DISCARD

TEXTBOOK OF ANATOMY AND PHYSIOLOGY IN RADIOLOGIC TECHNOLOGY

Textbook of anatomy and physiology in radiologic technology

CHARLES A. JACOBI
B.Sc., FNCRT (A.R.C.R.T.), R.T. (A.R.R.T.), M.T. (A.S.C.P.)

Professor and Chairman, Department of Radiologic Technology, University of Nevada, Las Vegas, Las Vegas, Nevada; member, Radiological Safety Board, University of Nevada System; member, Preprofessional Advisement Committee, University of Nevada, Las Vegas; member, Education Qualifications and Standards Committee, A.R.C.R.T.; previously Assistant Professor and Chairman, Medical Radiologic Technology Department of Medical-Dental Associates, and member of Radioisotopes Committee, Oregon Technical Institute, Klamath Falls, Oregon

SECOND EDITION

with 361 illustrations

THE C. V. MOSBY COMPANY
SAINT LOUIS   1975

Second edition

Copyright © 1975 by The C. V. Mosby Company

All rights reserved. No part of this book may be reproduced in any manner without written permission of the publisher.

Previous edition copyrighted 1968

Printed in the United States of America

Distributed in Great Britain by Henry Kimpton, London

Library of Congress Cataloging in Publication Data

Jacobi, Charles A
    Textbook of anatomy and physiology in radiologic technology.

    Includes index.
    1. Anatomy, Human. 2. Physiology. 3. Radiography.
I. Title. [DNLM: 1. Anatomy. 2. Physiology.
3. Technology, Radiologic. WN160 J16t]
RC78.J17   1975   616.07′572   74-20889
ISBN 0-8016-2390-1

VH/VH/VH   9 8 7 6 5 4 3 2 1

TO MY WIFE

a gracious lady with great understanding and patience

Foreword

It has been the experience of every surgeon that, during his medical school days, the subject of anatomy seemed ponderous and formidable; and, as the year crept on, a little bit dry. Later, upon taking up surgery, he reopens the dusty volume on this subject and studies it with a different eye. As he reviews the anatomy of the surgical approach the tissues become alive, interesting, and supremely important. He does not look at a maze of dry detail and facts but a *sine qua non* to the health and safety of his patient. He does not have to memorize, but finds the information he seeks, knowing he will put it to good use.

So it is with radiologic technology. The technician in training must, from the very start, be constantly reminded of the clinical correlation of his studies, thus gaining insight into what lies ahead. This will make his work more purposeful and help him to be a better technologist upon graduation.

During the many years that I have been privileged to know Mr. Charles A. Jacobi while he was Chairman of the Department of Radiologic Technology of Oregon Technical Institute, I have seen him carry out this policy of clinical correlation to the great benefit of the students and the department. In the present volume this principle has been continued, and I have attempted by means of some footnotes to tie in the section on skeletal anatomy with clinical practice of orthopedic surgery with which it is so closely allied.

Each year Mr. Jacobi has invited practicing physicians in the community to address the classes in the particular physician's field of interest and practice. A superior grade of technologist has been graduated from Oregon Technical Institute.

The Department of Radiologic Technology was organized by Mr. Jacobi at Oregon Technical Institute in 1952 with the assistance of Dr. J. M. Hilton, radiologist. In the sixteen years since its beginning, many outstanding graduate radiologic technologists have been educated. I am sure they will join me in saying that it is an honor and privilege to have worked with Mr. Charles A. Jacobi as teacher and friend.

A. M. Compton, M.D.
Diplomate, American Board of Orthopaedic Surgery

Preface TO SECOND EDITION

Increasing diversity of examination in departments of radiology dictates that future technologists possess greater understanding of both the anatomy and physiology in these new examinations. In many instances technologists have become specialists in the techniques of these newer studies; for example, in special procedures, ultrasonography, and thermography. The ultimate result is the need for a technologic specialist in routine radiography. With this in mind, the second edition has been *streamlined;* that is, the material presented follows the general pattern of routine examinations in a general department.

Some of the in-depth physiology of the first edition has been removed and saved for extensive treatment in later textbooks designed for specialized examinations. Many illustrations have been replaced and other illustrations have been added. All illustrations have the structures identified on photograph or drawing, rather than by number as in the first edition. The chapters treating each system have been rearranged as indicated in the previous paragraph.

Credits indicating our appreciation for various contributions are given to Mr. John Goad, of the University of Nevada's Audio-Visual Department, for the excellent photography; to Mrs. Doris Hinderliter, R.N., for the new illustrations; and to Desert Springs and Valley Hospitals of Las Vegas, Nevada, for several radiographs reproduced in the textbook.

Finally, and above all, my wife must receive the greatest credit for her tolerance and assistance during the preparation of this edition.

Charles A. Jacobi

Preface TO FIRST EDITION

With the speed of the advances in radiology there is an increase in the need for additional knowledge of anatomy and the physiology related to the assimilation and excretion of contrast media in radiologic technology.

A teacher can impart knowledge only to those students desirous of learning. The learning climate is enhanced in proportion to the interest of the subject material and the manner of presentation. Nothing is so interesting to ourselves as is a practical knowledge of our inner selves. Practical knowledge of anatomy and related physiology constitutes one of the principal necessities for the successful practice of radiologic technology.

The material presented in this text is that which we have found in the past sixteen years to be most successfully retained and utilized by our graduates. The anatomy and physiology are presented together in some chapters and separately in other chapters. This arrangement has proven itself to be the most popular with our students. The items of note on the illustrations have been identified (in most cases) by numbers rather than by names; this is to afford both students and instructors the opportunity of using the illustrations for review and test purposes. BNA and NA terminology is used in conjunction with the revised standard terminology relating to radiographic positions.

Actual skeleton photographs have been chosen in order to emphasize the variations seen in so many skeletons. Also, the large variety of mannikins and models available presents excellent facilities for the demonstration of nonskeletal anatomic structures.

In the numerous revisions of this material many wonderful persons have made valuable contributions. Dr. A. M. Compton, orthopedic surgeon and personal friend, has given most generously of his time and advice. He has added much to the chapter on the skeletal system and has helped with other chapters.

Dr. J. M. Hilton, radiologist and friend, has given much of his time and timely advice on the subject of heart margins, which is mainly found in the chapter on the circulatory system. Also, he has furnished many excellent radiographs. Other physicians who have taken valuable time from their busy schedules to review material that is directly within their specialty area are Dr. H. B. Currin, urologist; Dr. G. C. Miller, general practitioner; and Dr. G. J. Nicholson, vascular surgeon, who also supplied many fine radiographs. Mrs. Franc Bailey, physiology teacher, made numerous valuable contributions and suggestions.

Dr. G. R. Nicholson, pathologist, has been particularly helpful and has made available to me many slides from his teaching files. Drs. E. L. Lawson, A. S. Markee, R. H. Saul, and W. R. Stewart, radiologists, have given graciously of their time and have furnished some much needed radiographs. To each of the persons mentioned in this and the preceding paragraphs I extend my sincere thanks for generous efforts and continuing friendship.

Professor Don Q Paris, R.T., a good friend and highly valued colleague, has been most helpful and considerate during the development of this material. Dr. H. M. Hunt of our physics department was always available for advice and produced the drawing used as Figure 1-2. My sincere thanks is gladly extended to these two men.

The following companies and persons have been most kind in granting permission to use certain items: Clay-Adams Inc., General Biological Supply House, Dr. Justus F. Mueller and Ward's Natural Science Establishment, Denoyer-Geppert Company, Dr. J. A. G. Rhodin and The W. B. Saunders Company, and Mr. G. M. Larson and Miss B. Taber. To each of these I extend my sincere thanks.

Finally, and certainly at the top of the list of valued contributors to receive my wholehearted thanks I place my wife, for her understanding and help, for the many hours she has spent listening to various parts of the material, and for typing most of the material many, many times.

Charles A. Jacobi, R.T.

Contents

1 Introduction to anatomy and physiology, 1

    Terminology and anatomic location and direction, 2
    Basic units, 8
    Protoplasm, 11

2 Basic body structures and functions, 12

    Microscopic examination of cells, 12
    Cell structure, 14
        Nucleus, 14
        Cytoplasm, 16
    Cell division, 17
    Epithelial tissue, 22
        The skin, 23
    Muscle tissue, 24
        Striated muscle, 24
        Nonstriated muscle, 28
        Cardiac muscle, 29
    Nerve tissue, 29
    Connective tissue, 31
        Blood, 32
        Osseous tissue, 33
        Cartilage tissue, 34
        Adipose tissue, 34
        Areolar tissue, 34
        White fibrous tissue, 35
        Yellow fibrous tissue, 35
        Lymphoid tissue, 35
        Reticular tissue, 35
        Embryonal tissue, 35

Membranes, 35
    Dialysis, 36
    Diffusion, 37
    Filtration, 38
    Osmosis, 39
    Phagocytosis, 41
    Pinocytosis, 41
Homeostatic balance of fluids and electrolytes, 42
Homeostatic balance of acids and bases, 43
    Expression of hydrogen ion concentration in terms of pH, 43
    pH control mechanisms, 44

3 Embryologic survey, 46

Meiosis, 46
Fertilization, 50
Segmentation, 51
Embryo development, 52

4 The skeletal system, 54

Functions, 54
Anatomy, 55
    Typical bone, 55
    Types of bone, 57
    Markings on bone, 61
    Skeleton, 61
Bone physiology, 173
    Tissues derived from the mesenchyme, 173

5 The digestive system, 188

Function, 188
Anatomy, 189
    Buccal cavity, 189
    Oral cavity, 190
    Fauces, 194
    Pharynx, 195
    Esophagus, 195
    Abdominopelvic cavity, 197
    Accessory organs, 212
Digestion and absorption of food, 219
    Oral cavity, 219
    Deglutition, 220
    Stomach, 220
    Small intestine, 224
    Accessory organs, 224
    Large intestine, 229
    Portal circulation, 229
    Metabolism, 229
Radiographic interest, 231

6 The circulatory system, 235

General functions, 235
    Transportation, 235
    Water balance maintenance, 237
    Heat equilibration, 237
Anatomy, 237
    Heart, 237
    Arteries, 243
    Arterioles, 249
    Capillaries, 249
    Veins, 250
    Venules, 256
    Lymphatic system, 256
Blood, cells, fluids, and interrelated functions, 261
    Blood, 261
    Cells, 262
    Platelets (thrombocytes), 264
    Fluids, 264
    Interrelated functions, 264
    Lymph fluid, 266
    Interstitial fluid, 267
    Intracellular fluid, 267
Fetal circulation, 268
Circulatory system divisions, 270
    Coronary circulation, 270
    Pulmonary circulation, 270
    Systemic circulation, 271
Radiographic studies, 272
    Circulatory system radiography, 273
    The circulatory system as a transport mechanism, 274
    Heart margins, 277

7 The urinary system and the skin, 287

Urinary system, 287
    Functions, 287
    Anatomy, 288
    Urine formation, 296
Radiographic interest, 300
Skin, 302
    Functions, 302
    Anatomy, 303

8 The respiratory system, 305

Functions, 305
Anatomy, 305
    External structures, 305
    Internal structures, 312

Thoracic cavity, 314
    Pleura, 316
    Mediastinum, 316
Respiration, 317
    Mechanics of respiration, 317
    Physiology of respiration, 319
Radiography, 323
    Air contrast, 323
    Opaque media contrast, 324

9   The nervous system, 326

Function, 326
General anatomy, 326
    Central nervous system, 326
    Peripheral nervous system, 330
    Autonomic nervous system, 330
    Protective coverings, 331
Specific anatomy, 335
    Brain, 336
    Spinal cord, 339
    Peripheral nerves, 341
Physiology, 343
    Cerebrospinal fluid, 343
    Synaptic and neuromuscular impulse conduction, 344
Special senses, 346
    Smell, 347
    Sight, 347
    Hearing, 349
    Taste, 353
    Touch, 353
Radiographic studies, 353

10   The muscular system, 357

Muscle functions, 357
    Movement and heat production, 358
    Muscle activity, 361
    Posture, 362
Muscle anatomy, 362
    Skeletal muscles, 362
    Visceral muscles, 376
    Cardiac muscles, 377

11   The reproductive system, 378

Anatomy, 378
    Male, 378
    Female, 381

Structure and functions of the gonads, 384
    Male, 384
    Female, 386
Female mammary gland (breast), 388
Radiography, 390

12 The endocrine system, 394

    Specific endocrine glands, 394
    Other hormone sources, 394
    Function, 394
Endocrine glands, 395
    Pituitary gland, 395
    Thyroid gland, 398
    Parathyroid glands, 400
    Adrenal glands, 400
Other hormone sources, 402
    Compound glands, 402
    Glands of unknown function, 403
Radiographic interest, 403
General interest, 404

Glossary, 405

CHAPTER 1

Introduction to anatomy and physiology

Increasing sophistication of modern radiologic equipment along with corresponding advances in the manufacture of contrast media make possible visualization of anatomic structures and pathologic and other abnormal conditions not possible in earlier years. The radiologic technologist must possess the ability to visualize, in the finished radiograph, the requested structures as related to the particular examination. Concurrently, this technologist has a similar need for a broad understanding of the processes of assimilation and excretion of contrast media. Appreciation of contraindications for use of specific contrast media is of equal importance in the technologist's knowledge of the several physiologic processes.

The radiologic technologist consistently produces radiographs of diagnostic quality comparable to his knowledge of radiographic anatomy and related physiology.

Anatomy and physiology each are branches of biologic science. Studies of anatomy deal with both the architecture and the structure of the body. Studies of physiology deal with the function of lesser structures within the body and the integration of the several parts to effect the processes of life. The practice of radiologic technology employs practical application of each of these biologic sciences. Broad knowledge and understanding of skeletal anatomy, including articulations, and the overlying structures and protected viscera permit accurate positioning and radiographic demonstration of the individual structures. In all instances throughout this text, the student should consider that the skeleton, bone, part, or patient is facing him (the student); i.e., the left side of the skeleton, etc. is opposite the right side of the student.

The numerous anatomic charts, models, and other similar visual aids presently available can be of great assistance to the student of radiographic anatomy, especially when these aids are used with well-executed radiographs of the corresponding structures. A major portion, perhaps as much as 90 percent, of learning by college students takes place through the sense of sight.

An essential part of learning radiographic anatomy is the concurrent learning of the anatomic nomenclature and relationships of the body structures.

TERMINOLOGY AND ANATOMIC LOCATION AND DIRECTION

If one were to be placed suddenly in a foreign country, a very necessary facet of knowledge required for early success in that foreign country would be that of the language. A thorough study of anatomy and physiology places a new student in a situation comparable to living in a foreign-language–speaking country; i.e., the language of radiologic technology (and other allied health professions) is quite strange to him. It behooves the new student to begin immediately to learn and to comprehend the meaning and application of each new term. When reading each assignment, it is recommended that the student have both an English and a medical dictionary at hand and that he refer to the glossary, pp. 405-424; *he should make a habit of seeking the definition and application of each new term,* especially those terms applying to human anatomy and physiology.

The human body is an organized group of highly specialized tissues, which together perform the necessary functions contributing toward *homeostasis* (condition of status quo or relative uniformity of cellular environment). *A tissue is an aggregation of similarly specialized cells united in the performance of a particular function. A cell is any one of the minute protoplasmic structures that make up organized tissue;* the parts of a cell are threefold—the cell membrane, the cytoplasm (protoplasm), and the cell nucleus. The number of cells in the adult human body has been estimated to be from as few as approximately 30,000,000,000,000 to as many as more than 400,000,000,000,000 (four hundred trillion).

The cells of the human body function characteristically according to specialization* and are vital components of all other bodily structures, which are membranes, organs, glands, and systems. *A membrane is a thin layer of tissue that covers a surface or divides a space or organ. An organ is an assortment of tissues working together to perform a special function,* such as the heart, kidney, etc. *A gland is an aggregation of cells, specialized to secrete or excrete materials not related to their ordinary metabolic needs. A system is a series of interconnected or interdependent parts or entities that function together in a common purpose or produce results impossible of achievement by one of them acting or operating alone,* such as the digestive or nervous systems.

The human body contains nine systems—skeletal, muscular, circulatory, nervous, respiratory, digestive, urinary, reproductive, and endocrine. Each system is

---

*As an example of specialization, bone has a *matrix* of inert material that is *laid down* by the bone cells (osteoblasts), but the cells individually do not support the body weight. The matrix is the groundwork upon which anything is cast, or the basic material from which a thing develops.

in evidence in more than one anatomic region; some systems extend throughout the body and others tend to be located chiefly in a particular cavity.

The human body contains several cavities of different sizes, enclosed either partially or entirely by different structures. Some of the cavities are classed as being somewhat open; such are the orbital, nasal, buccal, and oral cavities. The air sinuses of the skull are entirely closed cavities.

The body cavity of the embryo is the *coelom*. The cavity is both extraembryonic (without—outside the embryo) and intraembryonic (within—inside the embryo). The principal cavities of the trunk arise from the intraembryonic portion of the coelom.

Collectively these cavities comprise the large ventral cavity, the body cavity lying in front of the vertebral column and composed of the mouth, throat, thorax, abdomen, and pelvis. Our principal concern in this chapter is with the thoracic and abdominopelvic divisions of the ventral cavity.

The large muscular diaphragm separates the thoracic cavity from the abdominopelvic cavity. The *thoracic cavity,* which is superior to the diaphragm, contains the pleural sacs and lungs, the pericardial sac and heart, and the mediastinal cavity and viscera. The *abdominopelvic cavity* (often called the abdominal cavity) contains the digestive viscera and the reproductive and urinary viscera, although some of the latter lie posterior and/or inferior to the peritoneum in this cavity. There is no membranous wall separating the abdominal part from the pelvic part of the cavity.

Dorsal (posterior) to the ventral cavity and contained within the dorsal body wall is the *dorsal cavity* (often named the spinal and cranial cavities), another very important cavity in radiographic procedures. The dorsal cavity consists of the cranial case, which contains the brain, and the spinal canal, which contains the spinal cord. The cranial case and spinal canal have direct communication with each other through the foramen magnum of the occipital bone in the floor of the skull.

New terms have been introduced in the last few paragraphs, and many more must be learned in succeeding pages. In describing the location, position, or direction of a part of the human body, it is necessary to have a standard point or condition of reference. The *normal anatomic position* of the human body is described as follows. The person stands erect (the heels together and the feet in a comfortable position), facing forward with the arms extended down the sides and the hands facing forward (the thumbs away from the body). If the person were lying on his back, he would be in the *supine* position, and his hands would be supinated, or in the supine position.

Among the many new terms to be learned are the following words having opposite reference: *superior* and *inferior, anterior* and *posterior, ventral* and *dorsal, palmar* and *dorsal, volar* and *dorsal, plantar* and *dorsal, proximal* and *distal, cephalad* and *caudad, craniad* and *caudad, medial* and *lateral, transverse* and *oblique, sagittal* and *coronal, supine* and *prone, erect* and *recumbent.* Most of these terms are illustrated in Fig. 1-1 and in Tables 1-1 and 1-2.

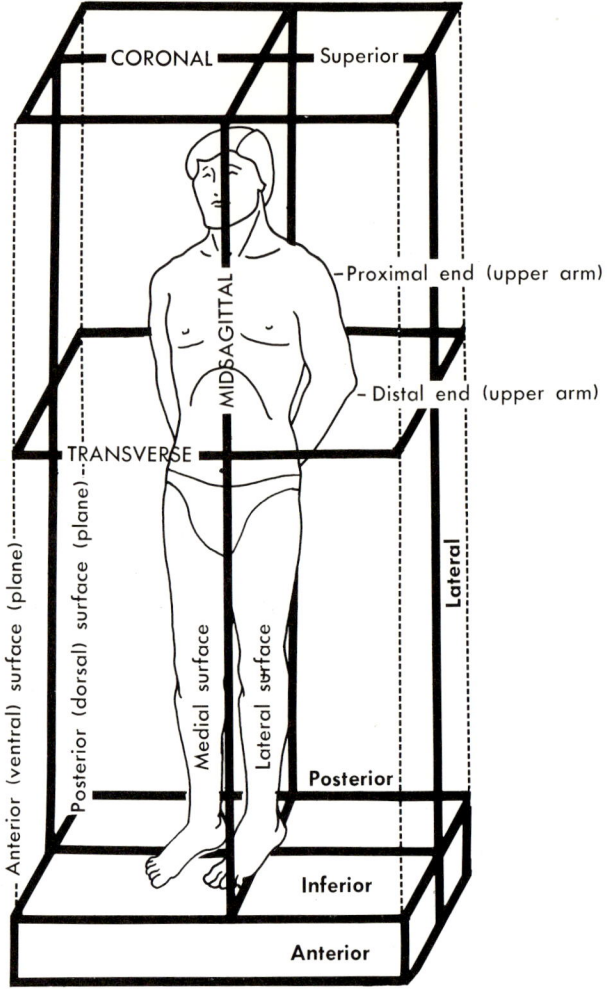

Fig. 1-1. Body planes. The figure illustrates the major and some minor planes and surfaces; it is drawn with the left foot parallel with the right to show the medial and lateral surfaces of the lower leg. The arms, hands, and left foot and leg are not in the true anatomic position.

The three major planes of the body are the sagittal, coronal, and transverse; each is at right angles to the other two. *A sagittal plane extends from ventral to dorsal and separates right from left.* A midsagittal plane separates the body into exact right and left halves. *A coronal plane extends from right to left and separates ventral from dorsal. A transverse plane extends from ventral to dorsal and from right to left and separates superior from inferior.* In numerous instances, two of these terms may be combined to describe a particular locus of trauma, disease, or anatomic reference. Such is the case with *anterolateral;* the last syllable of anterior is replaced with *o*, and the second term (lateral) is attached. The names of body planes and body positions are listed and explained in Tables

Table 1-1. Directional terms

Term	Opposite	Description
Superior	Inferior	Situated above, or directed upward; used in reference to the upper surface of an organ or other body structure
Inferior	Superior	Situated below, or directed downward; used in reference to the lower surface of an organ or other body structure
Anterior	Posterior	Situated in front of or in the forward part of, toward the head end of the body; used in reference to the ventral, or belly, surface of the body
Posterior	Anterior	Situated in the back of, or in the back part of, or affecting the back part of an organ; used in reference to the back, or dorsal, surface of the body
Ventral	Dorsal	The same as anterior in the human body
Dorsal	Ventral	The same as posterior in the human body; however, the word is used as opposite to three terms as seen in the following entries
Dorsal	Palmar	Applied to the hand, means the back of the hand
Dorsal	Volar	Applied to the forearm, means the back of the forearm
Dorsal	Plantar	Applied to the foot, means the upper surface of the foot, opposite the sole surface
Palmar	Dorsal	Pertaining to the palm surface of the hand
Volar	Dorsal	Pertaining to the flexor surface of the forearm, wrist, or hand
Plantar	Dorsal	Pertaining to the sole of the foot
Proximal	Distal	Nearest; closer to any point of reference
Distal	Proximal	Remote; farther from any point of reference
Cephalad	Caudad	Directed toward the head
Craniad	Caudad	Directed toward the head; toward the anterior or superior end of the body
Caudad	Cephalad	Directed toward a cauda or tail; used as meaning toward the feet
Medial	Lateral	Pertaining to the middle or meaning closer to the median plane, or midline, of a body or structure
Lateral	Medial	Denoting a position farther from the median plane, or midline, of a body or of a structure

Table 1-2. Body positions

Term	Opposite	Description
Transverse	Oblique	Placed crosswise; situated at right angles to the long axis of a part
Oblique	Transverse	Slanting; inclined; between a horizontal and a perpendicular direction; or between the coronal and sagittal planes
Sagittal	Coronal	Pertaining to situation in the direction of the sagittal suture
Coronal	Sagittal	Pertaining to situation in the direction of the coronal suture
Supine	Prone	Lying on the back with the face upward
Prone	Supine	Lying with the face down
Erect	Recumbent	Upright; having a vertical position
Recumbent	Erect	Lying down

Table 1-3. Body planes

Region	Name	Description
Skull	Aeby's	A plane through the nasion and basion and perpendicular to the midsagittal plane
	Auriculoinfra-orbital	A horizontal plane extending from lines extending between the lowest points on the margins of the eye orbits and the highest points on the margins of the auditory meati
	Axiobucco-lingual	A plane parallel with the bony axis of a tooth and passing through the buccal and lingual surfaces
	Axiolabio-lingual	A plane parallel with the long axis of a tooth and cutting its labial and lingual surfaces
	Axiomesio-distal	A plane parallel with the long axis of a tooth and cutting its mesial and distal surfaces
	Baer's	A plane through the upper border of the zygomatic arches
	Bite	The occlusal plane
	Blumenbach's	A plane parallel with the base of a dry skull that has had the lower jaw removed
	Broca's	A visual plane passing through the visual axes of the two eyes
	Coronal	Any vertical plane through the body parallel with the coronal suture, and at right angles to both the midsagittal and transverse planes
	Daubenton's	A plane passing through the opisthion* and the lower edges of the orbits
	Eye-ear and Frankfort-horizontal	See auriculoinfraorbital
	Horizontal	Any plane of the body (and head) that is (normally) parallel with the horizon. In dentistry, it is a plane passing through a tooth at right angles to its long axis.
	Labiolingual	See axiolabiolingual
	Listing's	A transverse vertical plane perpendicular to the anteroposterior axis of the eye, and containing the center of motion of both eyes. This plane contains the transverse and vertical axes of ocular rotation.
	Meckel's	A plane passing through the auricular and alveolar points
	Mesiodistal	The plane of a tooth that passes through its mesial and distal surfaces
	Morton's	A plane through the most projecting points of the parietal and occipital protuberances
	Nuchal	The outer surface of the occipital bone between the foramen magnum and the superior curved line
	Occipital	The outer curved surface above the superior curved line
	Occlusal	The imaginary surface (plane) between the upper and lower teeth in normal occlusion
	Orbital	The orbital surface of the maxilla, or a plane passing through the visual axes of both eyes
	Of regard	A plane that passes through the center of rotation and the point of fixation in the eye
	Sagittal	Any plane of the body (and head) that extends parallel with the midsagittal suture of the skull
	Temporal	The depressed area on the side of the skull below the inferior temporal line
	Tentorial	A straight line drawn (plane extended) through the tentorium

*The opisthion is the midpoint of the lower border of the foramen magnum.

Table 1-3. Body planes—cont'd

Region	Name	Description
Skull—cont'd	Tooth	Any imaginary plane (axial, horizontal, or vertical) made by section of a tooth
	Transverse	See horizontal
	Visual	See orbital
Trunk	Addison's	Plane used as a landmark in the thorax and abdomen
	Coronal	See under skull
	Hodge's	A plane passing through the second sacral vertebra and the upper border of the pubis, and parallel with the plane of the pelvic inlet
	Horizontal	See under skull
	Intertubercular	A transverse plane extending between the tubercles of the ilia
	Medial, median	See sagittal under skull
	Sagittal	See under skull
	Midsagittal	That plane parallel with the sagittal suture, and dividing the right from the left half of the body (or skull)
	Popliteal	The popliteal space
	Spinous	The horizontal plane extending between the iliac anterior superior spines
	Sternal	The anterior surface of the sternum
	Sternoxiphoid	A horizontal plane extending through the junction of the sternum and xiphoid
	Subcostal	A horizontal plane touching the inferior (subcostal) surfaces of the right and left tenth ribs
	Suprasternal	A horizontal plane through the suprasternal notch
	Thoracic	A transverse (horizontal) plane extending between the right and left fourth ribs anteriorly
	Transpyloric	A transverse (horizontal) plane extending through the pylorus
	Transverse	See horizontal under skull
	Oblique	Any plane of the trunk (or skull) not at right angles to the three major (coronal, sagittal, transverse) planes
	Umbilical	A transverse plane through the umbilicus (navel)
	Vertical	Any plane of the body perpendicular to the horizon
Outside the trunk	Anterior (frontal, ventral)	A vertical plane parallel with the anterior surface of the trunk
	Posterior (dorsal)	A vertical plane parallel with the posterior surface of the trunk (or buttocks)
	Lateral	Any vertical plane parallel with the midsagittal plane and contiguous with the left or right lateral surface

1-2 and 1-3. Further understanding of professional terminology is achieved when *standard* terms regarding patient, central ray, and radiograph are employed. The *patient* is said to be in a particular *position,* or is *positioned* (thusly). The *central ray* has *projection,* or is *projected* through the patient from anterior to posterior (etc.). The *radiograph* is a two-dimensional *view* of a projection of a positioned part.

Although anatomy and physiology are already familiar terms, understanding of the meanings of both anatomy and physiology (as terms) enables the student to progress in his studies. *Anatomy is the science of the structure of the animal body and the relation of its parts. Physiology is the science of the function of the living organism and its parts.* Other biologic sciences augment knowledge accumulated in studies of anatomy and physiology. Some of the more common contributing biologic sciences are biology, zoology, cytology, histology, and botany. Biology is the science that deals with the phenomena of life in general. Zoology is the biology of animals. Cytology is the scientific study of cells, their structure and functions. Histology is the science of the minute structure, composition, and function of tissues. Botany is the science of the plants, or the vegetable kingdom.

The preceding branches of biologic science deal with animate substance. *All substance, both animate and inanimate, consists of small physical particles.*

BASIC UNITS

A study of physics teaches that all matter exists invariably in one or more of three forms—*liquid, solid* and *gas*. The body is composed of cells, which are physical substance or matter. The major difference between the (animate) matter of our bodies and the (inanimate) matter of nonbiologic substance is that of life itself. Chemically, the human body can be separated into its elemental forms. There are presently at least 104 known elements. The human body includes thirteen of these in rather large quantities and several more of these elements in trace quantities. With a single exception, each of these is one of the lighter elements. The exception is iodine, which has an atomic number of 53; the other twelve elements found in the human body, including the atomic number of each, are as follows: hydrogen—1, carbon—6, nitrogen—7, oxygen—8, sodium—11, magnesium—12, phosphorus—15, sulfur—16, chlorine—17, potassium—19, calcium—20, and iron—26.

Each of the elements is classified in the periodic table* according to the increasing number of positively charged particles *(protons)* in the nucleus. An atom consists of its nucleus, containing the protons and *neutrons,* and the orbital *electrons* as diagrammed in Fig. 1-2. The nucleus is very dense and exerts an exceedingly strong electropositive attraction on each orbital electron. The electron is a minute particle having an electronegative charge. The three components mentioned, in addition to other unmentioned components, are referred to as subatomic particles. The neutron of the nucleus is of essentially the same mass as the proton but possesses no electrical charge; it is neutral and is, therefore, called the neutron. Stable atoms have one proton in the nucleus for each electron in the planetary orbits. There are as many as seven orbits surrounding the nucleus, and there is definite arrangement of the electrons in each orbit.

Hydrogen is the first element in the periodic table of the elements. Hydrogen has a nucleus consisting of a single proton and has a single orbital electron in the

---

*The periodic table is a table of the elements arranged according to the periodic law, which states (in modern terms) that most of the properties of the elements are periodic functions of their atomic numbers.

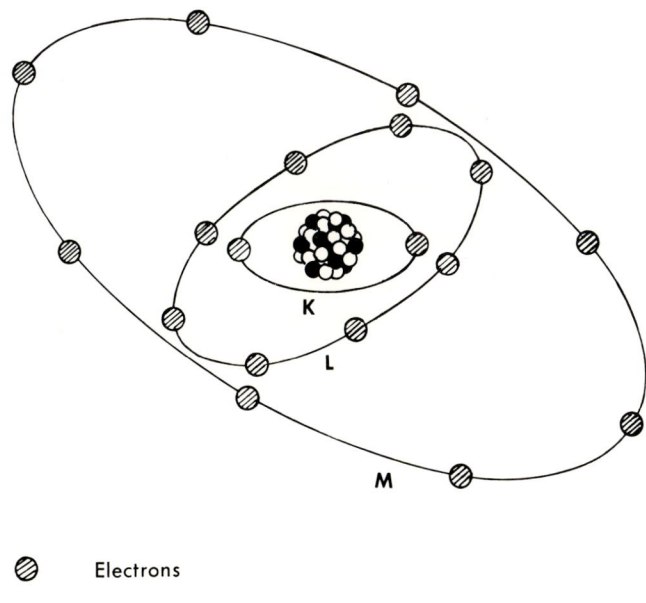

- ⊘ Electrons
- ● Protons ⎫
- ○ Neutrons ⎬ Nucleons
- K,L,M, ... Electron orbits

Fig. 1-2. Argon atom diagram, a modern concept.

K (first) orbit. The orbits, lettered from the nucleus outward, are K, L, M, N, O, P, and Q. Helium is the second element in the periodic table and its nucleus contains two protons and two neutrons. Since there are two protons in the nucleus there are two electrons, both of which are in the K orbit. Lithium is the third element in the periodic table. The lithium atom nucleus contains three protons and four neutrons. The lithium atom has two electrons in its K orbit and one electron in its L orbit.

The atoms are incredibly small, so small in fact that if all the space between the nucleons (disregarding the orbital electrons) in the atoms of a 150-pound adult were nullified and the remaining nuclear material were condensed in one place, the total mass of this nuclear material would probably be of insufficient size to cover the point of a pin! However, the nuclear material is very dense, so dense in fact that one cubic centimeter (1 cc.) of pure nuclear (material) substance would weigh in excess of 100,000,000 (one hundred million) tons! These statements are estimates based upon extensive research performed by many investigators. In this discussion of atomic structure and size no mention of the electrons is included because they are so small in comparison with the size of the nucleons (protons and neutrons). It has been estimated that if the nucleus of a tungsten atom ($^{184}_{74}W$) were enlarged to the size of a baseball, the orbital electrons would be in excess of 100 miles distant; this is a suggestion of the enormous relative space existing within the confines of a single atom including its electrons. An electron is approximately 1/1,840 the size of a proton.

The approximate weight of one cubic centimeter of nuclear material is mentioned to stimulate thought and to develop further a small degree of understanding of relativeness. Logic reveals that it is impossible to weigh a single nucleon or even a single atom. Therefore, the term "atomic weight" has meaning different from weight as we normally consider it. The atomic number of an atom, however, is based upon common numbers. *The atomic number of an atom is determined from the number of protons in the nucleus. The atomic weight of an atom includes both the protons and the neutrons and is the sum of the number of protons plus the number of neutrons in the atomic nucleus.* These are relative figures; i.e., figures having no relationship to actual weights as we know them.

The structural parts of matter can be compared with tangible parts; i.e., we can *think* of nucleons as *building blocks*. Protons represent one kind of building block and neutrons represent another (or second) kind. So many of each are required to form (build) a particular element. The protons, as previously indicated, possess positive electrical charges. The neutrons have no electrical charges and are neutral. The tiny electrons possess negative electrical charges. Each proton within an element attracts one electron (actually the closest electron at a given instant) that orbits about the element (atom) with great velocity. Certain forces *bind* the protons and neutrons in a relatively compact mass as the nucleus, and other forces prevent the electrons from falling into the nucleus and from leaving (flying away from) the atom.

Thus, it is seen that the nucleons, organized together in specific numbers, form the atoms of the elements. A simple change, either a gain or a loss, of one nucleon causes the atom to exhibit different characteristics. If the change is the gain or loss of a proton, the result is an atom of a different element. If the change is the gain or loss of a neutron, the result is an *isotope* of that element. Such changes relate to nuclear energy reactions (fission or fusion) and are not the subject of this text.

When an atom loses an electron, the atom has one positive charge in excess of its negative charges, and the atom is a *positive ion (cation)*. Conversely, when an atom gains an electron, the atom has an excess of negative charges and is a *negative ion (anion)*. Either condition results from the process of *ionization*. Ionization of these elements (atoms) occurs as a result of high-energy rays (or particles) interacting with the atoms in such a manner as to cause the loss or gain of one or more electrons. Roentgen rays (x rays), gamma rays, alpha particles, beta particles, and neutrons are among the forms of energy that cause ionization. Ionization occurs for other reasons not pertinent to this discussion.

From the foregoing paragraphs it is seen that nucleons comprise the atoms. In a similar manner the atoms are organized together in specific arrangement (with constant numbers of certain atoms) to form the molecules of substance. *A molecule is a very small mass of matter, an aggregation of atoms: specifically, a chemical combination of two or more atoms which form a specific chemical substance. Substance is the material constituting an organ or body.* Substance, then, is the material constituting protoplasm.

PROTOPLASM

Protoplasm is the only known form of matter in which life is manifested. Protoplasm is the very essence of all living matter and so is the essential "stuff" of the body cells. The fact that protoplasm is essential to life is universally accepted, yet a universally accepted definition of protoplasm has, so far, been elusive. However, understanding some of the intricacies of protoplasm is possible, especially if we study in areas in which we are somewhat knowledgeable.

Protoplasm is a semifluid substance, grayish to transparent in color, consisting of about 75 percent water and 25 percent protein. Protoplasm is without form, and when it is viewed under a high-power microscope, particles are seen to move through the protoplasm with little limitation. In addition to protein, protoplasm contains many essential minerals, such as calcium, magnesium, potassium, sodium, and others. Protoplasm also contains small amounts of sugar, starch, and fatty substance.

Protein molecules vary somewhat in composition from one to another but, for the most part, consist of atoms of carbon, hydrogen, nitrogen, and oxygen and small numbers of atoms of sulfur; some protein molecules contain atoms of phosphorus. Protein molecules are colloids. *A colloid is a mixture of microscopic particles so small that the energy of the molecules of the liquid portion is great enough to prevent the particles from settling out.* It follows that protoplasm is a colloidal solution and that protoplasm evidences characteristics of such solutions. An example is that protoplasm may change from its fluidlike (sol) state to a solid (gel) state under certain conditions.

The presence of carbon atoms in protein molecules is, at least to a large extent, responsible for the amazing versatility of protoplasm. Carbon atoms are capable of combining with other carbon atoms and with atoms of other elements to form great numbers of different chains and rings. The use of carbon in certain chemical experiments provides an insight into many of the phenomena fashioned from protoplasm during eons of time by the forces of nature.

The ameba, a single-celled animal, is a good example of protoplasm. The ameba is capable of carrying out all of the functions necessary for living, whereas the human and other higher animals depend upon the specialized functions of myriads of highly specialized cells to perform these functions.

REFERENCES

Best, C. H., and Taylor, N. B.: The human body, ed. 4, New York, 1963, Holt, Rinehart and Winston.

Kimber, D. C., Gray, C. E., Stackpole, M. A., Leavell, L. C., Miller, M. A., and Chapin, F. M.: Anatomy and physiology, ed. 15, New York, 1966, The Macmillan Co.

Langley, L. L.: Cell function, ed. 2, New York, 1968, Reinhold Publishing Co.

Langley, L. L., and Cheraskin, E.: Physiology of man, ed. 4, New York, 1971, Reinhold Publishing Co.

CHAPTER 2

Basic body structures and functions

We have learned that the body is composed of an extremely large number of cells: cells that are highly specialized and organized into the different body structures. It is next necessary to gain insight into the structure and function of individual cells.

No typical cell exists for uncomplicated examination of its basic parts and, subsequently, examination of the special function or functions of each of its parts. However, practical purpose is served if we examine a collection of results combined into a composite cell. Such is the approach to basic understanding in the following pages.

MICROSCOPIC EXAMINATION OF CELLS

Somatic cells, germ cells, erythrocytes, leukocytes, in fact all cells, vary considerably in size as will be discussed in the following section. Since the cells are generally too small for visualization and study by the naked eye, the cells and tissues must be examined under a microscope. For most of these microscopic examinations, the material to be examined must undergo a special preparation such as staining and sectioning. Fig. 2-1 illustrates a research microscope without camera. Properly equipped, it can be used to make pictures (photomicrographs) of tissue slides.

A photomicrograph is a photograph of a magnified image of a small object. The microscope, regardless of its complexity, uses two lenses to enable critical examination through the second lens (ocular or eyepiece) of the magnified image obtained through the first lens (objective).

Following is a brief description of the operation of the compound* micro-

---

*The compound microscope consists of an objective lens and the ocular lens mounted in a draw tube (to adjust the distance between the two lenses) and focused by means of screw arrangements.

Basic body structures and functions   13

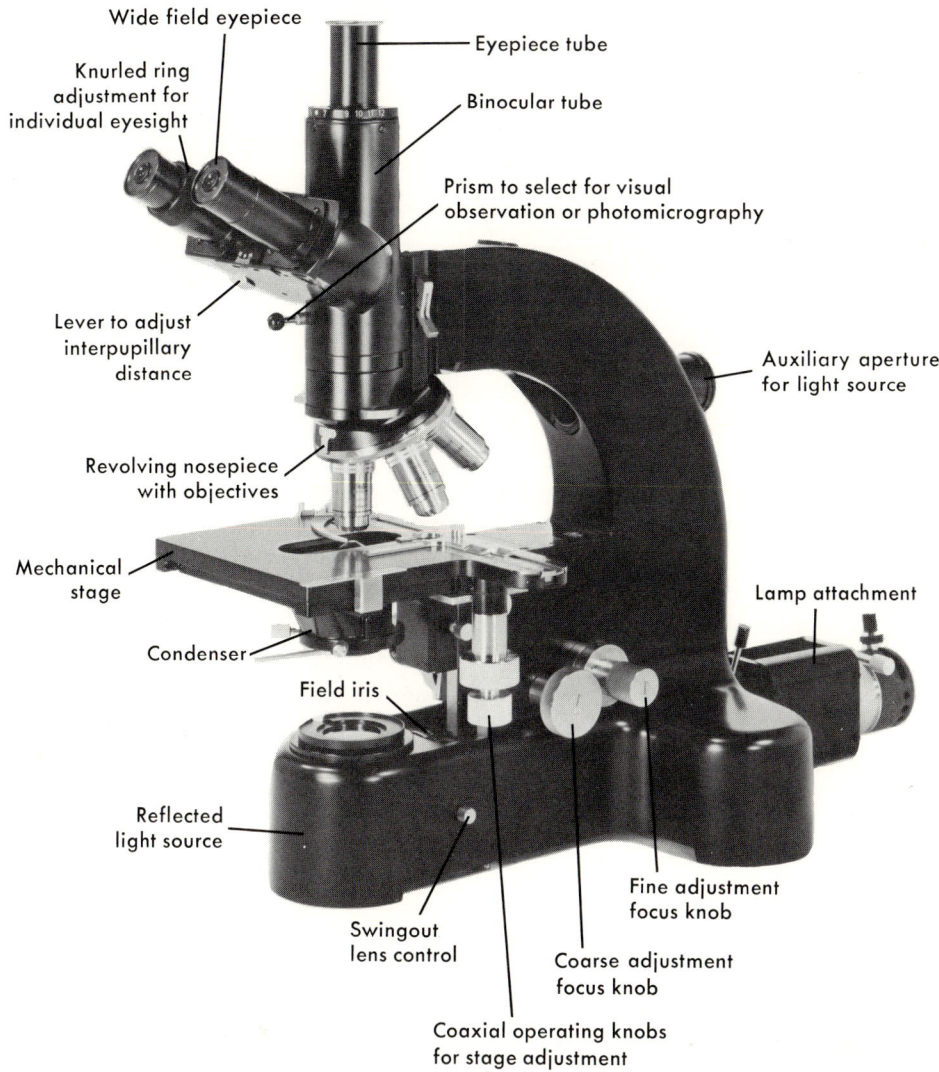

Fig. 2-1. Leitz Ortholux I Microscope. (Courtesy E. Leitz, Inc.)

scope. Rays, diverging from each point of the object being examined, are brought into sharp focus beyond the objective lens in a plane. This is determined by three factors: the relative object-to-lens distance, the lens-to-focal-plane distance, and the focal length of the lens. The rays again diverge beyond the plane of focus, and are again focused beyond the ocular lens. The resulting image size depends on the magnitude of divergence from the first lens and the magnifying power of the second lens. The usual laboratory microscope (often called student microscope) is equipped with three objective lenses and either 5× (five power) or 10× ocular lens (or lenses, depending on whether the microscope is monocular or binocular). The three objective magnifications are: low power—16 mm., giving

10 power magnification; high (dry) power—4 mm., giving 44 power magnification; and oil immersion—1.8 mm., giving 95 power magnification. The power of the ocular lens times the magnification of the objective equals the total magnification of the object of focus. More complex microscopes, such as the one illustrated, have a reducing factor incorporated in the system, thus allowing for less magnification than described.

CELL STRUCTURE

Because of the varied activities of the cells and because of anatomic structure and location, cell diameter is quite variable; i.e., some cells may be as small as 10 microns (μ) in diameter, and others may be as large as 100 microns in diameter. The inner parts of a cell receive the necessary nutriments and oxygen as a result of diffusion through the cell cytoplasm; if the cell diameter were too large for effective diffusion, the inner parts of the cell would be undernourished. Excretory processes of cellular wastes would be equally ineffective.

Although there is considerable variation in cellular anatomy and activity among the different kinds of cells, certain basic components are common to most cells. A cell usually contains a nucleus, which is surrounded by a form of protoplasm and is enclosed by a membrane. The direction for and control of cellular activity is a function of the nucleus. All body (somatic) cells consist of the protoplasmic substance, *cytoplasm,* and each cell is surrounded by a living membrane. Each cell has a nucleus, consisting largely of nucleic acids. Nucleic acids are substances of high molecular weight which, when combined with proteins, comprise those *nucleoproteins* vitally involved in cellular growth and reproduction and the transmission of hereditary traits. The two nucleic acids pertinent in the very brief discussion are DNA (deoxyribonucleic acid) and RNA (ribonucleic acid). The chromatin material exists as small granular particles along a fine, meshlike network. Continuing study of and research into the nature of cells and cell growth and metabolism provide the student of radiologic technology with information so necessary to his understanding of the uses of roentgen rays (x rays) and other high-energy radiations.

Nucleus

Transmission of hereditary traits is a known function of the cell nucleus, specifically the chromosomes. DNA, a super molecule, comprises the chromosomes and carries the hereditary traits. DNA molecules appear as a double helix* composed of ribose sugars and phosphates with crossbars of components of purines and pyrimidines. Because of the bundled arrangement, the fibers appear to be apportioned in a particular order at each cell division in mitosis. If such orderly arrangement were absent, new daughter cells would contain such a severely tangled collection of chromatin material that further cell division would

---

*The double-helix arrangement of the DNA molecule strongly resembles a spiral staircase; it contains the very numerous components arranged in a definite pattern throughout the entire structure.

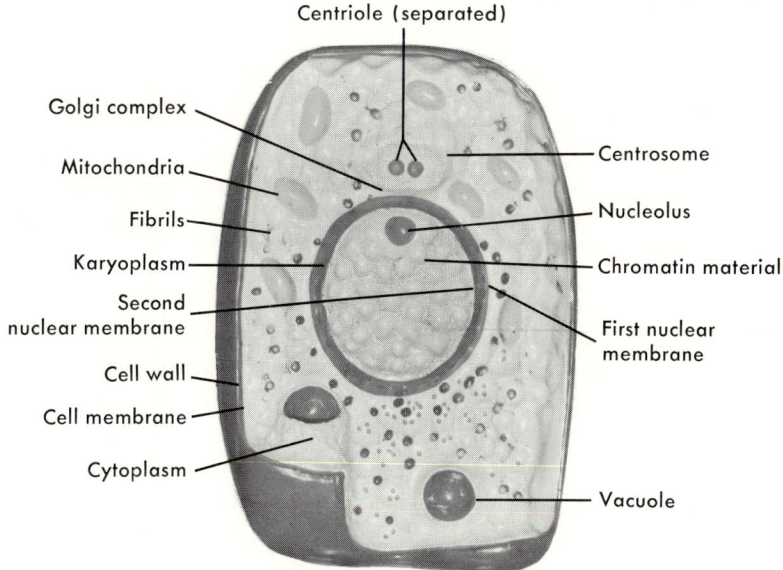

Fig. 2-2. Typical somatic cell. (Courtesy General Biological Supply House, Inc., Chicago, Ill.)

likely be impaired if not completely interrupted; therefore, cell division would be prevented.

The orderly arrangement and physical structure of the DNA fibers appear to make possible the conduction of *control messages* governing and regulating the vital functions of metabolism in the cells. Both DNA and RNA are found in the nucleus of each cell. There is a preponderance of DNA in the cell nucleus and a corresponding preponderance of RNA transfer* in the cell cytoplasm. Both DNA and RNA are composed of a series of nucleotides†; however, each of these acids yields somewhat different components when subjected to complete hydrolysis. Hydrolysis of RNA yields phosphoric acid, which is vitally important in muscular activity and muscle-cell metabolism. RNA messenger is vitally important in protein synthesis and in the determination of the amino acid sequence in proteins.

The fluid substance of the cell nucleus is the *karyoplasm,* which is usually more viscid than the cytoplasm. The nucleus is contained within a definite membranelike substance, which is probably a part of the karyoplasm. The true *nucleolus* and the *karyosome* (chromatin nucleolus) are contained within the nucleus (see Fig. 2-2).

---

*The RNA molecules are found in three forms: (1) RNA messenger, which is produced by the DNA molecule in the nucleus and which bears the genetic code; (2) RNA transfer, which picks up specific amino acids in the cytoplasm; and (3) ribosomal RNA, which is found in the ribosomes along the borders of the endoplasmic reticula. (With the aid of the electron microscope, the endoplasmic reticulum is seen as a very fine structural network in the cytoplasm of most cells.)

†A nucleotide is one of the compounds into which nucleic acid is split by the action of nuclease.

A nucleolus is a round body within the cell nucleus; it has poor, if any, staining characteristics. The nucleolus is often called a plasmosome. With ordinary histologic preparation, the nucleolus appears to be a distinct homogeneous body. Electron micrographs show the nucleolus as having no discrete boundary and appearing to be formed by clumps or skeins of microscopic particles. The particles are quite coarse and are somewhat darker than the remaining nuclear contents.

Cytoplasm

The cytoplasm, the fluid substance surrounding the nucleus, is limited on the outer periphery by a cortical layer, a plasma membrane, and a true membrane. The plasma membrane functions to *control* or *direct* the movements of various materials both inward to and outward from the cell, and the cytoplasm performs the major part of cellular work. The cytoplasm includes *organelles*,* which are capable of self-perpetuation, and *inclusions,* which are collections of nonliving substances. The organelles include centrosomes, the Golgi (apparatus) complex, fibrils, and mitochondria. The inclusions consist of collections of proteins, lipids (usually as fat droplets), carbohydrates, pigments, granules, and various crystals. The Golgi complex and the endoplasmic reticulum compose the two types of intracytoplasmic membranous systems of a cell. Strong evidence indicates continuity between the smooth-surfaced vesicles of the Golgi complex and the rough-surfaced vesicles of the endoplasmic reticulum.

The endoplasmic reticulum is a network of protoplasmic substance situated in the central portion of the cell cytoplasm. The smooth-surfaced reticulum may function importantly in glycogen synthesis and storage in the hepatic cells, in steroid hormone synthesis in the interstitial cells of the testis, and perhaps in striated-muscle impulse conduction.

Attached to the endoplasmic reticulum are minute granules, ribosomes, so small that they are seen only with the electron microscope. Ribosomes are largely nucleic acid in composition and are presumed to be sites of protein synthesis. These granules are frequently called RNA granules when in reality they are RNP (ribonucleoprotein). Ribosomes are often referred to as "protein factories."

Centrosomes have been found in most mammalian cells and are usually very close to the cell nucleus. The centrosome contains a spherically-shaped mass of clear cytoplasm in which is located a small centriole. It is possible that the centrioles initiate mitosis. Numerous fine bands, *astral rays,* surround the centrioles. During mitosis a spindle forms; the spindle is composed of many strands. The chromosomes attach themselves to these strands during the mitotic process.

The *Golgi complex* is usually associated with the centrosome and centriole

---

*An organelle is a specialized part of a cell, as a cilium, performing functions that are analogous to those of the organs of the multicelled animals.

and is found between the outer cell surface and the nucleus. The Golgi complex appears to increase greatly in size when the cell is secreting.

The *fibrils* are thin, protoplasmic threads extending throughout the cytoplasm. The fibrils probably function to give the cell structural support. The fibrils are highly developed in both the muscle cells and the nerve cells and function differently in these tissues.

Mitochondria (chondriosomes) are specific, minute fluid bodies containing granules; the bodies are of rodlike, oval, or spherical shape and are present in the cytoplasm of most cells. Many mitochondria are saclike, with an outer membrane and an inner membrane. The mitochondria function importantly in cell life; numerous chemical reactions occur within them. The energy for cellular work derives from these reactions. (Mitochondria are sometimes called the *power plants* of the living cells.)

Vacuoles appear in most cells as transparent or empty spaces of rounded and/or ovoid shape. These are probably openings through which waste substances, the products of metabolism, are collected and eliminated. The single living cell is capable of performing all of the necessary activities of life.

CELL DIVISION

In order to survive and propagate its kind, each cell must produce new cells. This is achieved in all life through one of three processes: *amitosis*—simple binary fission; *mitosis* (karyokinesis)—somatic cell division, which includes the four stages of *prophase, metaphase, anaphase,* and *telophase;* and *meiosis*—reduction division, which occurs in the maturation of germ cells.

In amitosis the cell simply divides into two equal parts, each containing exactly one-half of the original substance. Each half then increases to mature size and, after a given period of time, it divides into two equal halves.

Mitosis occurs in higher forms of animal life and is the process whereby somatic (body) cells multiply. Mitosis requires varying lengths of time according to cell and animal type, from about 30 minutes in some animals to several hours in other animals. During prophase, the first stage of mitosis, the chromatin material of the nucleus begins to shorten and thicken into distinct forms that resemble sausages. In a particular species there will always be a predetermined number of chromosomes in each cell, the number depending upon the species. The human somatic cell contains forty-six chromosomes (twenty-three pairs—one member of each pair from each parent). Immediately prior to chromosome formation, the centriole begins to separate and each half migrates toward opposite poles of the cell. Stretching between the centrioles are the forming strands to which the chromosomes later attach themselves.

During metaphase, the second stage of mitosis, the chromosomes begin to line up along the equatorial plane exactly halfway between the two centrioles. Each chromosome now splits longitudinally into two exact halves, each half appearing at one side of the equatorial plane. During this phase the greatest effect of ionization occurs in the cells. When ionizing radiations pass through matter, the radiation energy may be absorbed by the atoms, or radicals, of the matter. In such instances the absorbed energy may eject one or more electrons from the atom and leave it in the form of a cation. Subsequently, free electrons may attach to other atoms or radicals of the tissue so irradiated and may become anions. Since cancer cells undergo mitosis more rapidly than normal cells, metaphase occurs more frequently, thus permitting the ionizing rays to effect destruction of cancer cells more rapidly than the normal cells of surrounding tissue.*

During anaphase, the third stage of mitosis, the new chromosomes are drawn to each pole. Simultaneously, a small distinct indentation appears at each end of the equatorial plane.

---

*This simplified explanation of the treatment of cancer with ionizing rays cannot include the necessary and detailed discussions of therapeutic dose, recovery time of cells, etc. Such information will be given students of radiologic technology in courses of radiation physics and radiobiology.

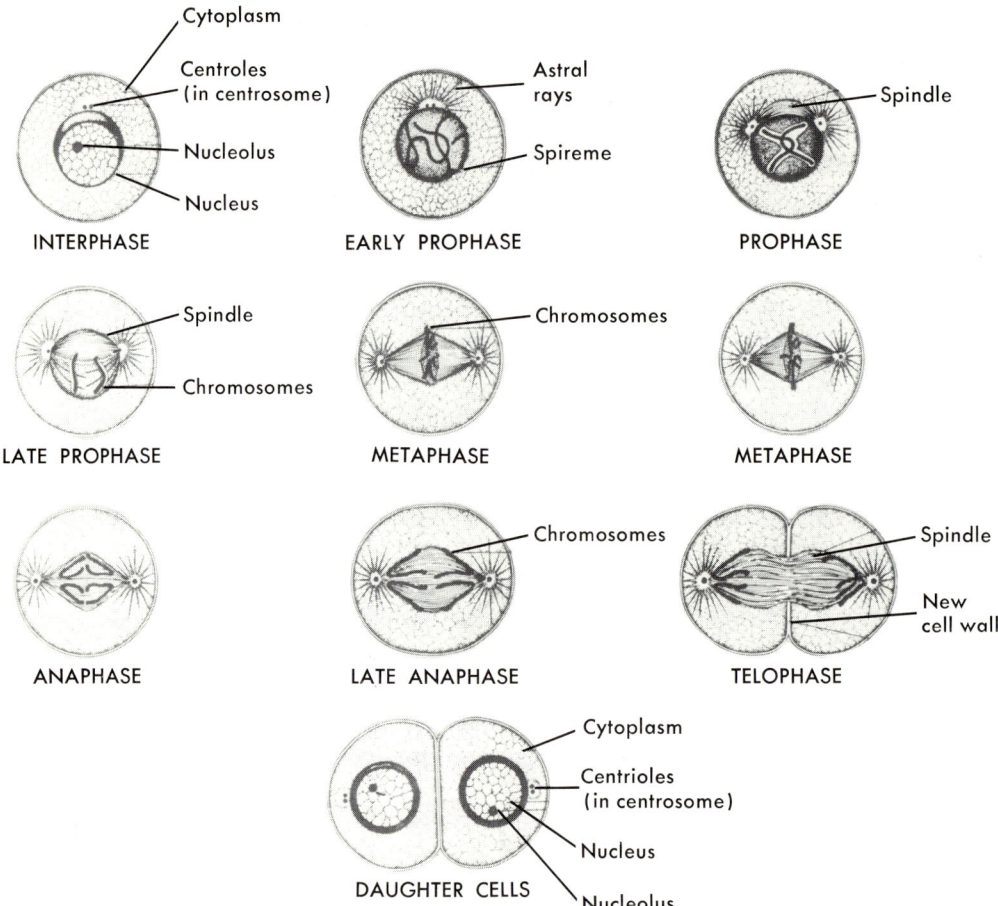

Fig. 2-3. Animal mitosis. (Courtesy General Biological Supply House, Inc., Chicago, Ill.)

Basic body structures and functions   19

In telophase, the fourth stage of mitosis, the indentations increase in depth and finally meet. There are now two new daughter cells, each possessing the same number of chromosomes as the parent cell. There now occurs a very brief period (interphase) in which the chromosomes of the new cells begin preparations for mitosis. During this interphase the DNA molecule replicates.* The principal steps of mitosis are demonstrated in Fig. 2-3.

Mitosis is the process by which all somatic cells mature and increase in number. The somatic cells include the four types of cells: *epithelial, muscle, nerve,* and *connective tissue.* These same four classifications serve as the names of the corresponding tissue types.

Meiosis is discussed in connection with a brief discussion of embryology on pp. 46 to 50.

Organizations of cells form the tissues. Tissues are structurally different and are therefore functionally different. The numbers, sizes, and shapes of cells along with the quantity and nature of intercellular substance determine the type of tissue.

*Replication is a turning back of a part so as to form a duplication.

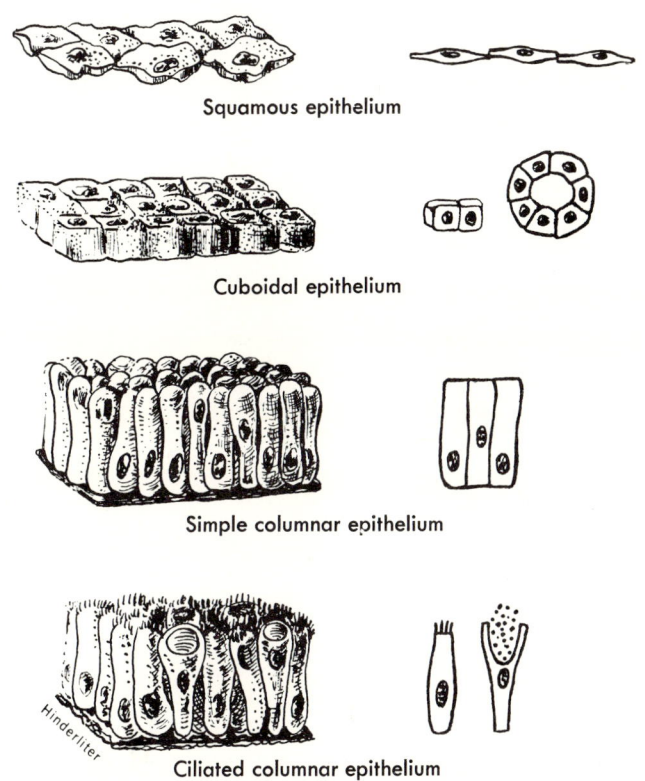

Squamous epithelium

Cuboidal epithelium

Simple columnar epithelium

Ciliated columnar epithelium

Fig. 2-4. Epithelial tissue.

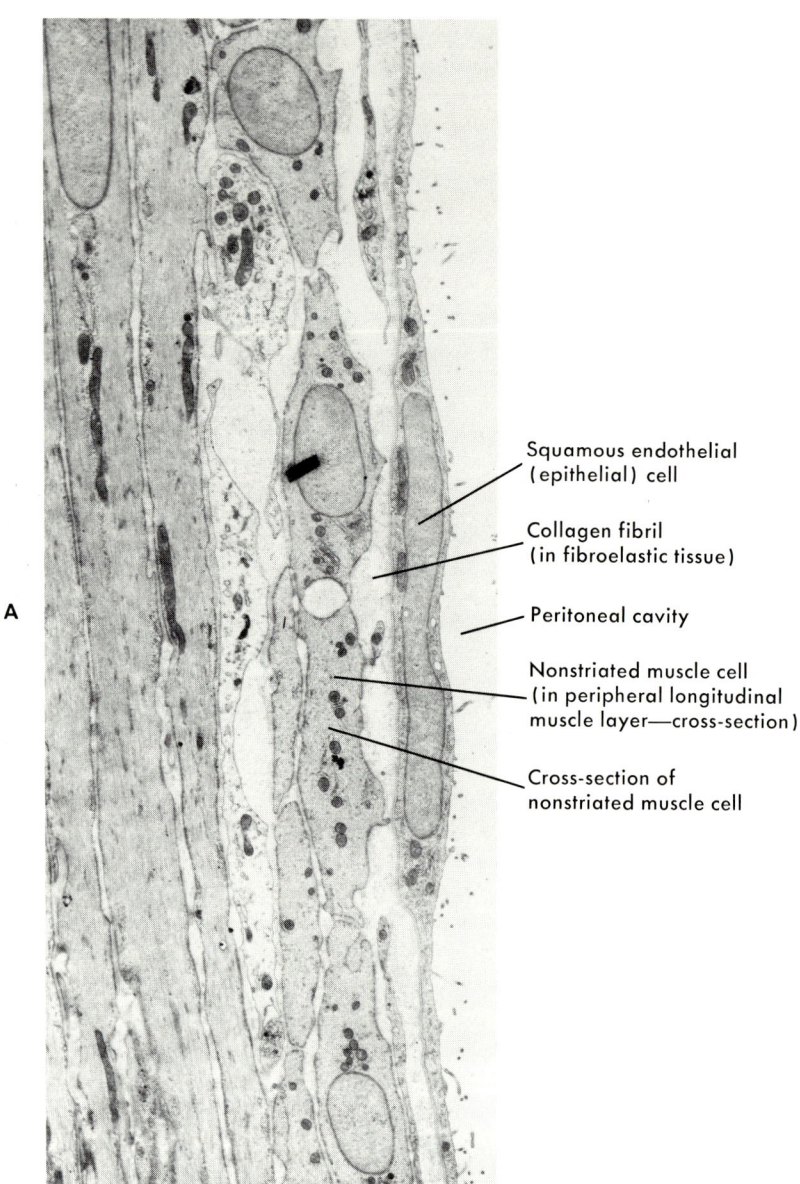

Fig. 2-5. Electron micrographs. **A,** Squamous epithelium of mouse intestine ($\times$ 5,000). **B,** Cuboidal epithelium from distal tubule of mouse kidney ($\times$ 2,000). **C,** Columnar epithelial cell of mouse jejunum ($\times$ 2,500). **D,** Ciliated columnar epithelium of human trachea ($\times$ 2,700). (From Rhodin, J. A. G.: An atlas of ultrastructure, Philadelphia, 1963, W. B. Saunders Co.)

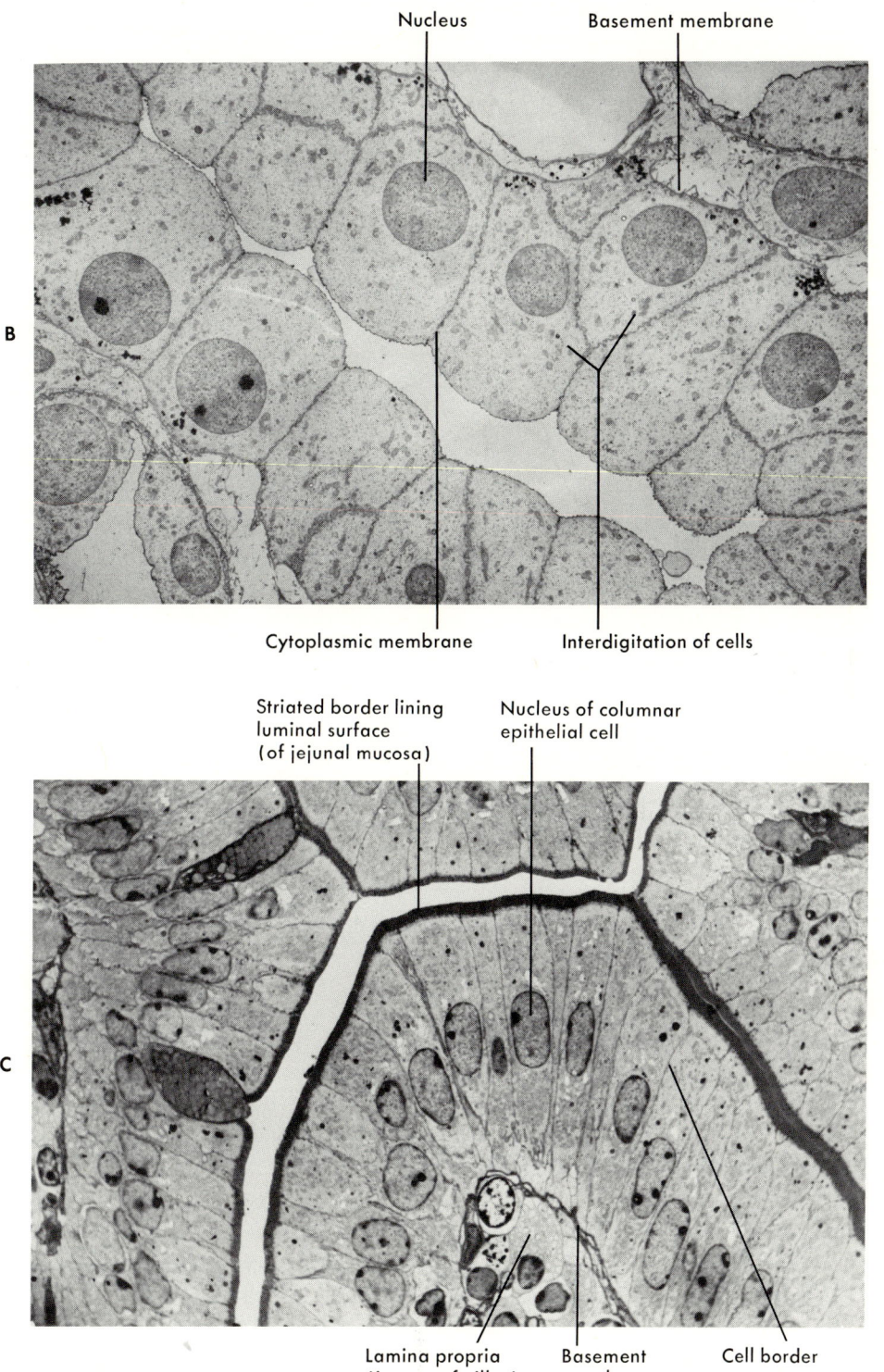

Fig. 2-5, cont'd. For legend see opposite page.

Continued.

22 *Textbook of anatomy and physiology in radiologic technology*

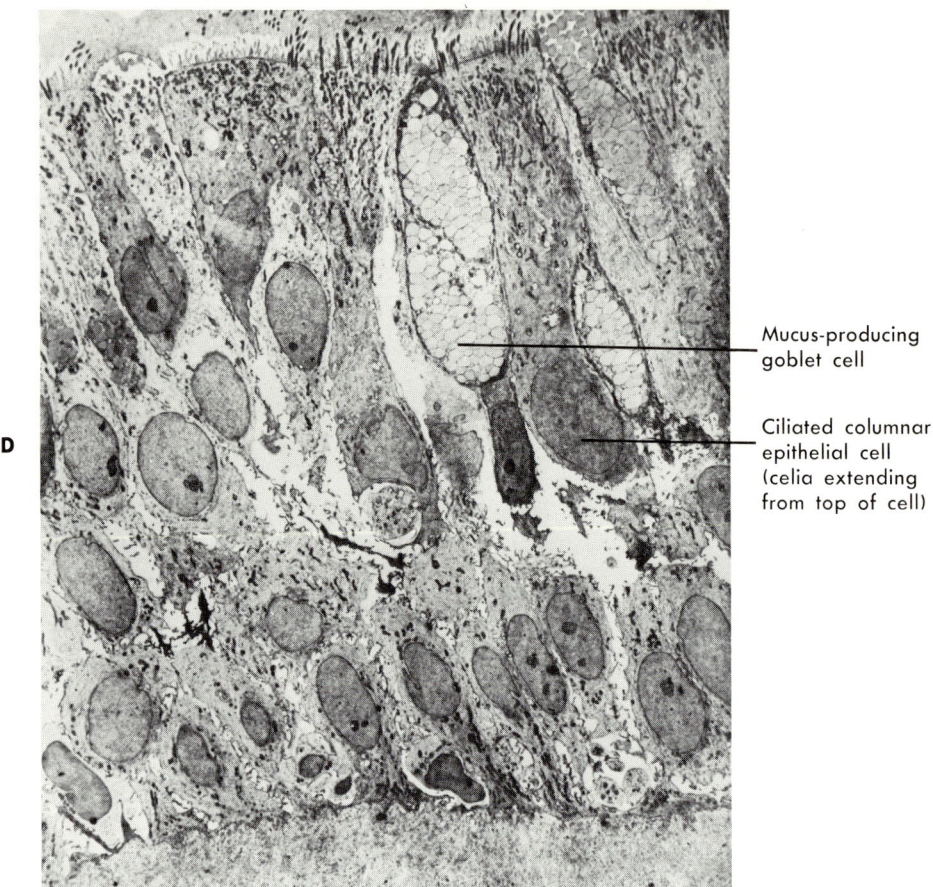

Fig. 2-5, cont'd. For legend see p. 20.

EPITHELIAL TISSUE

Epithelial tissue contains the least amount of intercellular substance and contains no blood vessels; the cells are nourished from the blood supply of the underlying connective tissue. The epithelial tissue consists of four subtypes of cells: *squamous*—flat and scalelike, *cuboidal*—shaped like a cube, *simple columnar*—coffin shaped, and *ciliated columnar*—the same shape as simple columnar with hairlike *cilia* on the upper surface (see Figs. 2-4 and 2-5).

Epithelial tissues are quite compact or dense, having the least amount of intercellular substance of all tissues. Cell outlines are fairly distinct and regular, and the nuclei are nearly always well defined. Most free surfaces, both within the body and on the outside of the body, are covered with some form or type of epithelium.

Epithelial tissues function to secrete* and to protect. In addition to secretion

---

*To secrete is to separate, elaborate, or emit as a secretion.

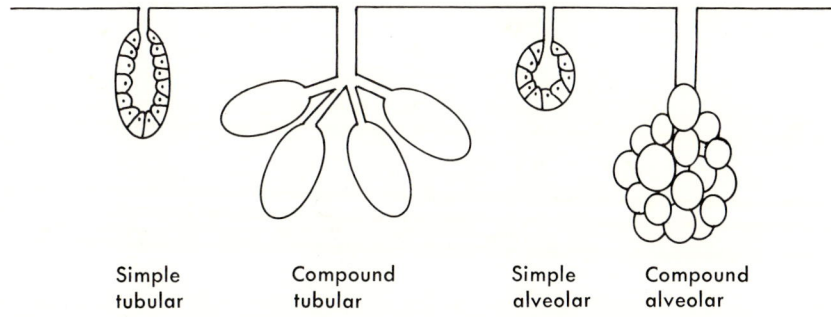

Fig. 2-6. Exocrine glands, tubular and alveolar.

and protection, ciliated columnar epithelium evidences a special function in that the cilia beat outwardly to move particles of material in one direction.

Of the four types of tissues, epithelial tissues have the ability to regenerate more rapidly following injury than the other tissues. Each of the glands in the body is composed of specialized epithelial cells. According to the method of excretion (or secretion) of their secreted substances, glands are either endocrine or exocrine. Endocrine glands secrete hormones directly into the blood as it passes through and flows in the gland. Exocrine glands are those of interest at this time, and expel their secretions via a duct or tube or onto a membrane. Exocrine glands may be simple tubular, compound tubular, simple alveolar, or compound alveolar as illustrated in Fig. 2-6.

The skin is composed of epithelium and serves as a protective covering for the entire body. Specialized epithelial cells form the mucous linings of many parts of the respiratory, urinary, digestive, and reproductive systems. The very special innermost linings of the circulatory system vessels are composed of *endothelium,* a variety of epithelium.

The skin

The skin is composed of two layers, the inner layer and the outer layer. The inner layer is the *dermis* (true skin, corium, or cutis vera), which is beneath the epidermis. The outer layer is the *epidermis.* (See Fig. 2-7.) The whole skin is called the *integument.* The dermis is the true, living tissue that sends nutritive materials into the innermost stratum of the epidermis.

The epidermis consists of four strata, which are, from inside to outside, the *stratum germinativum, stratum spinosum, stratum granulosum,* and *stratum corneum* (see Fig. 2-8). Each of these four strata consists of simple columnar epithelium. The stratum germinativum, like the dermis, is living tissue. The color of the skin is due to the presence of a pigment, *melanin,* which is formed in the stratum germinativum and which progresses outwardly through the other strata. The outer stratum (layer) of hard, flattened cells is in a constant state of sloughing and is replaced from below with cells of the next stratum. New cells form in the dermis and in the stratum germinativum; however, the cells formed in the

24 *Textbook of anatomy and physiology in radiologic technology*

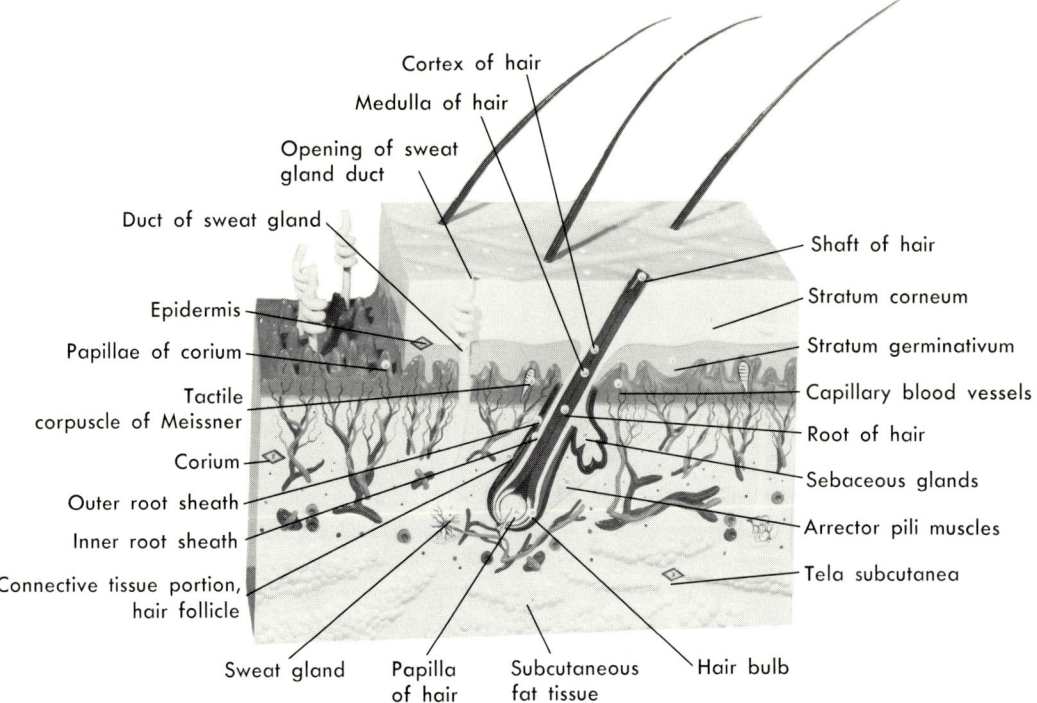

Fig. 2-7. Human skin (model), microscopic structures.

dermis do not contribute to the epidermis. The new cells of the epidermis push their way up to the surface from the stratum germinativum.

The skin functions to protect the body and aid in excretion of certain waste materials from the body. The excretory function is closely associated with the urinary system in maintaining the fluid and electrolyte balance of the body. A more detailed discussion of the skin appears in Chapter 7.

MUSCLE TISSUE

Muscle tissues consist of three subtypes of cells that differ in many ways from epithelial cells. The three types of muscle cells are as follows: *striated*—skeletal (voluntary); *nonstriated*—visceral (involuntary); and *cardiac*—branching, also involuntary. Muscle tissue functions in three ways: (1) to produce movement, (2) to maintain posture, and (3) to aid in the production of heat. The three types of muscle cells are demonstrated in Fig. 2-9.

Striated muscle

Striated muscle cells are multinucleate; i.e., they possess more than one nucleus. The nuclei are situated at random on the surface of the cell. The encasing membrane of striated muscle cells is the *sarcolemma*, and the cytoplasm of these cells is the *sarcoplasm*. The transverse striations are found throughout

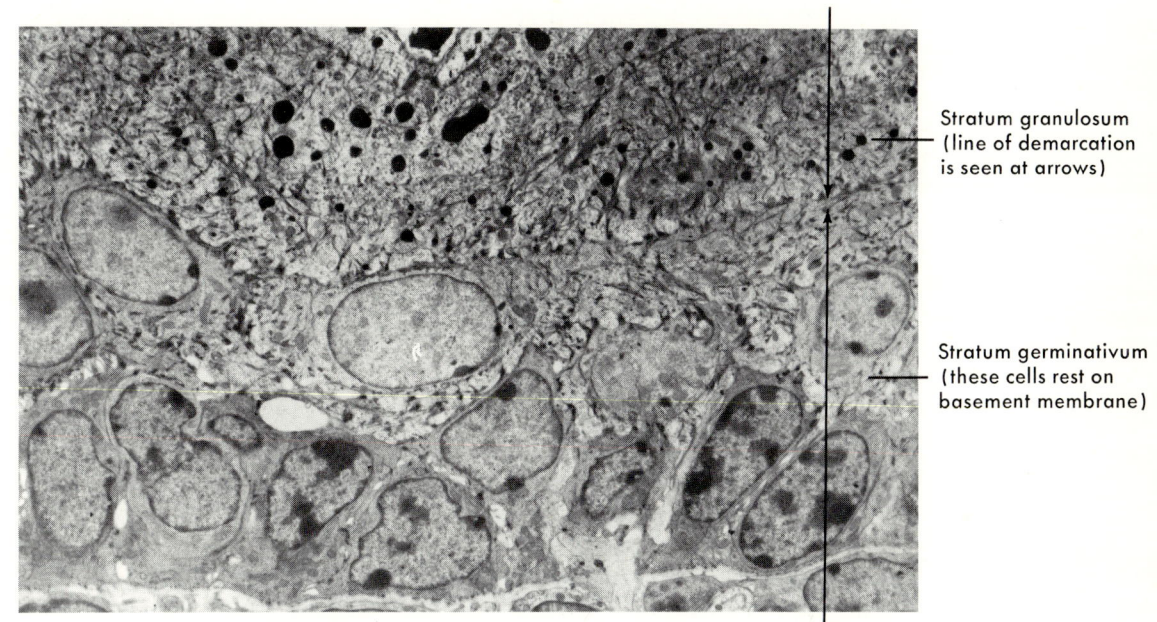
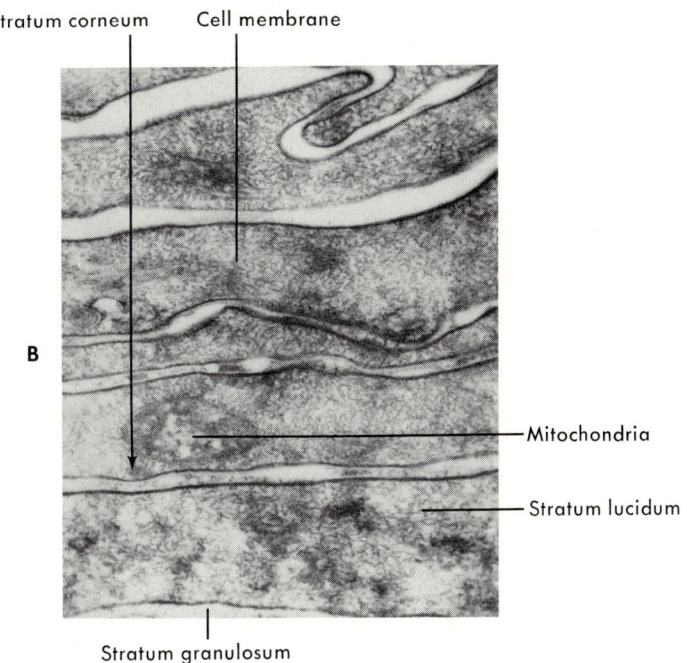

Fig. 2-8. Electron micrographs. **A,** Epidermis of newborn mouse (× 4,800). **B,** Stratum corneum (lower part) of mouse esophagus (× 66,000). (From Rhodin, J. A. G.: An atlas of ultrastructure, Philadelphia, 1963, W. B. Saunders Co.)

26 *Textbook of anatomy and physiology in radiologic technology*

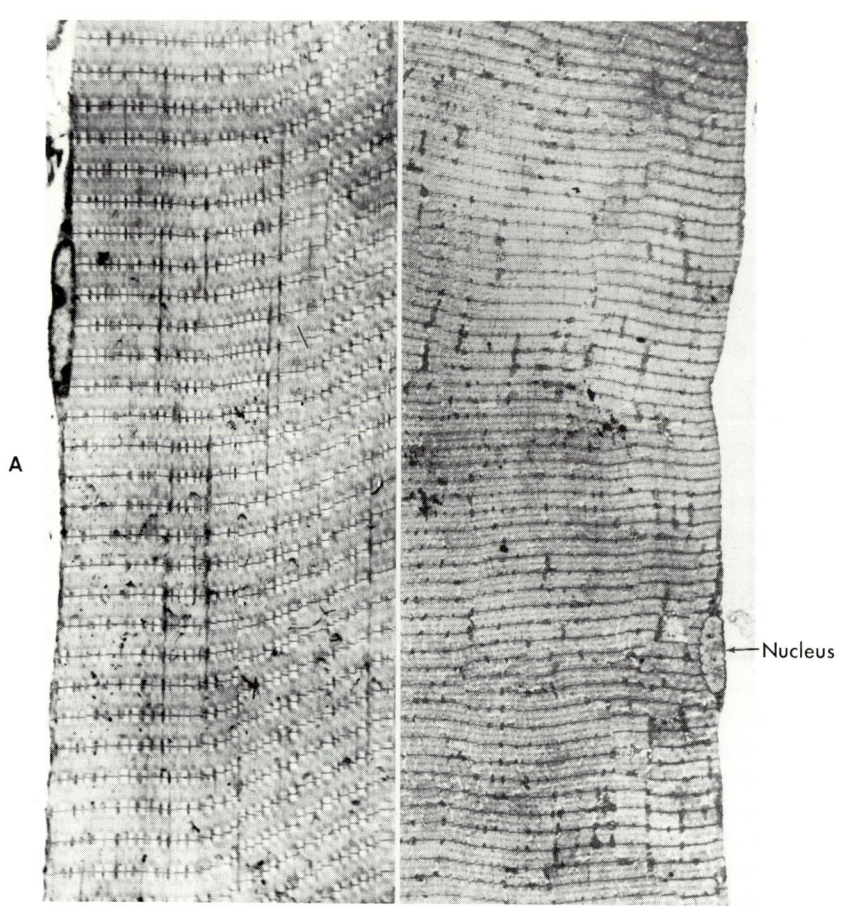

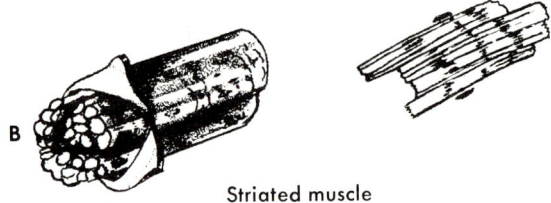

Striated muscle

Fig. 2-9. Electron micrographs. **A-B,** Striated muscle cell (in part) of mouse leg ($\times$ 2,800). In **A** the nucleus seen is one of several, usually located at the cell surface, of long striated muscle cell of a contracted muscle; the muscle in the left half of the illustration is relaxed. **C-D,** Non-striated muscle cell of mouse small intestine ($\times$ 1,200). In **C** the nucleus indicated is in a cell in the peripheral muscle layer; the inner circular layer is seen in the lower part of the illustration. **E-F,** Cardiac (branched) muscle cell of steer heart, ventricular septum ($\times$ 2,000). (**A, C, E** from Rhodin, J. A. G.: An atlas of ultrastructure, Philadelphia, 1963, W. B. Saunders Co.)

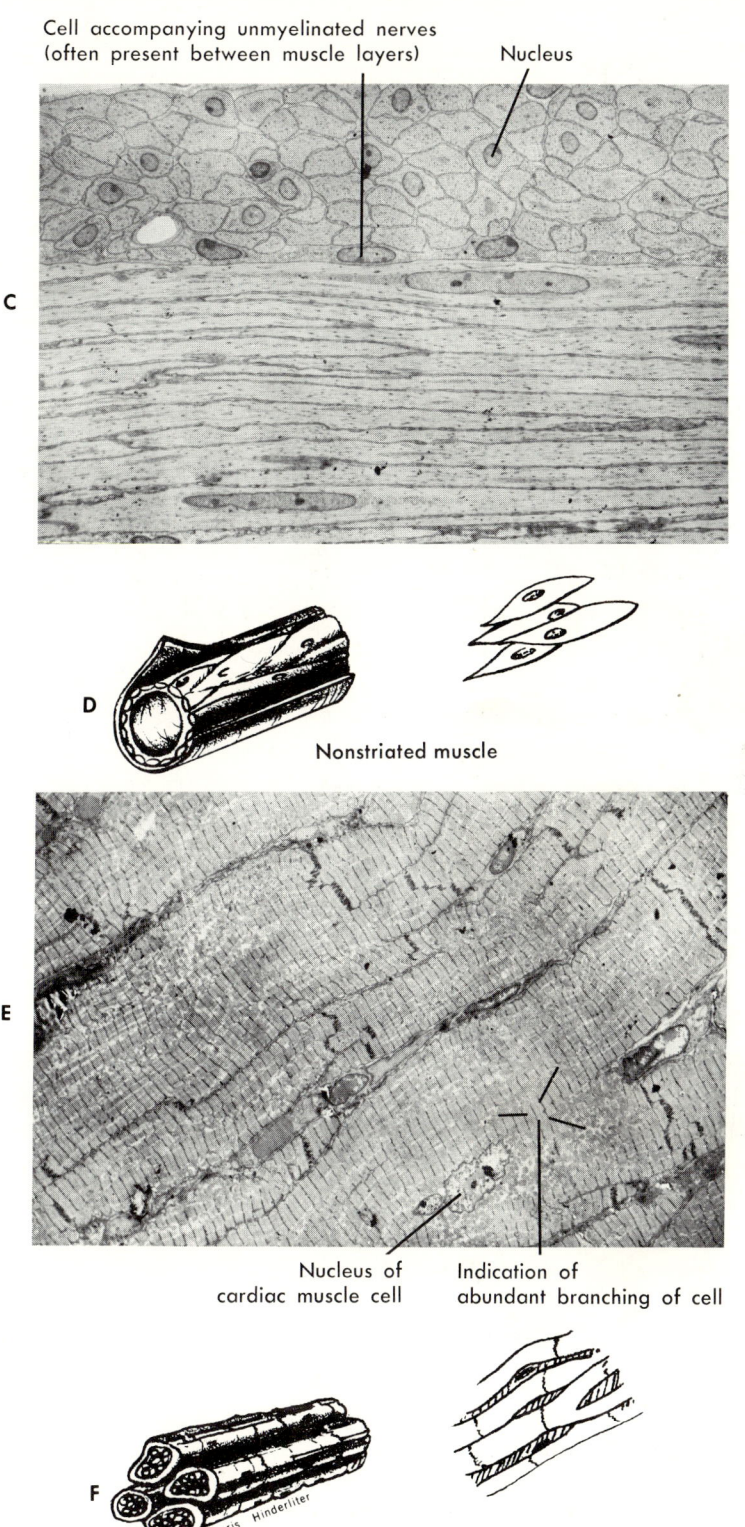

Fig. 2-9, cont'd. For legend see opposite page.

the length of each cell, and their presence accounts for the term "striated." These cells vary considerably in length, and in the larger muscles are bound together by connective tissue into bundles called *fasciculi*.

In most instances the striated muscles attach to bone, although some may attach to tendons and a few attach to skin or to cartilage. The chief attachment of any muscle is its *origin*, which, in the case of skeletal muscle, usually is the bone proximal to the joint. The opposite end of the muscle inserts (the *insertion*) into another bone, tendon, or ligament. The sites of attachment on bone usually are the roughened and/or raised areas: e.g., tuberosities, crests, etc. The origin of any muscle usually is the more stationary end of the muscle, and the insertion of this muscle usually is the more moveable end. In some instances a muscle will insert into a broad, flat tendinous sheet; this is an *aponeurosis* and is exemplified in the occipitofrontalis muscle and in the epicranial aponeurosis of the scalp. Movements of muscles usually apply across a joint.

Striated muscles are divided into three subtypes according to function as *prime movers, antagonists,* and *synergists.** A muscle responsible for the principal motion of a body part is the prime mover. A muscle that counteracts the action of another muscle is the antagonist of the second muscle. A muscle that assists another muscle in action is a synergist to the second muscle. Examples of the three muscle subtypes are as follows: prime mover—biceps of the upper arm in flexing the elbow; antagonist—triceps of the upper arm in extending the forearm; and synergist—extensor muscles of the dorsal surface of the wrist and forearm when the fist is clenched. (*Note:* A synergistic muscle may assist the prime mover to produce its movement, or the synergistic muscle may stabilize a body part to enable the prime mover to produce a more effective movement.)

Nonstriated muscle

Nonstriated muscle cells are mononucleate with the single nucleus in the center of the cell. In length, the nonstriated muscle cells are measured in thousandths of an inch as compared with striated muscle cells, many of which are about 1 inch in length. The ends of the nonstriated muscle cells exhibit considerable overlap of each adjacent cell. The nonstriated muscle differs from the striated muscle in the following comparisons:
1. Nonstriated muscle is considerably more sluggish in its contractions than striated muscle.
2. Nonstriated muscle is capable of remarkably greater stretching than striated muscle.
3. Nonstriated muscle exhibits much longer periods of sustained *tonus* than striated muscle.
4. Nonstriated muscle is innervated by two branches of the autonomic

---

*Another functional subtype is the fixation muscle, which is classed here with the synergists (see p. 365).

nervous system (from the sympathetic and parasympathetic divisions), whereas striated muscle is innervated by motor nerves.
5. Nonstriated muscle exhibits the phenomenon of rhythmic contractions as evidenced in the rhythmic peristaltic waves. This is not so in striated muscle.
6. Nonstriated muscle demonstrates a remarkably greater sensitivity to thermal and chemical stimuli than striated muscle.

Nonstriated (visceral) muscle tissue is found in the walls of the digestive tract, in the eyes, in the blood vessel and lymph vessel walls, in the urinary system, in parts of the reproductive system, and in the skin. The small nonstriated muscles in the skin that cause the condition of *goose flesh* to occur and raise the *hackles* on the necks of dogs are called the *arrector pili*.

Nonstriated muscle cells have no sarcolemma, they possess no cross striations, but they do possess very fine longitudinal markings.

Cardiac muscle

Cardiac muscle cells have a single nucleus and many branching fibers. The nucleus may be located near the center of the cell, or it may be in any one of the several branches. The branches intermingle with each other and appear to form a syncytium.* This type of muscle tissue is found exclusively in the heart (myocardium). Cardiac muscle cells have transverse striations, and cardiac muscle tissue is involuntary.

NERVE TISSUE

Nerve tissue functions to conduct messages (impulses) both to and from the central nervous system (CNS) along specific nerve pathways of nerve cells.

A conducting nerve cell is called a *neuron*. The body of a neuron may be any one of several shapes, such as stellate, oval, round, or pyramidal. The cytoplasm of the body extends in two opposing directions and forms two important processes, the *dendrons* (dendrites) and the *axons*. The axon usually is the longer of the two processes and always is single; it may possess small branches. *The axon always conducts the nerve message or impulse to the adjacent cell dendron and from the cell body.* The dendron may be single, or there may be several dendrites, each or all of which are in contact with the adjacent axon of another cell. *The dendrites conduct the messages to the cell body and from the adjacent axon.* A nerve cell, the body, and its processes, constitutes a oneway cell; i.e., the message must enter the cell via the dendrite, must pass through the cell body, and must be transmitted from the cell via the axon to the dendrites of the adjacent nerve cell. This pathway of impulse transmission may be remembered as being opposite in direction to the sequence of the alphabet: i.e., dendrite cell body to axon—DCBA.

*A syncytium is a multinucleate mass of protoplasm.

30   *Textbook of anatomy and physiology in radiologic technology*

Both axons and dendrites may vary in length from very short to quite long, depending upon location in the type of nerve fiber. Many nerve cells are bound together in bundles that resemble wire cables.

Large numbers of nerve fibers comprise these bundles, commonly considered as nerves. Considering a single nerve fiber, the *axis cylinder* (axon) extends from the cell body and is surrounded by a *myelin* (fatty) *sheath* shortly after leaving

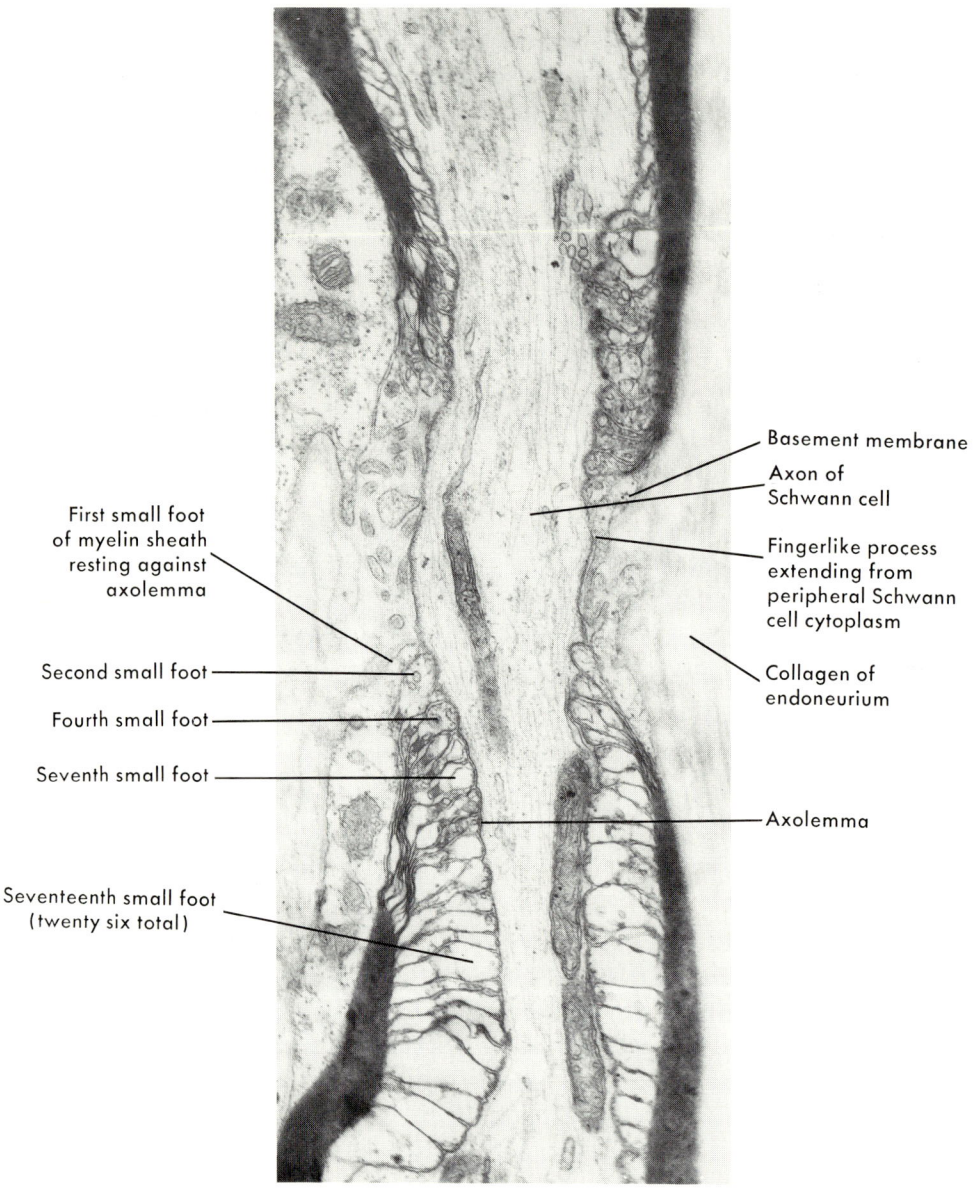

Fig. 2-10. Electron micrograph. Node of Ranvier of mouse sciatic nerve ($\times$ 32,000).

the cell body and prior to emerging from the spinal cord. The myelin sheath consists of both lipid and protein materials formed by the Schwann cells.* The myelin sheath serves the nerve fiber as a protective sheath and as an insulator. The outer covering of the myelin sheath and the axis cylinder is the *neurilemma*, a special membraneous covering. The neurilemma is constricted at regularly spaced intervals, the *nodes of Ranvier*. These nodes appear to assist propagation of the nerve impulses. Since the myelin sheath is much less evident (thinner) in the nodes (indentations), ionic movement between the inside and outside of the axon is facilitated in these parts. The propagation that occurs between the nodes of Ranvier is called saltatory† propagation. Fig. 2-10 is an electron micrograph of a node of Ranvier.

The passage of nerve impulses is invariably connected with a change in electrical potential. The frequency of nerve impulses may vary from as few as ten each second to as many as 1,000 each second. The conduction of nerve impulses is discussed in greater detail on pp. 344 to 346. Nerve impulses move with great velocity along the nerve fiber; the frequency of impulse transmission is a factor of the refractory period. Nerve impulses are transmitted with equal efficiency along all kinds of nerves.

Nerves are classified as either afferent or efferent. Afferent (sensory) nerves conduct messages toward the central nervous system while efferent (motor) nerves conduct messages from the central nervous system to the periphery of the body or to organs and/or muscles. Upon examination of the structure of both afferent and efferent nerve fibers, the axons and dendrites are found to be situated in opposite directions in relation to each nerve cell in afferent and efferent bundles.

CONNECTIVE TISSUE

Connective tissues are found interspersed extensively throughout all the other tissues of the body. Connective tissues are of several subtypes; one method of classification includes ten subtypes—*blood, bone, cartilage, adipose, areolar, white fibrous, yellow fibrous* (dense elastic), *lymphoid, reticular,* and *embryonal.*

In general, connective tissues have the dual functions of being the *filler* and *binder* substances of the body and of forming the *supporting framework* (skeleton) of the body and all the body organs. Connective tissues connect or bind certain tissues to each other, as with muscle to bone or as with bone to bone. Connective tissues form the basement membrane of most other tissues. In most tissues there is a predominance of cells, whereas connective tissues have but few cells: e.g., muscle tissues, which are composed mostly of muscle cells bound together by connective tissue, as compared with the cartilage and bone matrix, which contain but few cells. The predominant intercellular material of a particu-

---

*Schwann's cells are any of the cells of the neurilemma.
†Saltatory refers to a leaping motion by which the impulses move over the myelin sheath between the nodes.

lar connective tissue determines the general nature of the tissue, as in bone, cartilage, tendon, blood, etc.

Blood

Blood is fluid connective tissue that bathes, either directly or indirectly, each living cell in the body. Blood conveys all the nutritive substances to the cells; it conveys the toxic and exhausted materials from the cells to the excretory organs.

Blood consists of approximately from 42 to 47 percent cells and from 58 to 53 percent plasma (fluid). Blood comprises from 6 to 8 percent of the total body weight, the usual quantity varying between 5 and 6 liters, depending upon the size and health of the individual.

The blood cells include two major types: *erythrocytes* and *leukocytes*. The erythrocytes (red cells) number from 4,000,000 to 6,000,000 in each cubic millimeter of blood; the leukocytes (white cells) number from 5,000 to 10,000 in each cubic millimeter of blood. The *platelets* (thrombocytes), which are not true cells because they possess no nucleus, average between 250,000 and 400,000 in each cubic millimeter of blood. These values vary considerably between sexes and with age and with disease. The mature erythrocyte, which appears in peripheral

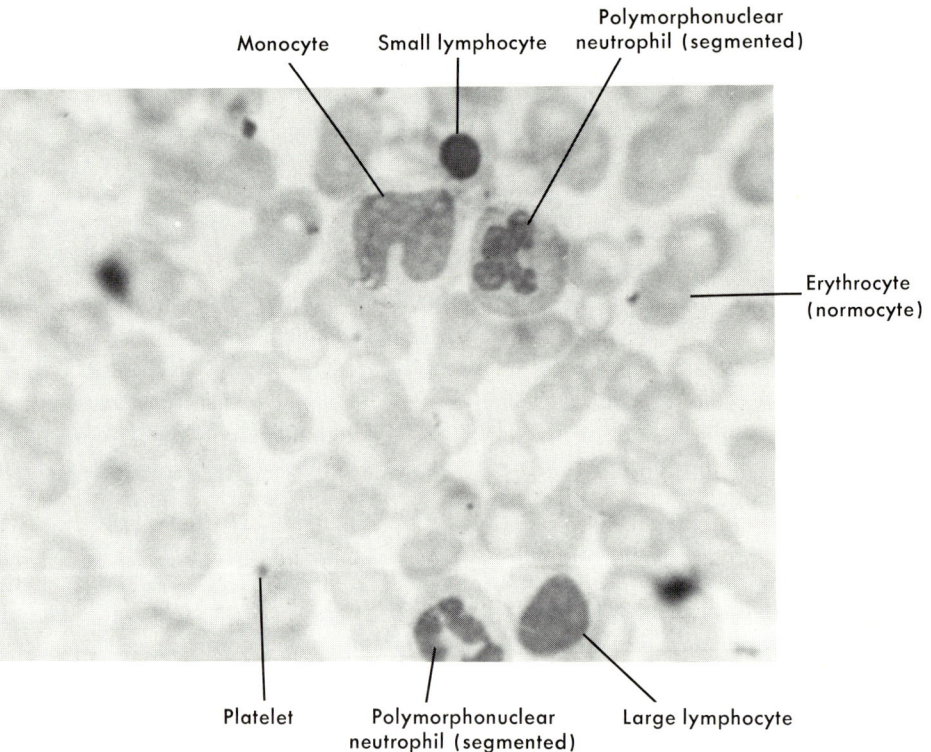

Fig. 2-11. Normal peripheral blood smear, Wright's stain. (Courtesy Dr. G. R. Nicholson.)

blood, usually has no nucleus and averages approximately 7 microns* in diameter. The leukocyte is considerably larger, has a definite nucleus, and includes three major subtypes: the *neutrophils, lymphocytes,* and *monocytes.* The platelets are quite small, averaging between 2 and 4 microns in diameter. Fig. 2-11 is a photomicrograph of a normal blood smear stained with Wright's stain.

Osseous tissue

Bone tissue is the substance of all bones and contains the least amount of water of all the tissues. In embryonic life the deposition of calcium begins in the cartilage at varying stages of growth and development. Calcification of cartilage in the growing ends of the long bones continues throughout adolescence until the bones have attained their predetermined length or until injury or disease terminates growth. Bones grow by means of organized haversian systems and by unorganized interstitial lamellae as discussed on pp. 178 to 186. *Osteoblasts* are the immature bone cells, and *osteocytes* are the mature bone cells.

There are two types of bone tissue—*cancellous* (spongy) and *cortical* (true, compact, or hard). Cancellous bone is found in the diploë of the skull, in the ribs and other flat bones, in the body of the sternum, and in the ends of the long

*One micron = 1/1,000 mm. = 1/25,000 inch.

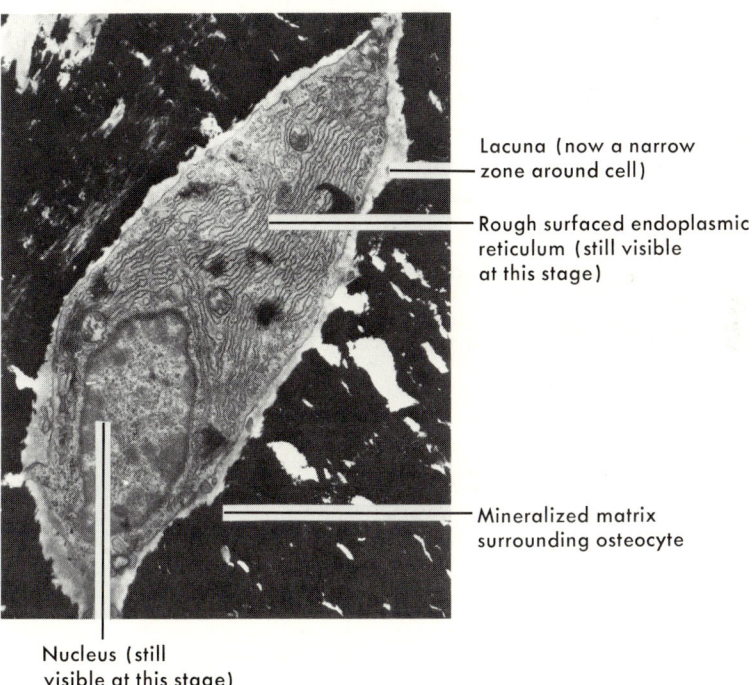

Fig. 2-12. Electron micrograph. Osteocyte (bone cell) of newborn mouse tibia (× 10,000). (The light triangular space in upper left corner is a part of a canaliculus interconnecting the lacunae.) (From Rhodin, J. A. G.: An atlas of ultrastructure, Philadelphia, 1963, W. B. Saunders Co.)

bones. Cortical bone comprises the hard substance of bones. Cortical bone is covered by a tough outer layer of connective tissue, the *periosteum*.

The functions of bones and the skeletal system are the same. Bones function as *sites of attachment* for muscles and tendons, *support* for the body, *protection* for much of the vital viscera, and *sites for cytogenesis** and *hemopoiesis*.

Bones are described in four classes: *short, long, flat,* and *irregular*. The structure of bone and the composition of the skeletal system are of such importance to radiologic technologists as to warrant detailed description. For this reason many parts of osteology are excluded in this chapter and are contained in Chapter 4.

The cancellous bone in the epiphyses and toward the epiphyseal ends of long bones appears grossly and radiographically as a fine network of trabeculae.

Cartilage tissue

Cartilage is subdivided into three lesser subtypes by older classifications but is considered to be a single type with certain modifications by more modern classification methods. The older method of cartilage classification included *hyaline*—found in the nasal septum, larynx, trachea, bronchi, and sternal ends of the ribs and covering the articular surfaces of bones; *fibrous*—found in the symphysis pubis and forming the discs between the vertebral bodies; and *elastic*—forming the cartilaginous portions of the external ears and the walls of the auditory tubes. The modern concept is that all cartilage is hyaline and that each subtype is a variation or modification of hyaline cartilage.

In the embryo and during the growth years, cartilage functions to form the *model* for the future skeleton (except in the cranial vault). In the adult, cartilage functions to act as a shock absorber in many places and to furnish a smooth surface for articulation in many other places. (*Note:* The latter two functions exist in the growth years and in the adult.)

Cartilage is distinguished from other connective tissues in that *cartilage possesses no vascular network;* it derives its nourishment from the *perichondrium,* a dense connective tissue membrane.

Adipose tissue

Adipose (fatty) tissue appears in most other tissues. Each cell of this tissue contains a rather large droplet of fat in the cytoplasm. Adipose tissue cells are present in very small quantity when compared with other cells. Adipose cells are packed quite closely and possess a rather small quantity of intercellular material between each two cells.

Areolar tissue

Areolar tissue is very thin and loosely arranged. It contains many of the different types of connective tissue cells scattered infrequently throughout the

---

*Bone cells and most subtypes of leukocytes form in certain bone tissues.

viscid intercellular substance. Areolar tissue contains elastic fibers and collagenous fiber bundles.

Viewed under the microscope, areolar tissue appears to contain numerous gaps (holes) in its surface.

White fibrous tissue

Viewed macroscopically, white fibrous tissue has a distinct sheen and tends to glisten when struck by light. It contains very few cells and many collagenous fibers. It constitutes the substance of ligaments and tendons.

Yellow fibrous tissue

Yellow fibrous tissue is similar to white fibrous tissue, but it is yellowish in color. Yellow fibrous tissue contains numerous closely packed, elastic fibers, few cells, and small quantities of ground substance; matrix exemplifies yellow fibrous tissue.

Lymphoid tissue

Lymphoid tissue is contained in a fluidlike ground substance and contains numerous lymphocytes tightly packed within a dense network of reticular fibers. Lymphoid tissue comprises the lymph nodes, including the tonsils and the adenoids, and is the site for lymphocyte manufacture.

Reticular tissue

Reticular tissues function as supporting frameworks for many other tissues; e.g., the lymphoid tissue. Reticular tissues are simple, rather dense networks of reticular fibers.

Embryonal tissue

Embryonal tissue exists in the embryo as mesenchyme and in the form of mucous connective tissue. It is superabundant in the embryo and almost nonexistent in the adult.

MEMBRANES

A membrane is a limiting protoplasmic surface. Membranes are composed of one or more of the four major tissues. Understanding the composition and functions of membranes is most important in the study of physiology because all fluids, gases, and solids that pass into or out of the blood must pass through one or more membranes.

Membranes are very thin sheets of tissue furnishing protective and/or excretory linings for much of the body. The mechanism of fluid (solvent) and solid particle (solute) passage through membranes (physiochemical processes) is explained briefly in the following paragraphs.

All substances must be reduced into their elemental components, either molecular or atomic, to effect passage across membranes. These processes are discussed in greater detail in Chapters 5 to 8. "Decomposition" and "hydrolysis" are terms applied to such chemical actions. Decomposition is the process of decomposing or the state of being decomposed or distintegrated. Hydrolysis is a chemical process of decomposition involving addition of the elements of water. In general, food substances must be hydrolyzed into their simplest molecular forms in preparation for passage through the membrane. Certain physical phenomena govern the passage of atoms and molecules through membranes. Membranes are either (completely) permeable or semipermeable. *An atom is the smallest quantity of an element that can exist and still retain the chemical properties of that element. A molecule is an aggregation of atoms or a chemical combination of two or more atoms that forms a specific chemical compound.* In addition to the phenomena of decomposition and hydrolysis is the phenomenon (process) of *electrolysis*.

Electrolysis is the act or process of conduction of an electric current by an electrolyte* through the presence of charged particles, called ions, in the electrolyte. When the ions reach the electrodes, chemical changes occur if the applied potential is sufficient. The new substances may be deposited at the electrodes or liberated as gases, or they may react chemically with the liquid to form new ions.

Thus through decomposition, hydrolysis, or electrolysis, the food particles are reduced to their elemental forms in preparation for absorption into the blood stream. Also, certain harmful and toxic wastes are eliminated from the blood stream for eventual discharge from the body. Absorption and excretion result from other physical phenomena. Those pertinent in this text are dialysis, diffusion, filtration, osmosis, phagocytosis, and pinocytosis.

Dialysis

In some respects dialysis can be compared with chemical filtration. *Dialysis is the separation of crystalloids and colloids in solution by means of their unequal diffusion through certain natural or artificial membranes.* Crystalloids diffuse readily; colloids, not at all or very slowly. Crystalloids are smaller than colloids and are thus permitted to pass through the animal membrane, whereas colloids are retained inside the membrane sac or on the initial side of the membrane. This is exemplified by placing a starch solution inside a dialysis bag and placing the dialysis bag in an iodine solution. The iodine solution will pass through the bag wall (membrane) into the starch solution; the starch will be retained inside the dialysis bag. The following reactions are noted:
1. As the iodine solution enters the starch solution, the typical blue color or starch-iodine reaction appears in the starch solution.
2. No color change occurs in the iodine solution outside the dialysis bag (see Fig. 2-13).

---

*An electrolyte is an electric conductor in which passage of current is accompanied by liberation of matter at the electrodes.

Basic body structures and functions 37

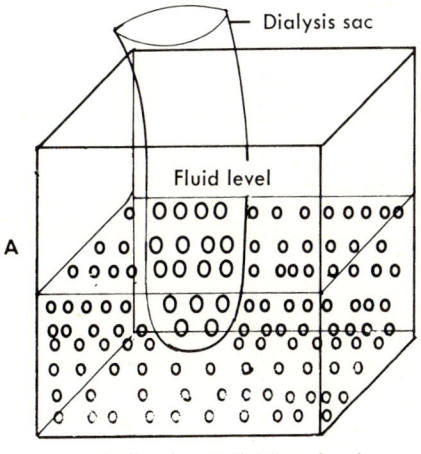 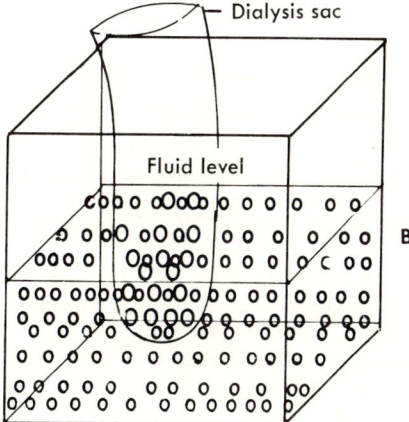

o = Iodine (crystalloid) molecule
0 = Starch (colloid) molecule

Fig. 2-13. Dialysis separating colloids from crystalloids. **A,** The dialysis sac containing a starch solution, which is colloidal, has been placed in a glass vessel containing an iodine solution, which is crystalloidal. **B,** The dialysis membrane permits penetration of the crystalloid-sized iodine molecules but prevents outward penetration of the colloid-sized starch molecules. As a result, the starch and iodine molecules mix inside the dialysis sac.

Diffusion

Diffusion is the property of a gas or liquid to distribute itself in equal concentration throughout a second gas or liquid by action of the kinetic motion[*] of the gas molecules. In the process of diffusion, the diffusing substance moves from an area of higher concentration to one of lower concentration. Diffusion occurs when two liquids of different concentration are placed in the same container; this is true unless the liquids are immiscible.[†] Diffusion occurs between two gases[‡] of different concentration, between two liquids of different concentration, and through (or across) permeable membranes. Diffusion of solute particles and solvent molecules occurs in both directions through permeable membranes.

Solutes diffuse through a nonliving membrane into a solution (solvent molecules) of lower concentration, whereas water, the solvent, diffuses in the opposite direction. The kinetic motion of both water (solvent) molecules and sugar (solute) molecules causes these molecules to strike a membrane placed between (for example) a volume of 10-percent sugar solution and an equal volume of water. If the membrane is permeable to both sugar and water and separates

[*]The kinetic theory of gases assumes that the particles of gas move in straight lines with high average velocity, continually encountering one another and hence changing their individual velocities and directions, and that the pressure of the gas is due to the impact of the particles against the walls of the containing vessel.
[†]Immiscible liquids are incapable of mixing or of being mixed.
[‡]The gas laws are discussed on p. 319.

38 *Textbook of anatomy and physiology in radiologic technology*

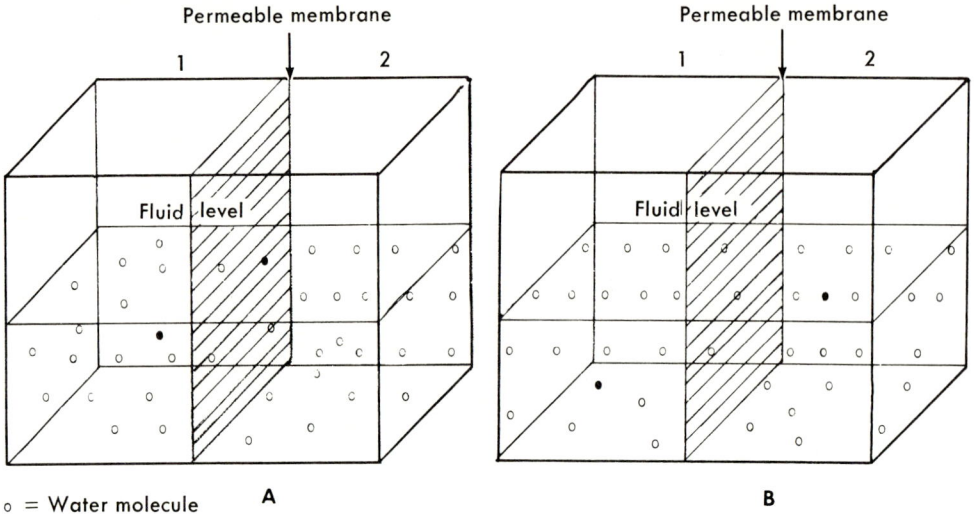

o = Water molecule
• = Sugar molecule

Fig. 2-14. Diffusion across a membrane. **A,** Side **1** contains a solution of 10 percent sugar and 90 percent water; side **2** contains a solution of 100 percent water. The two sides are separated by a permeable membrane. **B,** Side **1** and side **2** contain solutions of 5 percent sugar and 95 percent water. Equilibrium in concentration has been achieved on each side of the permeable membrane.

equal volumes of solutions, the sugar molecules will pass from the sugar solution into the 100-percent water solution at the same time that the water molecules pass from the 100-percent water volume into the 10-percent sugar–90-percent water volume. After a sufficient period of time elapses, each volume will be 5 percent sugar and 95 percent water. The sugar molecules pass first into the pure water and later in both directions across the membrane. Water molecules pass in both directions across the membrane. The net passage (diffusion) of sugar molecules is from the higher concentration to the lower concentration: i.e., from the 10-percent solution into the pure water solution. The net passage (diffusion) of water molecules is from the higher concentration (100 percent water) into the lower concentration (10 percent sugar and 90 percent water). The cells comprising such membranes do not participate in the process of diffusion; therefore, diffusion (like dialysis) is said to be a passive transport mechanism (see Fig. 2-14).

Filtration

If a mixture or suspension of barium sulfate in water is poured into a filter paper that is folded and placed in a filter funnel, the water passes through the filter paper and the barium sulfate remains spread over the upper surface of the filter paper. The water has been filtered and is called the *filtrate. Filtration is the process of passing through (or as through) a filter (percolation), hence diffusion.*

The force of gravity acts upon the water to draw (attract) the water through

Basic body structures and functions 39

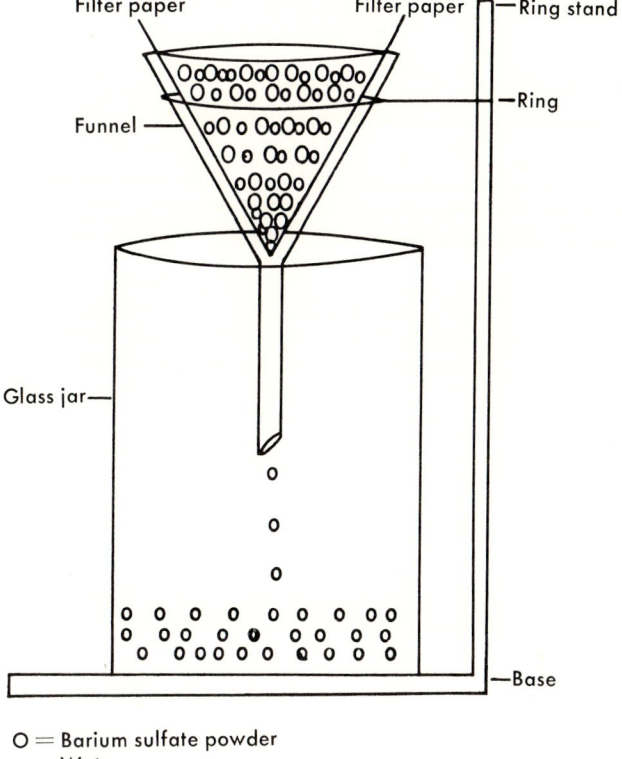

O = Barium sulfate powder
o = Water

Fig. 2-15. Filtration. The mixture of barium sulfate and water within the filter paper settles in the filter paper as the water (filtrate) passes (filters) through the filter paper. The barium sulfate (in powder form) is too large to pass through the filter paper, and so it is filtered from the mixture. (The barium sulfate is *held back* by the filter paper.)

the filter paper. The gravity force is sufficient to overcome the static condition imposed upon the water by the filter paper. This is an example of hydrostatic pressure. *Hydrostatic pressure is that pressure resulting from the weight of the liquid and exerted uniformly and perpendicularly to its surface.*

In some instances both the solvent molecules and the solute particles are so small that each passes through the filter or membrane. This occurs when the hydrostatic pressure is higher on one side of the membrane than on the other side. Hydrostatic pressure is the force pushing both solvent and solute through the permeable membrane. When permeable membranes separate solutions having different hydrostatic pressures, the solvent and solute particles having the greater hydrostatic pressure pass through the membrane into the solution (filtrate) with the lesser hydrostatic pressure (see Fig. 2-15).

Osmosis

Osmosis is the passage of a pure solvent from the lesser concentration to the greater concentration when the two solutions are separated by a membrane that selectively prevents the passage of solute molecules but is permeable to the solvent. In general, the net movement will be from a solution of higher (solvent) concentration into a solution of lower (solvent) concentration. A permeable membrane will permit passage of water and many small solute particles. Some membranes permit passage of water only. A semipermeable membrane permits movement of all molecules (including water) that are smaller than the pores of the membrane. A selectively permeable membrane may permit passage of some substances more readily than passage of other substances.

If a membrane is permeable to water and impermeable to sucrose (one kind of sugar), the membrane is said to be a semipermeable membrane. If such a membrane separates a 10-percent solution of sucrose in water from a solution of 100 percent (pure) water, the pure water (being of higher concentration than the 90-percent water–10-percent sucrose solution) will pass through the membrane into the 90-percent water–10-percent sucrose solution; there will be no passage of sucrose into the 100 percent water. The net osmotic pressure increases in the sucrose solution because the net movement of water is into the sucrose solution. The water moves into the sucrose solution, causing the *osmotic pressure* that results from the process of diffusion. In this example, the sucrose solution is *hypertonic* to the water, and the water is *hypotonic* to the sucrose solution. If equilibrium of the two solutions is achieved as discussed on pp. 37 and 38, the two solutions would be *isotonic* to each other (see Fig. 2-16).

Using erythrocytes and blood fluid as examples, hypertonic, hypotonic, and

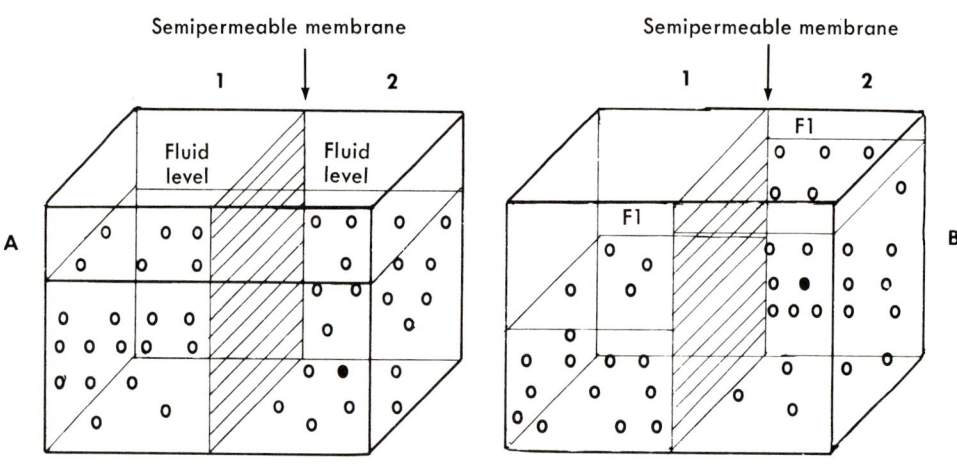

o = Water molecule
● = Sucrose molecule

Fig. 2-16. Osmosis. **A,** Side **1** contains a solution of 100 percent (pure) water; side **2** contains an equal volume, but the solution is 10 percent sucrose and 90 percent water. **B,** Side **1** contains a solution of 100 percent (pure) water; side **2** contains a solution (of sucrose and water) of greater volume than side **1**. The water molecules have bombarded the semipermeable membrane on both sides, but the net (osmosis) movement of water molecules has been from side **1** into side **2**, i.e., from higher water concentration to lower water concentration. **Fl,** Fluid level.

isotonic solutions are explained in the following statements. In a hypertonic solution, the diffusion (osmotic) pressure is greater than that of the plasma or of the intracellular content. Therefore, the erythrocytes lose water to the plasma and the cell membrane shrinks; the cell becomes crenated. In a hypotonic solution, the opposite conditions exist. The diffusion (osmotic) pressure is less than that of the plasma or intracellular content. Therefore, the erythrocytes "take in" water from the plasma and swell. If the quantity of water taken in exceeds that which can be contained successfully, the cell membrane will rupture (the cell undergoes lysis). When applied to the erythrocyte, the proper term is hemolysis; i.e., the erythrocyte membrane ruptures and permits the hemoglobin to escape into the plasma.

Under normal conditions the blood plasma is isotonic with the erythrocyte contents, and neither crenation nor hemolysis occurs.

Phagocytosis

In the preceding four methods of fluid passage through membranes, passive physical forces have been involved. Both phagocytosis and pinocytosis employ active physiologic forces; i.e., the membrane cells actively participate in the passage of substances through the membrane.

Phagocytosis is the ingesting or engulfing and usually destroying of microorganisms and other foreign bodies by phagocytes. *A phagocyte is any cell that ingests microorganisms or other cells and substances.* The phagocyte often digests the ingested material. Phagocytes may be fixed cells of the reticuloendothelial* system, or they may be motile cells of the circulatory system. Of the leukocytes found in peripheral blood, both monocytes and polymorphonuclear neutrophils are phagocytic. Macrophages, large wandering mononuclear cells originating in the tissues, are known to be phagocytic. Small interstitial cells of the nerve tissue, called microglia, migrate within the nerve tissue and are phagocytic for waste products of the nervous system.

In the process of phagocytosis a small segment of the cell membrane surrounds a small particle of foreign matter outside the cell wall. This segment of membrane next separates from the cell and then moves into the cell to resemble a vacuole or vesicle with no external opening. Excessive quantities of ionizing radiations are known to decrease and sometimes to terminate phagocytic activity.

Pinocytosis

Pinocytosis differs from phagocytosis in the material engulfed. Pinocytosis involves the engulfing of fluid particles. The processes of phagocytosis and pino-

---

*The reticuloendothelial system consists of the cells of the body having both endothelial and reticular attributes and showing a common phagocytic behavior toward dyestuffs, including cells of the spleen and lymph nodes, Kupffer cells of the liver, the reticuloendothelium of the bone marrow, and the clasmatocytes; this system is concerned in blood cell formation and destruction, storage of fatty materials, and the metabolism of iron and pigment and plays a defensive role in inflammation and immunity.

cytosis are similar. Much remains to be learned regarding what makes these processes possible.

The various physiochemical processes discussed in this section are related to the integration of body functions and to maintaining homeostasis. Homeostasis includes the very important responsibility of maintaining the body's fluid and electrolyte balance.

HOMEOSTATIC BALANCE OF FLUIDS AND ELECTROLYTES

From 80 to 95 percent of the total volume of substances taken into the body through the mouth are liquid, mostly water. Also, some water forms within the body from metabolism of food. The body eliminates water and waste fluid via the digestive, respiratory, and urinary systems and the skin. *A cardinal rule of homeostasis is that the fluid output must equal the fluid intake and, conversely, the fluid intake must equal the fluid output.* All body fluids are distributed among the three fluid compartments of the body as follows:
1. Intracellular fluid is within the cells.
2. Interstitial fluid is in the interstitial spaces (between the cells and tissues).
3. The blood fluid is contained within the circulatory system.

These three compartments and their fluid contents comprise the *structural distribution of fluid; functional distribution* places all the fluid volume either in the cell (intracellular) or outside the cell (extracellular). In order for homeostasis to exist, the total volume of fluid in the body must remain relatively constant and the volume of fluid in each of the three fluid compartments must remain relatively constant.

The homeostatic condition of the body fluids is interdependent upon the same condition of body electrolytes.* It is not possible for the body's fluid balance to be stable without its electrolyte balance also being stable. The ratio of fluid intake to fluid output is partially regulated through the lungs, skin, and fecal excreta. The principal regulatory structures in maintaining the proper relation of fluid intake to fluid output are the kidneys. There is automatic adjustment within the kidneys between fluid intake and fluid output. The mechanisms of regulation of fluid intake to fluid output are discussed in Chapter 7, which treats those specific structures; see p. 305 for a discussion of water eliminated through the lungs; see pp. 302 and 303 for a discussion of water eliminated through the skin; and see pp. 211 and 229 for discussions of the volume of water reabsorbed through the walls of the large intestine. Chapter 7 presents additional discussion of fluid and electrolyte elimination and reabsorption in the kidneys.

In addition to an obviously necessary balance of fluids and electrolytes for homeostasis, there is an equally necessary balance between acids and bases for

---

*An electrolyte is an electrical conductor in which passage of current is accompanied by liberation of matter at the electrodes; also, a substance such as an acid, base, or salt becomes this type of a conductor when dissolved in a suitable solvent. The current is carried by charged particles, ions. It is believed that the ions exist in the substance and are rendered mobile by the solvent or by heat.

homeostasis. The balance between acids and bases in the body refers to the hydrogen ion concentration of body fluids.

HOMEOSTATIC BALANCE OF ACIDS AND BASES
Expression of hydrogen ion concentration in terms of pH

All solutions of which water is a constituent (or is the solvent) contain free ions of H and OH. H ions are positive and are written as H^+. OH ions are negative and are written as OH^-. When the number of H ions exceeds the number of OH ions, the solution is said to be *acid*. Conversely, when the number of OH ions exceeds the number of H ions, the solution is said to be *base*, or *alkaline*. When the H^+ and OH^- ions are equal in number, there is a balance and the solution is said to be *neutral*. A change in one group of ions is accompanied by an inverse and proportional change in the other group of ions.

At neutrality a liter of water contains 0.0000001 gram of ionized hydrogen and an equivalent quantity of OH^- ions. This concentration of H^+ (C_H) at neutrality may be expressed as follows:

$$C_H = 0.0000001 = \frac{1}{10,000,000} = \frac{1}{10^7}$$

The actual number of grams of H^+ per liter is a cumbersome figure. A more convenient method of expressing the H^+ concentration is to use the *numerical value of the logarithm of the reciprocal of the H^+ concentration*. This mathematic expression is called pH.

Thus:

$$pH = \log \frac{1}{C_H}$$

and at neutrality: $C_H = \frac{1}{10^7}$

Then, substituting:

$$pH = \log \frac{1}{\frac{1}{10^7}}$$
$$pH = \log 10^7$$
$$pH = 7.0$$

The comparative difference in pH values is presented in Table 2-1.

As is evident in the table, pH 6 is ten times more acid than pH 7 (neutrality); pH 8 is ten times more alkaline than neutrality. At pH 0 the acidity is 10 million times greater than neutrality, while pH 14 is 10 million times more alkaline than pH 7.

Remember that the values representing pH are logarithms; therefore, the system conflicts with our mental habits, since we usually think in terms of arithme-

Table 2-1. Acid-base relation at fifteen pH values

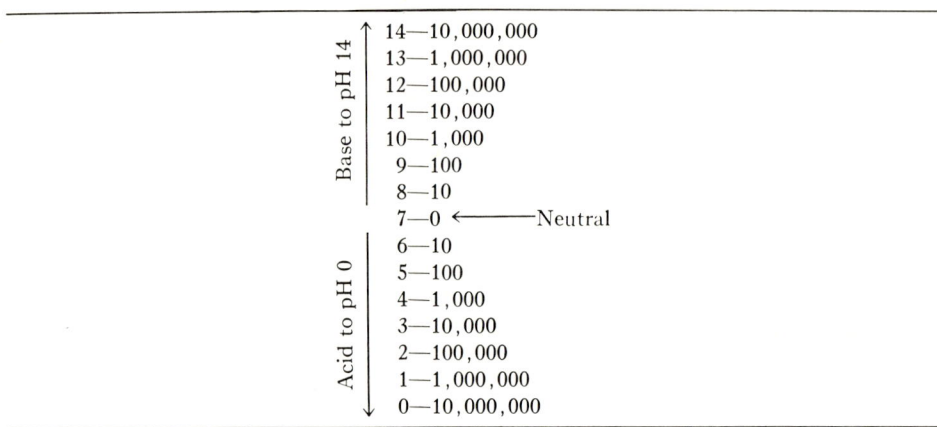

tic progressions. The following three statements are presented for the sake of ready comprehension.
1. As H^+ concentration increases, pH values decrease and vice versa.
2. Alkalinity is expressed in terms of H^+ concentration; pH values greater than 7 are increasingly more alkaline.
3. Pure water is neutral in reaction; pure water has pH 7.

pH control mechanisms

The variety of foodstuffs and fluids taken into the digestive system contain, accordingly, various acids and bases. The necessity of neutralizing and/or eliminating these substances to maintain blood pH at a constant value (homeostasis) is obvious. Three mechanisms within the body function to maintain the exceedingly critical value of blood pH between 7.35 and 7.45. These three mechanisms are buffers in the blood, respirations, and urinary excretions.

Buffers are substances that prevent great changes in the pH of a solution when an acid or a base is added to that solution; the buffers consist of a weak acid (or its acid salt) and a basic salt of that weak acid.

Chemically, two kinds of substances comprise a buffer or buffers: therefore the term "buffer pairs." Several pairs of buffers are found in body fluids, among which are as follows:
1. Sodium bicarbonate ($NaHCO_3$) and carbonic acid (H_2CO_3)
2. Potassium bicarbonate ($KHCO_3$) and carbonic acid
3. Sodium proteinate and proteins (weak acids)
4. Potassium hemoglobin ($K \cdot Hb$) and acid hematin (Hb)
5. Potassium oxyhemoglobin ($K \cdot HbO_2$), and oxyhemoglobin (HbO_2 – Hb and HbO_2 are weak acids)
6. Disodium phosphate (Na_2HPO_4) and sodium diphosphate (NaH_2PO_4) (Na_2HPO_4 is basic phosphate; NaH_2PO_4 is acid phosphate.)

Proteins may combine with either acids or bases and, therefore, must be classed with the salts of the acids or bases. The *alkali reserve* of the blood is equivalent to an approximate 0.5-percent solution of sodium bicarbonate.

Carbonic acid is one member of at least two of the buffer pairs; carbonic acid *exists only in solution and is the only free acid in the blood*. Carbonic acid reacts with the bases to form carbonate; this acid dissociates into H^+ and HCO_3^- and readily breaks up into H_2O and CO_2.

Respirations consist of inspirations and expirations. Oxygen is inhaled, and moisture and carbon dioxide are exhaled. If more CO_2 is exhaled, as in hyperventilation, less CO_2 exists in the blood to combine with H_2O to form H_2CO_3. Less H_2CO_3 in the blood permits less dissociation of this acid into H^+ and HCO_3^-, which results in an increase in pH.

Urinary excretions assist in maintaining blood pH by a continuous process of excreting and reabsorbing both Na^+ and K^+ to replace NH_3^+. Nitrate ions combine with hydrogen ions to form ammonia, which is eliminated in the urine.

The combined result of these three mechanisms is the maintenance of blood pH. When blood pH exceeds 7.5, the patient is suffering from alkalosis. Conversely, when blood pH falls below 7.3, the patient is suffering from acidosis. In either case, the patient may die if the condition worsens or continues without proper medication. Both acidosis and alkalosis have pH values above 7.0, neutrality. All blood is alkaline, so acidosis is actually a condition of reduced alkalinity. (See pp. 261 and 262 for further discussion of pH control.)

REFERENCES

Anthony, C. P.: Textbook of anatomy and physiology, ed. 9, St. Louis, 1975, The C. V. Mosby Co.

Bard, P.: Medical physiology, ed. 11, St. Louis, 1960, The C. V. Mosby Co.

Best, C. H., and Taylor, N. B.: The human body, ed. 4, New York, 1963, Holt, Rinehart and Winston.

Francis, C. C, and Martin, A.: Introduction to human anatomy, ed. 7, St. Louis, 1975, The C. V. Mosby Co.

Gardner, E.: Neurology, ed. 5, Philadelphia, 1968, W. B. Saunders Co.

Jacobi, C. A., and Paris, D. Q: Textbook of radiologic technology, ed. 5, St. Louis, 1972, The C. V. Mosby Co.

Schottelius, B. A., and Schottelius, D. D.: Textbook of physiology, ed. 17, St. Louis, 1973, The C. V. Mosby Co.

Wallace, B., and Dobzhansky, T.: Radiation, genes and man, New York, 1963, Holt, Rinehart and Winston.

Windle, W. F.: Textbook of histology, ed. 4, New York, 1969, McGraw-Hill Book Co.

CHAPTER 3

Embryologic survey

A very important part of the knowledge to be learned by students of radiologic technology deals with the effects of ionizing radiations on living tissues. Just as a baby learns to crawl before it learns to walk, an understanding of embryology is most helpful in (and almost necessary for) understanding the adolescent and adult anatomy, especially the effects of radiation on growth and metabolism. The previous comparison not only justifies, but requires that radiologic technology students understand embryology prior to studying human anatomy.

Embryology is the science that deals with the development of the embryo; it is correct to consider that embryology is a study of the development of the individual from the egg to the adult stage. The developmental history of an individual organism is called *ontogeny*, whereas the evolution of a race or genetically related group of organisms (such as a species, family, or order) is called *phylogeny*. The factors both affecting and effecting a new individual are too numerous to treat in this text. However, much of the related knowledge is both necessary and interesting to the student and to the practitioner of radiologic technology. Since sex cells mature by a different process than somatic cells and since a new individual begins with the union of two mature sex cells, it is logical that this chapter begin with a discussion of sex cell maturation, *meiosis*.

MEIOSIS

Meiosis is the sequence of complex nuclear changes that result in the production of cells (usually gametes*) with half the number of chromosomes present in the original cell; meiosis is also called *reduction division* or *maturation*. When gametes form from the original totipotent† cells, the entire process is that of maturation.

*A gamete is a matured sex cell or germ cell.
†A totipotent cell may mature into a somatic cell, or it may mature by meiosis into a germ cell.

In the male gonads (testes) the potential sex cells are called *spermatogonia;* in the female gonads (ovaries) the potential sex cells are called *oogonia*. In each sex the potential sex cells divide frequently by mitosis and every so often small groups of the totipotent cells undergo a process leading to the formation of *spermatozoa* in the male and *ova* in the female.

A new individual forms following the union (fertilization) of a mature ovum with a mature spermatozoon; the union of two gametes forms a *zygote*. In the somatic cell the nucleus contains the chromosomes. This is also true in the gametes; each ovum and each spermatozoon contains its respective chromosomes in its nucleus. *The chromosome number for any species remains constant from one generation to the next.* This is evidence that an *exact process* exists to maintain the constant chromosome number, thus precluding the possibility of doubling the chromosome number with each subsequent generation. This is the process of reduction division—meiosis.

Chromosomes are so small as to require observation through a microscope. When reduction division is seen through the microscope, the first indication of beginning meiosis is the rapid enlargement of the spermatogonium or oogonium. (For the sake of simplicity, the spermatogonium is used as the example of this process since oogonium maturation [oogenesis] includes the same steps until the final step, which is explained on p. 49.)

Following enlargement, the spermatogonium is called a *primary spermatocyte*. Completion of the enlargement process is followed by a very short, quiet (resting) stage, after which the slender, coiled threads (chromosomes) appear essentially as in prophase of mitosis (see pp. 17 to 19).

Next, the two chromosomes of each pair undergo *synapsis*, a condition wherein the two chromosomes become closely applied to each other throughout their entire lengths. During this very intimate association each chromosome splits longitudinally to produce two strands *(chromatids)*; the four chromatids remain quite closely associated. The two chromatids of a single chromosome comprise a *dyad*, and the two dyads comprise a *tetrad*. After synapsis is completed the four chromatids extend outwardly by pairs into loops and present a very tortuous appearance.

The extended chromatids next shorten and thicken. Each dyad is aligned parallel with and near the equatorial plane and becomes attached by the centromere (spindle fibers) from its end of the cell. The two dyads of each tetrad migrate toward the corresponding end of the cell. The cell indents at each end of the equatorial plane to form two new cells, each of which contains one set of the duplicated homologues of the primary spermatocyte. The two newly formed cells are the *secondary spermatocytes*.

Each of the secondary spermatocytes divides along the equatorial plane, and in this second meiotic division each chromosome separates along the longitudinal split, which occurred during tetrad formation. The final result is the formation of four *spermatids*, each possessing one-half of the number of chromosomes of the original spermatogonium. No additional divisions occur; the spermatids ex-

48 *Textbook of anatomy and physiology in radiologic technology*

trude a very slender tail that tapers to a sharp point. The head of each new *spermatozoon* consists almost entirely of the nucleus and functions in fertilization; the tail forms from the cytoplasm and functions in the movement of the spermatozoon. The scheme of meiosis is diagrammed in Fig. 3-1.

During the close application of a pair of chromosomes undergoing synapsis, each chromosome splits in its longitudinal axis. Following synapsis, each mem-

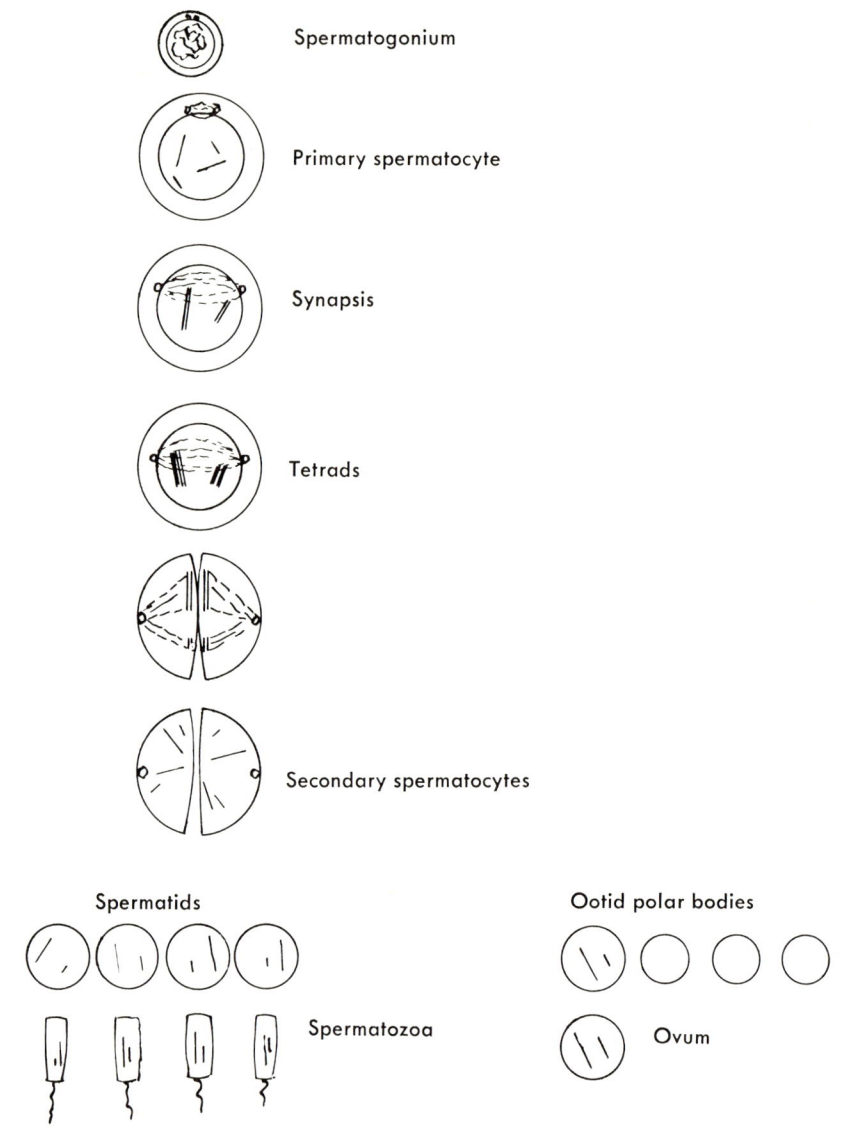

Fig. 3-1. Meiosis. The principal steps of meiosis are diagrammed to demonstrate the haploid number of chromosomes in each gamete. To avoid confusion, one long line and one short line are used to indicate two of the twenty-three chromosomes found in human gametes.

ber of a pair separates from the other member in a peculiarly crossed manner called a chiasma.* Present understanding of this phenomenon indicates that the chiasma occurs as a result of the exchange of segments between the two threads of different members of the pair. The exchange of segments is quite important in heredity; at this stage in meiosis the greatest effects of ionizing radiations upon new gametes become apparent.

As mentioned previously the ovum develops from the oogonium in a manner identical with the development of the male spermatozoon until the final step. Following formation of the secondary oocytes, *one of the four newly formed bodies in the female becomes an ootid;* the other three become *polar bodies* and are reabsorbed by the ovarian tissue. (Following synapsis in the primary oocyte, the first meiotic division produces one secondary oocyte and one, smaller first polar body. The first polar body receives one-half of the total number of the disjoined, reduplicated homologues of the primary oocyte and receives considerably less than one-half of the cytoplasm. The second and third polar bodies form from division of the first polar body.)

Since the secondary oocyte contains most of the cytoplasm, it, therefore, contains most of the nutritive material. As a result, the first polar body cannot mature as an oocyte. Next, the new *ootid*, formed from the original oogonium, becomes considerably larger than most cells of the body and is now an *ovum;* the ovum matures following extrusion from the graafian follicle as described on pp. 386 to 388. The exchange of segments (chiasma) occurs in the same developmental stage of the ovum as in development of the spermatozoon; therefore, ionizing rays have potentially the same effects upon both male and female gametes.

High-energy rays of any kind may contact living tissue. If, in the contact, there is absorption of the energy by the tissue cells, biologic changes must occur.

The energy of any ionizing rays directly controls the possible quantity of biologic effect. Roentgen rays (x rays) generated at 200 kVp, gamma rays, and beta rays have similar biologic effects when compared with each other. When compared with any of the preceding energy forms, neutrons are capable of causing about ten times the biologic effect and alpha rays are capable of causing about twenty times the biologic effect.

Ionizing effects are both immediate and future; cell changes have been observed for periods from a few minutes to several years following irradiation. The changes may occur in either or both the nucleus and the cytoplasm. As a result, the effects may be immediate to somatic cells, or the effects may be delayed and become evident in future generations because of gamete changes.

Ionizing radiations may prevent fertilization; they may cause changes in the immediate offspring; or they may cause changes that will occur in the embryo and prevent normal birth.

---

*A chiasma is a crosswise fusion.

FERTILIZATION

Since the tail of the spermatozoon gives it motility, the spermatozoon migrates from the vaginal tract, where it was deposited during coitus, into and through the cervix of the uterus and into the uterine lumen. From the latter, the spermatozoon migrates into either one of the two cornua of the uterus and into the corresponding uterine tube. The spermatozoon then ascends to approximately the middle third of the uterine tube; where the ovum has migrated following expulsion from the ovary. Fertilization usually occurs at this point, and the new *zygote* migrates into the uterus. Failure of the zygote to migrate into the uterus at this time is the cause of *ectopic pregnancy.**

The graafian follicle of the ovary expels the ovum, which is received by the fimbria of the uterine tube. Following this, the ovum migrates down the uterine tube to its midportion where fertilization occurs.† The normal migration process through the uterine tube and into the uterine lumen occurs in approximately 72 hours. If the ovum has become fertilized, it will lodge in the endometrial lining of the uterus. If no fertilization occurs, the ovum disintegrates in the uterus prior to (or during) the next menses.

Viable spermatozoa have been found in the cervical canal as long as 48 hours following coitus. The normal period of time during which an ovum can become fertilized has been estimated to be from 18 to 24 hours following ovulation. As a result, the probability of conception depends upon these two time factors in addition to viability of the gametes. Various authorities have placed the time limit for conception as varying from 6 to 36 hours following ovulation.

There are several theories regarding penetration of the ovum wall by the spermatozoon. Regardless of the cause (or method), the spermatozoon does penetrate the *zona pellucida* (the tough, transparent membrane enclosing the ovum), and the surface of the ovum becomes refractory to penetration by other spermatozoa. Subsequently, the ovum completes the final stage of maturation division in which the second polar body is cast off.

At this stage the spermatozoon loses its tail. The male *pronucleus* forms from the persisting nuclear elements; the female pronucleus forms from the persisting nuclear elements of the ovum nucleus. Cytoplasmic granules surround the male pronucleus radially, similar to the spokes of a wheel, as the entire mass migrates farther into the center of the ovum. The male and female pronuclei move toward each other, fuse, and form the *segmentation nucleus.* The segmentation nucleus forms as the male and female pronuclei break down, each into its constituent chromosomes near the center of the cell (ovum).

If the ovum is fertilized, its initial cleavage will probably occur within the first 24-hour period. It has been estimated that the second cleavage (into four cells) will occur in not more than an additional 12 hours. A third cleavage into

---

*Ectopic pregnancy is gestation elsewhere than in the uterus.
†The ovum can migrate into the uterus and become fertilized in the uterine lumen.

eight cells within the next 24 to 30 hours is probable, thus permitting a *morula**
of eight cells to enter the uterus approximately 72 hours following ovulation.

Many authors suggest that the term "morula" be applied to the new individual (organism) after it has entered the uterus; however, the morula is said to exist with cleavage, and cleavage follows the first mitotic division of the nucleus into the two new cells.

SEGMENTATION

During the development of the morula and the blastocyst (modified blastula† of mammals), the progressing processes of segmentation can be noted. Segmentation is said to begin with the original cleavage. The rate of mitosis differs in these cells, and no particular size increase occurs from that of the ovum during the first few cleavages. During cleavage it is highly probable that the different protoplasmic stuffs incompletely segregate into separated groups of cells.

Following the entry of the morula into the uterine lumen, the morula grows as a result of an increased rate of mitosis in its cells. During this stage of the morula, true segregation becomes discernible; the *trophoblast,* an outer layer of cuboidal cells and an inner cell mass, forms. The nourishing membranes of the embryo form from the trophoblast along with the attachment membranes. The embryo develops from the inner cell mass. The blastocyst stage begins with the secretion of fluid from the trophoblast into the inner portion of the morula.

The trophoblastic fluid is secreted under pressure, causing the entire morula to enlarge into the blastocyst. The innermost of the two lining membranes of the ovum is the zona pellucida. Cleavage of the trophoblastic cells causes them to flatten against the zona pellucida, which eventually disappears. One pole of the blastocyst retains attachment to the inner cell mass. Following this is the expansion of the *amnion,* the innermost fetal membrane, which forms both the sac that encloses the fetus and the sheath that encloses the umbilical cord.

The blastocyst continues to grow, and upon reaching approximately 0.5 mm. in diameter it attaches to the epithelial lining of the uterus by the embryonic pole of the blastocyst. After attachment, the blastocyst commences to flatten and to *digest* the uterine tissues that lie immediately beneath. The uterine epithelium soon envelops the blastocyst so that it is implanted within the endometrial lining. Openings into the maternal blood vessels from the endometrium occur as a result of continuing rapid digestion and invasion of the uterine tissues. These forced openings permit the trophoblast spaces to fill with maternal blood. The intervillous space forms as a result of the pressure of the maternal blood in the intercommunicating trophoblastic spaces. The trophoblast and the inner cell mass comprise the implanted blastocyst. The intervillous space contains nu-

---

*A morula is a globular mass of cells (blastomere) formed by cleavage of the zygote.
†A blastula is the usually spherical structure produced by cleavage of a fertilized ovum, consisting of a single layer of cells (blastoderm) surrounding a fluid-filled cavity (blastocoele).

merous strands, which anchor to the uterine wall to form the *chorionic villi*. Beyond this stage, the trophoblast is called the *chorion*, and its cavity is called the *chorionic cavity*. The amnion forms on the inner surface of the trophoblast over the *embryonic disc*, the inner cell mass.

Chorionic growth progresses steadily, and the amnion expands to contact the chorion, thus filling the chorionic cavity. Growth of the chorion forces the enveloping *decidua capsularis* (that part of the decidua covering the ovum) into contact with the *decidua vera* (that part of the decidua lining the remainder of the uterine body). As a result of this forced contact, the entire lumen of the uterus is now occupied.

The *placenta* connects the fetus to the uterine wall and functions as the organ of nutrition, respiration, and excretion for the fetus. Both fetal and maternal portions comprise the placenta. *There is no direct mixing of fetal and maternal blood.* Fetal blood circulates through the chorionic villi, which extend into the intervillous space, to contact the maternal blood. Exchanges of oxygen, carbon dioxide, nutritive substances, and waste products occur through the very thin, specialized membrane of the villi walls.

EMBRYO DEVELOPMENT

Late in the stage of morula development the inner cell mass becomes distinguishable. At this time the embryonic disc consists of a layer of columnar cells higher in the center than at the periphery, and the disc contributes cells to the embryo. The yolk sac develops from the inner cell mass on the ventral surface of the embryonic disc.

Five important parts of the future embryo (later called the fetus) develop from the embryonic disc as follows:
1. The *primitive streak* gives rise to the *primitive mesoderm*.
2. The *primitive groove* contains the primitive streak.
3. The *primitive node* is formed by cells from the embryonic disc and lies anterior to the primitive streak.
4. The *notochord* develops at the anterior end of the primitive node prior to the appearance of the *neural fold*.
5. The *prochordal plate* extends anteriorly from the notochord.

Close association of the prochordal plate with the *entoderm* probably gives rise to formation of the *mesoderm*. At this particular stage of embryo development, the blastoderm consists of three layers. From inside to outside these layers are the *entoderm, mesoderm,* and *ectoderm*.

The entoderm gives rise to the epithelial lining and the glands of the entire digestive and respiratory systems.

The mesoderm gives rise to the epithelium of the urinary and reproductive systems; the linings of the peritoneal, pleural, and pericardial cavities; a major part of the striated muscles; the myocardium; and the mesenchyme. The mesenchyme gives rise to all the nonstriated muscles and the various connective tissues.

The ectoderm gives rise to the brain, spinal cord, lining of the buccal cavity, epidermis, and parts of the sense organs.

Growth and development of the zygote from the morula to the blastocyst, the embryo, and the fetus follow a definite plan with control of the complete process the probable responsibility of the zygote nucleus and its contents as explained on pp. 14 to 16. One such variation, with interest to radiologic technologists, is that of Meckel's diverticulum. A brief discussion of its development follows.

The *body stalk* connects the embryonic disc to the chorion. The amnion develops on one side of the body stalk and the yolk sac develops on the opposite side. The chorionic mesoderm fixes the embryo to the inner wall of the chorion and envelops both the yolk sac and the amnion. The *allantois,* a tubular diverticulum of the posterior part of the yolk sac, grows into the body stalk. The yolk sac appears as a small, pear-shaped vesicle near the end of the fourth week of pregnancy, and opens into the digestive tube via the *vitelline duct,* the narrow tube connecting the umbilical yolk sac with the abdominal cavity of the embryo. The duct normally undergoes complete obliteration, usually by the end of the seventh week of pregnancy (some authors state from the third to the seventh week). However, some researchers report the existence of the vitelline duct in from 2 to 3 percent of cadavers examined. *Meckel's diverticulum is a continuation of the vitelline duct persisting in the embryo as a diverticulum from the ileum;* it occurs at varying distances proximal to the ileocecal valve—from a few centimeters to as much as 4 feet. Like the appendix, Meckel's diverticulum may fill with barium sulfate suspension during routine fluoroscopic or radiographic examinations of the digestive tract (barium enema, usually). Radiographic demonstration is quite rare in the case of the diverticulum; it is less rare in the case of the appendix.

REFERENCES

Goss, C. M.: Gray's anatomy, ed. 27, Philadelphia, 1959, Lea & Febiger.

Holmes, G. W., and Schulz, M. D.: Therapeutic radiology, Philadelphia, 1950, Lea & Febiger.

Lindsey, A. W.: Principles of organic evolution, St. Louis, 1952, The C. V. Mosby Co.

Snyder, L. H.: The principles of heredity, ed. 4, Boston, 1951, D. C. Heath and Company.

Wallace, B., and Dobzhansky, T.: Radiation, genes, and man, New York, 1963, Holt, Rinehart and Winston.

CHAPTER 4

The skeletal system

FUNCTIONS

The skeletal system is the bony framework of the body; it functions in the five following ways to assist in maintaining homeostasis:

1. The skeleton *supports* the body and gives the body its *shape*. The skull bones are arranged in a generally globular shape, thus giving the skull (superior to the mandible) this shape. The shoulder girdle bones and the ribs generally form the top of the thoracic cavity. The ribs between the thoracic vertebrae posteriorly and the sternum anteriorly encompass the thoracic viscera as a cage might encompass or contain them. The pelvic girdle articulating posteriorly with the sacrum gives shape to the lower part of the abdominopelvic cavity. The muscles, tendons, and other tissues of the extremities are arranged about, and generally exhibit the shape of, the bones of these extremities. In all the preceding, the bony structures give support to the related viscera.

2. The skeleton *protects* the vital viscera. The bones of the cranium protect the brain. The articulated vertebral bones and segments protect the spinal cord and cauda equina. The ribs, sternum, and thoracic vertebrae protect the thoracic viscera. The lumbar vertebrae and several muscles protect the abdominal viscera. The bones of the pelvic girdle along with the sacrum protect the abdominopelvic viscera. The shoulder girdles assist in protection of the upper thoracic viscera.

3. The bones of the skeleton serve as sites of *attachment* of tendons and muscles (origins and insertions). As muscles contract, force is exerted, principally on the insertion site, thus causing or enabling motion of a part in a certain direction.

4. All parts of the skeleton produce erythrocytes (red blood cells) and act in *hemopoiesis* to a varying degree. Hemopoiesis is quite active throughout the embryonic skeleton and continues in extrauterine life, although this activity declines in much of the cortical bone throughout the adult skeleton. Erythrocyte production increases as a result of proper stimulus (or demand) when homeo-

stasis exists, and at other times under certain conditions. Conversely, erythrocyte production may decline or fail completely as a result of certain pathologic conditions.

5. *Calcium storage* is a function of the skeletal system. The deposit of calcium is a dynamic process, the calcium being deposited when it is in excess in the body and removed from the bone when calcium is demanded by the body. Nature provides the greatest calcium deposits in the heaviest and strongest bones, which are subject to strain; the reverse is also true—the calcium is removed when a bone is not used. This is seen in the bones of an arm when the cast is removed following prolonged immobilization for fracture. The calcium is deficient. It is rapidly replaced, however, as soon as the body part is used. This calcium storage and buildup is closely akin to Wolff's law, which is, simply stated, that bone forms in response to stress and strain. Orthopedic surgeons make use of this fact and deliberately apply a walking cast to place strain on a fracture site. It is then noticed that union is often achieved, whereas nonunion was imminent prior to weight bearing.

ANATOMY
Typical bone

A typical (long) bone consists of two ends (epiphyses) and a central shaft (diaphysis). Hyaline cartilage covers the articular ends, while periosteum covers the remainder of the outer surface of the bone. Correspondingly, endosteum covers the inner surface of the cortical bone, forming the lining of the medullary canal (see Fig. 4-1).

Periosteum. Periosteum is a tough, fibrous covering membrane formed from the local mesenchyme. Two indistinctly separated layers form the periosteum in adults. The outer layer is a network of dense connective tissue that contains blood vessels, while the inner layer, adjacent to the bone, consists of loosely arranged collagenous bundles. Some of these bundles curve archlike, to form *Sharpey's fibers,* which penetrate the bone. *Fibroblasts* and a network of thin elastic fibers are contained within the inner layer. From the outer layer, blood vessels extend through the inner layer and pass via Volkmann's canals to the haversian canals.

Normally, the adult periosteum has no osteogenetic function and contains no osteoblasts. The embryonic properties of the periosteum return following fractures of the bone; this is due to the presence of fibroblasts and reticular cells, which, under special conditions, will reproduce the active osteoblasts. The latent osteogenetic property of the periosteum manifests itself following transplantation into other sites in the same organism, where it may form bone; this is particularly true if spicules of the old bone are transplanted with the periosteum. When certain conditions prevail, cartilage may form from the periosteum.

Bone marrow. The cavities of all bones contain only red marrow in the embryo and the newborn infant. In certain bones, the red marrow is gradually

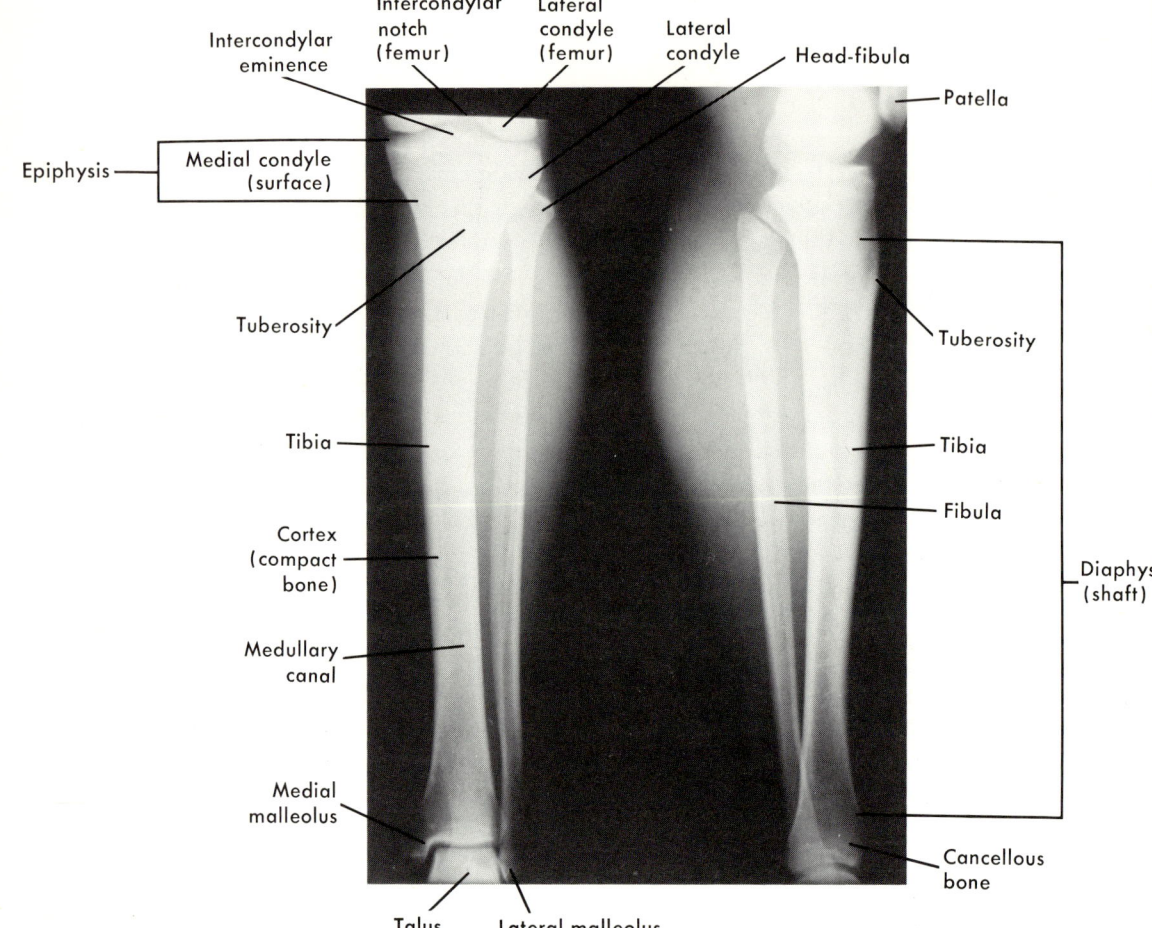

Fig. 4-1. Left tibia and fibula, posterior and lateral views (showing some bone markings). Endosteum lines the medullary canal, and periosteum covers the cortical bone, except on the articular surfaces.

replaced with yellow marrow by the twenty-fifth year of life. Red marrow persists in the adult vertebrae, ribs, sternum, diploë of the skull bones, and the proximal epiphyses of the humeri and femora. The primary function of red marrow is the manufacture of erythrocytes and granulocytes. Extreme anemia may stimulate some of the yellow marrow to revert to red marrow to aid in the production of erythrocytes.

Bone marrow is richly supplied with blood from branches of the nutrient (medullary) artery.* These arteries penetrate cortical bone to supply both

---

*One or more nutrient arteries supply each bone and pass into the bone via nutrient canals. The nutrient canals usually point toward the end of the bone in which the first union of epiphysis to shaft occurs. The canals point away from the growing end of the bone.

bone marrow and cancellous bone. Multiple small arterial branches from the periosteum supply cortical bone with blood. Extensive anastomoses exist between the medullary canal and cortical bone vessels. Veins corresponding to the arteries return blood from the bones to the circulatory system. The nerves supplying the bones accompany the arteries and veins of the bones.

Endosteum. The endosteum is a thin, connective tissue layer lining the walls of the larger bone cavities, which are usually filled with yellow marrow. The endosteum resembles the periosteum and is the condensed peripheral layer of the *stroma* (framework) of the bone marrow, where it is in contact with the bone. Like the periosteum, the endosteum is capable of forming bone under certain circumstances.

Types of bone

Bones are classified in four types: *long, short, flat,* and *irregular*. The humerus and femur (and many other bones) are typical long bones. In this classification of bones (long) the following items are demonstrated:
1. Epiphyseal ends containing the red marrow in the cancellous portions
2. Cortical bone in the diaphysis covered with periosteum and enclosing the medullary canal, which contains the yellow marrow, nerves, and blood vessels

The endosteum contains the medullary canal and forms the inner lining of cortical bone (see Fig. 4-1). The cancellous bone in the epiphyses is covered with a thin, peripheral cortex of cortical bone. The cavities in the epiphyses are direct continuations of the medullary canal of the diaphysis. Other bones classified as long bones include the metacarpals, metatarsals, fibulae, tibiae, clavicles, radii, ulnae, and phalanges. In a typical long bone there is a single center of ossification in the diaphysis and one in each of the two epiphyses.

Under the classification of short bones are the carpals and tarsals. Short bones are usually composed of cancellous material covered with a layer of cortical bone. There are no epiphyses and no diaphyses in short bones. Each of the carpals and tarsals, except the calcaneus, ossifies from a single center. The calcaneus ossifies from two centers—the primary center in the body and a secondary center in the posterior epiphysis.

The flat bones of the skull are the frontal, parietals, temporals, and occipital; other flat bones in the body are the scapulae, ribs, and sternum. In the flat bones of the calvarium (skullcap), there are two relatively thick layers of cortical bone sandwiching a thin layer of cancellous bone, the *diploë*. The diploë is the nutrient layer for all of the *inner table* and the inner portion of the *outer table* of each of these bones. In the Caucasian race the tables are each approximately 2 mm. thick; other races may have, characteristically, inner and outer tables of either greater or lesser thickness. In persons having increased thickness of either of these tables, an additional 2 to 4 kilovolts may be required for adequate penetration. Increased aluminum filtration compensates for the very slight increase in R-dose resulting from the increased kilovoltage. Sometimes both tables

Table 4-1. Bone projections

Type	Name	Location
Condyle (a rounded projection on a bone, usually articular)	Capitulum Femoral Mandibular Occipital Trochlea	Humerus Femur Mandible Occipital bone Humerus
Crest (a projection or projecting structure, or ridge, especially one surmounting a bone or its border)	Iliac Intertrochanteric Maxillary Supraorbital	Ilium Femur (posterior) Maxilla Frontal bone
Eminence (a prominence or projection, especially one upon the surface of a bone)	Frontal Intercondylar* Parietal	Frontal bone Tibia Parietal bone
Epicondyle (an eminence upon a bone, above its condyle)	Lateral Medial	Femur and humerus Femur and humerus
Head (the upper, anterior, or proximal extremity of a structure or body)	Femoral Humeral Mandibular Radial Scapular Ulnar	Femur Humerus Mandible Radius Scapula Ulna
Line (a narrow ridge; often an imaginary line connecting different anatomic landmarks)	Intertrochanteric	Femur (anterior)
Process (a prominence or projection from a bone)	Acromion Alveolar Clinoid Coracoid Coronoid Crista galli Frontal Glenoid Greater wing Hamular Lesser wing Mastoid Maxillary Palatine Perpendicular plate Petrosal Petrous Pterygoid	Scapula Mandible and maxilla Sphenoid bone Scapula Mandible and ulna Ethmoid bone Maxilla Scapula Sphenoid bone Hamate, lacrimal bone, and sphenoid bone Sphenoid bone Temporal bone Zygoma Maxilla Ethmoid bone Sphenoid bone Temporal bone Sphenoid bone

*The intercondylar eminence and the tibial spine are two names for the same projection.

Table 4-1. Bone projections—cont'd

Type	Name	Location
Process—cont'd	Spina angularis	Sphenoid bone
	Spinous	Vertebra
	Styloid	Fibula, radius, temporal bone, and ulna
	Sustentaculum tali	Calcaneus
	Temporal	Zygoma
	Transverse	Vertebra
	Xiphoid	Sternum
	Zygomatic	Maxilla and temporal bone
Protuberance (a projecting part, or prominence)	External occipital	Occipital bone
	Malleolus	Fibula and tibia
Spine (a thornlike process or projection)	Ethmoid	Maxilla
	Frontal	Ethmoid bone
	Iliac	Ilium
	Ischial	Ischium
	Nasal (anterior and posterior)	Maxilla
	Pubic	Pubis
	Scapular	Scapula
	Tibial	Tibia
Trochanter (either of the two processes below the neck of the femur)	Greater and lesser	Femur
Tubercle (a nodule)	Anterior and posterior	Atlas
	Anterior	Tibia
	Greater and lesser	Humerus
Tuberosity (an elevation or protuberance)	Calcaneal	Calcaneus
	Deltoid	Humerus
	Femoral	Femur
	Ischial	Ischium
	Occipital	Occipital bone
	Radial	Radius
	Tibial	Tibia

plus the diploic layer will measure as much as 15 mm. without the presence of pathology.

Irregular bones include the vertebrae, sphenoid, ethmoid, sacrum, coccyx, ilia, ischia, pubes, hyoid, mandible, auditory ossicles, and the facial bones of the skull. These, like the short bones, consist of a cancellous portion covered with a thin layer of cortical bone.

From the healing standpoint, most of the irregular bones are classified as flat bones. By definition the flat bones include two cortical tables, separated by a layer of bone marrow and cancellous bone. Healing of fractured flat bones is excellent, and a nonunion of a flat bone is almost unknown. Flat bones form from membranes and are sometimes called membranous bones. Although con-

Table 4-2. Bone depressions

Type	Name	Location
Acetabulum (the large cup-shaped cavity on the lateral surface of the pelvic bone (os coxa) in which the head of the femur articulates)	Acetabulum	Junction of ilium, ischium, and pubis
Canal (a relatively narrow tubular passage or channel)	Medullary	All long bones
	Spinal	Vertebral column
Fissure (any cleft or groove, normal or otherwise)	Superior orbital	Middle cranial fossa
Foramen (a natural opening or passage)	Caecum	Anterior cranial fossa between ethmoid and frontal bones in midsagittal plane
	Transverse	Transverse processes of cervical vertebrae
	Incisive	Maxilla
	Infraorbital	Maxilla
	Intervertebral	Vertebra (between each two)
	Jugular	Posterior cranial fossa
	Lacerum	Middle cranial fossa
	Magnum	Occipital bone
	Mental	Mandible
	Obturator	Pelvis, between ischium and pubis
	Optic	Sphenoid bone
	Ovale	Middle cranial fossa
	Parietal	Parietal bone
	Rotundum	Middle cranial fossa
	Spinosum	Middle cranial fossa
	Supraorbital	Frontal bone
	Vertebral (spinal)	Vertebra
Fossa (a trench or channel)	Coronoid	Humerus
	Glenoid	Scapula
	Infraspinous	Scapula
	Intercondylar	Femur
	Mandibular (temporal)	Temporal bone
	Olecranon	Humerus
	Radial	Humerus
	Supraspinous	Scapula
Groove (sulcus) (a shallow linear depression)	Bicipital (intertubercular)	Humerus
Meatus (an opening to some passageway in the body)	Auditory	Temporal bone

Table 4-2. Bone depressions—cont'd

Type	Name	Location
Notch (an indentation or depression)	Ethmoidal	Frontal bone
	Intercondylar	Femur
	Mandibular	Mandible
	Radial	Ulna
	Semilunar (sigmoid)	Ulna
	Suprasternal	Sternum and clavicles
	Ulnar	Radius
Sinus (a cavity or hollow space)	Ethmoidal	Ethmoid bone
	Frontal	Frontal bone
	Mastoid	Temporal bone
	Maxillary (antrum of Highmore, nasal antrum)	Maxilla
	Sigmoid	Temporal bone
	Sphenoid	Sphenoid bone

trary to general classification as a long bone, the clavicle forms from membranes and, as a result, heals very rapidly following fracture.

Markings on bones

The varied markings on the bones represent a rather large variety of functions to which the markings are specifically adapted. The markings include all of the various projections and depressions. Smooth projections may be articular; roughened projections usually are sites for attachment of muscles and/or tendons. Both nerves and blood vessels pass through bone substance via small holes named foramina. Shallow, smooth depressions (fossae) are often articular; roughened depressions usually are sites of attachment.

The projections on bones include all elevations, from the body or ends of the bone, that are visible radiographically or palpable on the skeleton; some of the more prominent projections are palpable through the skin. Some of the more distinct and slightly elevated lines classify as projections. The projections are listed in Table 4-1.

The classification of depressions of bone markings includes both deep and shallow depressions, notches, grooves (sulci), and holes (either through a single bone or part thereof or formed by the incomplete junction of two or more bones). The depressions are listed in Table 4-2.

The projections and depressions listed in Tables 4-1 and 4-2 represent those of greatest importance to radiologic technologists. As will be noted, many of these bone markings are either a part of or contribute to the formation of the various articulations.

Skeleton

The entire skeleton of the human body consists of 210 bones radiographically or 206 bones anatomically; the fact that each pelvic (hip) bone (os innomi-

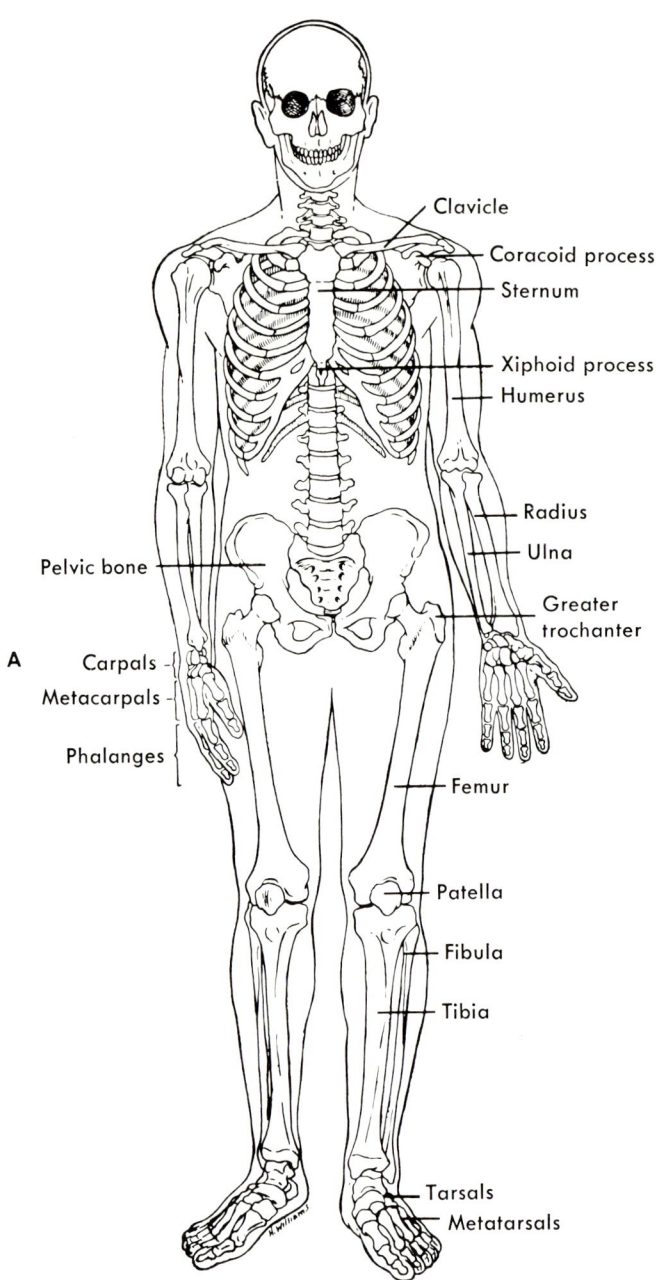

Fig. 4-2. Skeleton. **A,** Anterior view. **B,** Posterior view. (Modified from Anthony, Catherine Parker: Textbook of anatomy and physiology, ed. 7, St. Louis, 1967, The C. V. Mosby Co.)

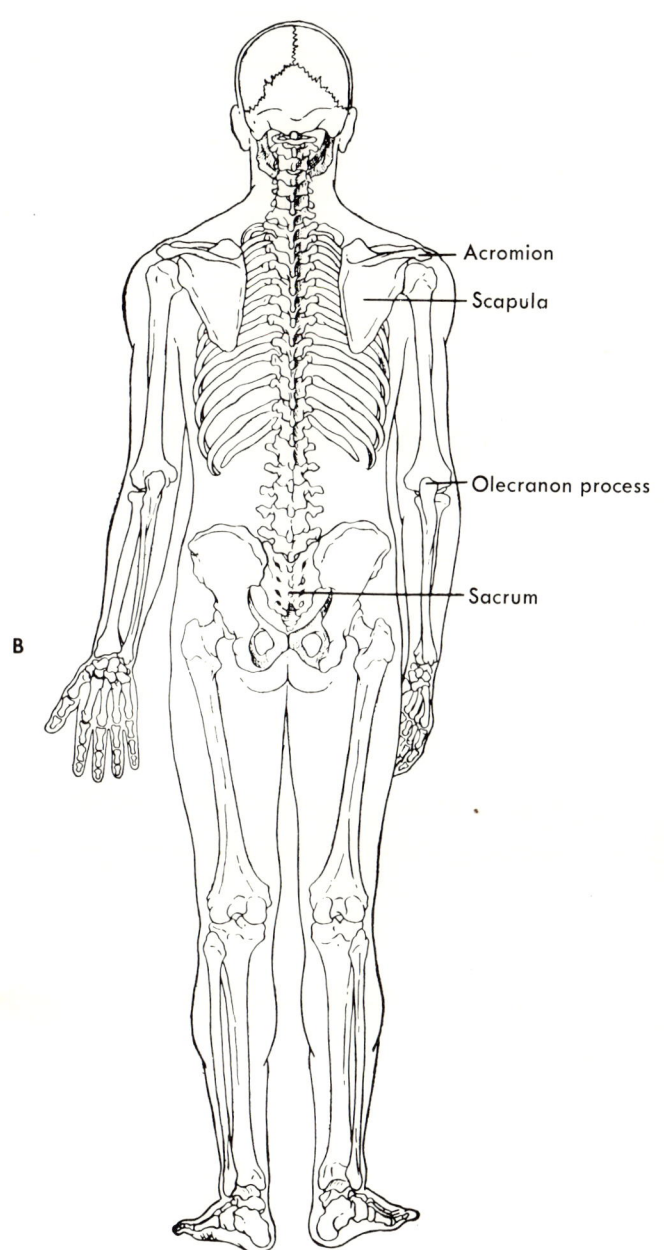

Fig. 4-2, cont'd. For legend see opposite page.

Table 4-3. Classification of bones by region—cont'd

Region	Section	Name	Number
Skull	Cranial	Ethmoid	1
		Frontal	1
		Occipital	1
		Parietal	2
		Sphenoid	1
		Temporal	2
	Facial	Inferior concha	2
		Lacrimal	2
		Mandible	1
		Maxilla	2
		Nasal	2
		Palatine	2
		Vomer	1
		Zygoma (malar)	2
	Temporal	Incus	2
		Malleus	2
		Stapes	2
Neck	Anterior spine	Hyoid	1
		Cervical bodies C-1 to C-7	7
		Atlas	1
		Axis	1
		C-3 to C-6	4
		Vertebra prominens	1
Thorax	Spine	Thoracic bodies, T-1 to T-12	12
	Rib cage	Left ribs, 1 to 12	12
		Right ribs, 1 to 12	12
		Sternum	1
	Shoulder	Clavicle	2
		Scapula	2
Lumbar	Spine	Lumbar bodies, L-1 to L-5	5
Pelvis		Coccyx	1 (counted as 1 although there are from 3 to 5 segments, averaging 4)
		Ilium	2
		Ischium	2
		Pubis	2
		Sacrum	1 (5 segments fused)
Extremity, upper	Upper arm	Humerus	2
	Forearm	Radius	2
		Ulna	2
	Wrist	Capitate	2
		Hamate	2
		Lunate	2
		Pisiform	2
		Scaphoid	2
		Trapezium	2
		Trapezoid	2
		Triquetrum	2
	Hand	Metacarpals	10
	Fingers	Proximal phalanx	8
		Middle phalanx	8
		Distal phalanx	8

Table 4-3. Classification of bones by region

Region	Section	Name	Number
Extremity, upper—cont'd			
	Thumb	Proximal phalanx	2
		Distal phalanx	2
Extremity, lower	Thigh	Femur	2
	Knee	Patella	2
	Lower leg	Tibia	2
		Fibula	2
	Ankle	Talus	2
	Foot	Calcaneus	2
		Cuboid	2
		Cuneiform (intermediate)	2
		Cuneiform (lateral)	2
		Cuneiform (medial)	2
		Metatarsal	10
		Navicular	2
	Toes	Proximal phalanx	10
		Middle phalanx	8
		Distal phalanx	10
Total			210

natum) consists of three bones (ilium, ischium, and pubis) accounts for the difference. Other differences, *in consideration only,* include the radiographic and anatomic variations in the members of the appendicular skeleton.

The skeletal organization presents two anatomic divisions: axial and appendicular. The axial division includes the skull, vertebral column, ribs, sternum, hyoid, and auditory ossicles. The appendicular division includes the clavicles, scapulae, humeri, radii, ulnae, carpals, metacarpals, phalanges (both manual and pedal), ilia, ischia, pubes, femora, patellae, tibiae, fibulae, tarsals, and metatarsals.

Because the problems of radiographic penetration and demonstration are similar in the bones of the shoulder and hip (pectoral and pelvic girdles) to the bones of the axial skeleton in that particular region, the radiologic technologist usually considers these appendicular skeleton bones as being members of the axial skeleton (see Fig. 4-2). Other variations from true anatomic classification are included to emphasize the radiographic demonstration of the particular part.

Table 4-3 lists the bones according to region of the body, regardless of classification either as axial or appendicular.

Upper extremity (excluding the pectoral girdle)

Since the clavicles and scapulae are considered radiographically as members of the axial skeleton, it is proper to state that the upper extremity begins with the humerus.

The humerus, the long bone of the upper arm, is usually ossified by the twenty-first year of life. Proximally the humerus consists of the *head*, which articulates with the glenoid fossa of the scapula to form the shoulder joint; the *greater* and *lesser tubercles*, between which is the *bicipital groove* (intertubercular sulcus); and the *surgical neck*. A shallow groove, the *anatomic neck*, separates the greater and lesser tubercles from the head. The greater tubercle is lateral; the lesser tubercle is in the approximate center of the anterior surface of the humerus. The greater tubercle is best demonstrated radiographically when the arm is fully extended with the arm and hand in external rotation (see Fig. 4-82, A and C). The bicipital groove extends lengthwise with the shaft in a somewhat oblique direction from its origin between the tubercles to its terminus below the greater tubercle and toward the lateral surface of the shaft. The tendon of the long head of the biceps brachii muscle moves through the bicipital groove. Synovitis often occurs in this site. The long head tendon forms a right angle at the top of the humerus to facilitate its attachment to the supraglenoid tuberosity, which is the origin of this head of the muscle. The surgical neck rings

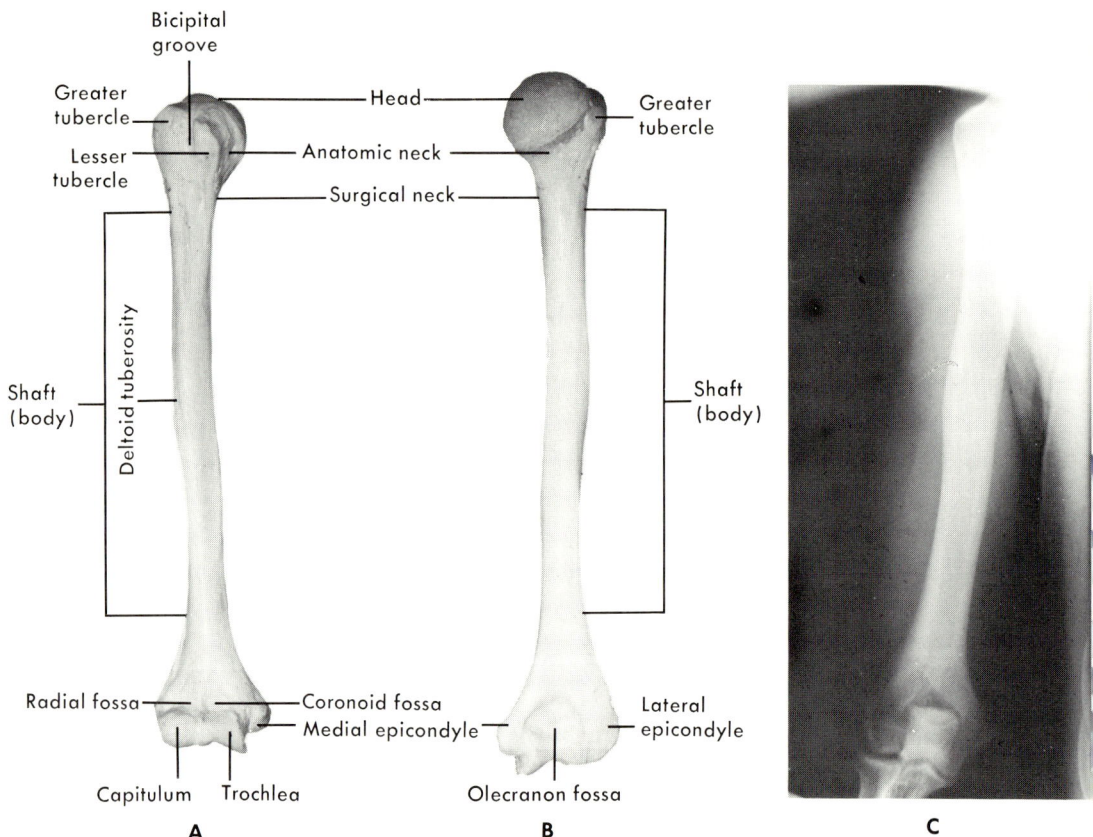

Fig. 4-3. Right humerus. **A**, Anterior surface. **B**, Posterior surface. **C**, Posterior view.

the proximal end of the shaft immediately inferior to the elevation of the lesser tubercle.

For the most part the *shaft* is straight and round with a tendency to flatten out as it approaches the distal end of the humerus. The *deltoid tuberosity* is a slightly raised, roughened area in approximately the middle third of the shaft; it is somewhat lateral on the anterior surface. The deltoid muscle inserts into the tuberosity. Below the tuberosity is a spiral groove in which lies the radial nerve.

Distally on the anterior surface of the lateral side of the humerus is the *radial fossa,* which receives the radial head when the elbow is flexed. On the same level with the radial fossa and somewhat medial to it is the *coronoid fossa,* which receives the prominence of the coronoid process of the ulna when the elbow is flexed. On the lateral surface below the radial fossa is the roughened *lateral epicondyle.* A common tendon arising from this epicondyle gives origin to all the extensor tendons of the forearm. Under prolonged or severe strain the tendon becomes frayed and produces the chronic condition, *epicondylitis.* The *capitulum* (capitellum) is the smooth, rounded articular surface inferior to the radial fossa on the extreme distal end of the lateral side of the humerus, against which the radial head articulates. The epiphysis appears quite early in skeletal ossification and is most susceptible to injury. The medial epicondyle is a roughened prominence extending medially from the coronoid fossa. Immediately inferior to the coronoid fossa on the extreme distal end of the humerus is the *trochlea* for articulation with the semilunar (sigmoid) notch of the ulna. The capitulum is the lateral condyle and the trochlea replaces the medial condyle of the humerus. As ossification progresses the capitulum and trochlea form a single epiphysis. The trochlea seldom, if ever, causes trouble in elbow dislocations, but the capitulum is often displaced from the joint and must be replaced surgically. If the capitulum is not replaced, the elbow assumes an excessive *carrying angle;* the normal carrying angle is 15 degrees. Posteriorly in the midportion of the distal end of the humerus and immediately superior to the trochlea is the *olecranon fossa,* which receives the prominence of the olecranon process of the ulna when the forearm is extended (see Fig. 4-4). When the olecranon process is in complete extension, it fits into the corresponding fossa. Following injury the fossa frequently fills with scar tissue, preventing full extension of the joint to its normal 180-degree angle.

The elbow is a hinge (ginglymus) joint. The following statements describe its movements. The semilunar notch of the ulna articulates on the trochlea of the humerus in flexion and extension of the forearm. The capitulum of the humerus articulates in the fossa in the superior surface of the radial head during flexion and extension of the forearm. The head of the radius rotates on the capitulum and in the radial notch of the ulna in pronation and supination of the forearm and hand. For radiography of the elbow in true lateral position, the joint is flexed 90 degrees.

In the true anatomic position, the radius is lateral and the ulna, medial in the forearm. The anterior surface of the elbow is named the antecubital space. Both

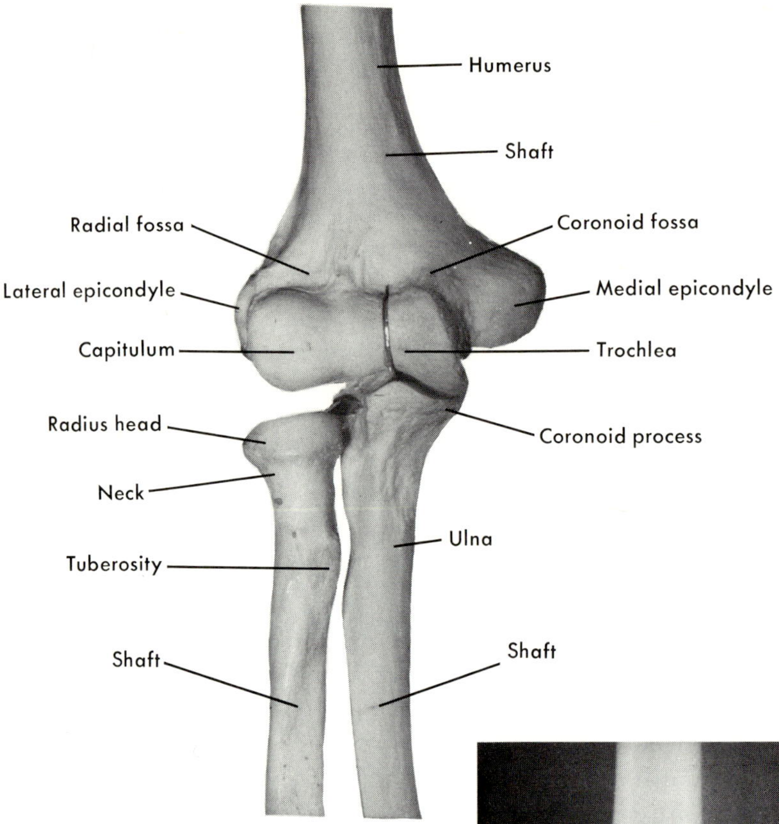

68 *Textbook of anatomy and physiology in radiologic technology*

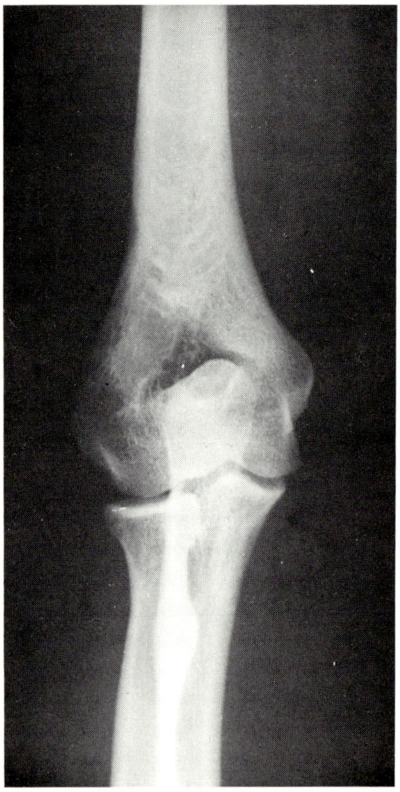

Fig. 4-4. Right elbow. **A**, Anterior aspect. **B**, Posterior view.

The skeletal system 69

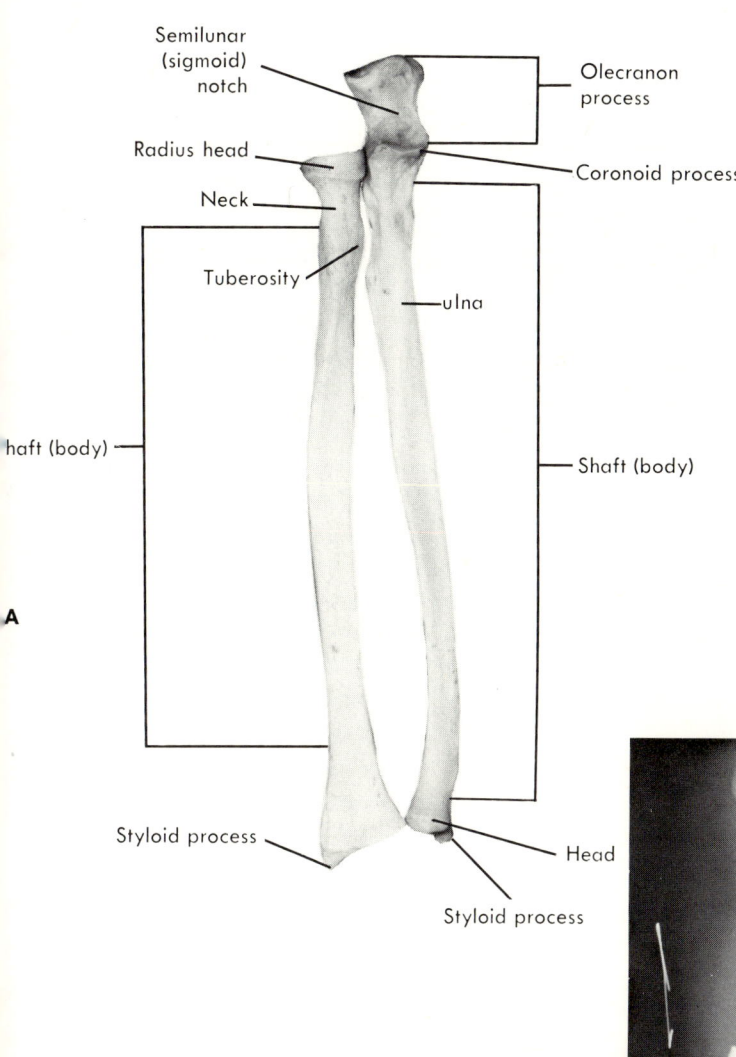

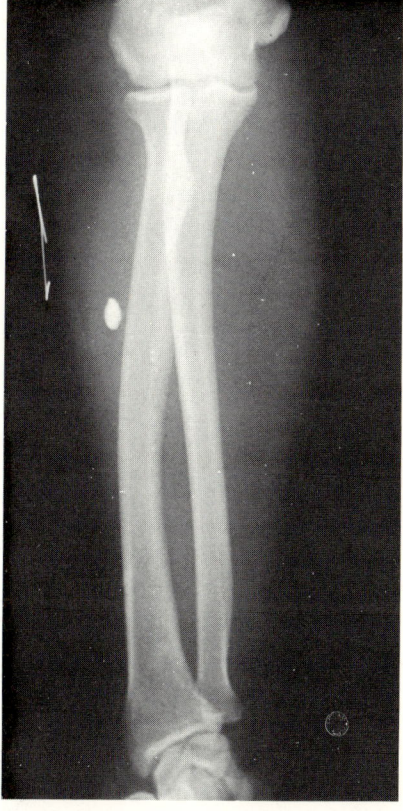

Fig. 4-5. Right forearm. **A,** Anterior surface. **B,** Oblique view. A foreign body in tissues next to radius and a marker on lateral surface of forearm required that an oblique position rather than true posterior position be used. (**B,** Courtesy Dr. J. M. Hilton.)

the radius and ulna are usually ossified by the twenty-fifth year of life. The ulna is quite large proximally and quite small distally. The radius is considerably larger in its distal end than in its proximal end.

The proximal end of the ulna consists of the large and prominent *olecranon process* extending superiorly from the posterior surface; the olecranon process contains the *semilunar notch* in the anterior surface. Immediately inferior to the olecranon process is the roughened *coronoid process,* which supports the prominent lower lip of the semilunar notch. The coronoid process contains the *radial notch* in its lateral surface. The *shaft* tapers to a rather small *head* at the distal end of the ulna. The head supports a sharp *styloid process* medially and posteriorly (see Fig. 4-5).

The radius begins proximally with the *head,* a rounded eminence containing a shallow fossa in its extreme superior end. The head of the radius rotates in the annular ligament. Any roughness of the head prevents free rotation in this closely fitting area; thus the motions of both pronation and supination of the forearm are either limited or prevented. Full rotation is restored by surgery. A true *neck* rings the radius immediately inferior to the head. The neck continues into the *shaft,* which becomes increasingly larger in its distal third. A raised and roughened area, the *radial tuberosity,* is approximately 1 inch inferior to the head and is on the medial surface somewhat posteriorly. The tuberosity is the site of insertion of the biceps brachii tendon. The distal end of the radius is broadened; it supports a *styloid process* laterally on the posterior surface. Medially on the distal end of the radius is a small notch, the *ulnar notch;* the notch is often nonevident, although it is for articulation with the ulnar head. The extreme distal end of the radius forms the major portion of the proximal articulation of the carpals since there is no direct articulation between the carpals and the distal end of the ulna.

The carpals are classified as short bones; there are eight carpal bones arranged in a proximal row and a distal row. Proximally from lateral to medial, the

Table 4-4. Articulations of the carpus

Bone	Number of articulations	Articulations
Scaphoid	5	Radius, trapezium, trapezoid, capitate, and lunate
Lunate	5	Radius, capitate, hamate, triquetrum, and scaphoid
Triquetrum	3 and disc	Lunate, pisiform, hamate, and triangular-shaped cartilage
Pisiform	1	Triquetrum
Trapezium	4	Scaphoid, trapezoid, and first and second metacarpals
Trapezoid	4	Scaphoid, trapezium, capitate, and second metacarpal
Capitate	7	Scaphoid, lunate, trapezoid, hamate, and second, third, and fourth metacarpals
Hamate	5	Lunate, triquetrum, capitate, and fourth and fifth metacarpals

carpal bones are the *scaphoid* (navicular), *lunate* (semilunar), *triquetrum* (triangular), and *pisiform*. Distally, from lateral to medial, the carpal bones are the *trapezium* (greater multangular), *trapezoid* (lesser multangular), *capitate*, and *hamate*. The carpal bones are described as a group since radiographic requests are for the carpus as a whole rather than for individual bones. However, certain conditions may cause requests specifically for the scaphoid or for the pisiform and the hamulus process. The carpal bones are usually ossified by the tenth year of life.

The scaphoid is the most troublesome carpal bone from the standpoint of nonunion. It has a single, small artery entering the bone from the radial side.*

*The small depression in the skin and tissues distal to the scaphoid on the radial side is called the anatomic "snuffbox."

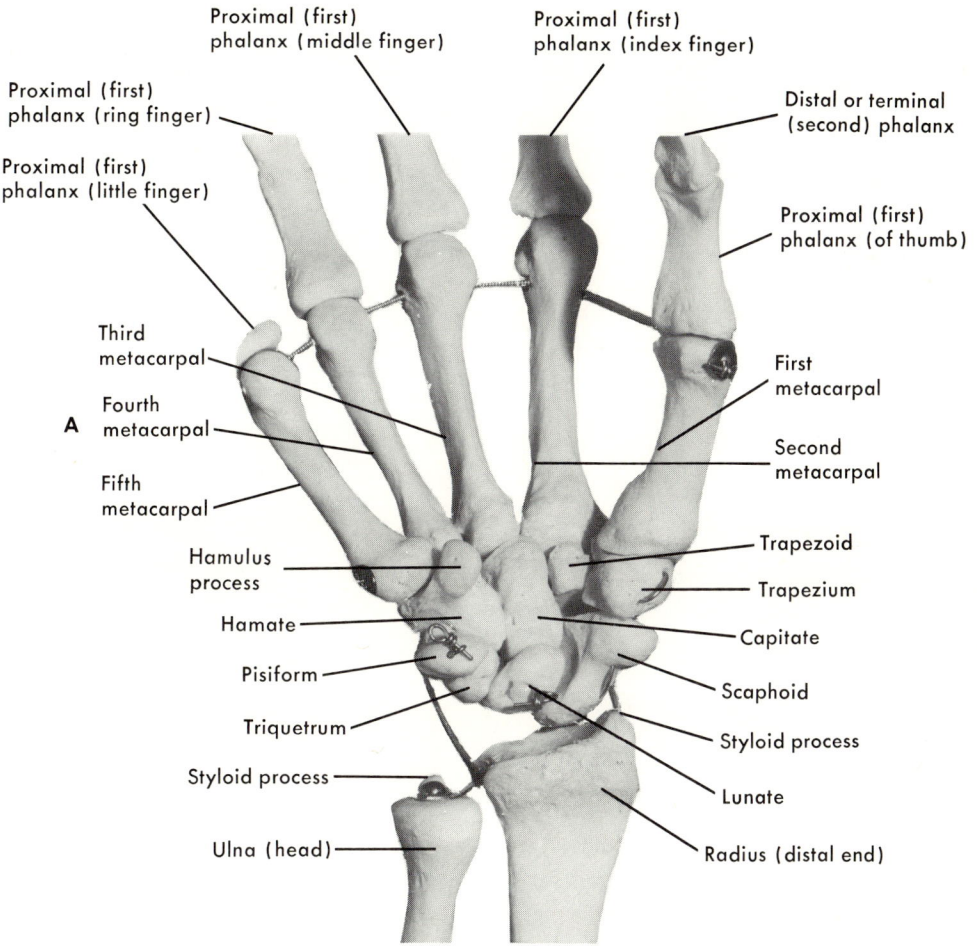

Continued.

Fig. 4-6. A, Right carpals and metacarpals, anterior surface. **B,** Right hand, anterior view. **C,** Right hand in ulnar flexion.

A fracture of the scaphoid in its narrow portion causes the ulnar half to be without circulation. Such a condition usually causes death of the bone part and requires removal. The scaphoid is best demonstrated radiographically with the hand prone and the wrist in ulnar flexion.*

The carpals articulate proximally with the radius and distally with the metacarpals to form the wrist joint. The articulations of the carpals are found in Table 4-4. For radiography of the carpals, the wrist and hand may be either prone or supine; however, if information is desired regarding the distal ends of the radius and ulna, supination is the preferred position. In supination, the radiograph will be a posterior (dorsal) view. The carpals are included routinely in radiographs of the hand (see Fig. 4-6).

The metacarpals form the bony part of the hand; there are five metacarpals, numbered from one through five from the lateral to the medial side. These long bones are usually ossified by the nineteenth year of life. As a group they articu-

---

*Ulnar flexion places the hand and wrist flexed *toward* the ulna (medial surface of the forearm) and *away from* the radius. Ulnar deviation places the hand and wrist flexed *toward* the radius (lateral surface of the forearm) and *away from* the ulna. (See Fig. 4-6, C.)

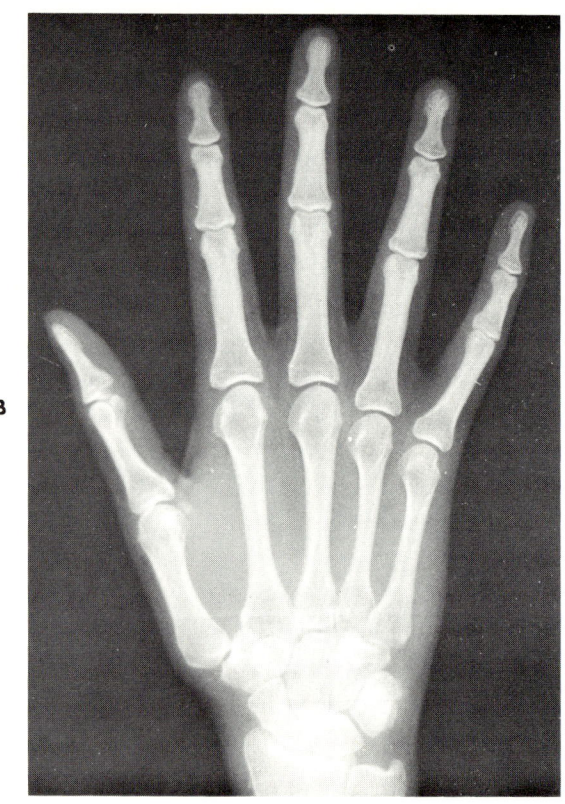

Fig. 4-6, cont'd. For legend see p. 71.

The skeletal system **73**

late proximally with the distal row of carpals (see Fig. 4-6 and Table 4-4). Each metacarpal articulates distally with the proximal end of the corresponding proximal phalanx.

The phalanges are long bones that form the thumb and fingers. The thumb (first digit), index or first finger (second digit), middle or second finger (third digit), ring or third finger (fourth digit), and little or fourth finger (fifth digit) comprise the five digits of the hand. The manual phalanges are usually ossified by the nineteenth year of life. There are fourteen phalanges in each extremity: two in the thumb and three in each of the remaining four fingers. The thumb has a proximal and a distal (terminal) phalanx; each finger has a proximal, a mid-

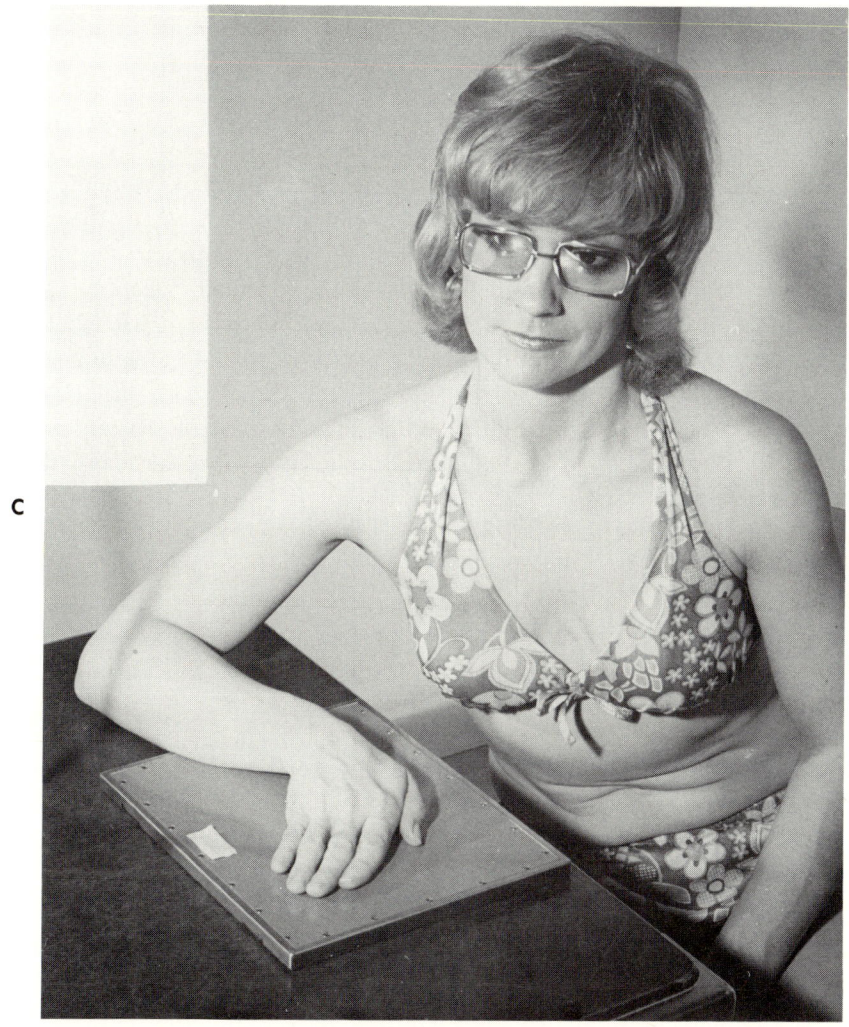

Fig. 4-6, cont'd. For legend see p. 71.

dle, and a distal (terminal) phalanx. The broad, flattened ends of the distal phalanges are named the *tufts* (see Fig. 4-6, *B*).

Lower extremity (excluding the pelvic girdle)

As in the upper extremity, the bones of the pelvic girdle (ilium, ischium, and pubis) are considered radiographically as members of the axial skeleton. As a result, we may state that the lower extremity begins with the femur.

The femur, the bone of the upper leg (thigh), is the largest single bone in the body. It is usually ossified by the eighteenth year of life. The femur begins proximally with the *head,* followed by the *anatomic neck, greater* and *lesser trochanters,* and *surgical neck.* The *intertrochanteric line,* for attachment of the iliofemoral ligament and origin of the upper part of the vastus medialis muscle, begins with the *tubercle of the femur* and extends obliquely medialward between the trochanters to terminate in the linea aspera. The tubercle of the femur is the common junction of the gluteus minimus, vastus lateralis, tendon of the obturator internus, and superior and inferior gemelli muscles. Posteriorly and extending obliquely medialward from the greater trochanter, the *intertrochanteric crest* (spinal line) terminates in the lesser trochanter. The upper part of the quadratus femoris muscle attaches to the intertrochanteric crest. Medially at the base of the greater trochanter is the deep *trochanteric fossa,* which is the site of insertion of the obturator externus tendon. The head of the femur articulates in the acetabulum of the pelvic bone to form the hip joint, a ball-and-socket joint. The *fovea capitis* is an oval-shaped depression situated somewhat below and behind the center of the head. The fovea capitis is the site of attachment for the ligamentum teres.

The shaft of the femur is nearly cylindric and noticeably flattened in its distal end. The shaft angles toward the midline in both sexes, although the angle is greater in females than in males.

The posterior surface of the femoral shaft presents five lines and two foramina. The *linea aspera,* a distinct ridge, serves as attachment and origin sites posteriorly in the midline of the femoral shaft for several muscles. It continues both above and medially with the *spiral line* and the *pectineal line.* A roughened, raised ridge, the *gluteal tuberosity,* extends upward on the lateral side from the linea aspera to the base of the greater trochanter. A part of the gluteus maximus muscle attaches to the gluteal tuberosity. The pectineal line extends to the inferior surface of the lesser trochanter and is the site of attachment of the pectineus muscle. The linea aspera continues below with the *lateral epicondylic line* to the lateral epicondyle and with the *medial epicondylic line* to the medial epicondyle. The *popliteal surface* lies between the medial and lateral epicondylic lines; the popliteal artery crosses this surface.

Located on the posterior surface and approximately at the junctions of the proximal third with the middle third and the middle third with the distal third of the femoral shaft are two *nutrient foramina,* through which pass branches from the femoral artery to supply the medullary canal of the femur.

The distal end of the femur presents two *epicondyles, lateral* and *medial,* and two *condyles, lateral* and *medial.* The *adductor tubercle* lies at the summit of the medial condyle and is the site of insertion of the tendon of the adductor magnus muscle. Anteriorly and between the condyles is the smooth articular area for the patella—*patellar surface.* Posteriorly and between the condyles is the in-*intercondylar notch.* The *intercondylar fossa* is the depression in the deepest part of the bony margin of the notch. The condyles are the smooth, rounded prominences on the extreme distal end of the femur to articulate with the corresponding condyles of the tibia and form the knee joint, a hinge joint (see Fig. 4-7). The posterior surface of the knee is named the popliteal surface.

The femur may be compared with the humerus in the following points:

Femur	Humerus
Head	Head
Anatomic neck	Anatomic neck
Greater trochanter	Greater tubercle
Lesser trochanter	Lesser tubercle
Intertrochanteric line	Bicipital groove (intertubercular sulcus)
Intertrochanteric crest	
Surgical neck	Surgical neck
Shaft	Shaft
Epicondyles	Epicondyles
Medial condyle	Trochlea
Lateral condyle	Capitulum
Patellar surface	
	Radial fossa
	Coronoid fossa
Intercondylar fossa	Olecranon fossa

Trauma and disease necessitate extensive radiographic study of the femur. Although nongrid (Bucky) radiographs are often satisfactory, grid radiographs are often made because of the overall thickness of the femur and surrounding tissues. Lateral views of the upper third of the shaft must include the head and anatomic neck; these views are easily obtained when the grid technique is employed. Obtaining a lateral view of the head and anatomic neck of the femur is imperative during surgery because the surgeon cannot close the wound until he is absolutely certain that the nail he has inserted does not perforate the head, damage the socket, or contribute to other deleterious conditions. In order to obtain lateral radiographs of the femoral head and neck, the technologist must recognize the patient's ability to endure movement of this leg as well as other parts of his body. Either the *frog-leg* (bilateral) or the *unilateral* view is satisfactory when there is no recent trauma or suspected pathology. If trauma has occurred recently or if pathology is suspected, the *lateral* (transverse hip) view is the position of choice (see Fig. 4-8). Polaroid technique is distinctly advantageous since it permits very rapid inspection of the radiograph and supplies the answer to the very critical question of the surgeon, "Is the nail in place?" Modern mobile x-ray machines with image intensifying (and viewing) units offer even more satisfactory answers to this question.

76 Textbook of anatomy and physiology in radiologic technology

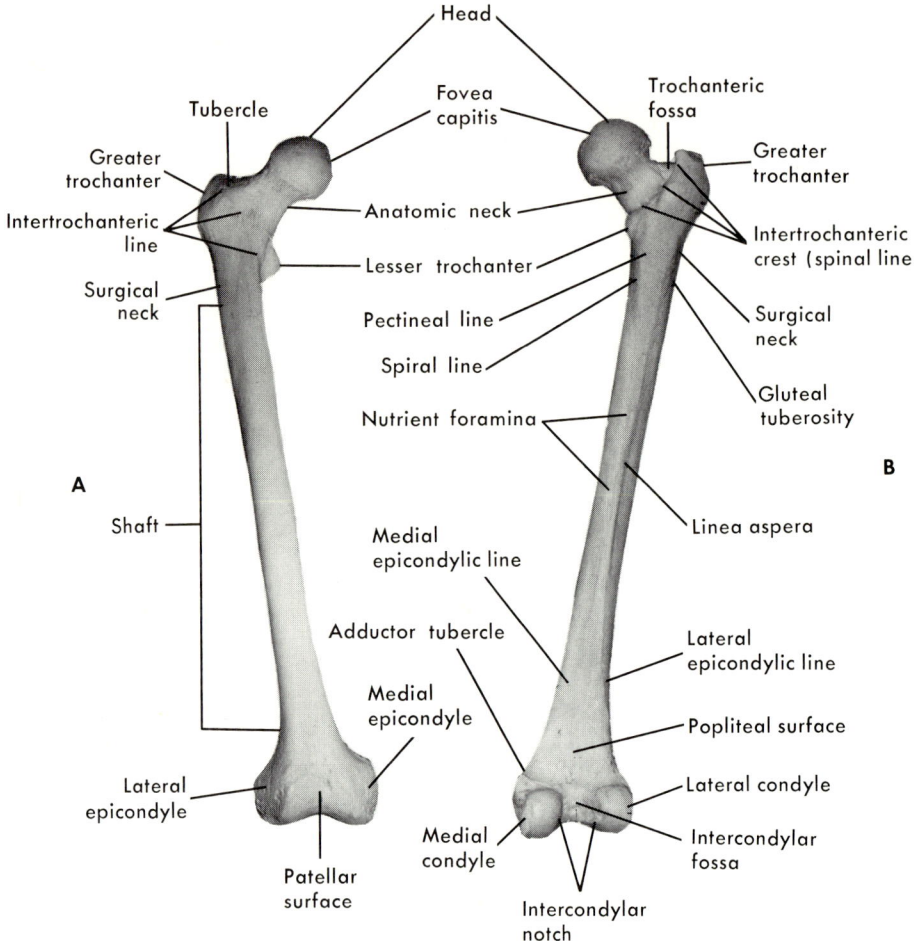

Fig. 4-7. Right femur. **A,** Anterior surface. **B,** Posterior surface.

A single radiograph of the complete femur is difficult to obtain on one 14 × 17 inch film. It is often useful to project the head, neck, and three-fourths of the femoral shaft on a 14 × 17 inch film, and the lower one-fourth of the shaft and the distal end of the femur on a 10 × 12 inch film. Regardless of the method, it is usually necessary to demonstrate the entire femur when routine exposures and/or surveys are requested.

The patella (knee cap) forms the cap of the knee joint and is the only sesamoid bone of the body classed as essential. The patella is rounded and presents some resemblance to a biscuit. The anterior surface of the patella is somewhat convex; the posterior surface is similarly concave. The patella articulates only with the femur (see Fig. 4-9). Ossification of the patella is usually completed by the thirteenth year of life.

Although the patella is essential in normal skeletal anatomy, it may be re-

The skeletal system 77

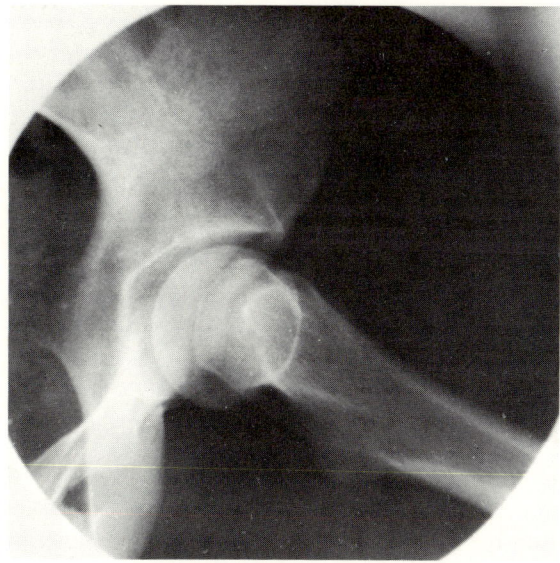

Fig. 4-8. Left hip, lateral view.

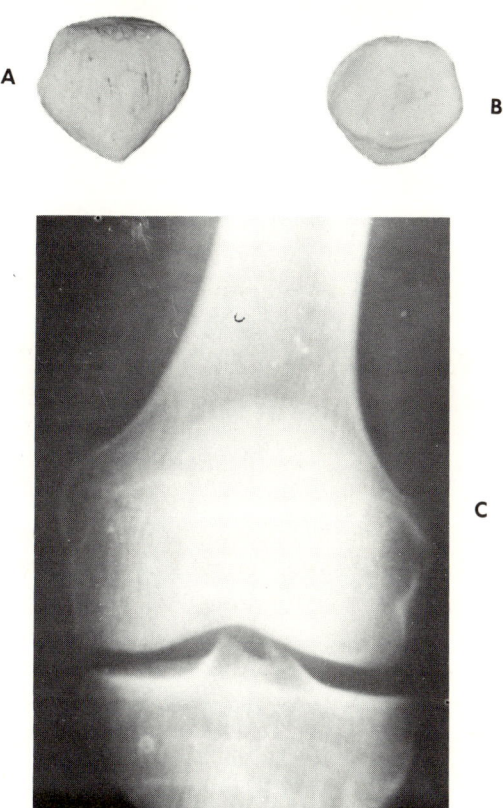

Fig. 4-9. Right patella. **A,** Anterior surface. **B,** Posterior surface. **C,** Anterior view; patella is outlined through the femur.

moved completely and leave a normally functioning knee joint. Surgical removal is often necessary following fracture when the posterior surface, which is articular, becomes roughened. During extension of the knee the patella presses firmly against the femur, and the slightest roughness and/or displacement of the patella is not tolerated.

Routine radiography of the patella includes the following views: anterior (P-A), posterior (A-P), lateral, and tangential (sunrise).

The lower leg includes two bones: the tibia and fibula. The tibia is quite large, especially in its proximal third, and is the medial bone of the lower leg.

The tibia articulates proximally with the femur in the knee joint. The knee joint consists of the femur, patella, and tibia. Distally the tibia articulates with the talus to form a major portion of the ankle joint. At both proximal and distal ends the tibia articulates laterally with the fibula.

The superior surface of the tibia presents two shallow, smooth depressions, the *medial* and *lateral condyles;* the condyles of the tibia articulate with the corresponding condyles of the femur. The tibial condyles are separated by a raised, double-pointed projection, the *intercondylar eminence* (tibial spine). Extending from anterior to posterior and between the two points of the tibial spine is a large depression, which is divided into the *anterior* and *posterior condyloid fossae.* In several loci in these fossae are attachments of the *menisci* and *cruciate ligaments.*

The medial and lateral condyles and the intercondylar eminence make up the *tibial plateau.* Inferior to the tibial plateau and anterior to the tibial spine, the medial and lateral condyles join. The *tibial tuberosity* is the raised and roughened area projecting from the center of the junction of the condyles to serve as attachment site for the ligament of the patella. Posterolaterally on the lateral condyle and inferior to the tibial plateau is a flat, articular facet for the head of the fibula.

The head of the tibia (proximal end) includes the condyles, intercondylar eminence, tibial tuberosity, and facet for the head of the fibula. The neck of the tibia is not well defined.

The tibial shaft extends distally from the neck of the tibia and is shaped somewhat like a triangle in its proximal portion; the base of the triangle is in an anteromedial oblique plane. The shaft flattens considerably in the lateral surface and narrows in the distal third. In males the tibial shaft is more nearly vertical than in females, in whom the shaft angles outwardly from the knee. Accompanying the greater inward angle of the female femur is the greater outward angle of the female tibia, which causes a *knock-kneed* condition in most females.

Posteriorly on the tibia is the *popliteal (soleal) line,* a slightly elevated ridge to serve as attachment site for the soleus muscle. The popliteal line extends obliquely between the neck laterally and the center of the tibial shaft medially. Parallel with the approximate middle third of the popliteal line and somewhat lateral to it is the nutrient foramen, through which the nutrient artery passes to the medullary canal. The nutrient artery arises from the posterior tibial artery.

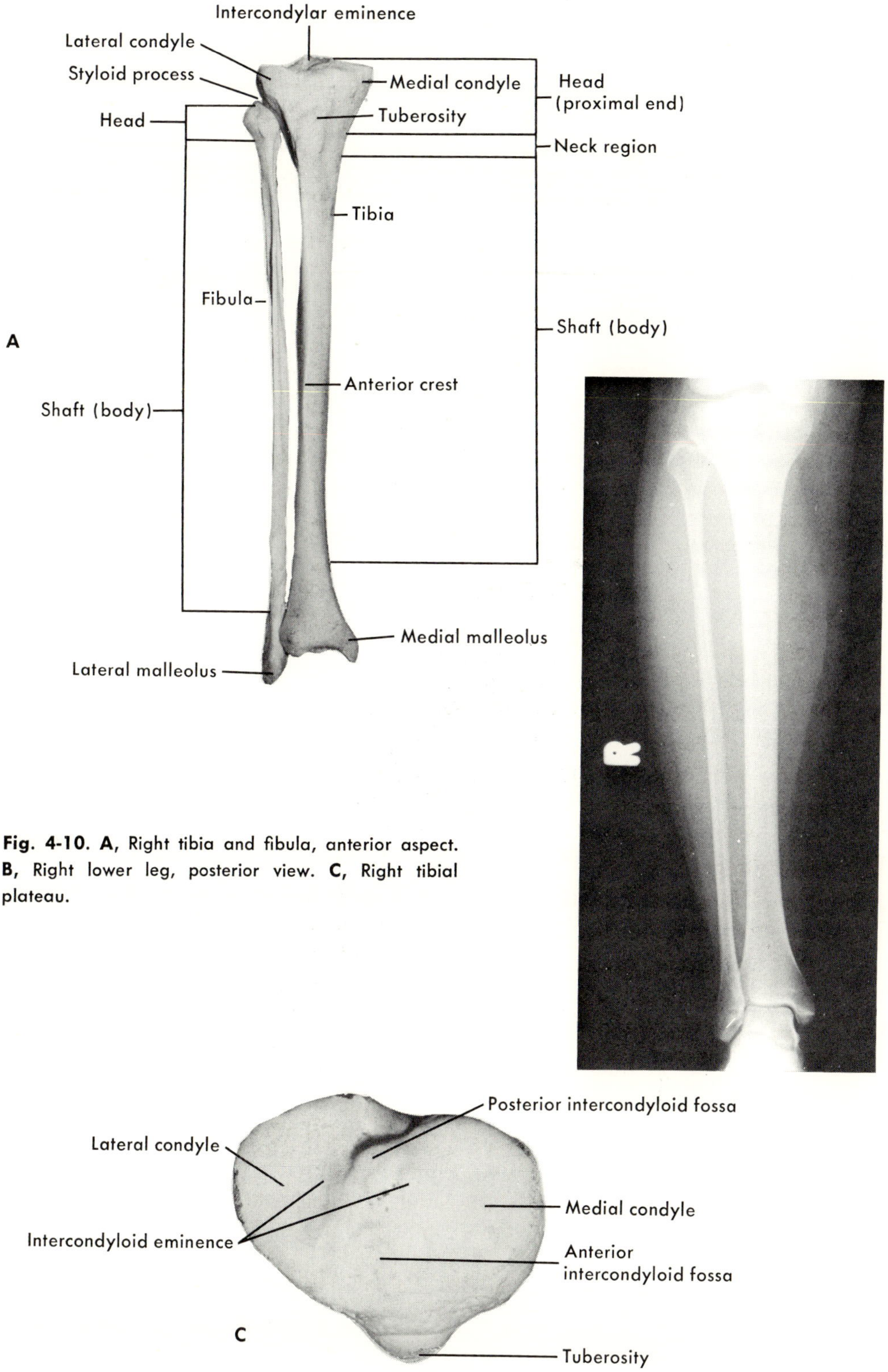

The skeletal system 79

Fig. 4-10. **A**, Right tibia and fibula, anterior aspect. **B**, Right lower leg, posterior view. **C**, Right tibial plateau.

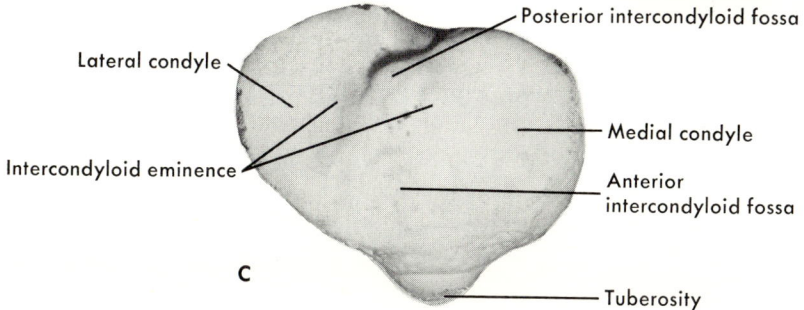

The distal end of the tibia is quite large and is very strong. A large projection, the *medial malleolus,* extends well below the remainder of the tibia to articulate laterally with the talus and to form the medial prominence of the ankle. A large area of the inferior surface of the tibial shaft articulates with the superior surface of the talus, forming the principal ankle articulation. There is limited articulation laterally with the fibula near the distal end of the tibia. Certain injuries of the ankle necessitate radiographic visualization of the space between the lower parts of the tibia and fibula. A tangential view of the triangle-shaped lower third of the tibia often demonstrates pathology of this surface, otherwise not routinely visible. The tibia is usually ossified by the twentieth year of life.

The fibula, the lateral bone of the lower leg, is quite slender. The *styloid process,* a roughened and pointed projection, extends superiorly from the lateroposterior surface of the head; it serves as attachment for the tendon of the biceps femoris muscle and for the fibular collateral ligament of the knee. The head and neck are included in the proximal portion of the fibula.

The fibular shaft is parallel with the tibial shaft and terminates in the enlarged *lateral malleolus,* which extends slightly below the inferior tip of the medial malleolus. The lateral malleolus forms the lateral prominence of the ankle joint.

The fibula articulates with the tibia both proximally and distally. The medial surface of the lateral malleolus articulates with the talus. The fibula is usually ossified by the twenty-third year of life (see Fig. 4-10).

Routine radiographs of the lower leg are made to visualize both the tibia and fibula. The routine views include posterior (A-P), internal oblique, and lateral views. All the body weight is carried by the tibia; none is carried by the fibula. This is visualized radiographically when the ankle mortise is viewed from the anterior aspect.

Following a fracture of the ankle the entire ankle mortise must be demonstrated radiographically. To make such demonstration possible, the toes must be turned slightly inward (about 30 degrees). If the ankle mortise is widened, the physician will return the mortise to natural spacing by lateral pressure, which is maintained by application of padded sponge rubber, a cast, and other appliances. Positioning of this body part is most important in radiography to assure direction of the central rays exactly through the mortise.

The bony part of the foot includes the seven tarsal and five metatarsal bones. The seven tarsal bones include the talus (astragalus), calcaneus (os calcis), cuboid, navicular, medial (first) cuneiform, intermediate (second) cuneiform, and lateral (third) cuneiform. The tarsals and metatarsals are held together by ligaments, tendons, and muscles to form two distinct arches, the transverse and longitudinal. These arches permit a "springing" action of the foot while standing, walking, and running.

The arrangement of the five metatarsals and the distal row of tarsals forms the transverse arch. The longitudinal arch consists of lateral and medial portions. The

Table 4-5. Articulations of the tarsal bones

Bone	Number of articulations	Articulations
Talus	5	Lateral and medial malleoli, distal end of tibia, navicular, and calcaneus
Calcaneus	2	Talus and cuboid
Navicular	4(5)	Medial, intermediate and lateral cuneiforms, talus, and sometimes cuboid
Cuboid	4(5)	Calcaneus, lateral cuneiform, fourth and fifth metatarsals, and sometimes navicular
Medial cuneiform	4	Navicular, intermediate cuneiform, and first and second metatarsals
Intermediate cuneiform	4	Navicular, medial and lateral cuneiforms, and second metatarsal
Lateral cuneiform	6	Navicular, medial cuneiform, cuboid, and second, third, and fourth metatarsals

calcaneus, cuboid, fourth metatarsal, and fifth metatarsal form and shape the lateral longitudinal arch. The calcaneus, talus, navicular, three cuneiforms, first metatarsal, second metatarsal, and third metatarsal form and shape the medial longitudinal arch. Decline of tension and strength of the ligaments, tendons, and muscles of both the transverse and longitudinal arches results in the condition known as "flatfoot" or "flatfeet."

The talus is the major weight-supporting bone of the body; it is the second largest tarsal bone. Although the talus is less easily visualized radiographically, successful demonstration is achieved in posterior and lateral views of the foot. In the posterior view, the central ray is directed from 5 to 10 degrees toward the calcaneus; the increased density of structure requires an increase in kilovoltage of approximately 10. The talus, tibia, and fibula form the ankle joint. The articulations of the talus and other tarsal bones are listed in Table 4-5.

The talus ossifies from one center and is usually completely ossified by the twelfth year of life.

The calcaneus is the largest of the tarsal bones and forms the heel of the foot. The calcaneus is shaped somewhat like an elongated cube. A prominence, the *sustentaculum tali*, articulates with and supports the talus medially. The sustentaculum tali is medial on the calcaneus near the toe end of the bone. The articular facet on this prominence consists of two portions, the medial portion being the larger. Superior on the calcaneus and near its center is a facet for articulation with the talus. These three articular facets present a combined total surface of not more than 1 square inch, a surface that must support the entire weight of the body. This explains why fractures of the calcaneus are notorious producers of foot pain at a later date. Approximately 50 percent of all calcaneal fractures cause severe pain in the subtalar (subastragalar) joint. Fusion of this joint is obtained by a triple arthrodesis surgical procedure. Laterally on the calcaneus and nearly opposite to the sustentaculum tali is the rather prominent

82 *Textbook of anatomy and physiology in radiologic technology*

trochlear process, an important palpation point. A large, roughened prominence, the *calcaneal tuberosity*, is situated posteriorly near the superior surface of the calcaneus and serves as attachment for the abductor digiti quinti, abductor hallucis, and flexor digitorum brevis muscles and the plantar aponeurosis. The calcaneal tendon (of Achilles), the common tendon of the gastrocnemius and soleus muscles, inserts into the midportion of the posterior surface of the calcaneus. For the most part, the anterior surface of the calcaneus articulates with the cuboid bone.

Routine radiographs of the calcaneus include either the posterior view or anterior and lateral views. It is essential in the posterior and anterior views that sufficiently high kilovoltage be used to demonstrate the *talocalcaneal joint*.

The calcaneus ossifies from two centers, one for the body and one for the posterior epiphysis. The epiphyseal center does not appear until about the tenth year of life. The epiphysis attaches to the body shortly after puberty.

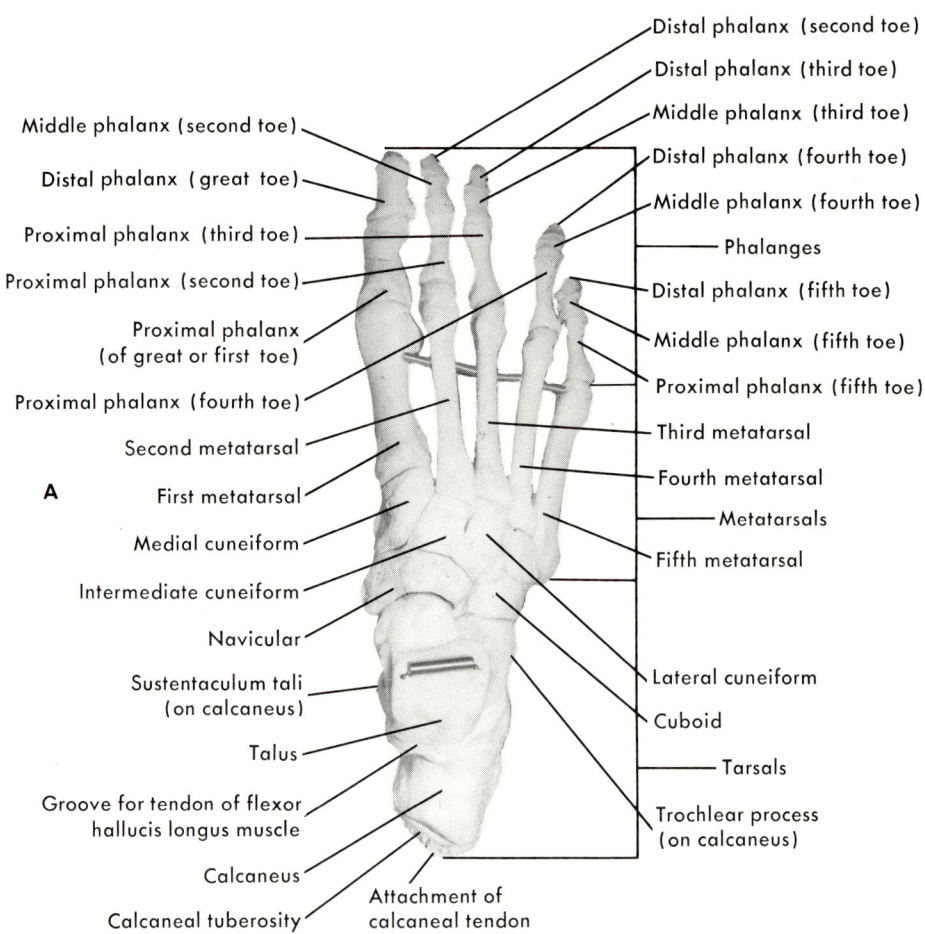

Fig. 4-11. Right foot. **A,** Superior aspect of bones. **B,** Dorsoplantar view.

The navicular bone is situated medially in the foot between the talus posteriorly and the cuneiform bones and cuboid bone anterolaterally.

The cuboid bone derives its name from its similarity in shape to a cube. It is situated laterally in the foot and sandwiched between the fourth and fifth metatarsals anteriorly and the calcaneus posteriorly.

The dorsal surface of the cuboid bone is roughened and is the attachment site of several ligaments. The posterior surface of the cuboid bone articulates with the calcaneus; the anterior surface articulates with the fourth and fifth metatarsals. The medial surface articulates with the lateral cuneiform bone and sometimes with the navicular bone.

The cuboid bone may be demonstrated radiographically in the dorsoplantar view of the foot when the central ray is directed about 10 degrees toward the calcaneus and in the oblique view when either the dorsal or plantar surface is nearest to the film.

The cuboid, the navicular, and the three cuneiform bones each ossify from a single center. The three cuneiform bones are named from medial to lateral—the medial, intermediate, and lateral. The cuboid bone is lateral to the lateral cuneiform bone. These four tarsals form a *distal* row of tarsal bones and are

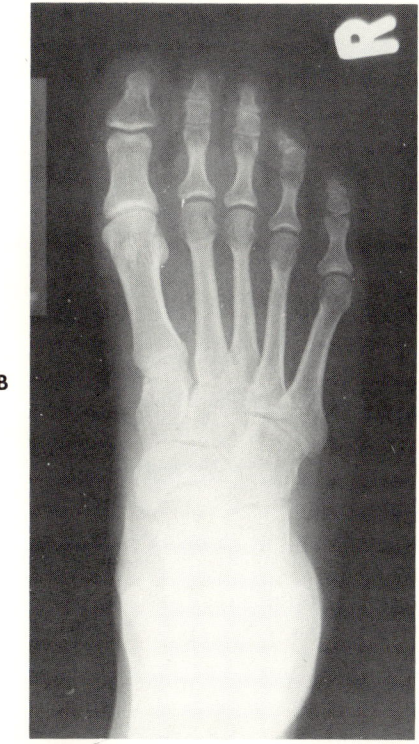

Fig. 4-11, cont'd. For legend see opposite page.

rarely requested for singular radiographic demonstration, but they are always included in views of the foot and are often included in views of the ankle.

There are five metatarsals, numbered from one through five from the medial to the lateral side. They compare in shape with the metacarpals. The foot is formed by the seven tarsals, five metatarsals, and fourteen phalanges. The highest prominence on the dorsum of the foot is over the proximal end of the second metatarsal (see Fig. 4-11).

The proximal articulations of the metatarsals are listed in Table 4-5. Each metatarsal ossifies from two centers. The ossification centers in the second, third, fourth, and fifth metatarsals are in the bodies and in the heads of the bones, which are anterior (toward the toes). The ossification centers in the first metatarsal bone are in the body and in the base, which is toward the tarsals.

The pedal phalanges comprise the toes. In the toes, as in the thumb and fingers of the hand, there are two phalanges in the first (great) toe and three phalanges in each of the remaining four toes. As in the thumb, the two bones in the first toe are the proximal and distal phalanges. The phalanges of the other four toes are the proximal, middle, and distal.

Each phalanx ossifies from two centers, one in the body and one in the base. Ossification progresses from each center toward the other, which causes each phalanx to be fully ossified, usually, during the eighteenth year of life.

Radiography of the foot must include the toes. The difference in thickness between the toes and the metatarsal region may be compensated for by either of the two following methods:
1. Use of the heel effect and angling the central ray slightly toward the calcaneus
2. Use of one 1 mm. thick aluminum filter between the film holder and the metatarsal region of the foot and use of two 1 mm. thick aluminum filters between the film holder and the toes

Vertebral column

The vertebral column (spine) consists of thirty-three segments, twenty-four of which are true vertebrae. Structurally, the vertebral column has five divisions, the cervical, thoracic, lumbar, sacrum, and coccyx. The articulated vertebrae support the trunk and protect the spinal cord (medulla spinalis). Viewed from the side, the vertebral column presents four curvatures as follows:
1. Cervical (anteriorly convex)
2. Thoracic (anteriorly concave)
3. Lumbar (anteriorly convex)
4. Sacrococcygeal (anteriorly concave)

The curvatures are illustrated in Fig. 4-12.

In addition to these normal curvatures, three abnormal conditions are frequently found. It is our responsibility to know and recognize these conditions so that we may demonstrate them radiographically. These conditions are: kyphosis, lordosis, and scoliosis.

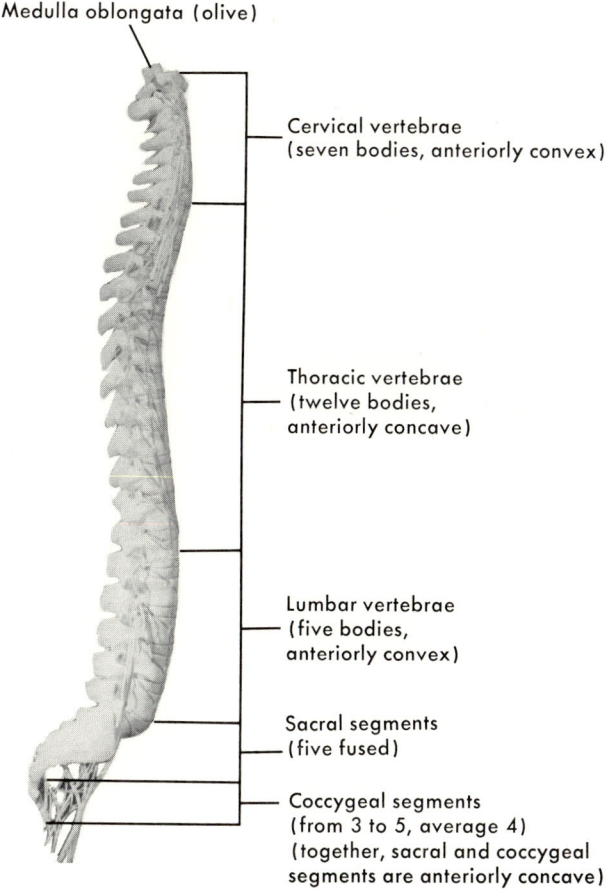

Fig. 4-12. Spine model, right lateral aspect. (Taken together, the sacral and coccygeal segments are anteriorly concave.)

Kyphosis (humpback) is abnormal anterior concavity of the thoracic or thoracolumbar vertebrae with dorsal prominence of the vertebral column. This condition occurs in the sagittal plane, and may result from trauma, degenerative disease of the vertebrae, or developmental abnormalities.

Lordosis is abnormal curvature of the vertebral column in the sagittal plane with anterior convexity, and often with abnormal or exaggerated prominence of the lumbosacral region. The latter condition causes a "hollow back" and may be called lumbar lordosis.

Scoliosis is abnormal lateral curvature of the vertebral column, a sideways curvature. This condition may include rotation of one or more vertebrae about the long axis of the column (torsion).

Both lordosis and scoliosis may occur as a result of the conditions described under kyphosis, although scoliosis seldom results directly from trauma (see Fig. 4-13).

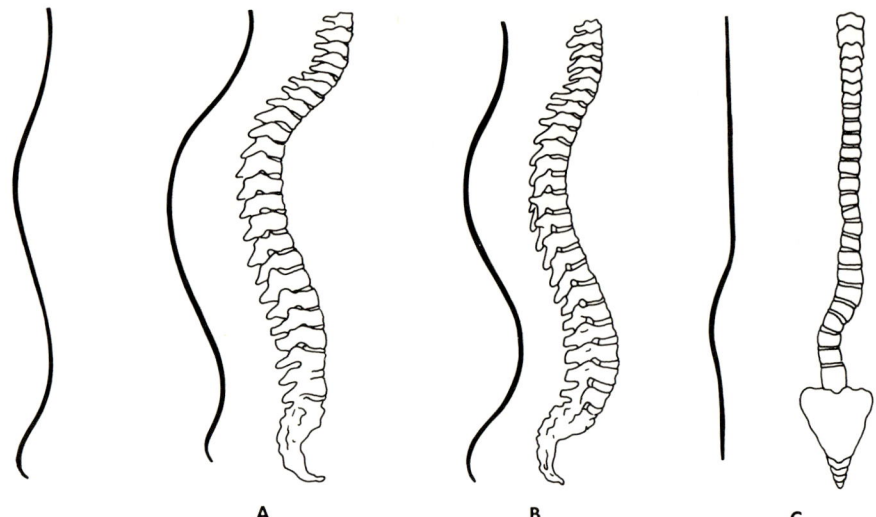

Fig. 4-13. A, Kyphosis, note the comparison of curved black line with the normal curvature line to the far left of the illustration. **B,** Lordosis, note the comparison of the curved black line with the normal curvature line to the far left of the illustration. **C,** Scoliosis, note the marked deviation from true vertical as evidenced by the curved black line.

Typical vertebrae: common aspects. A typical vertebra consists of the *body* and the *neural arch*. The *vertebral* (spinal) *foramen* is formed by the body anteriorly and the *pedicles* and *laminae* posteriorly. The pedicles and laminae form the neural arch. When the vertebral bodies are articulated, the several vertebral foramina form the *spinal canal*. Between each two vertebral bodies is an *intervertebral disc* of tough fibrocartilage, which acts as a shock absorber and hinge. The twenty-three intervertebral discs perform a very important function. Rupture of one disc may permit the superior and inferior surfaces of adjacent vertebrae to exert pressure on one or more of the spinal nerves and sometimes on the spinal cord. Radiographic examination to demonstrate this condition and other pathologic conditions is named *myelography* (see Fig. 4-14, *A*). Radiographic examination following opacification of the nucleus of an intervertebral disc is named *discography* (see Fig. 4-14, *B*).

The intervertebral discs are circular fibrocartilages. The discs function importantly to facilitate motion of the vertebral column. The motion possible between any two vertebral bodies is severely limited. However, the vertebral column is capable of considerable rotation, flexion, and extension, with constant protection being afforded the spinal cord.

The intervertebral discs function as hinges; each allows a few degrees of angulation between the adjacent two vertebrae. *Rupture* of a disc is more often a gradual degenerative process punctuated by attacks of back pain than the result of a single traumatic experience. Each attack terminates quickly. Following a prolonged period of time (from 6 to 7 years) a much more severe attack is likely;

The skeletal system 87

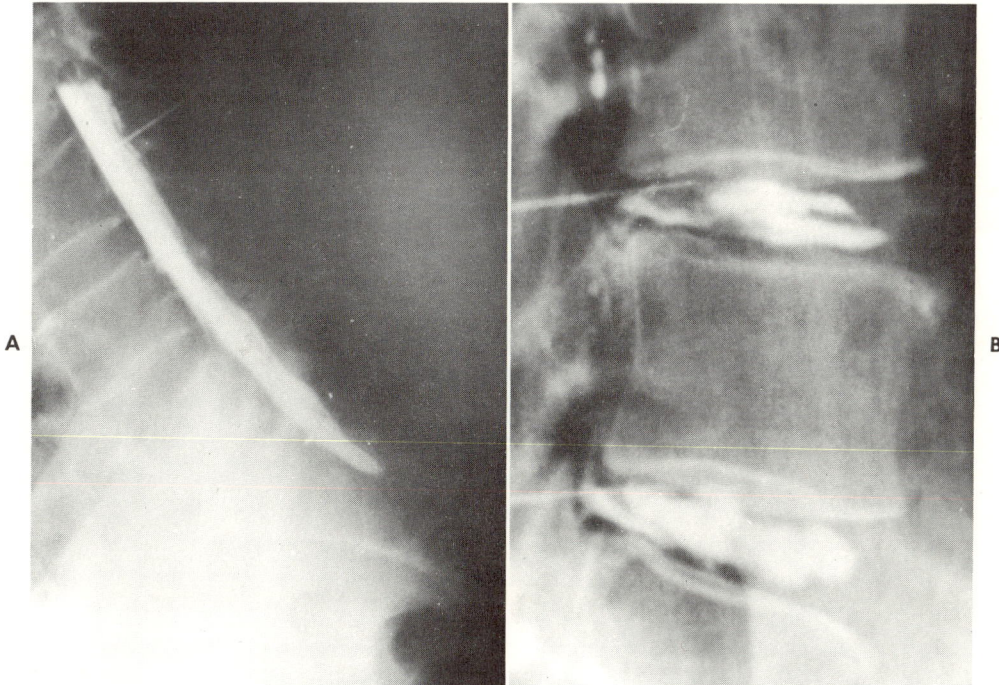

Fig. 4-14. A, Myelogram, lateral view. B, Discogram, lateral view.

this attack presents the typical symptoms of pain radiating down the leg accompanied by numbness. This is the actual time at which disc protrusion occurs and causes nerve root pressure. About 85 percent of degenerated discs occur between the fifth lumbar and first cervical vertebrae, and about 17 percent occur between the fourth and fifth lumbar vertebrae. All regions are examined radiographically.

Some mention of the relation of the skeletal and nervous systems is pertinent at this time, although growth and development of the nervous system is presented in Chapter 9. The notochord appears prior to the neural fold and remains, in part, as the pulpy mass in the center of each intervertebral disc; the pulpy mass is the *nucleus pulposus*. Prolapse of the nucleus pulposus into an adjoining vertebral body causes a nodule called *Schmorl's nodule*, which can be visualized in radiographs of the vertebral column.

Each vertebral body supports seven processes of which four are articular, two are transverse, and one, the spinous, extends posteriorly in the midline. Each articular process contains an articular facet. A typical vertebral body contains three primary centers of ossification and five secondary centers of ossification. There is one primary center in the vertebral body and one in each side of the neural arch. There is a secondary center of ossification in the tip of the spinous process, one in the tip of each transverse process, and one in each epiphyseal ring on the superior and inferior surfaces of each body. The primary centers are present at birth; the secondary centers usually appear at puberty.

BODY. The biscuit-shaped body (centrum), the largest single part of a vertebra, is composed almost entirely of cancellous bone. The inferior and superior surfaces of the body are flat and roughened and are depressed inwardly from the margin, thus forming a distinct rim on each surface. A compression fracture consists of a telescoping of the anterior rim. A lateral view demonstrates the condition as a decrease in anterior thickness when compared with the two adjacent vertebrae.

PEDICLES. The pedicles are two strong, bony processes that project bilaterally and posteriorly from the sides of the vertebral body. Each pedicle extends from the junction of the posterior and lateral surfaces of the vertebral body. The *vertebral notches* are the depressions found superior and inferior to the pedicles. When the vertebral bodies are articulated, the inferior vertebral notch of one body is situated above the superior vertebral notch of the body below, thus forming the *intervertebral foramen,* which is illustrated in Fig. 9-4. Since there are two pedicles situated bilaterally on each vertebral body, there are two intervertebral foramina between each two bodies. The spinal nerves and branches and the nutrient vessels pass through these foramina.

LAMINAE. The laminae extend posteriorly and medially from the pedicles. The laminae fuse posteriorly in the midline to complete the neural arch. Failure of the laminae to close the neural arch results in the condition called *spina bifida,** a common condition that is considered asymptomatic. When the condition is patently visible, it is a true spina bifida; when the condition can be visualized only by means of radiography, it is called a *spina bifida occulta,* hidden spina bifida, as demonstrated in Fig. 4-15.

PROCESSES. The *spinous process* of each vertebral body can be palpated easily in most persons. It projects posteriorly and inferiorly from the fusion of the laminae. Spinous processes serve as attachment sites for several ligaments and muscles. The processes are often fractured as a result of trauma; lateral radiographs of the particular area of the spine demonstrate these fractures. Because of the great density of the vertebral bodies compared with the lesser density and thickness of the spinous processes, it is frequently necessary in radiography to compensate for this variation in thickness and density by inserting a 1 mm. thick aluminum filter between the film holder and the patient in such a manner that *only* the spinous processes are over the filter.

The pedicles join the laminae bilaterally to form parts of the neural arch. Two *superior articular processes* and two *inferior articular processes* arise from these junctions. The superior articular process extends posteriorly and presents the *superior articular facet.* The inferior articular process extends anteriorly and presents the *inferior articular facet.* The plane of articulation between the su-

---

*Spina bifida means cleft spine, a congenital malformation in which the spinal column is cleft at its lower portion and the membranes of the spinal cord project as an elastic swelling from the gap thus formed.

The skeletal system 89

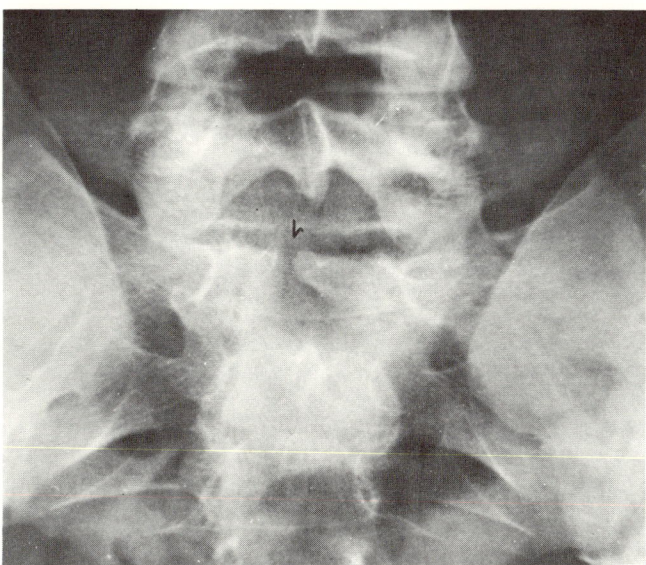

Fig. 4-15. Spina bifida occulta, posterior view. Arrow points to incomplete closure of neural arch. (Courtesy Dr. J. M. Hilton.)

perior and inferior articular facets varies in the different sections in the vertebral column from almost horizontal to almost vertical and from nearly coronal to nearly sagittal. The inferior articular facets of the vertebral body above articulate with the superior articular facets of the vertebral body below.

The junction of the laminae with the pedicles is below the base of the superior articular process and above the base of the inferior articular process. The transverse processes extend from this site. The transverse processes present roughened sites for attachment of several ligaments and muscles. Like the spinous processes, the transverse processes are often fractured as a result of trauma. Posterior and anterior radiographs of the particular region of the spine demonstrate these fractures. Compensation for the variation in density between the vertebral bodies and the transverse processes may be achieved by use of aluminum filters, properly positioned between the patient and the film holder.

Cervical vertebrae. The seven cervical vertebrae are the smallest of the true vertebral bodies. The first vertebra is the smallest of the cervical bodies; the size increases with progression toward the last. The cervical section is unique for several reasons. It contains the three named vertebral bodies: C-1* *(atlas)*, C-2 *(axis)*, and C-7 *(vertebra prominens)*. The axis supports the *dens epistropheus* (odontoid process). Early in intrauterine life the body of C-1 fails to ossify to the ring of C-1 and subsequently fuses to the superior surface of C-2. As a result, considerable rotation of the head is permitted by virtue of the turn-

---

*C-1 is the accepted form of writing the first cervical vertebra. C is used for cervical, T is used for thoracic, L is used for lumbar, S is used for sacral, and Co is used for coccygeal.

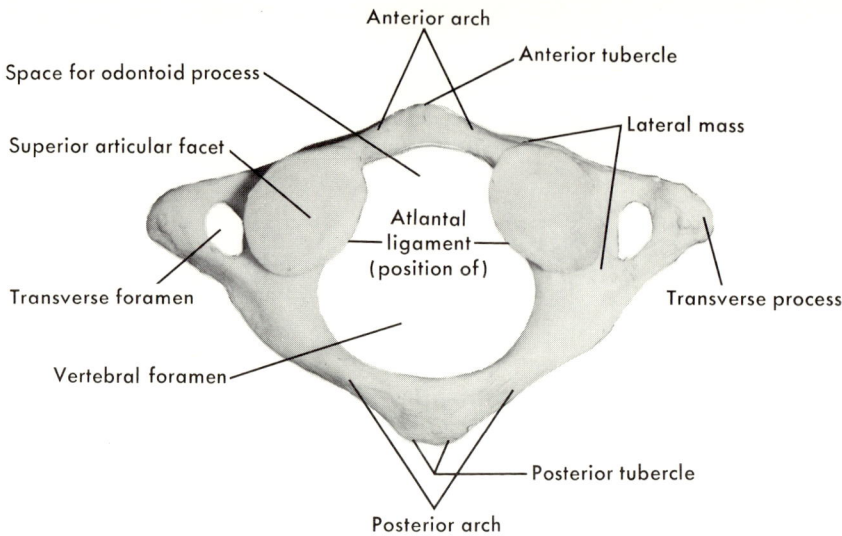

Fig. 4-16. Atlas, superior surface.

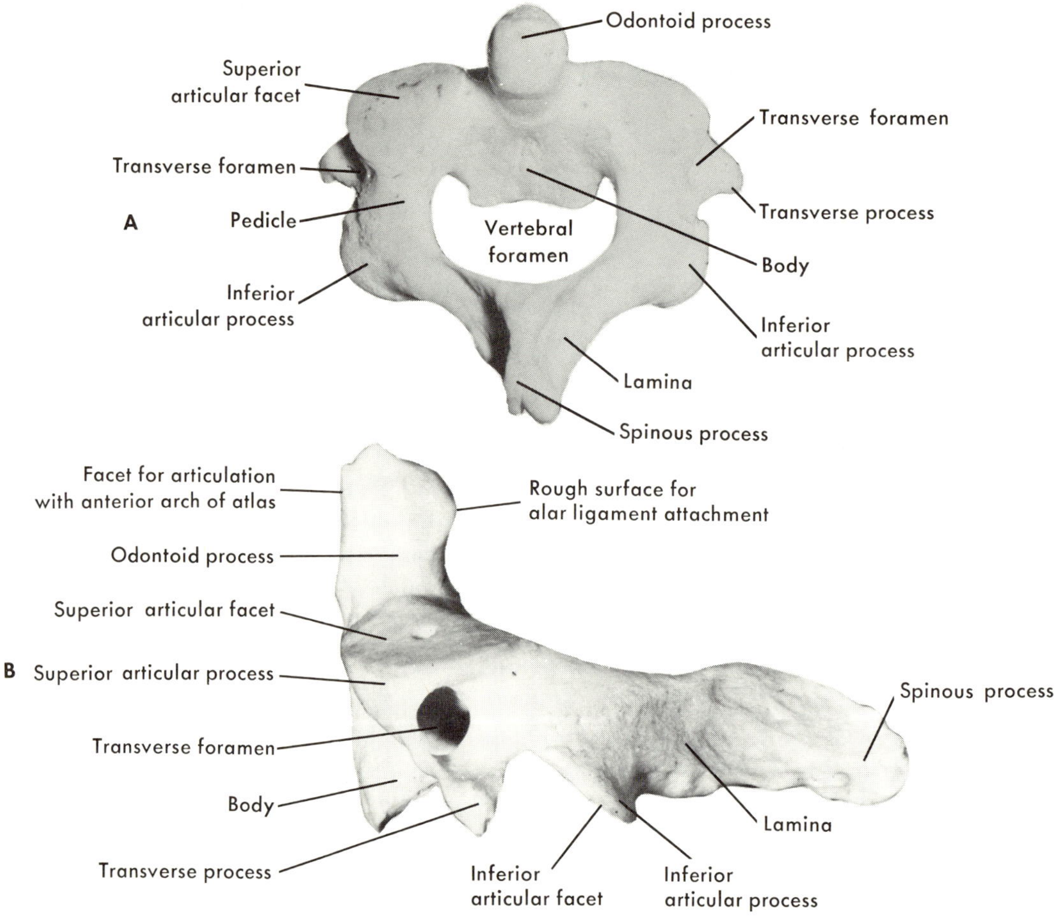

Fig. 4-17. Axis. **A,** Superoposterior aspect. **B,** Left lateral surface.

ing or rotating capability of C-1 about the odontoid process and upon the cartilage (intervertebral disc) between C-1 and C-2. The cervical vertebrae contain in their transverse processes the *transverse foramina* (one foramen in each transverse process), through which the vertebral arteries ascend from their origins to the base of the brain. The tips of the spinous processes of C-2 through C-6 are usually bifid.*

ATLAS. The head articulates with a forward and backward rocking motion upon the ring-shaped atlas, which has no body. In prenatal life the body fuses to the superior surface of the axis to form the odontoid process, which projects superiorly through the ring of the atlas. The atlas supports no spinous process, having instead either a single or double *posterior tubercle*. Two *lateral masses* separate its *anterior* and *posterior arches* and present the two *superior articular facets* for the occipital condyles. Posteriorly in the center of the anterior arch is a facet for articulation with the odontoid process. The transverse *atlantal ligament* extends between the lateral masses to prevent pressure of the odontoid process upon the spinal cord. The *inferior articular facets* articulate with the superior articular facets of the axis (see Fig. 4-16).

AXIS. The axis is unique in that it supports the odontoid (toothlike) process; its name derives from its similarity to a tooth projecting from the mandible. In other respects, the axis is similar to any typical cervical vertebra, i.e., C-3 through C-6 (see Fig. 4-17).

VERTEBRA PROMINENS. The most prominent spinous process of the vertebral column is C-7; for this reason C-7 is named the vertebra prominens. The tip of its spinous process is *monid*, not bifid. Its transverse processes are larger than those of the other six cervical bodies (see Fig. 4-18). Certain conditions cause

*Bifid means divided into two equal parts or lobes by a median cleft.

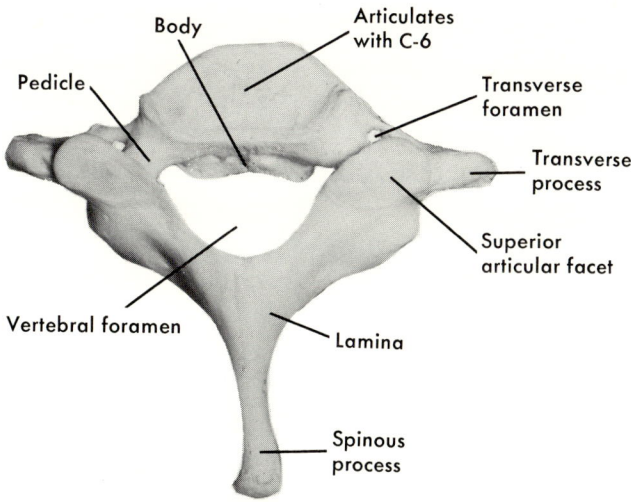

Fig. 4-18. Vertebra prominens, superoposterior aspect.

92 Textbook of anatomy and physiology in radiologic technology

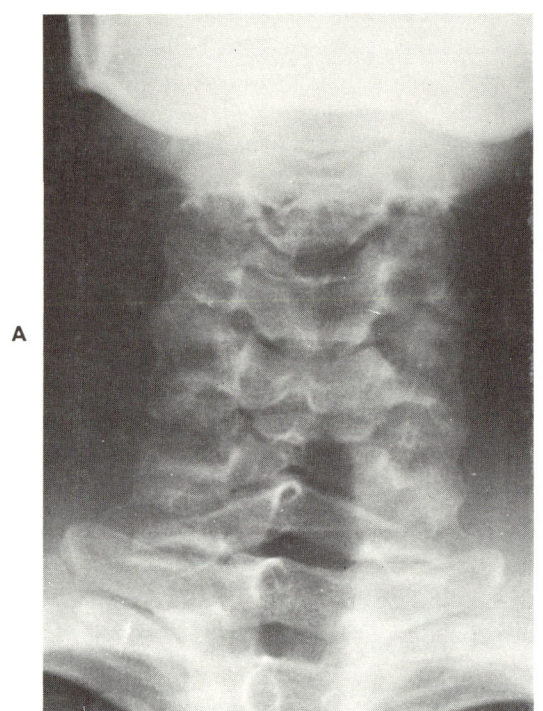

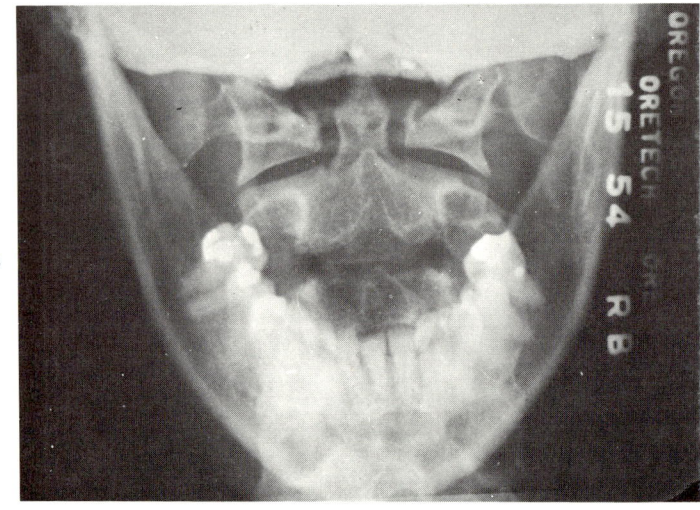

Fig. 4-19. Cervical spine. **A** and **B,** Posterior views. **B,** Open mouth for atlas and axis. **C,** Lateral view.

excessive growth of one or both transverse processes of C-7. When a patient has this condition, he is said to have a *cervical rib* (or cervical ribs).

Routine radiography of the cervical spine requires that posterior views, one with the mouth closed and one with the mouth open, be made in conjunction with lateral and oblique views. The open-mouth view demonstrates the odontoid process and the *very important interspace between C-1 and C-2*. The routine posterior view with the mouth closed demonstrates very little, if any, of the intervertebral spaces because the anterior convexity of the cervical spine places these spaces at angles remarkably different than those of the roentgen rays (x rays); placing the patient in a prone position and using an anterior projection of the x-ray beams usually overcomes this problem. An anterior view takes advantage of the natural curvature of the cervical spine (see Fig. 4-19). Another method of obtaining useful radiographs of the cervical spine employs two exposures on one film. A routine exposure on a 10 × 12 inch film is made with the mouth closed, the patient in supine position, and the head and neck in a normal position; without moving the patient or the film, a second exposure, using ten more kVp, is made in which the central ray* is directed through the open mouth to the atlas. A fully extended cone is used in the second exposure. The mandibular shadows are quite indistinct (see Fig. 4-20).

*Central ray may be written C-R.

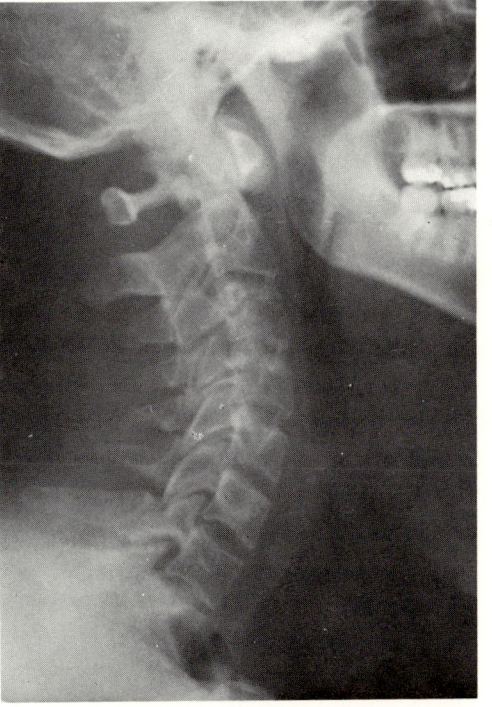

Fig. 4-19, cont'd. For legend see opposite page.

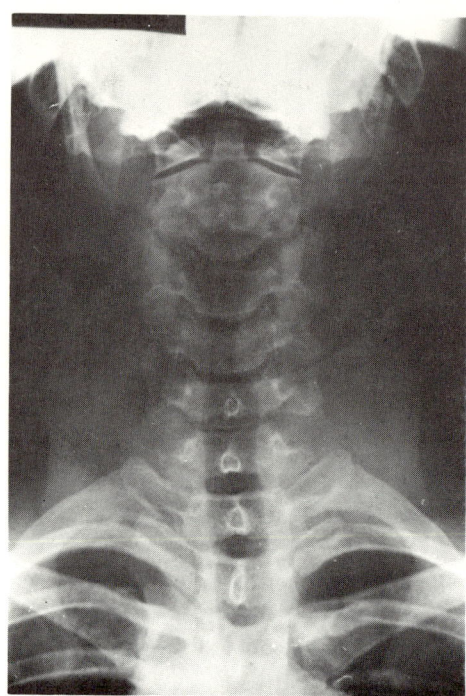

Fig. 4-20. Complete cervical spine, posterior view.

The lateral view of the cervical spine demonstrates pathology better than anterior and posterior views, except for a fractured odontoid process. Arthritic degeneration of the intervertebral disc between C-5 and C-6 often forms *without injury*. Such conditions are found in a high percentage of persons over 40 years of age. These persons are quite susceptible to whiplash injuries of the neck.

A fracture of the odontoid process, not visualized in the original radiograph, may frequently be demonstrated in radiographs made subsequently (about 4 weeks later) due to absorption; in such conditions a line may be demonstrated.

Following fracture of the mandible and/or maxilla, repair is frequently followed by immobilization of the jaws, often achieved by wiring the jaws together. There may be subsequent requests on such patients for radiographic demonstration of the odontoid process. This may be achieved by rotating the head from 4 to 8 degrees from vertical and directing the central ray through the upper nasal antrum to the odontoid process.

Among the less common radiographic requests are those for "pillar" views of the vertebrae, especially the cervical and upper thoracic regions. These are views, both posterior and posterior oblique, to demonstrate the articular processes and their facets. The term, pillar, refers to a firm upright support for a superstructure, such as the head upon the atlas. Other explanations for use of this term include: a column or shaft standing alone, and a natural pillar-shaped formation or mass.

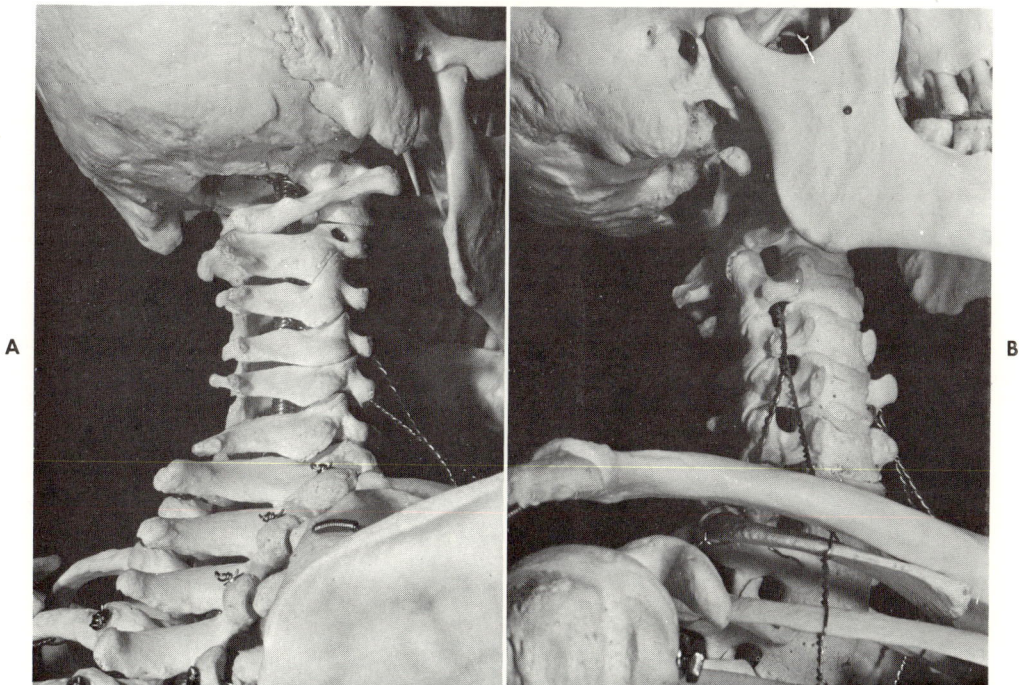

Fig. 4-21. Cervical vertebrae, articulated skeleton. **A,** Right postero-oblique aspect. **B,** Right antero-oblique aspect. In each aspect, the surface shown is that nearest the film during radiography.

These views are obtained with the patient supine, the head rotated either to the right or the left, and the central ray angled caudad from 25 to 35 degrees.

Oblique views of the cervical spine "open" the intervertebral foramina; it is necessary to direct the central ray through the foramina in a plane at right angles to the plane of the foramina (the principal foramen is positioned in the center of the central ray). If the position of the patient is antero-oblique, the central ray must be angled caudad; if the position of the patient is postero-oblique, the central ray must be angled cephalad (see Fig. 4-21). In all instances it is necessary that the central ray be parallel with the plane of the articular facets to "open" these articular spaces in the radiograph.

Lateral views with the neck in extreme flexion and in extreme extension demonstrate certain pathologic and abnormal conditions.

Thoracic vertebrae. Dual nomenclature is common in relation to the thoracic section of the vertebral column. This section is frequently called the dorsal spine rather than the thoracic spine, although the entire spine is dorsal in the body.

The twelve thoracic vertebrae are each larger than the largest of the cervical vertebrae. Similar to the cervical bodies, the thoracic bodies become progres-

96 *Textbook of anatomy and physiology in radiologic technology*

sively larger from the first one downward. The *costal* (rib) facets* are found only on the thoracic vertebrae. From the first through the tenth thoracic vertebrae, additional facets articulate with the rib tubercles; these are found on the transverse processes and are called the *costotransverse facets.* The articulating facets for the ribs are on the vertebral bodies and are called the *costovertebral facets.* When there are two facets, one above the other on a single side of the vertebral body, the facets are the *upper costal facet* and the *lower costal facet.*

*The rib facets are often incomplete and actually are demifacets. The demifacets are sometimes missing.

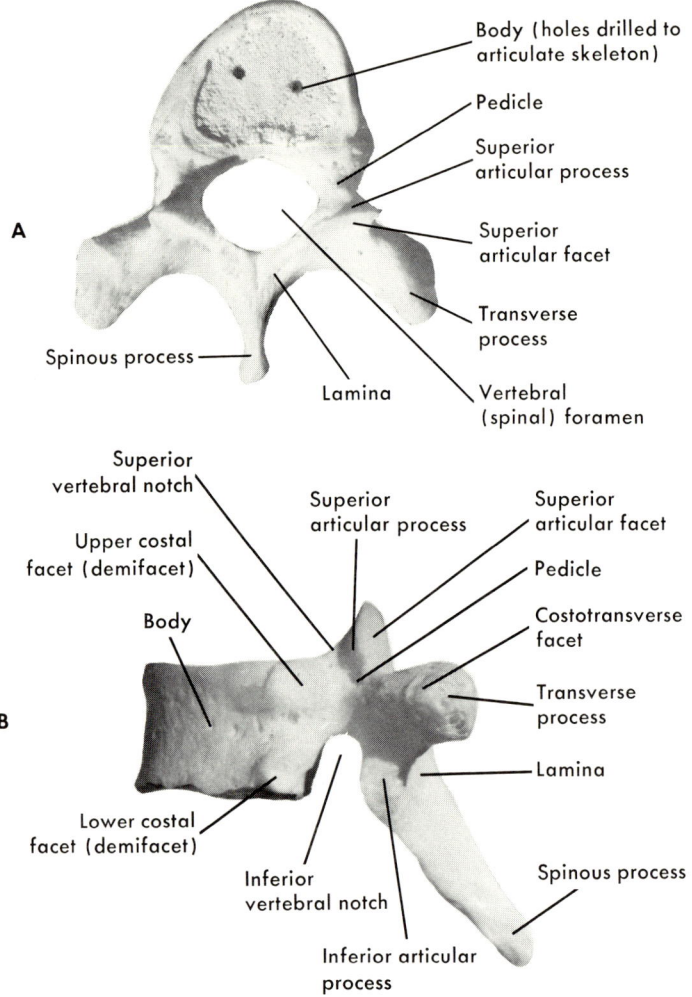

Fig. 4-22. A, Thoracic vertebra, superior aspect. **B,** Thoracic vertebra, lateral aspect. **C,** Thoracic spine, posterior view. **D,** Thoracic spine, semioblique view. **E,** Patient position for oblique view of the first, second, and third thoracic vertebrae. **F,** Oblique view, first, second, and third thoracic vertebrae.

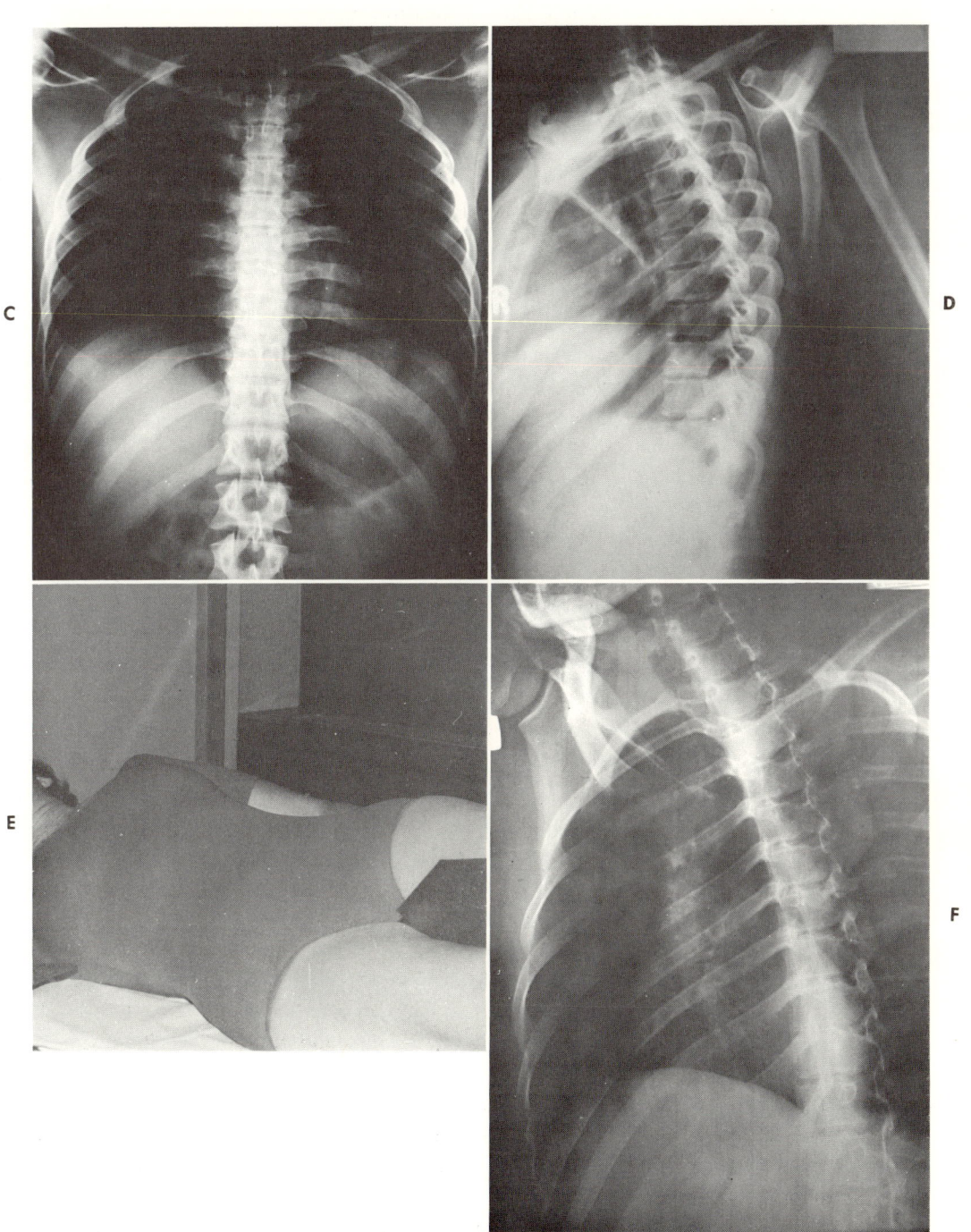

Fig. 4-22, cont'd. For legend see opposite page.

There are notable differences in the number and location of the various rib facets on the thoracic bodies. The differences are listed as follows:
1. T-1 has, on each side, a costovertebral facet and a lower costal facet. The transverse process of T-1 has a costotransverse facet.
2. T-2 through T-8 have, on each side, both an upper costal facet and a lower costal facet. The transverse process of each of these seven bodies has a costotransverse facet.
3. T-9 and T-10 each have, on each side, a single costovertebral facet. The transverse process of both T-9 and T-10 presents a costotransverse facet.
4. T-11 and T-12 each have, on each side, a single costovertebral facet. There are no costotransverse facets on the transverse processes of T-11 and T-12.

A posterior projection of the thoracic spine takes advantage of the anterior concavity of this section of the vertebral column, and the roentgen rays (x rays) easily penetrate the intervertebral spaces. If the radiographic request necessitates cone-down exposures on specific thoracic bodies, these bodies can be located easily with the patient lying in the supine position: T-1 is opposite a point 1.5 inches superior to the suprasternal notch; T-3 is directly opposite the suprasternal notch (or just below this notch in some persons); T-10 is directly opposite the xiphoid process.

A lateral view of the thoracic spine visualizes the fourth through the twelfth thoracic bodies; the first, second, and third thoracic bodies have the scapulae, humeri and/or clavicles superimposed over them in true lateral views. To demonstrate the first three thoracic bodies in a position other than prone or supine, rotate the left side (or right side) of the patient upward from a supine position with the lower arm pulled forward and the upper arm pulled backward; pull the upper shoulder backward and toward the table. Direct the central ray at right angles to the film through a point in the upper breast just inferior to the upper clavicle and opposite T-2. The central ray should emerge from the patient after passing obliquely through the first thoracic body at a point in the dorsal surface opposite the medial angle of the lower scapula and just medial to the vertebral border of the lower scapula. This is a semioblique (Swimmer's) projection of the first three thoracic bodies (see Fig. 4-22, *C* to *F*).

The principal pathology demonstrated in lateral views of the thoracic spine is the compression fracture. This usually occurs in the region of the maximum (anterior) concavity involving one or more of the bodies of T-5, T-6, or T-7. This is the region of maximum weight bearing, and fractures occur during forward, or jack-knife, injuries.

Juvenile epiphysitis occurs in the lower thoracic vertebrae sometimes causing back pain in a person of the late teen-age group. The condition often predisposes the rounded back, called by orthopedic surgeons the *dorsum rotundum* (round back).

Lumbar vertebrae. The five lumbar vertebrae are remarkably larger than the thoracic vertebrae and, like the thoracic bodies, increase in size from the

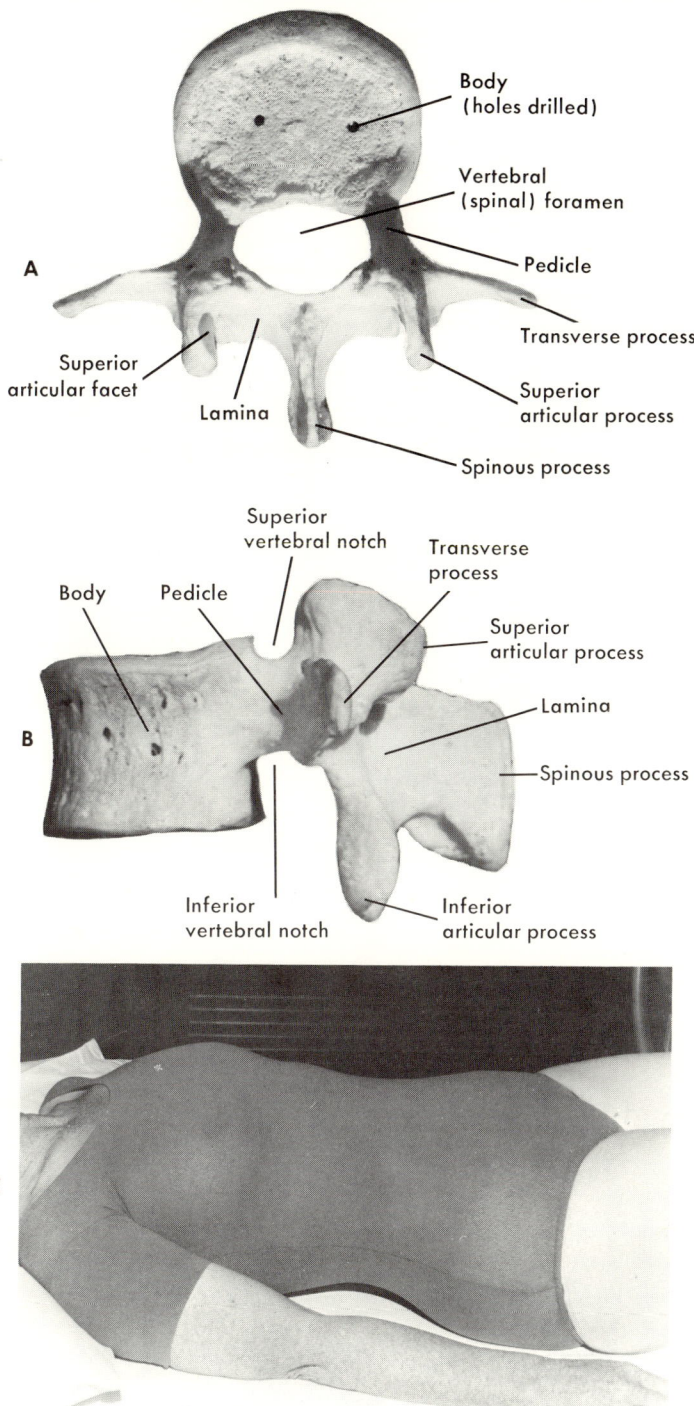

Fig. 4-23. A, Lumbar vertebra, superior aspect. **B,** Lumbar vertebra, lateral aspect. **C,** Patient position for posterior view of lumbar spine. **D,** Posterior view of lumbar spine. **E,** Patient position for anterior view of lumbar spine. **F,** Anterior view of lumbar spine. Compare the intervertebral spaces with those in **D**; the radiographs are of the same female. **G,** Lateral view of lumbar spine. **H,** Lumbar spine, R-P-O view.

100   Textbook of anatomy and physiology in radiologic technology

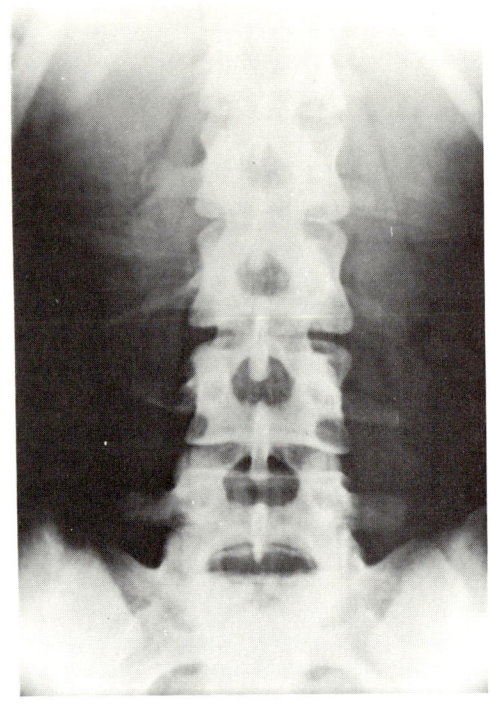

D

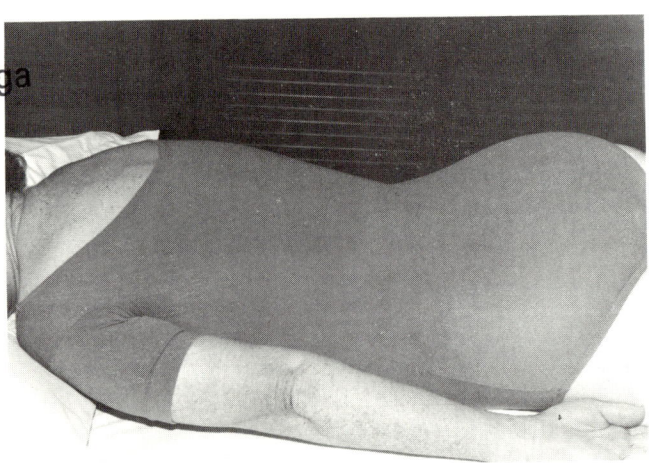

E

Fig. 4-23, cont'd. For legend see p. 99.

The skeletal system 101

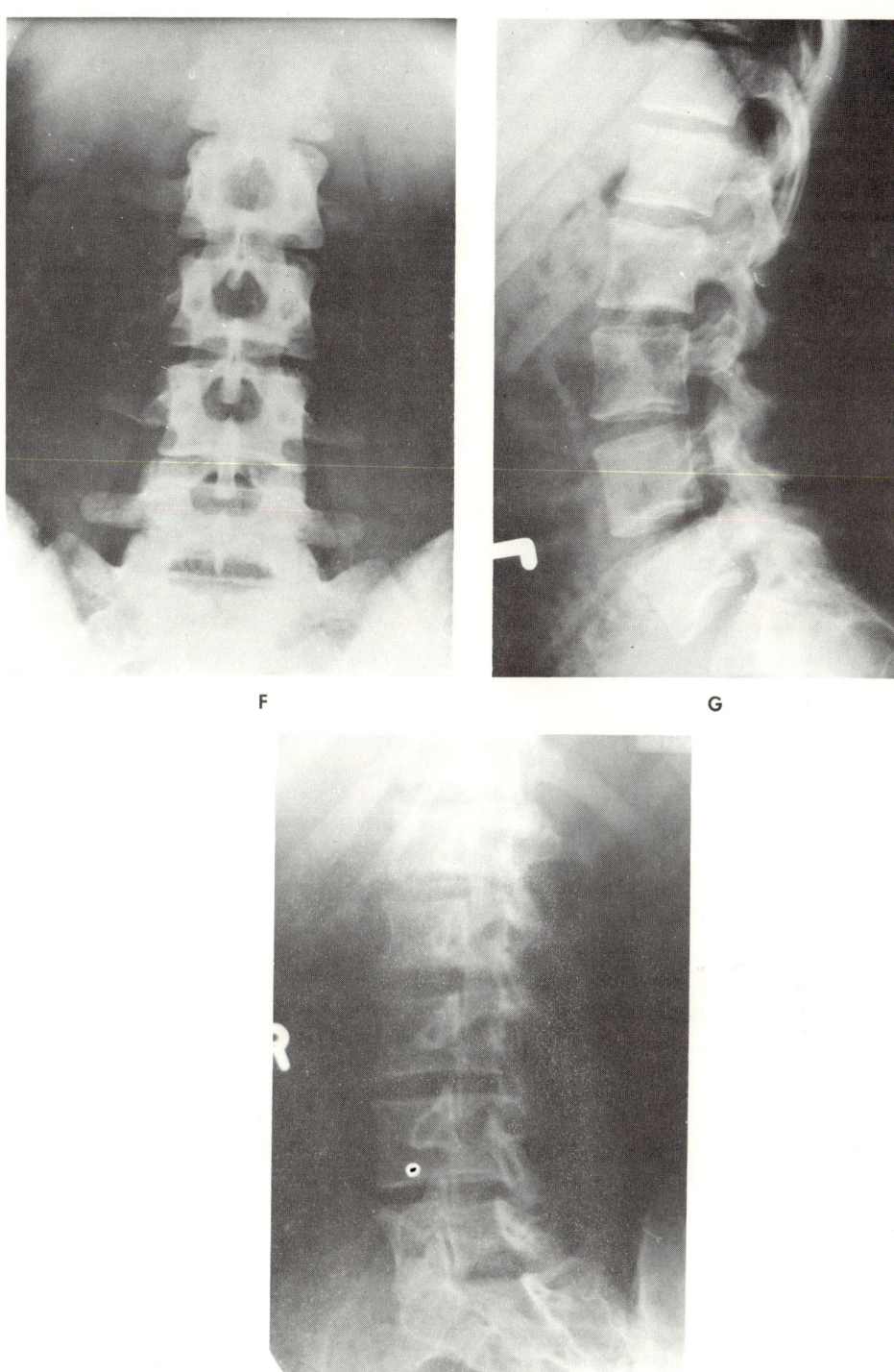

Fig. 4-23, cont'd. For legend see p. 99.

first downward. The transverse processes are more prominent radiographically; they are longer and more slender than those of other vertebral sections. The processes of the first three lumbar bodies are horizontal, while the processes of the fourth and fifth lumbar bodies project somewhat superiorly (see Fig. 4-23). The anterior (ventral) thickness of L-5 is considerably greater than the posterior (dorsal) thickness; this accounts for the prominence of the *lumbosacral articulation*. The spinous process of L-5 is smaller than the spinous processes of L-1 through L-4. The transverse processes of L-5 are thicker than the transverse processes of L-1 through L-4. Also, the inferior articular facets of L-5 are spaced farther apart than the inferior articular facets of the first four lumbar bodies. The most frequent site of spina bifida occulta is the fifth lumbar vertebra (see also Fig. 4-15).

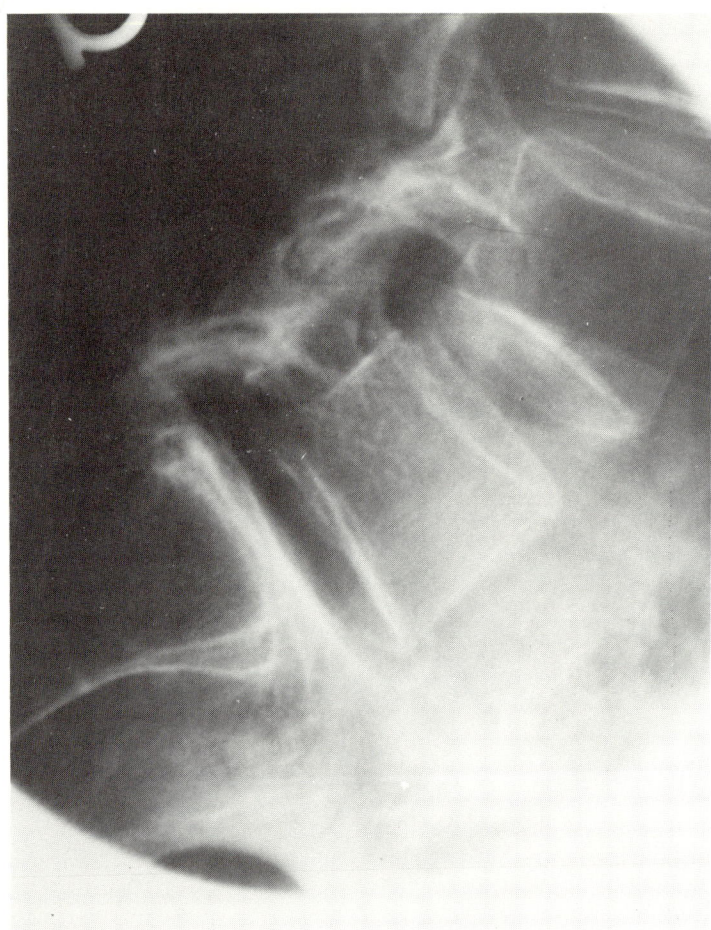

Fig. 4-24. Spondylolisthesis. (Courtesy Dr. Frank S. Cavallaro and Valley Hospital, Las Vegas, Nevada.)

The articulations of the lumbar vertebrae differ from the articulations of the thoracic vertebrae only in the absence of rib articulations. The very important articulation of L-5 with S-1 is the previously mentioned lumbosacral articulation. Radiographic examination of this joint is frequently requested. Traumatic injury may cause L-5 to slip forward on S-1 (the first sacral segment). This is the condition called *spondylolisthesis* (see Fig. 4-24). Another kind of pathology of the fifth (last) lumbar body and the first sacral segment is called lumbarization (see Fig. 4-25). When this condition is present the transverse processes of L-5 are either wholly or partially coalesced with the segment of S-1. Coalescence is the blending or fusion of parts.

The anterior curvature of the lumbar spine, like that of the cervical spine, is convex. Posterior projections of this region do not adequately show the inter-

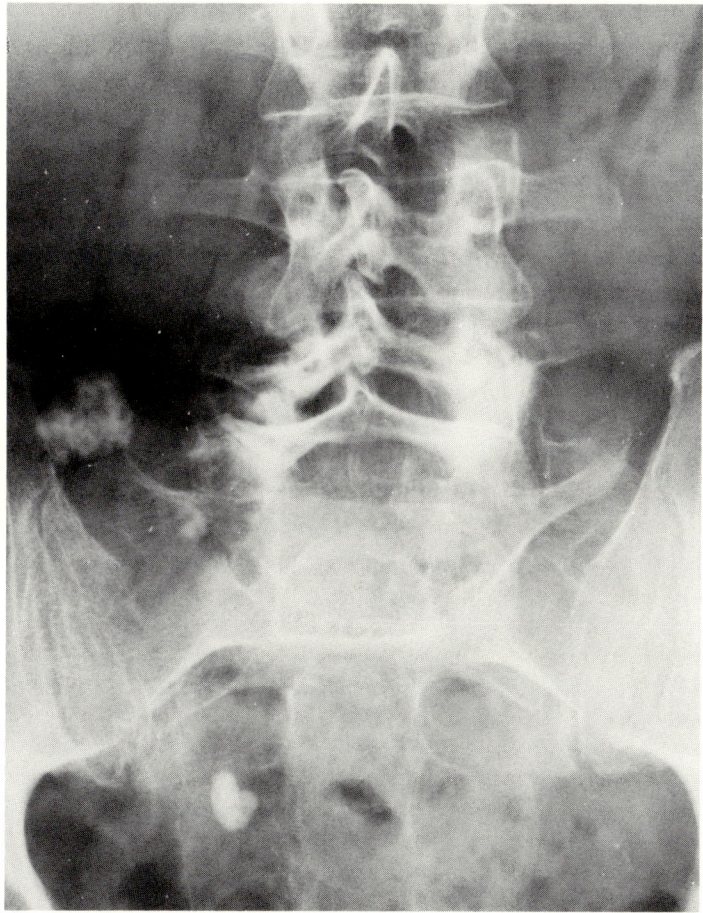

Fig. 4-25. Lumbarization. (Courtesy Dr. Frank S. Cavallaro and Valley Hospital, Las Vegas, Nevada.)

vertebral spaces. An anterior view of this section takes advantage of the natural curvature to demonstrate the intervertebral spaces (see Fig. 4-23, C to H).

When special emphasis of the lumbosacral joint is requested, *cone-down* views, both posterior and lateral, usually satisfy. In posterior views, the central ray must be angled 20 degrees cephalad to take advantage of the angle of the intervertebral disc. When one makes a cone-down view of any of the lumbar bodies, it is useful to remember certain landmarks; L-1 is opposite a point halfway between the xiphoid process and the umbilicus, L-3 is opposite the umbilicus, and L-5 is between the iliac crests (see pp. 124 and 125).

In addition to anterior and cone-down views of the lumbar spine and lumbosacral (L-S) joint, lateral and oblique views are routine. Oblique views open the intervertebral foramina for visualization. In this view, the facets resemble the head of a Scotch terrier dog when the facets are properly superimposed over the center of the vertebral body (see Fig. 4-23, H).

The lateral view of the lumbosacral joint is extremely important for diagnosis. The body is widest at this level, and superimposed upon the vertebral bodies are the two iliac crests and the posterior spines of the ilia. The physician needs to see the interspaces between L-5 and S-1 and between L-4 and L-5. He also needs to see the condition of the laminae, which form the neural arch.

Proper lateral views of the lumbar spine require that there be no unnatural, lateral curvature resulting from improper patient positioning. Place a radiolucent block of plastic material (foam) beneath the lumbar spine and between the rib cage and the hip so that, upon palpation, the spinous processes of the lumbar bodies are in a straight line and parallel with the table top, as illustrated in Fig. 4-26.

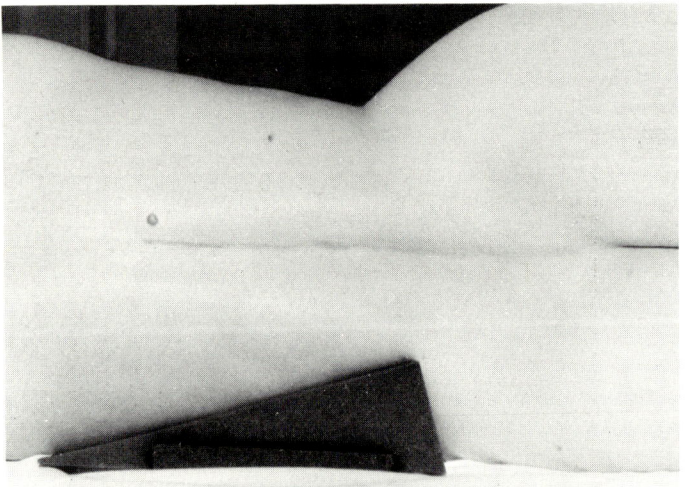

Fig. 4-26. Patient position for lateral view of lumbar spine.

The skeletal system **105**

Pectoral girdle

The pectoral (shoulder) girdle consists of the scapula and the clavicle. Neither of these bones attaches to the vertebral column. Anatomically, both bones are members of the appendicular skeleton. However, these bones are in the trunk of the body and therefore are radiographed with x-ray energy greater than that needed for an extremity to effect adequate penetration of the trunk.

The lateral end of the clavicle articulates with the acromion process of the scapula to form the shoulder girdle. The scapula supports the glenoid fossa in which the humeral head articulates to form the shoulder joint (see Figs. 4-78, *B* and 4-82).

A thorough examination of the shoulder includes two degrees of penetration, one for the structures of the principal articulation and a lighter (less dense) exposure for the acromioclavicular articulation. (Proper placement of a 1.0 mm thick aluminum filter achieves satisfactory results in a single exposure.)

Frequent radiographic requests are made for posterior views of the shoulder with both external and internal rotation of the humerus. These views present different surfaces of the humeral head and neck for visualization (see Fig. 4-82, *C* and *D*). A complete shoulder examination may include the following views:

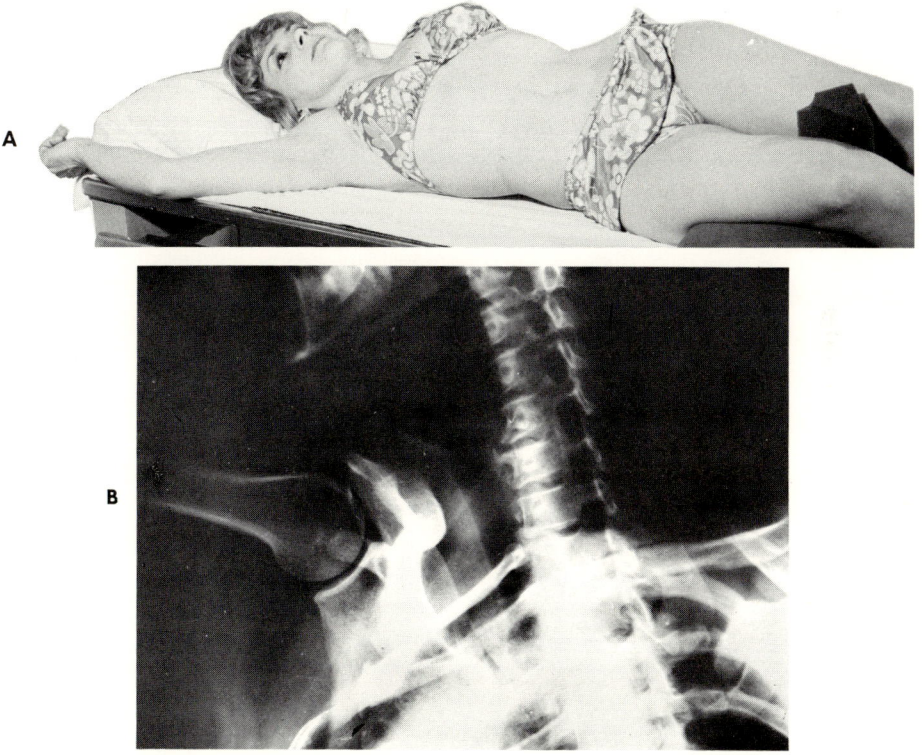

Fig. 4-27. A, Right shoulder. Patient is in position for a view of the articular space. **B,** Articular space between the humerol head and glenoid fossa of scapula.

1. Posterior in external rotation with low penetration (light technique)
2. Posterior in external rotation with high penetration (heavy technique)
3. Posterior with internal rotation

An excellent position to demonstrate the articular space between the humeral head and the glenoid fossa is the same as that described on p. 98 for an oblique view of T-1, T-2, and T-3. The direction of the central ray in the shoulder radiograph differs only in that the central ray is directed to a point opposite the humeral head as illustrated in Fig. 4-27.

Clavicle. The anterior part of the shoulder girdle is formed by the clavicle, which resembles a shallow S. The clavicle is slightly superior to the first rib anteriorly and extends almost transversely across to the sternum. Palpation of the clavicle is quite simple throughout its entire length from medial to lateral. The anteromedial part of the clavicle contains the convex curvature while the anterolateral part contains the concave curvature. The clavicle articulates laterally with the acromion process of the scapula to form the *acromioclavicular joint;* the clavicle articulates medially with the manubrium of the sternum to form the *sternoclavicular joint.* The clavicle articulates medially and inferiorly with the cartilage of the first rib (see Fig. 4-28).

The clavicle consists mostly of cancellous tissue and is the last bone to ossify completely. It is easily fractured and, under normal circumstances, heals easily. It ossifies from a medial and a lateral center for the body and from another center for the sternal end. The medial and lateral ossification centers are primary; the other is a secondary center.

Certain injuries to the acromioclavicular joint are such that the shoulder is said to be knocked down. If the ligaments fastening the clavicle to the coracoid process remain intact, the clavicle will be displaced slightly upward. However, if all four ligaments are torn, the clavicle will extend upward beneath the skin, which permits easy diagnosis. To demonstrate lesser degrees of displacement, or subluxation, *both* shoulders are radiographed on the same film, usually with the

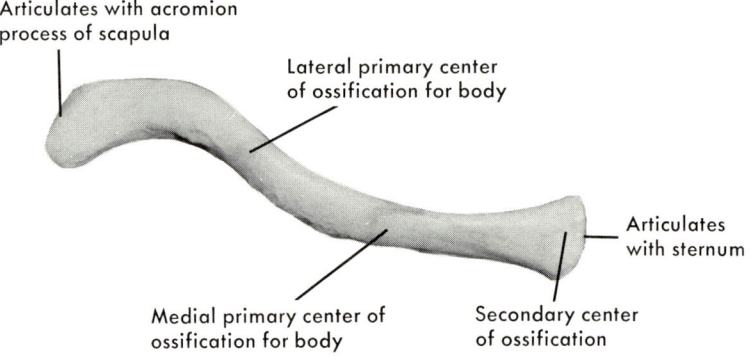

Fig. 4-28. Right clavicle, superior surface.

patient erect and holding equal weights in each hand. Such procedure demonstrates the pathology.

A frequent (over 90 percent) site of clavicular fracture is the middle third of the body of the bone. Radiographic examination is performed with the patient in either prone or supine position on the table, or the patient may be erect with either his ventral or dorsal surface against the cassette holder. These are the routine anterior or posterior views. The anterior view enables less magnification of the clavicle than the posterior view because of the decreased object-film distance. Lateral (more properly called axial) views are quite difficult to achieve and always cause great magnification because of the great object-film distance; however, this view is useful in certain cases. The tangential view is very useful as the second view (see Fig. 4-29).

Scapula. The scapula is superior and lateral in the thoracic cage. The scapula

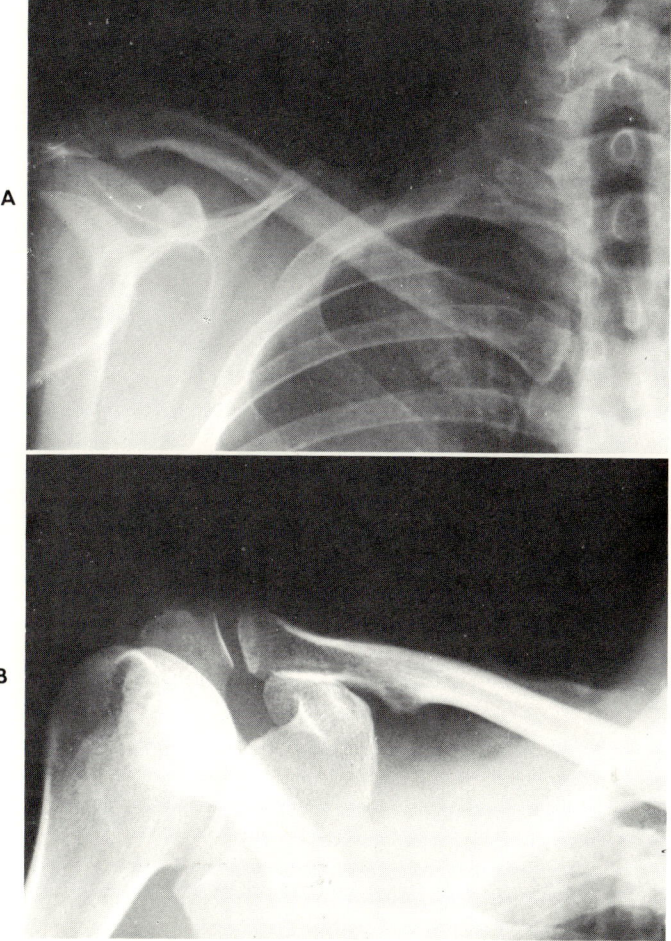

Fig. 4-29. Right clavicle. **A,** Anterior view. **B,** Tangential view. (**B,** Courtesy Dr. J. M. Hilton.)

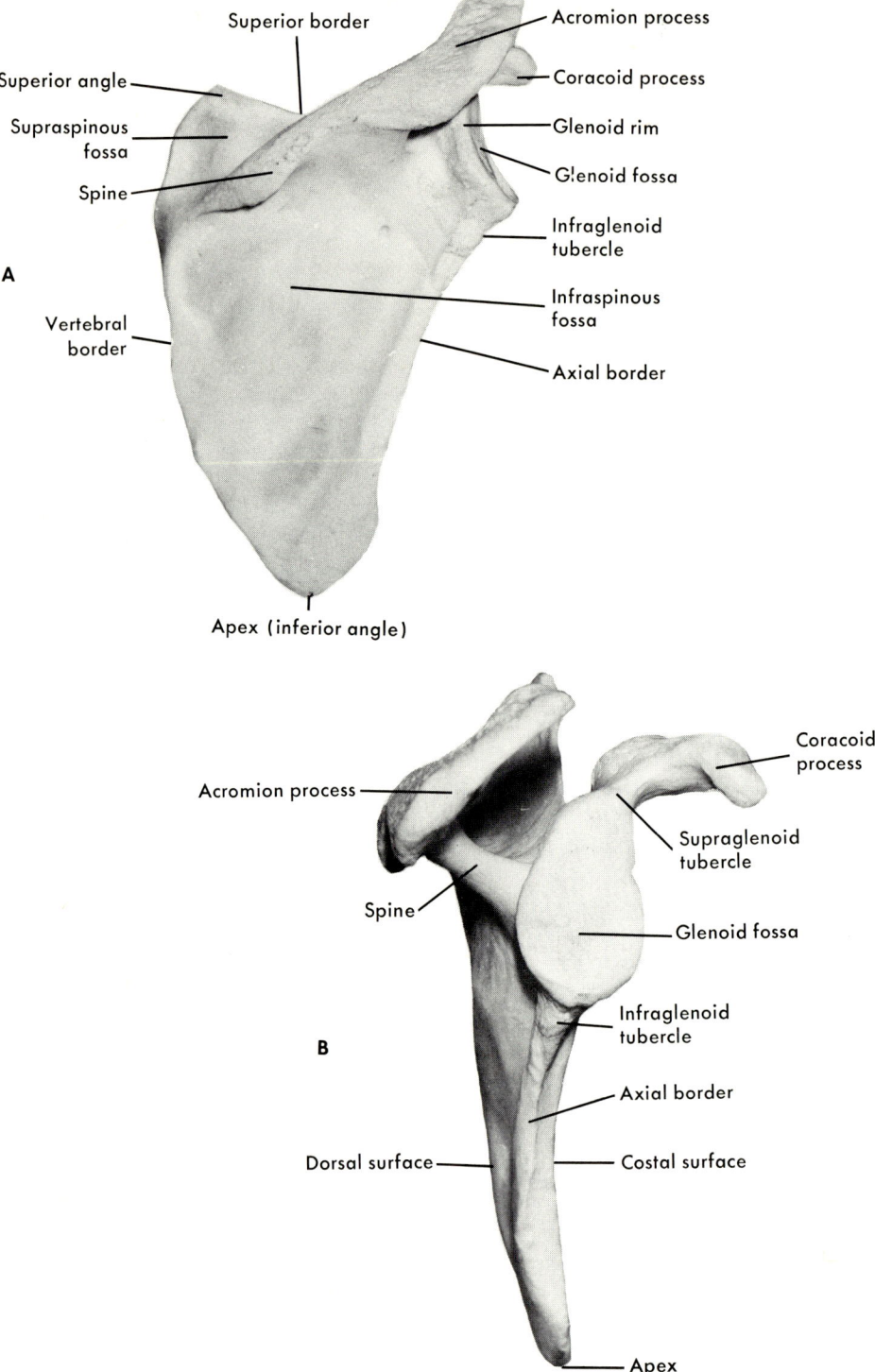

Fig. 4-30. Right scapula. **A,** Dorsal surface. **B,** Lateral aspect.

is a triangular, flat bone having three angles, three borders, three processes, and two surfaces. It supports a *spine,* which extends obliquely upward from the upper vertebral border on the dorsal surface, and terminates in the *acromion process,* which extends anteriorly above the glenoid cavity. The *vertebral border* is adjacent to the vertebral column. The *axial border* is adjacent to the axial or lateral margin of the thoracic cage. The *superior border* forms the *base* of the scapula and extends between the superior ends of the axial and vertebral borders. The axial and vertebral borders join inferiorly to form the *apex* (inferior angle) of the scapula. The vertebral and superior borders join superiorly and medially to form the *medial angle.* The axial and superior borders join superiorly and laterally to form the *lateral angle.* The lateral angle presents the *glenoid process,* which supports the *glenoid fossa* for articulation with the head of the humerus. A slight narrowing medial to the glenoid fossa forms the *neck* of the scapula, and the thickened portion medial to the fossa and lateral to the neck is the head of the scapula. The entire structure of the glenoid process includes the neck, head, and glenoid fossa. The *coracoid process,* a strong bony structure, arises medially and superiorly behind the neck and extends anteriorly and laterally over the glenoid cavity. The scapular spine divides the dorsal surface into two unequal divisions, the spinous fossae. The *supraspinous fossa* lies above the spine and is the smaller

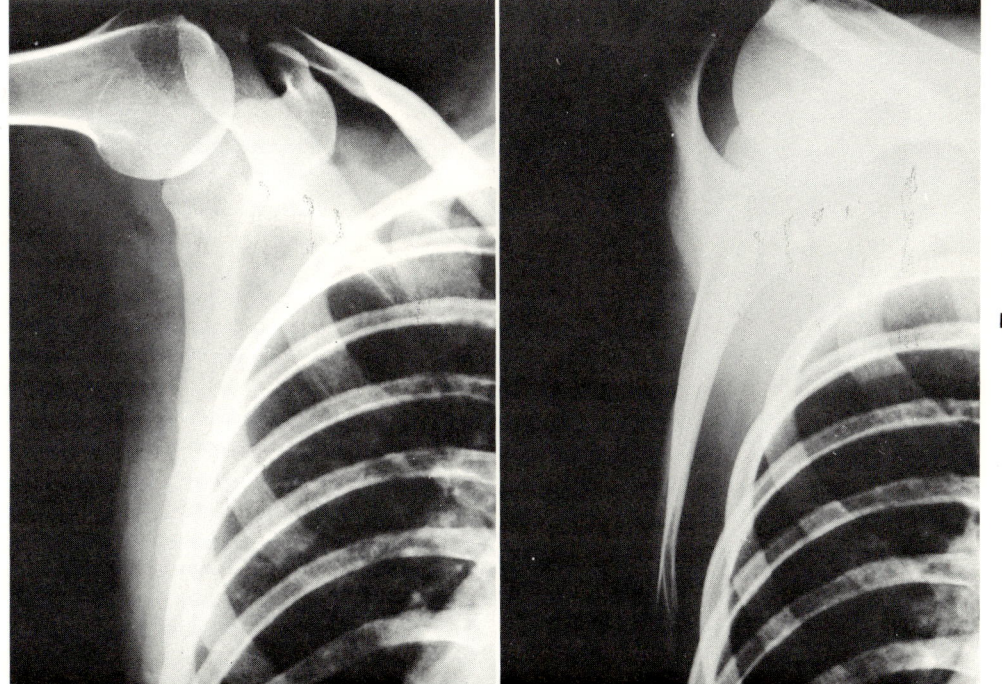

Fig. 4-31. Right scapula. **A,** Posterior view. **B,** Lateral view.

of the two. The *infraspinous fossa* lies below the spine and is the larger of the two. The *costal* (ventral) surface is adjacent to the ribs (see Fig. 4-30).

The most frequent injury of the scapular neck is fracture with inward telescoping. Since such a fracture is not easily reduced, the usual treatment is support of the arm in a sling.

Routine radiography of the scapula includes posterior and lateral views. The posterior view is made with the patient in the supine position, rotated slightly on the affected side. The lateral view may be made from either the prone or supine position; in either case, the patient is rotated sufficiently to present the scapula at right angles to the table top (see Fig. 4-31).

Thoracic cage

The thoracic cage includes the ribs and sternum; it provides bony protection for the lungs, mediastinal viscera, and heart. There are twelve pairs of ribs. Seven pairs of ribs attach posteriorly to the vertebral column and anteriorly to the sternum; these are the *true ribs*. The remaining five pairs of ribs attach posteriorly to the vertebral column but have no direct sternal attachment; these are the *false ribs*. The last two pairs of ribs have no anterior attachment and are called *floating ribs*. Cervical ribs are discussed on pp. 91 and 93 in conjunction with the cervical vertebrae. Some persons have one or more ribs extending from the sides of the first lumbar vertebrae. There is no difference in the number of pairs of ribs due to sex (see Fig. 4-32).

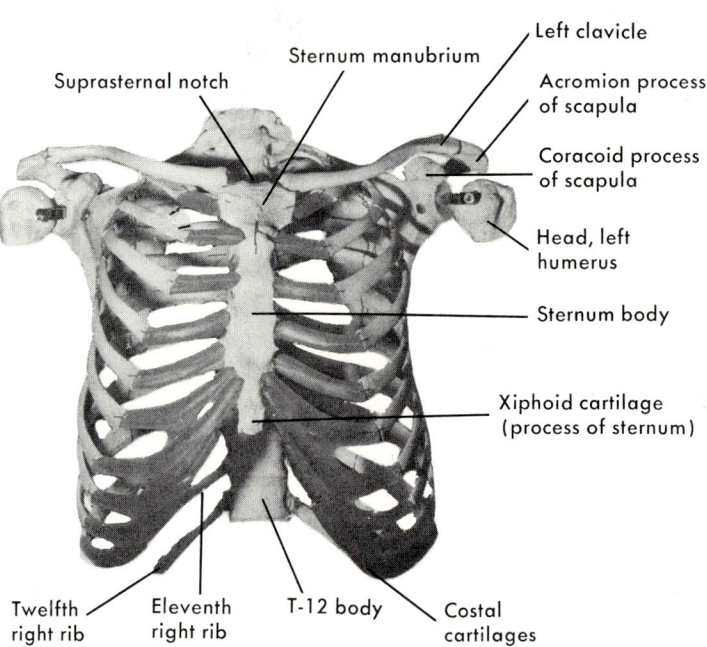

Fig. 4-32. Thoracic cage, anterior aspect.

The skeletal system

Ribs. A rib is a flat bone having two extremities, vertebral and sternal, which are separated by the body. The vertebral extremity consists of a head, neck, and tubercle. The head contains the facets for articulation with the thoracic vertebrae, forming the costovertebral joint. The neck is the narrowed portion immediately lateral to the head. The tubercle has both articular and nonarticular parts.

A typical rib extends obliquely downward to the midaxillary line, from which it continues, downwardly and anteriorly, to its sternal articulation forming the costochondral joint. The spaces between each rib, the *intercostal spaces,* contain two layers of intercostal muscles. The angle formed by the lateral border of the lower ribs with the diaphragm is the *costophrenic angle.* The ribs articulate as described in the following paragraphs.

On each side the first rib articulates with the costotransverse facet and the costovertebral facet of the first thoracic vertebra. The second rib articulates with the costotransverse facet and with the upper costal facet of the second thoracic vertebra and with the lower costal facet of the first thoracic vertebra.

The third rib articulates with the costotransverse facet and with the upper costal facet of the third thoracic vertebra and with the lower costal facet of the second thoracic vertebra. The rib articulations continue in this manner through the eighth rib.

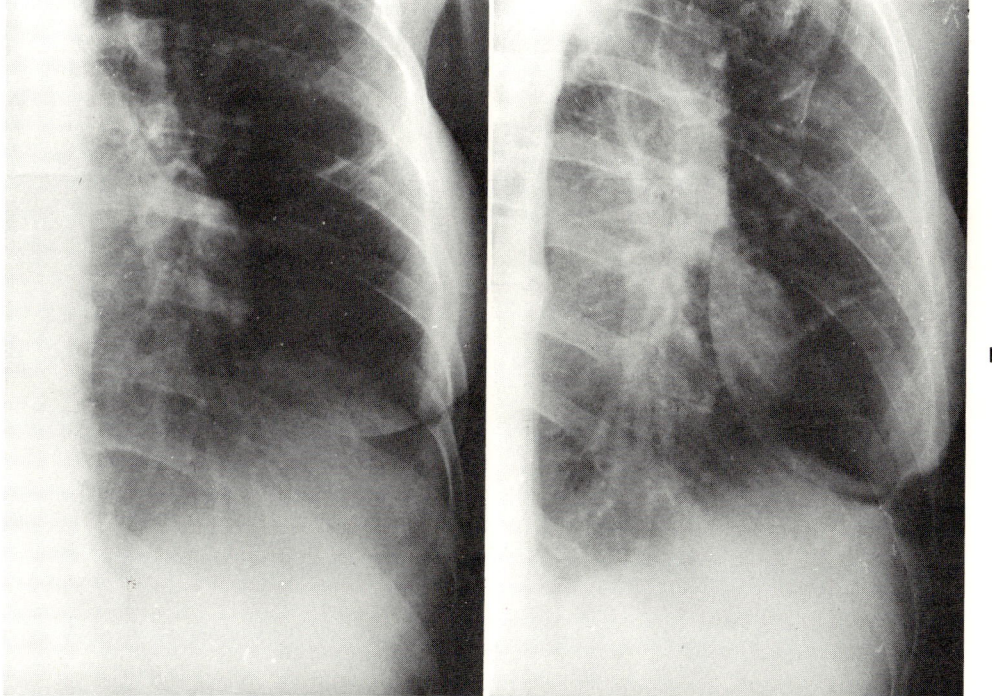

Fig. 4-33. Ribs. **A,** Posterior view. **B,** Oblique view. (Courtesy Dr. J. M. Hilton.)

112 *Textbook of anatomy and physiology in radiologic technology*

The ninth and tenth ribs articulate with the costotransverse and costovertebral facets of the corresponding vertebrae.

The eleventh and twelfth ribs articulate with the costovertebral facets of the corresponding vertebrae.

Rib fractures are often tangential and are so situated as *not to be seen* in radiographs. Approximately 50 percent of rib fractures are visualized in routine views. For radiographic examination, the ribs are considered in two groups: those above the diaphragm and those below the diaphragm. This is because of the differing requirements for penetration with roentgen rays (x rays) of the upper ribs contrasted against the lungs and the lower ribs contrasted against the diaphragm. Routine views include posterior or anterior and oblique views above and/or below the diaphragm. Some routines require that the entire side (right or left) be radiographed rather than the ribs above or below the diaphragm (see Fig. 4-33).

Sternum. The sternum is a flat bone consisting chiefly of cancellous tissue covered with a layer of compact tissue (cortical bone). It contains three sections: *manubrium, body,* and *xiphoid process.* The manubrium is superior, and the xiphoid process is inferior. The manubrium is the thicker part of the sternum. The xiphoid process usually does not ossify until quite late in life and then never ossifies completely.

The manubrium articulates bilaterally with the clavicles to form the sterno-

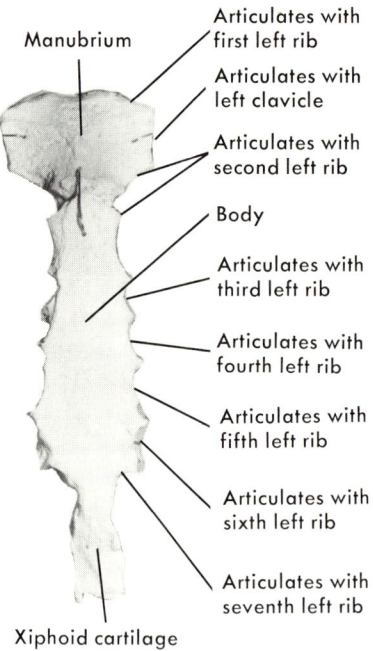

Fig. 4-34. Sternum, anterior surface.

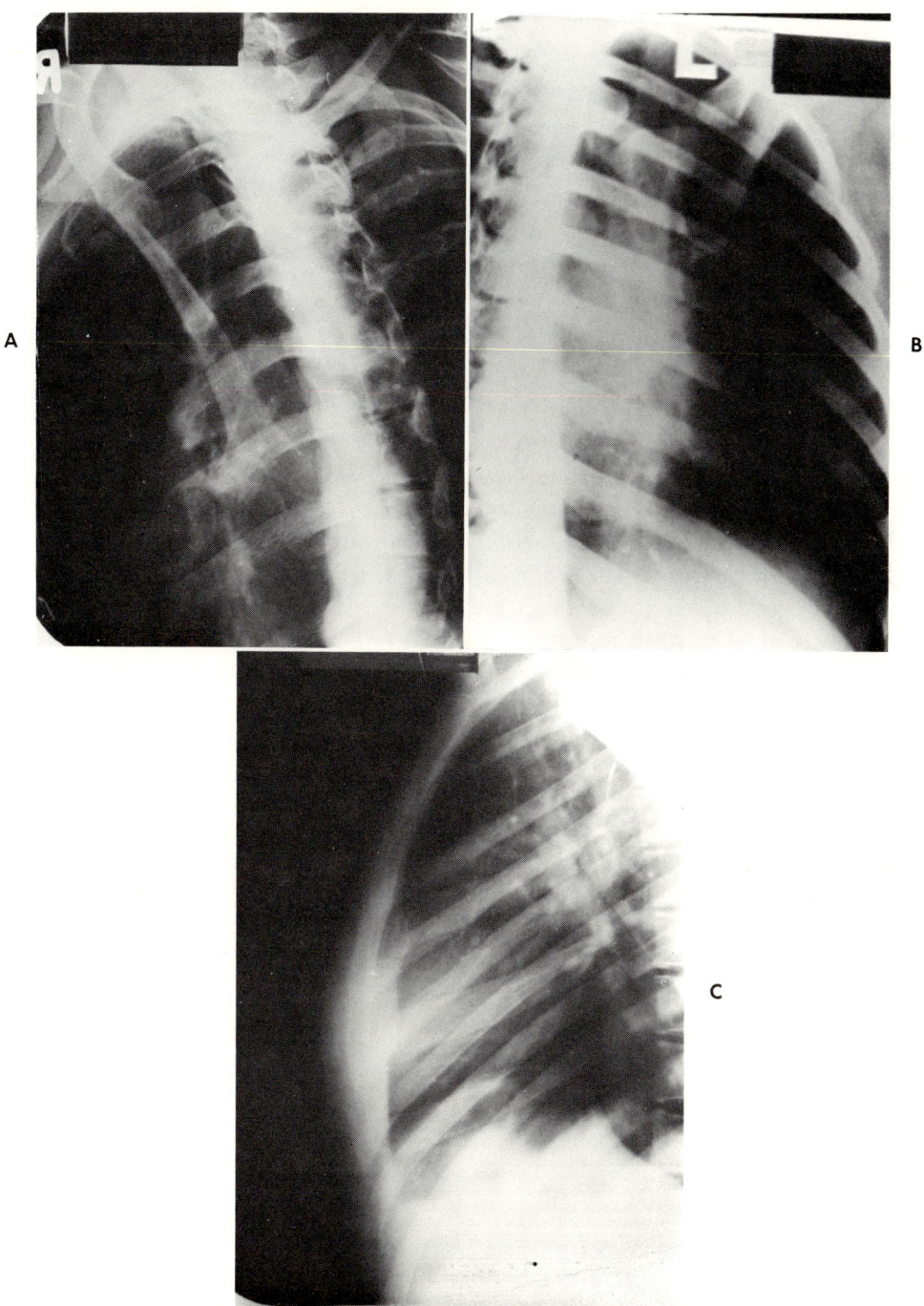

Fig. 4-35. Sternum. **A,** R-A-O view. **B,** L-A-O view. **C,** Lateral view.

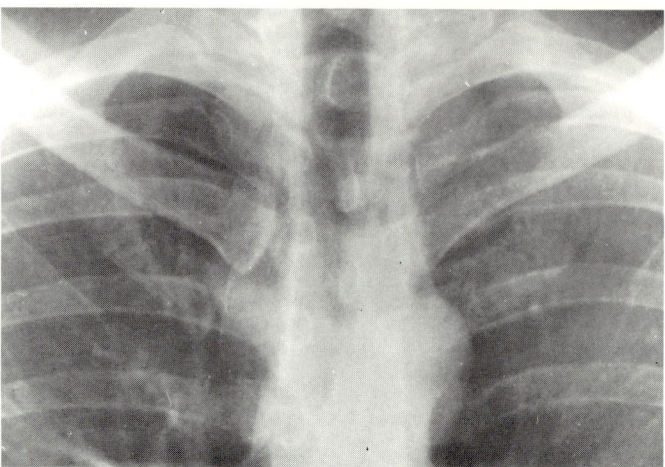

Fig. 4-36. Sternoclavicular joints, anterior view.

clavicular joints. The space between the medial ends of the clavicles and the superior margin of the manubrium is the *suprasternal notch;* it is opposite T-2 to T-3. The manubrium also articulates with the sternal ends of the first two pairs of ribs. The second pair of ribs articulates partially with the body of the sternum.

The body of the sternum articulates with the second through the seventh pairs of ribs, with the manubrium superiorly, and with the xiphoid process inferiorly (see Fig. 4-34).

The xiphoid process articulates slightly with the seventh pair of ribs and is opposite T-10.

The sternum is a frequent site of puncture for bone marrow studies in the clinical laboratory. The outer covering is quite thin; the inner substance (diploë) is composed of cancellous bone—thus an excellent source of erythropoietic samples.

Routine radiographic examinations of the sternum include lateral views, both oblique views, anterior sternoclavicular joint views, and both oblique sternoclavicular joint views. Patients who have suffered trauma in this region usually can be seated at the chest board or stood erect at the chest board or the upright radiographic table. Trauma to this region usually precludes successful positioning in a prone or semiprone position (see Figs. 4-35 and 4-36).

Fractures of the sternum are often not demonstrated in anterior and oblique views; as a result the lateral view is of great importance. It is of the greatest importance that this fracture be diagnosed because of the vital structures (heart, aorta, and mediastinum) beneath it. Excessive volume of blood in the mediastinum can cause death of the patient.

Pelvic girdle

The pelvic girdle includes the two pelvic bones (os innominata). Each pelvic bone is formed from three bones—the ilium, ischium, and pubis. The bodies of these bones fuse to form the *acetabulum,* the articular fossa for the head of the femur. The two pelvic bones are separated posteriorly by the sacrum (see Fig. 4-37). The two pelvic bones articulate anteriorly in the articulation of the two pubes, forming the *symphysis pubis.*

Fractures of the pelvis are usually anterior and involve the anterior and pos-

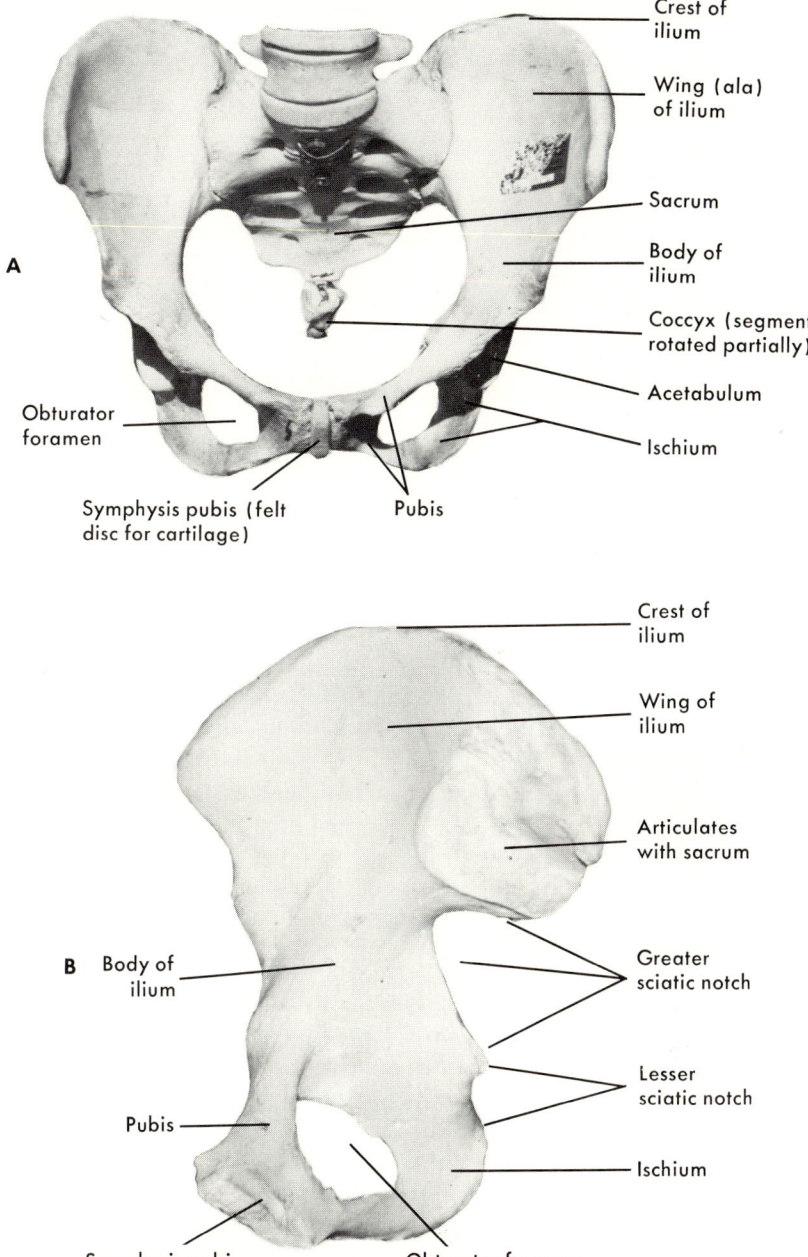

Fig. 4-37. A, Pelvis, anterior aspect. **B,** Right pelvic bone, medial aspect.

116   *Textbook of anatomy and physiology in radiologic technology*

terior rami. A fracture of the anterior ramus usually requires a corresponding fracture of the posterior ramus (and vice versa) due to the ring (surrounding the obturator foramen) structure of the rami.

Routine radiography includes only a posterior view; unless there has been trauma to the pelvis or unless there is suspected disease involving pelvic bones or in the pelvic viscera, other views are of little value (see Fig. 4-38).

Radiographic examination of the pelvis of a pregnant woman is called *pelvimetry* and is performed to determine if normal delivery can occur. Accurate positioning for this examination is extremely important both to demonstrate the necessary landmarks and to prevent repeated exposures, thus causing unnecessary

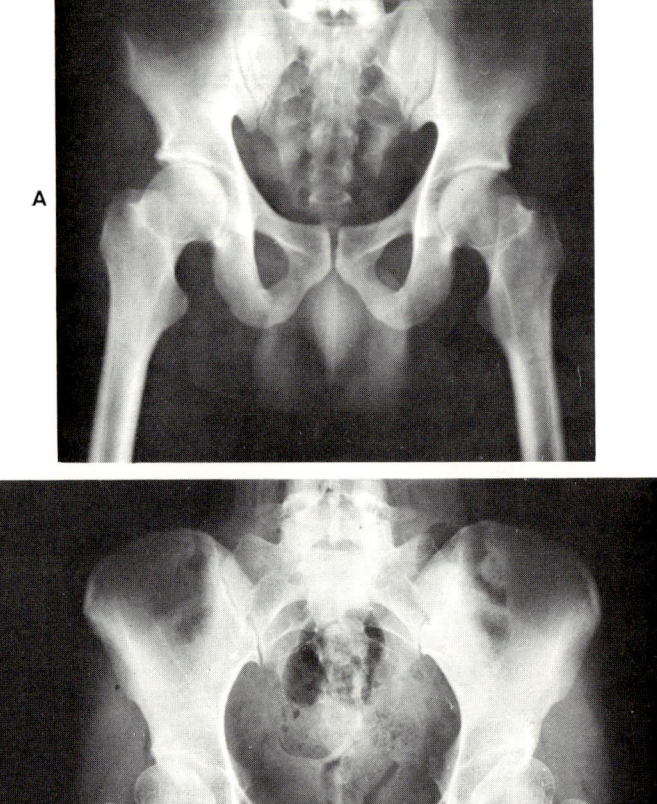

Fig. 4-38. A, Male pelvis, posterior view. **B,** Female pelvis, posterior view.

The skeletal system 117

exposure to the fetus and the female gonads. The landmarks are used in measurement of the various pelvic diameters.

The inferior pelvic strait is the plane of the pelvic outlet. The antero-posterior diameter extends from the tip of the coccyx to the subpubic ligament and is 9.5 cm. long. The transverse diameter extends between the ischial tuberosities and is 11 cm. long.

The superior pelvic strait is the plane of the pelvic inlet. The anteroposterior (conjugate) diameter extends from the superior border of the first sacral segment to a point ⅛ inch inferior to the superior border of the symphysis pubis and is 11 cm. long. The transverse diameter at the widest locus is 13.5 cm. long (see Fig. 4-40). During parturition (birth) the fetus must pass through both the inlet and the outlet of the pelvis.

A true posterior view of the pelvis is obtained when the two ischial spines are visualized equally. For this view it is necessary to place both feet together

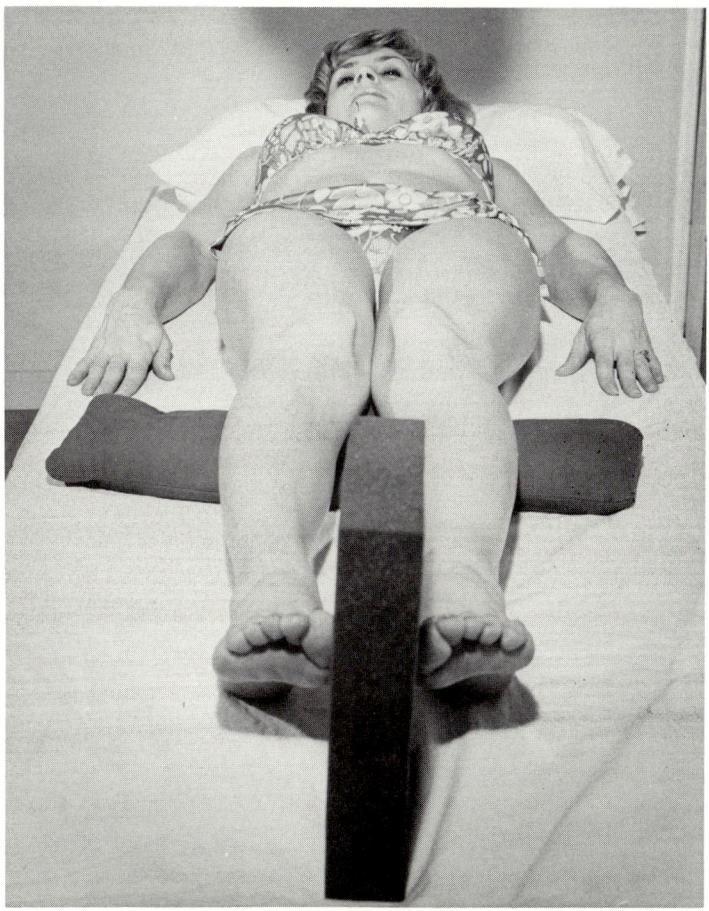

Fig. 4-39. Patient in position on radiographic table for correct posterior view of pelvis.

with the medial side of each foot forming an approximate 5-degree angle from the vertical (see Fig. 4-39). This position assures that the two hip joints are visualized with equal rotation of the femoral heads.

Sacrum. The sacrum is treated in the pelvic girdle because it is thus shown radiographically and because it is *not* a part of the true vertebral column. The following structural variations characterize the fixed sacrum. In the embryo the sacrum begins as five separate segments, which fuse into a single bone in early childhood. The neural arch of each segment is evident in a modified form, and the sacral foramina, both dorsal and ventral, serve for passage of the several sacral nerves from the filum terminale.

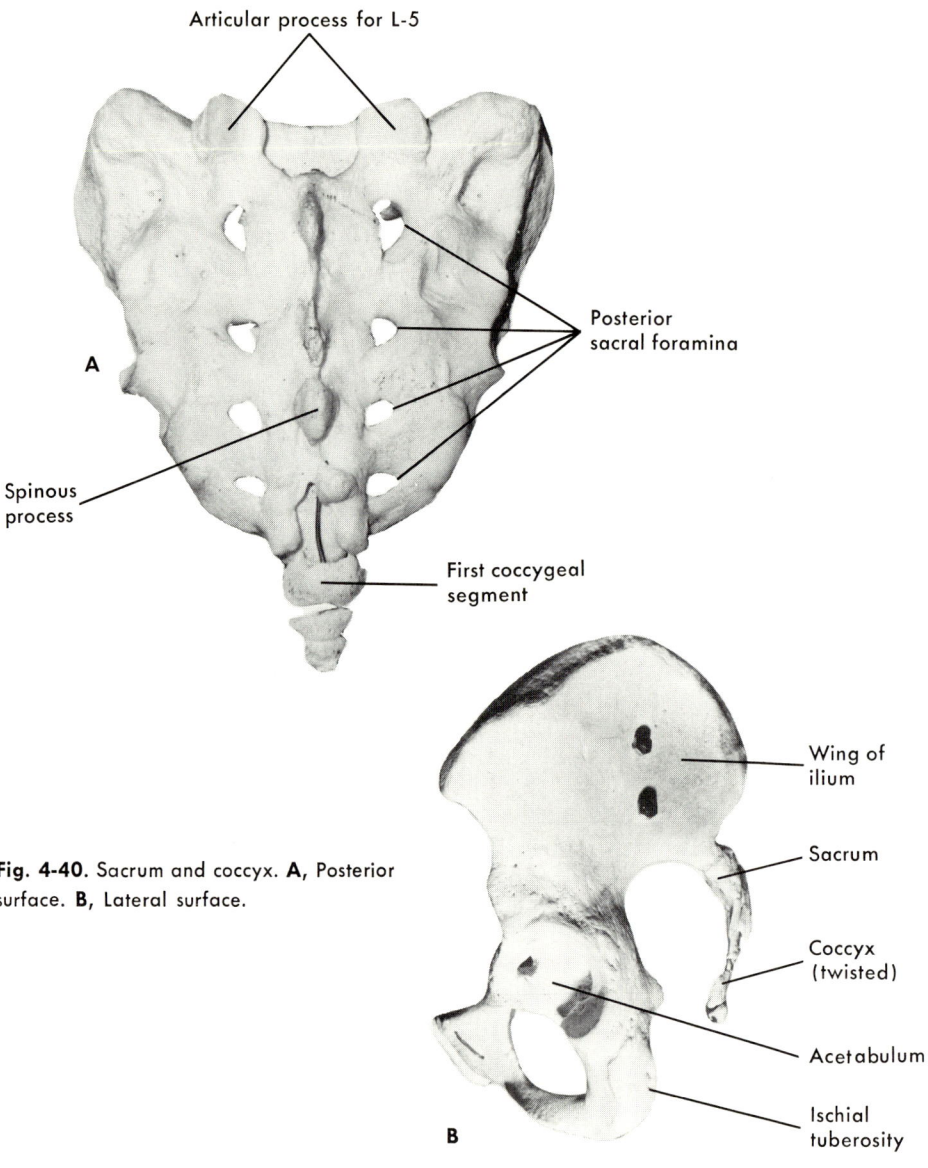

Fig. 4-40. Sacrum and coccyx. **A,** Posterior surface. **B,** Lateral surface.

The sacrum is the triangular bone situated between the two ilia and upon which the twenty-four true vertebral bodies rest. The first sacral segment articulates superiorly with L-5 to form the *lumbosacral joint* and bilaterally with the two ilia to form the *sacroiliac joints*. The apex of the sacrum articulates inferiorly with the first coccygeal segment. The sacrum forms the major portion of the mid-dorsal region of the pelvis. The male sacrum is both longer and narrower than the female sacrum. The angle between the upper and lower halves of the sacrum is greater in the female than in the male, thus increasing the volume of the pelvic cavity.

Fractures of the sacrum are infrequent, the usual injury being a separation of the sacroiliac joint treated with traction.

In routine views of the lumbar spine, the first sacral segment must be included. Since the sacrum has an anterior concavity and angles dorsally from the vertebral column, the central ray must be angled cephalad (when the patient is in the supine position) to demonstrate this bone most effectively. Lateral views of the sacrum visualize both the first sacral body per se and the sacrum in relation to the pelvic cavity (see Figs. 4-41 to 4-43).

Coccyx. The coccyx usually consists of four segments, extending inferiorly and anteriorly below the sacrum and forming part of the dorsal boundary of the pelvic cavity. Like the sacrum, the coccygeal segments are not true vertebral bodies; unlike the sacrum, the coccygeal segments are not fixed.

The coccygeal segments may vary in number from person to person, usually from three to five. Sometimes the neural arch is found in the first segment, but it is never found beyond this point. The segments are completely devoid of pedicles, laminae, and spinous processes; therefore, the nomenclature "segment" instead of "body."

The coccyx extends inferiorly from the sacrum and angles toward the ventral surface of the body, thus giving anterior concavity to the combined sacrum and coccyx. Posterior views of the coccyx are best obtained when the central ray is angled 10 degrees caudad, the patient lying in the supine position. Lateral views of the coccyx demonstrate pathologic deviations from the normal angle and fractures and/or dislocations of one or more segments. Oblique views are of no particular value. The technique employed to demonstrate the coccyx with the patient in the lateral position is to place a 1 mm. aluminum filter between the coccyx and the film to prevent overpenetration of the coccyx (see Figs. 4-40, *A*, and 4-43).

Ilium. The ilium consists of a body and a wing (ala). The body is inferior and contributes to the formation of the acetabulum. The wing is quite large and extends upward, laterally and medially. It forms the lateral boundary of the pelvis and derives its name from its relation with the flank. The crest of the ilium is convex and rough. The thickened anterior and posterior parts terminate in the *anterior* and *posterior superior spines*. The anterior superior spine is quite prominent and easily palpated; it is an important radiographic landmark. The *anterior* and *posterior inferior spines* are separated from the superior spines of the

120 *Textbook of anatomy and physiology in radiologic technology*

same surface by notches. The anterior notch is the site of origin of the sartorius muscle. Inferior to the posterior inferior spine is the *greater sciatic notch*. The anterior superior spine is an important radiographic landmark (see Fig. 4-44).

Ischium. The ischium is the most inferior of the pelvic bones, and it is the strongest pelvic bone. It extends inferiorly from the acetabulum to the large

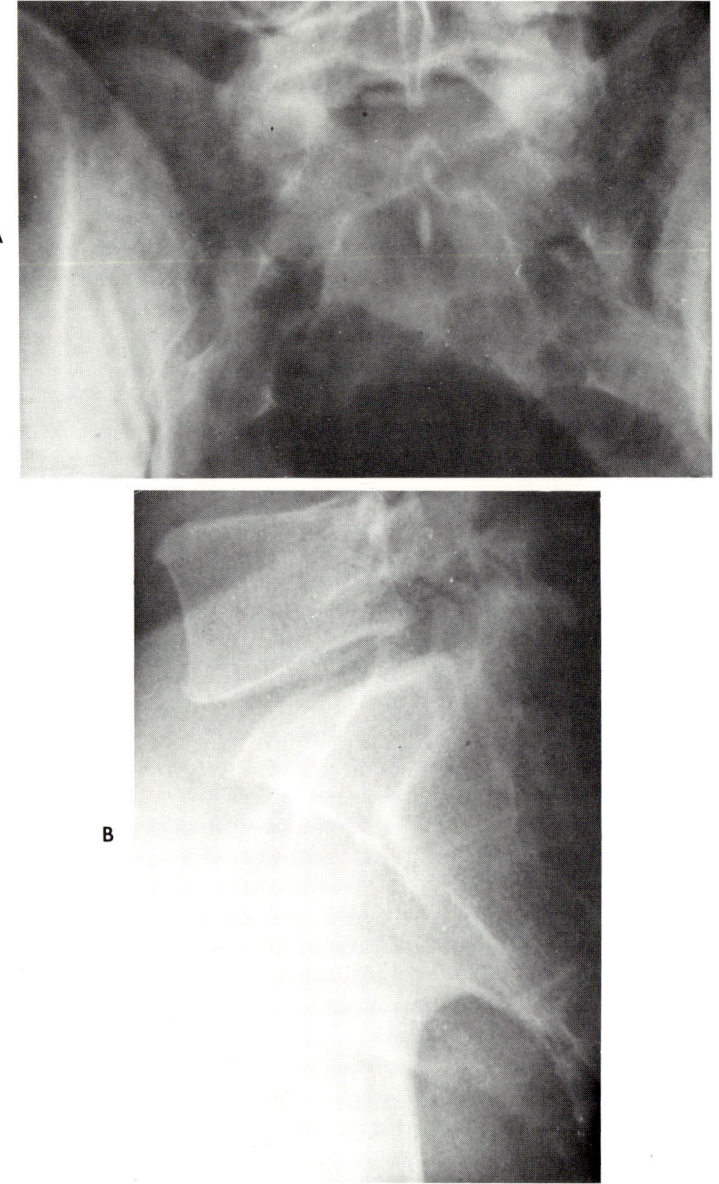

Fig. 4-41. Lumbosacral joint. **A,** Cone-down posterior view. **B,** Cone-down lateral view. (**B,** Courtesy Dr. J. M. Hilton.)

The skeletal system 121

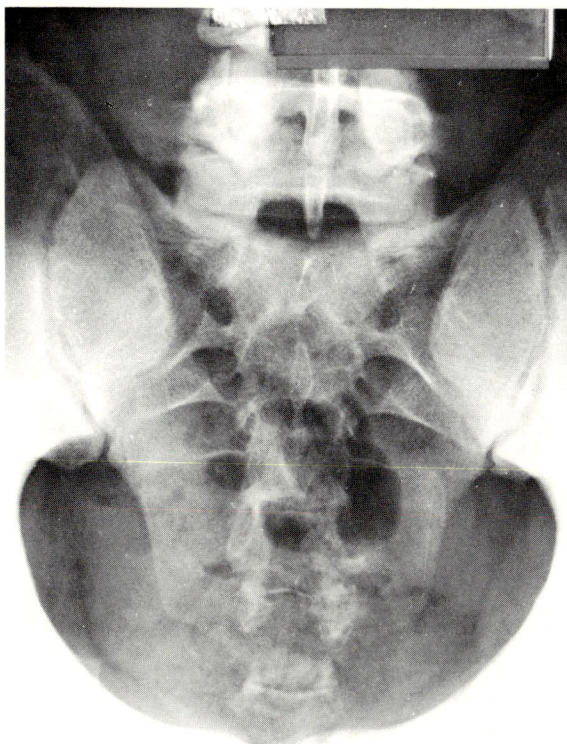

Fig. 4-42. Sacrum, posterior view.

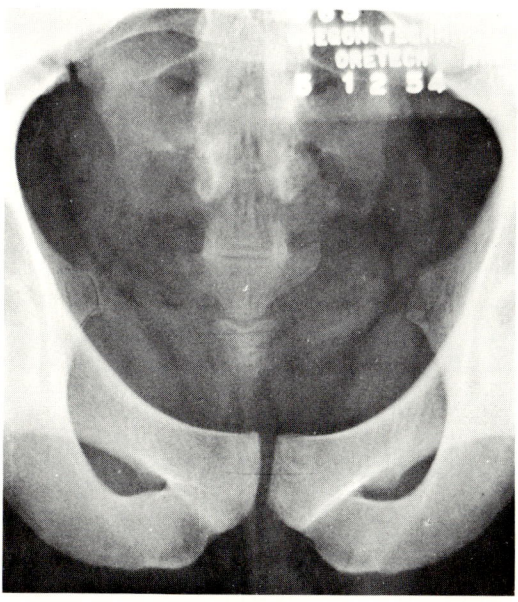

Fig. 4-43. Coccyx, posterior view.

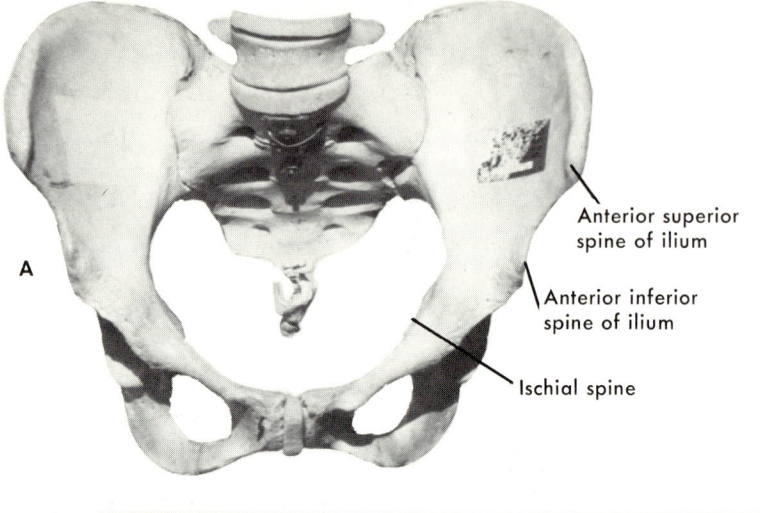

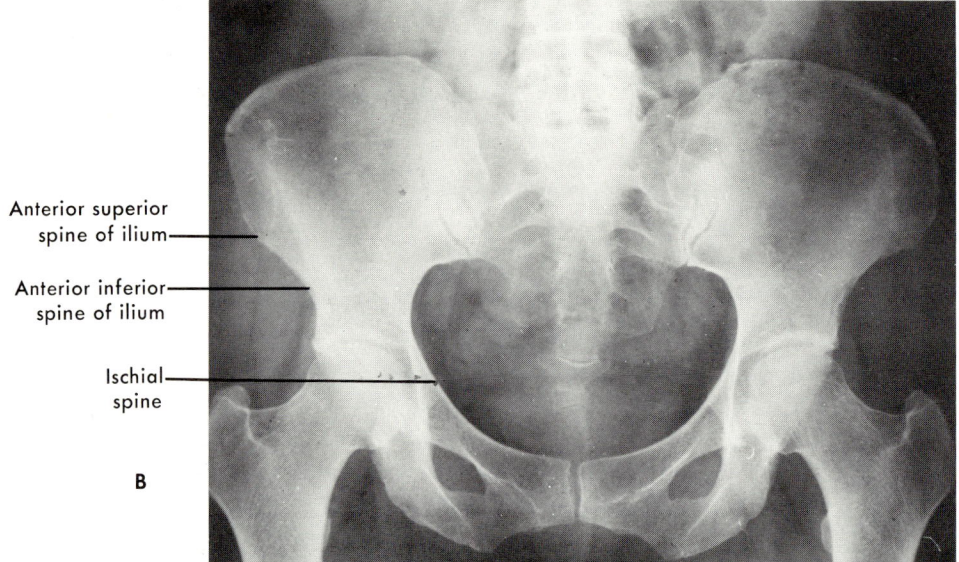

Fig. 4-44. **A,** Anterior aspect of pelvis showing spines. **B,** Posterior view of female pelvis showing spines.

ischial tuberosity on the posterior and inferior surfaces and then continues and curves anteriorly where it forms, with the pubis, the *obturator foramen*. The ischium consists of a body, which contributes to the formation of the acetabulum, and a *ramus*, with descending and ascending portions. The ischial tuberosity arises inferiorly from the angle formed by the descending and ascending rami below the obturator foramen. The ramus descends inferiorly and posteriorly from the acetabulum to below the obturator foramen and ascends anteriorly and

The skeletal system **123**

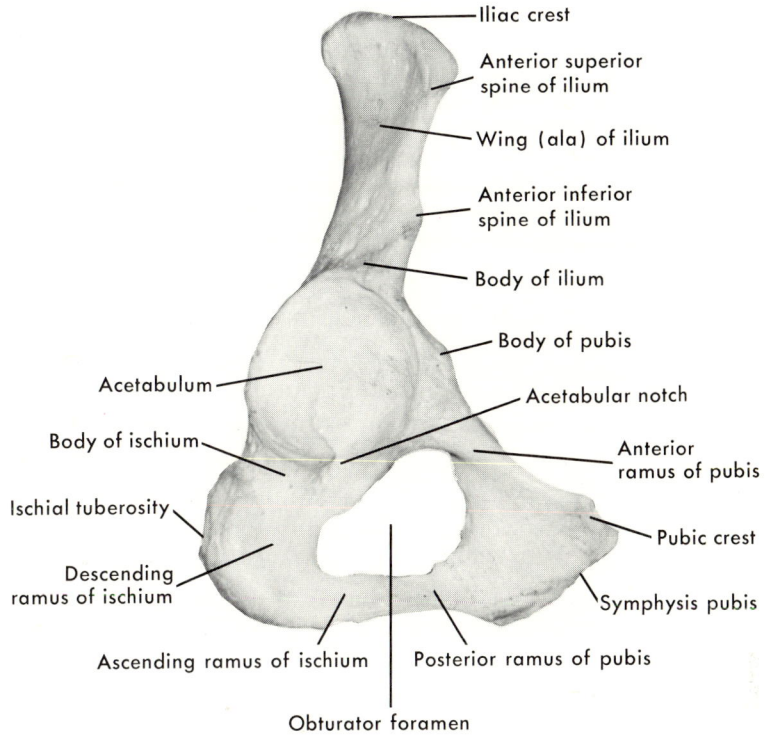

Fig. 4-45. Right pelvic bone, lateral aspect.

medially from this point to join the posterior ramus of the pubis. The *ischial spine*, an important radiographic landmark, extends medially and superiorly from the posterior border of the body (see Figs. 4-37 and 4-45).

Pubis. The pubis lies inferiorly and medially to the acetabulum. The body of the pubis contributes to the formation of the acetabulum. The *anterior ramus* extends anteriorly and medially from the body to articulate with the opposite pubic anterior ramus to form the *symphysis pubis*. The *posterior ramus* extends inferiorly, posteriorly, and laterally from the symphysis pubis to join with the ascending ramus of the ischium, forming the inferior border of the obturator foramen (see Figs. 4-37 and 4-45).

Acetabulum. The acetabulum (hip socket) is formed by the fusion of the bodies of the ilium, ischium, and pubis. The pubic body contributes one-fifth to the surface of the acetabulum, and the ilial and ischial bodies each contribute approximately two-fifths to the surface of the acetabulum (see Fig. 4-45).

The acetabulum is a deep fossa facing inferiorly, anteriorly, and laterally. It has a thick, strong, uneven rim superiorly to which the cotyloid ligament attaches. This ligament (the glenoidal labrum) contracts the orifice of the acetabulum and thus deepens the articular space. The *acetabular notch* is hemispherically shaped and is situated both inferiorly and medially to the rim.

The borders of the acetabular notch are the site of attachment of the *ligamentum teres*. In the center of the acetabular notch is a rough depression, the *acetabular fossa*. The fossa is surrounded by the *lunate surface,* the articular surface for the head of the femur. The other attachment for the ligamentum teres is in the fovea capitis femoris in the center of the medial surface of the femoral head. The ligamentum teres functions to hold the femur in the acetabulum.

Occasionally the acetabulum is displaced inwardly toward the urinary bladder, resulting from a force striking against the greater trochanter of the femur. Thus the head of the femur acts like a battering ram to shove the acetabulum inward. The condition responds well to external pull and manipulation.

Obturator foramen. The obturator foramen is formed by several structures of two of the pelvic bones. The anterior ramus of the pubis forms the superior border. The union of the anterior and posterior rami of the pubis forms the medial and anterior borders of the obturator foramen. The descending ramus of the ischium forms the lateral and posterior borders of the obturator foramen. The ascending ramus of the ischium joins the posterior ramus of the pubis to form the inferior border of the foramen (see Fig. 4-45).

The obturator foramina function to pass the obturator vessels and nerves out of the pelvis.

Comparison of the size and shape of the two foramina in a posterior view of the pelvis aids the technologist in his determination of accurate and correct positioning.

Radiographic interest of the trunk

This section includes many of the important landmarks and planes used in radiographic procedures. The anatomic landmarks are useful for all persons; however, anatomic variation between individuals makes these landmarks true on the majority of persons only. The anatomic planes hold true on all persons.

Anatomic landmarks

T-1 Opposite a point 2 inches superior to the suprasternal notch

suprasternal notch The space superior to the manubrium and between the the sternal ends of the clavicles

T-3 Opposite or slightly below the suprasternal notch

T-10 Opposite the xiphoid process

L-1 Opposite a point approximately halfway between the xiphoid process and the umbilicus

umbilicus In the approximate center of the abdominal wall

L-3 Opposite the umbilicus

L-4 Superior to the interiliac line

L-5 Between and slightly below the iliac crests

ischial spine Medial to and on a level with the superior margin of the femoral head

symphysis pubis On a level with the greater trochanter

Body planes

sagittal Any plane of the body that extends between ventral and dorsal and that is parallel with the median plane through the sagittal suture; it separates left from right (see Figs. 1-1 and 4-46)

parasternal A sagittal plane on either side of and next to the sternum (see Fig. 4-46)

coronal Any plane of the body extending between the right and left sides and parallel with the coronal suture; it separates ventral from dorsal (see Fig. 1-1)

oblique Any body plane not coincident with or parallel with the sagittal, coronal, and/or transverse planes (see Fig. 1-1)

transverse Any plane of the body that passes at right angles to both the sagittal and coronal planes, each plane being at right angles to the other two (see Figs. 1-1 and 4-46)

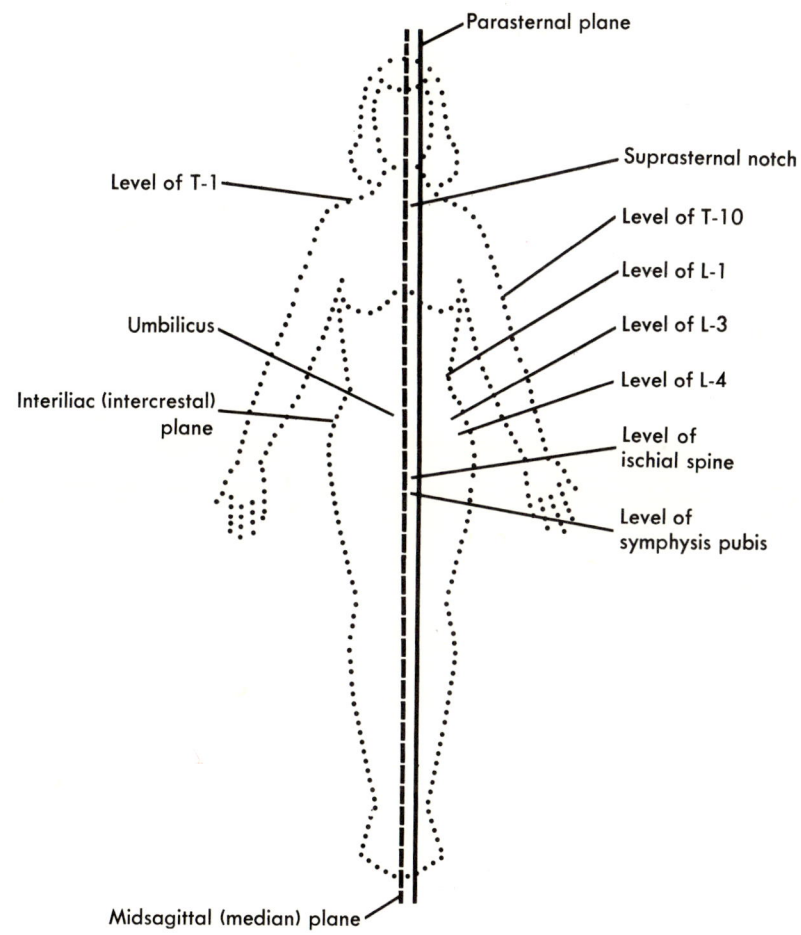

Fig. 4-46. Some radiographic landmarks and planes.

126 *Textbook of anatomy and physiology in radiologic technology*

interiliac (intercrestal) A transverse plane extending between the iliac crests (see Fig. 4-46)

Surgical landmarks. The femoral head can be located by drawing an imaginary line between the anterior superior iliac spine and the superior border of the symphysis pubis. A second imaginary line bisects the first line at right angles in the exact center of the first. The center of the femoral head is approximately 1 inch below the point of intersection (Note: In many instances the femoral head may be slightly lateral) (see Fig. 4-47). For posterior exposures during surgery, it is permissible to request that the surgeon (or other person wearing sterile dress) place his finger on the aforementioned location. During surgery the area is covered with sterile drapes, which renders the usual landmarks invisible.

Skull

The skull consists of two sections: the cranial section with eight bones and the facial section with fourteen bones. The twenty-two bones of the skull include neither the six auditory ossicles nor the single hyoid bone.

Since the skull is shaped somewhat spherically, superimposition of bones from one side or from within the inner structures over bones or structures of the side being examined (radiographically) is inevitable. Thorough knowledge of the anatomic relationship of all skull bones enables the radiologic technologist to demonstrate the required structures (see Fig. 4-48). Suture lines and other normal shadows are easily recognized by the well-informed technologist. Fig. 4-48, *B* is a photograph of a skull having a persisting metopic (frontal) suture extending vertically with the entire surface of the frontal bone. In both prenatal life and early infancy this suture may exist in part (or totally as illustrated) and

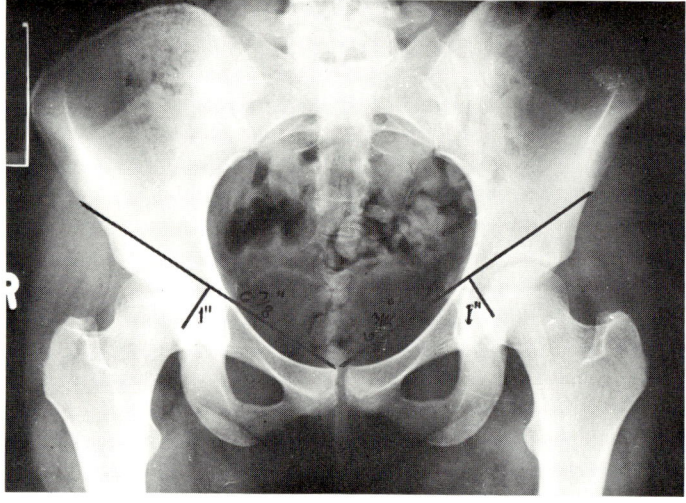

Fig. 4-47. Pelvis, femoral head localization, posterior view.

The skeletal system   **127**

separate this bone into two parts. The condition may persist throughout life as in this skull. The suture-shadow can cause technologists concern regarding the resulting radiograph.

The superior surface of the skull presents an ovoid shape; marked variations are common. The *skull vertex* is the highest prominence of the *crown* (top of the head). The *calvarium,* also called the cranial vault,* is the skullcap and includes parts of four bones—both parietals, the frontal, and the occipital. The average skull is *mesocephalic;* i.e., the cranial cavity is of medium capacity and the skull has a cephalic index† of between 76.0 and 80.9. A broad skull is *brachycephalic;* i.e., the transverse diameter is greater than eight-tenths (0.8) of its anteroposterior diameter and its cephalic index is between 81.0 and 85.4. Normally, the greatest transverse diameter exists between the two parietal bones. An elongated skull is *dolichocephalic;* i.e., the diameters are not the same as

*A vault is an arched or dome-shaped anatomical structure.
†The cephalic index is 100 times the maximal head breadth divided by the maximal head length.

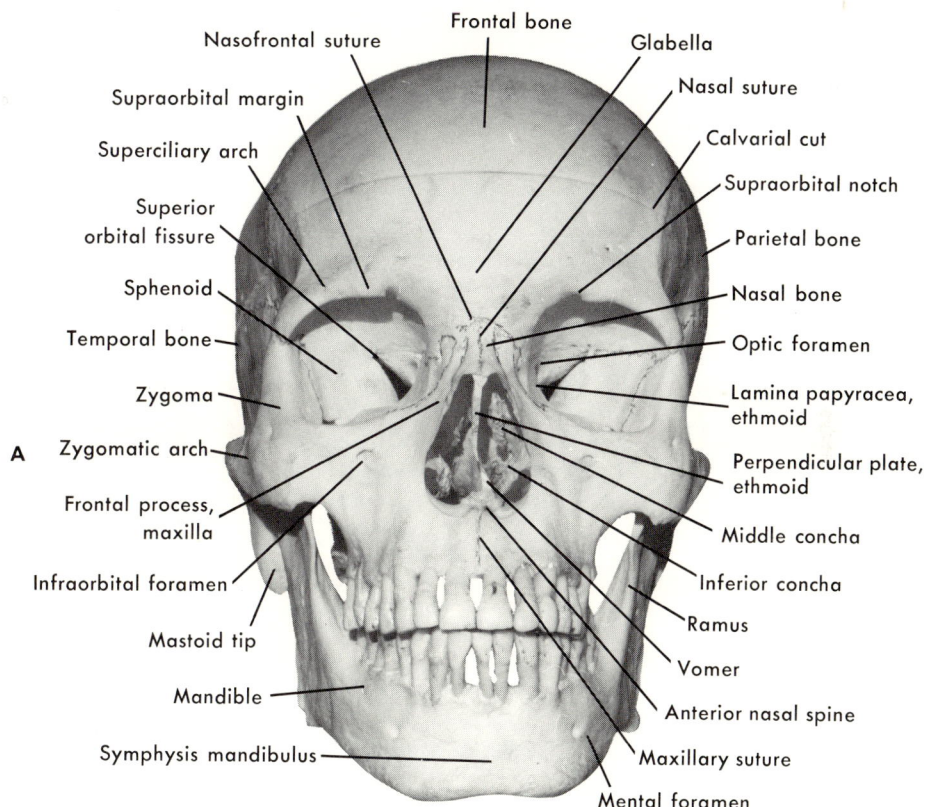

Fig. 4-48. Skull. **A,** Anterior aspect. **B,** Photograph of skull with persisting metopic (frontal) suture. **C,** Lateral aspect. **D,** Superior aspect. **E,** Posterior aspect. **F,** Inferior aspect.

Continued.

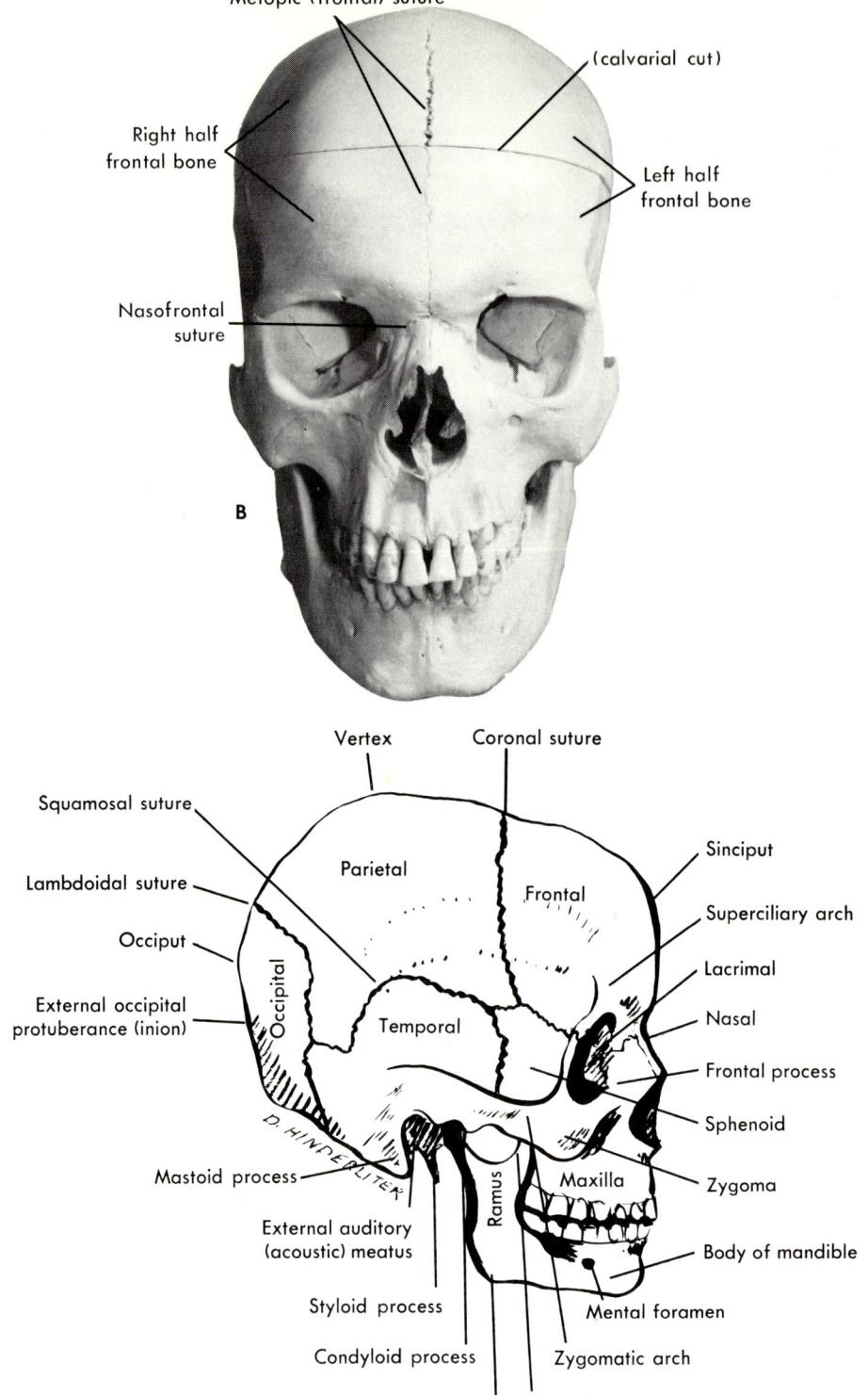

Fig. 4-48, cont'd. For legend see p. 127.

The skeletal system

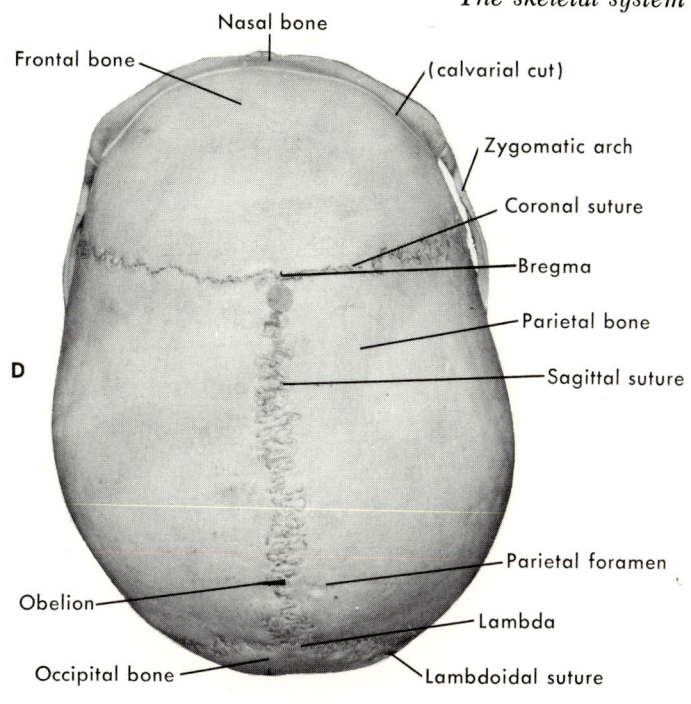

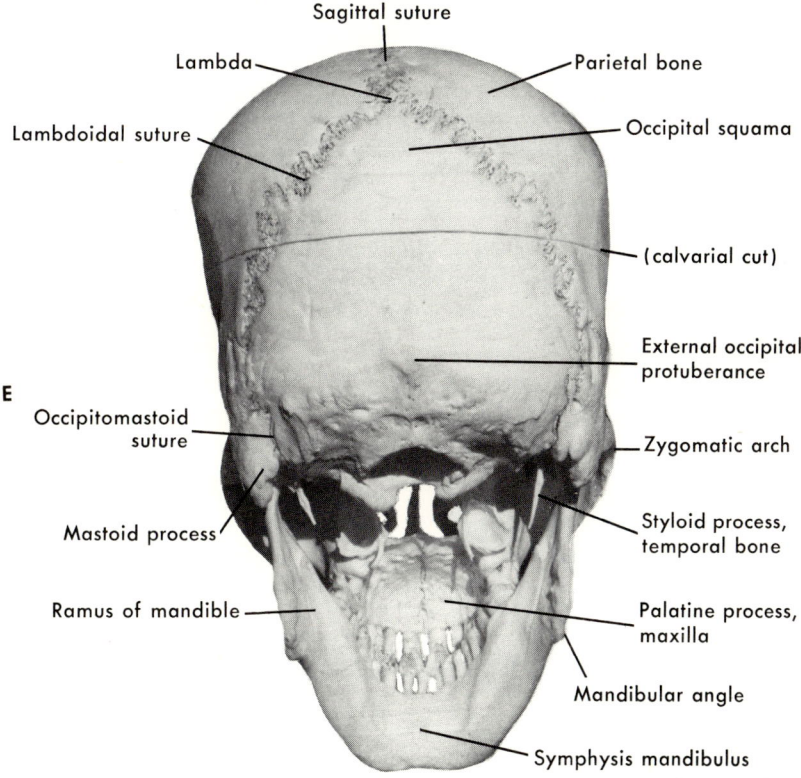

Fig. 4-48, cont'd. For legend see p. 127.

Continued.

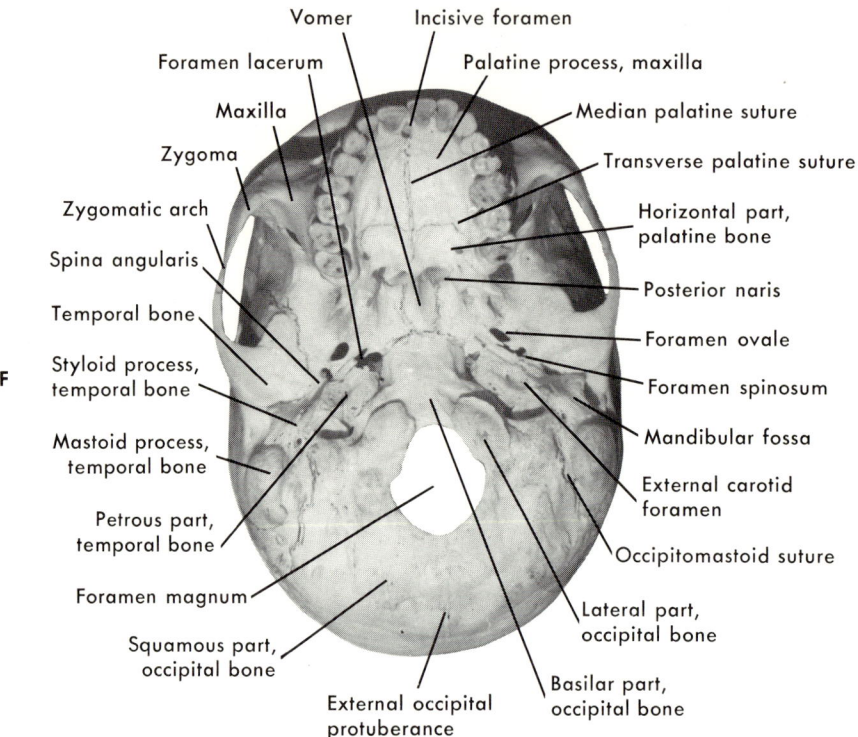

Fig. 4-48, cont'd. For legend see p. 127.

either the mesocephalic or brachycephalic skull and its cephalic index is 75.9 or less.

Knowledge of certain abnormalities of the skull is of considerable importance to radiographers. Among these are *microcephalic* and *hydrocephalic* skulls.

A *microcephalic* skull has abnormal smallness, specifically a cranial capacity less than 1,350 cc. as in some aboriginals. The term, *microcephalous*, is used to describe this skull or person. A *microcephalus* is a person or monster with an abnormally small head; this condition is usually associated with mental retardation.

Hydrocephalus is a condition characterized by an abnormal increase in the amount of cerebral fluid accompanied by dilatation of the cerebral ventricles. The disease is marked by enlargement of the head, with prominence of the forehead, atrophy of the brain, mental weakness, and convulsions.

The superior surface of the skull presents three suture lines: *sagittal*, between the two parietal bones and extending between the coronal and lambdoidal sutures; *coronal*, extending transversely between the two parietal bones posteriorly and the frontal bone anteriorly; and the *lambdoidal*, also extending transversely and between the two parietal bones anteriorly and the occipital bone posteriorly. The lambdoidal and coronal sutures are parallel with each other and are at right angles to the sagittal suture. The *bregma* is a point on the surface of the skull at the junction of the sagittal and coronal sutures. In the newborn infant this region is noncalcified and is called the *anterior fontanelle*. The *lambda* is a point at the junction of the sagittal and lambdoidal sutures. In the newborn infant this region also is noncalcified and is called the *posterior fontanelle*.

Other suture lines of the skull include the *sphenotemporal*, between the sphenoid and temporal bones; the *occipitotemporal*, between the occipital and temporal bones (sometimes called the occipitomastoid suture); and the *squamosal*, between the temporal and parietal bones.

Routine radiographs of the skull may include single or stereographic views, anterior or posterior; single or stereographic views, left and/or right lateral; occipital views; and sometimes verticosubmental or submentovertex views.

Skull landmarks and important lines

CRANIAL BONES. Except for the ethmoid and sphenoid bones, the cranial bones or parts thereof are flat and have an *inner* and an *outer table* of cortical bone. Between these tables is the highly nutritive layer of spongy bone, the *diploë*. The diploic layer consists of the nerve, blood, and lymph supplies for all of the inner table and the inner half of the outer table.

The eight cranial bones (one frontal, one ethmoid, one sphenoid, one occipital, two parietal, and two temporal) form the cranial case. The petrous part of each temporal bone contains the corresponding three auditory ossicles, which are the malleus, incus, and stapes.

There are many useful landmarks, lines, and points illustrated in Fig. 4-49, *A* and *B*. These are not all the lines used in the profession of radiography. However, the more common ones are indicated.

Frontal bone. The frontal bone consists of the horizontal and vertical parts. The horizontal part lies between the brain cavity and the eye orbits and forms a part of the cranial floor. The horizontal part consists of the two *orbital plates*, which are separated by the *ethmoidal notch*. The vertical part is the *squama*, which curves toward its perimeter from the anterior portion of the orbital part. This forms what is commonly called the forehead. The *frontal eminences* are on the anterior surface of the squama above the *superciliary arches*, bilateral to the midsagittal line. The *glabella* is raised and smooth and lies between the superciliary arches. Inferior to the superciliary arches on each side are the supraorbital margins; each is a prominent bony protection for the corresponding eye. At the end of the medial third of this margin lies the *supraorbital notch* (sometimes a foramen) for the passage of the supraorbital nerve and vessels.

The frontal bone articulates inferiorly with both nasal bones to form the *nasofrontal suture*. The frontal processes of the maxillae and the nasal bones are above the nasal process, which terminates in the *nasal spine*. The supraorbital margin is the line of demarcation between the squama and the orbital part (see Fig. 4-50). The ethmoidal notch, between the orbital plates, receives the cribriform plate of the ethmoid bone. In addition to forming the bony part of the superior portion of the eye orbits, the orbital plates (with the cribriform plate and

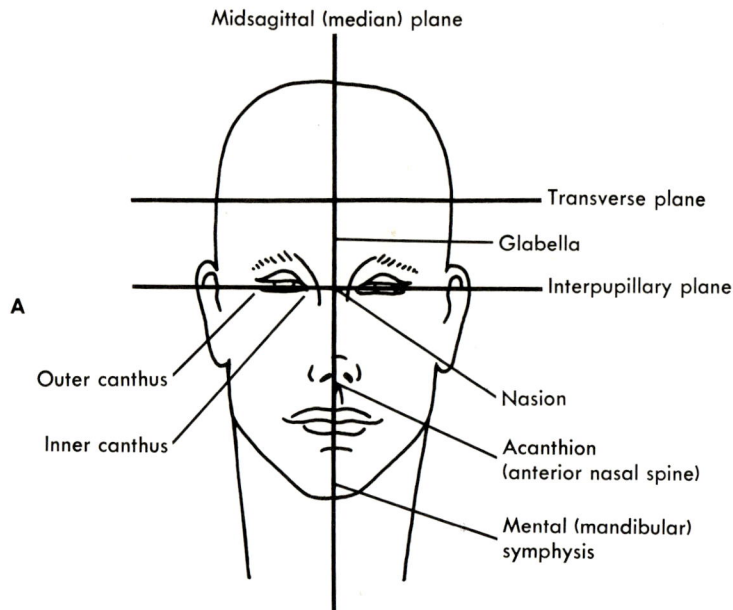

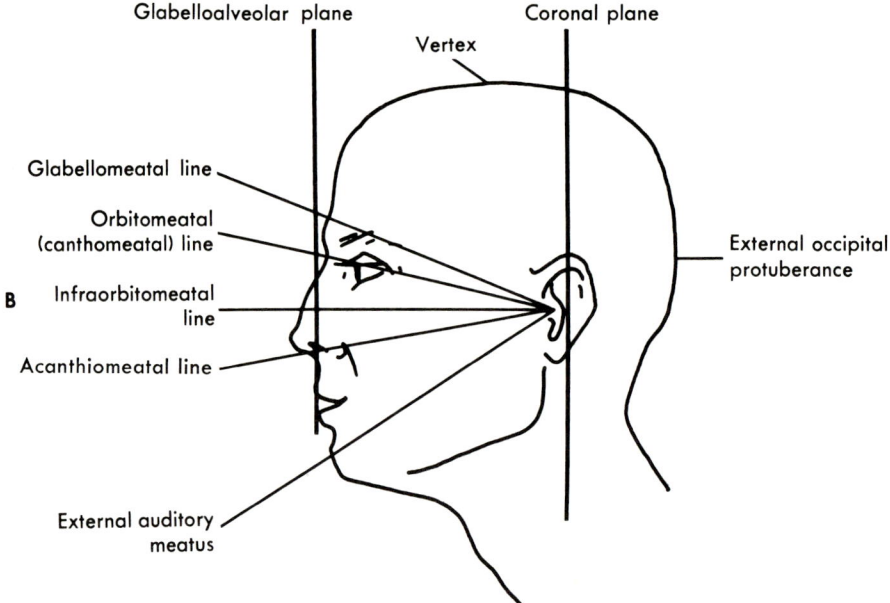

Fig. 4-49. A, Important landmarks, face—anterior surface. **B,** Important landmarks, face—lateral surface.

crista galli of the ethmoid bone) form most of the anterior cranial fossa. The frontal lobes of the brain lie within the anterior cranial fossa. The *frontal sinuses* lie directly behind the glabella. In embryonic life the frontal bone forms from two equal parts that join in the midsagittal plane, usually with a septum separating the frontal sinuses into two parts; these parts are often quite different in size, one occasionally being absent.

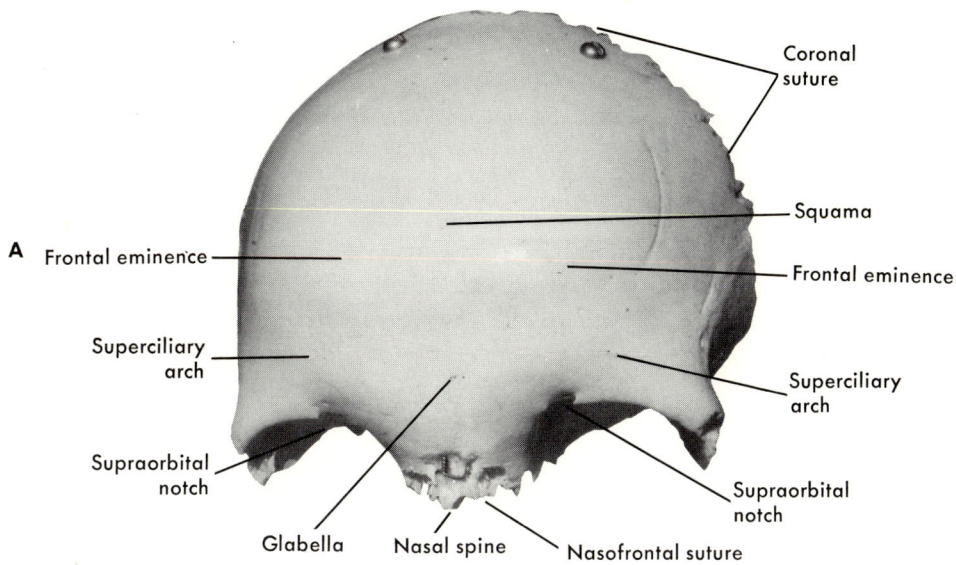

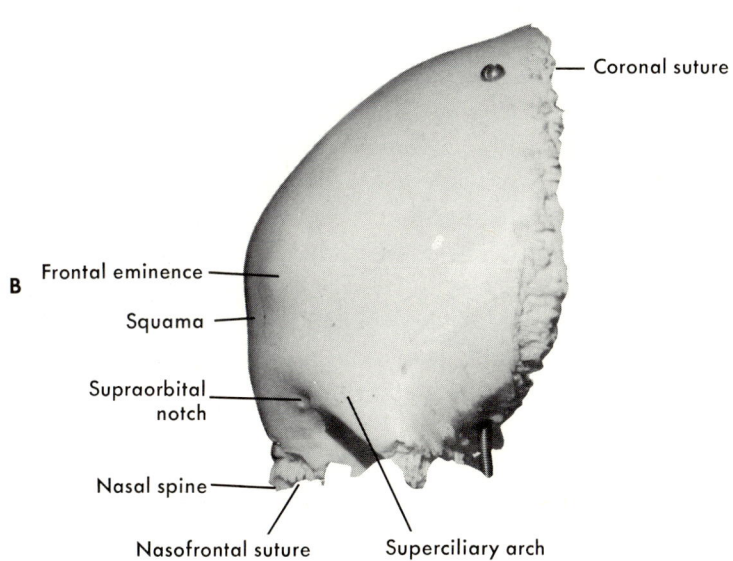

Fig. 4-50. Frontal bone. **A,** Antero-oblique surface. **B,** Left lateral surface.

The frontal sinuses and the squama are demonstrated radiographically with the patient in the frontal position and the central ray angled from 15 degrees to 23 degrees caudad. The frontal bone articulates with the ethmoid bone, the sphenoid bone, both maxillae, both zygomatic bones, both parietal bones, both nasal bones, and both lacrimal bones. The eye orbit is formed by the frontal bone, zygoma, sphenoid, ethmoid, lacrimal, and palatine bones, and maxilla.

Ethmoid bone. The ethmoid bone occupies the space surrounded by the frontal bone, both maxillae, the sphenoid bone, and the vomer. The *cribriform plate* is horizontal between the orbital plates of the frontal bone; this plate, being quite porous, serves as a site for communication between the frontal sinuses and the nasal cavity. The *perpendicular plate* extends downward and forms the major part of the *nasal septum*. The *laminae papyraceae,* one on the lateral surface of each *lateral mass,* articulate in the eye orbit in the space surrounded by the maxillae, palatine bones, lacrimal bones, sphenoid bone, and frontal bone. The *crista galli* extends superiorly from the midportion of the superior surface of the cribriform plate and serves as the attachment site for the falx cerebri of the brain. The *superior and middle conchae* are parts of the ethmoid bone and are more cartilaginous than osseous. The superior conchae contain the *ethmoid cells* (ethmoidal sinuses). The ethmoid cells are visualized in the frontal and verticosubmental views. These same views visualize the body of the ethmoid bone.

The ethmoid bone articulates with the frontal bone, sphenoid bone, both sphenoidal conchae, both nasal bones, both maxillae, both palatine bones, both lacrimal bones, the vomer, and both inferior conchae (nasal turbinates) (see Fig. 4-51).

Sphenoid bone. The irregularly shaped sphenoid bone comprises the major part of the floor of the middle cranial fossa. This bone occupies the space anterior to the occipital and temporal bones and posterior to the frontal and ethmoid bones. Two *great wings,* two *small wings,* and two *pterygoid processes* project from the sphenoid body, which is situated in the midline of the floor of the skull (see Fig. 4-52). A sagittal septum divides the hollow body into the two *sphenoidal sinuses,* which are inferior to the *sella turcica.* The sella turcica is the saddle-shaped part of the sphenoid bone that contains the *pituitary (hypophyseal) gland.* The cribriform plate of the ethmoid bone articulates with the *ethmoid spine,* which extends above the body of the sphenoid bone. The smooth and slightly raised *optic (chiasmatic) groove* is posterior to the ethmoid spine and anterior to the *tuberculum sellae.* The tuberculum sellae is the small, bony elevation that forms the posterior limit of the optic groove. The sella turcica is directly behind and below the tuberculum sellae. The sella turcica is bounded anteriorly by the *middle* and *anterior clinoid processes* and posteriorly by the *dorsum sellae,* which terminates superiorly in the *posterior clinoid processes.** The dorsum sellae is the square, bony plate that projects superiorly at the

---

*In a true lateral radiograph of the skull the left and right clinoid processes are superimposed in the shape of a saddle.

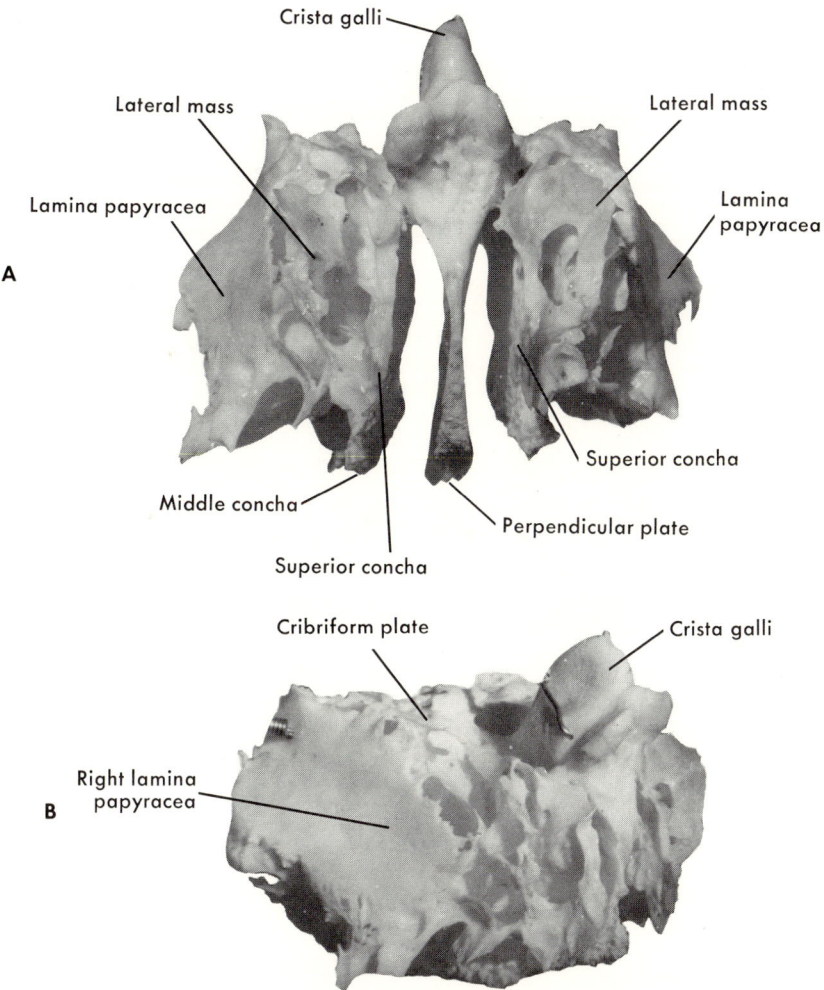

Fig. 4-51. Ethmoid bone. **A**, Anterior aspect. **B**, Right supero-oblique aspect.

posterior margin of the sella turcica. The *petrosal process* on the side of the dorsum sellae is inferior to the notch for the abducent nerve. The apex of the petrous part of the temporal bone articulates with the petrosal process. The *clivus* is the downward-sloping, shallow groove posterior to the dorsum sellae, and it articulates posteriorly with the basilar part of the occipital bone. The perpendicular plate of the ethmoid bone articulates posteriorly with the *sphenoidal crest*, which joins the *rostrum* of the sphenoid bone. The sphenoidal rostrum is a triangular spine on the inferior surface of the body. The rostrum is situated between the alae of the vomer. The lower lateral margins of the anterior surface of the sphenoidal body form parts of the eye orbits. The extreme lateral part of each great wing forms a small part of the corresponding lateral wall of the skull.

136 *Textbook of anatomy and physiology in radiologic technology*

This part of each great wing fits into the lateral wall between the temporal and frontal bones and between the parietal and zygomatic bones. The *spina angularis* (sphenoidal spine) is an inferior projection of the sphenoid bone adjacent to both the squamosal and petrous parts of the temporal bone. The pterion (pp. 138, 166) is usually located at the common junction of the frontal, temporal, and parietal bones and the great wing of the sphenoid bone about 3 cm. posterior to the

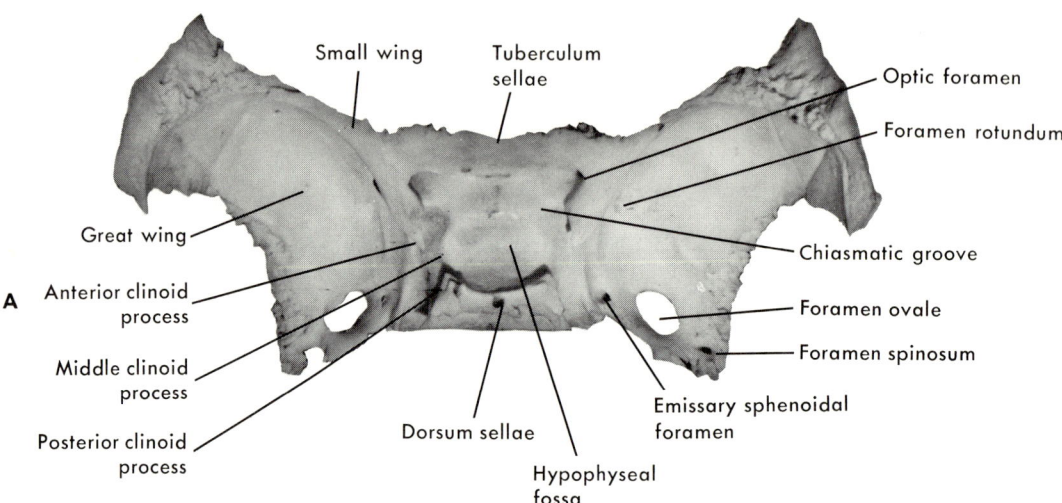

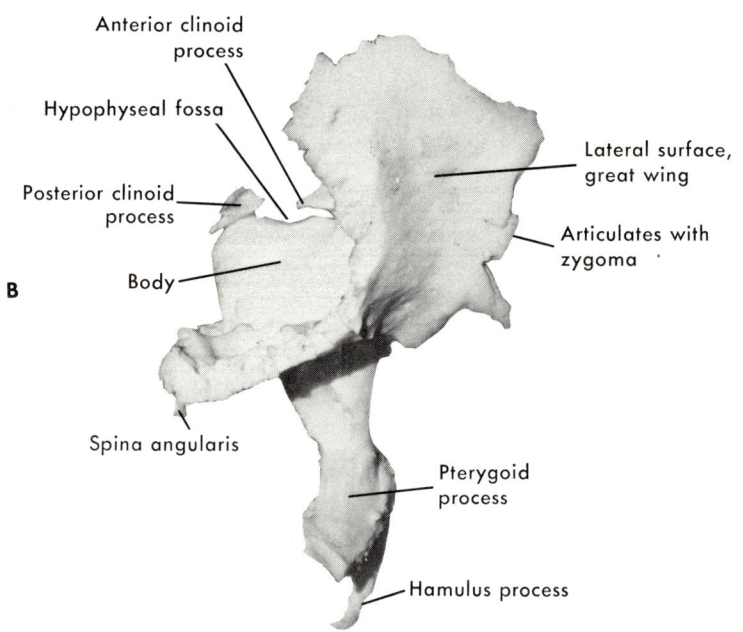

Fig. 4-52. Sphenoid bone. **A,** Superior aspect. **B,** Lateral aspect. **C,** Anterior aspect.

The skeletal system   137

external angular process of the eye orbit. However, this point may be located at the common junction of the parietal, temporal, and sphenoid bones in the many skulls wherein this part of the sphenoid bone is broad and separates the frontal bone from the temporal bone.

Each small wing of the sphenoid bone is superior to the body and extends laterally from the midsagittal plane. It contains the *optic foramen,* through which the second cranial (optic) nerve and ophthalmic artery pass. The flat superior surface of each wing supports part of the frontal lobe of the brain, and the opposite (inferior) surface contributes to the formation of the eye orbit. This surface also constitutes the upper boundary of the *superior orbital fissure,* the communication between the cranial cavity and the eye orbit. The superior orbital fissure is the passageway for the oculomotor, trochlear, and abducent nerves; the three branches of the optic division of the trigeminal nerve; the orbital branch of the middle meningeal artery; and other branches of nerves, arteries, and veins. The lateral fissure of the brain receives the smooth and rounded posterior borders of the small wings; the medial ends of these borders terminate in the anterior clinoid processes. In some skulls, on either side a small bony spicule may join the anterior clinoid process to the middle clinoid process and so convert the terminal end of the internal carotid groove into the caroticoclinoid foramen.

The paired *pterygoid processes* extend inferiorly, one on each side, from the junction of the body with the great wing. There are two *sphenoidal conchae* between the nasal cavity and the sphenoidal sinuses; the conchae permit passage of air between the cavity and the sinuses.

Each great wing arises from the side of the sphenoidal body as a strong bony process curving upward, laterally and posteriorly. The cerebral surfaces of the two great wings contain several depressions to receive the convolutions of the temporal lobes of the brain. The *foramen rotundum* is the circular opening sit-

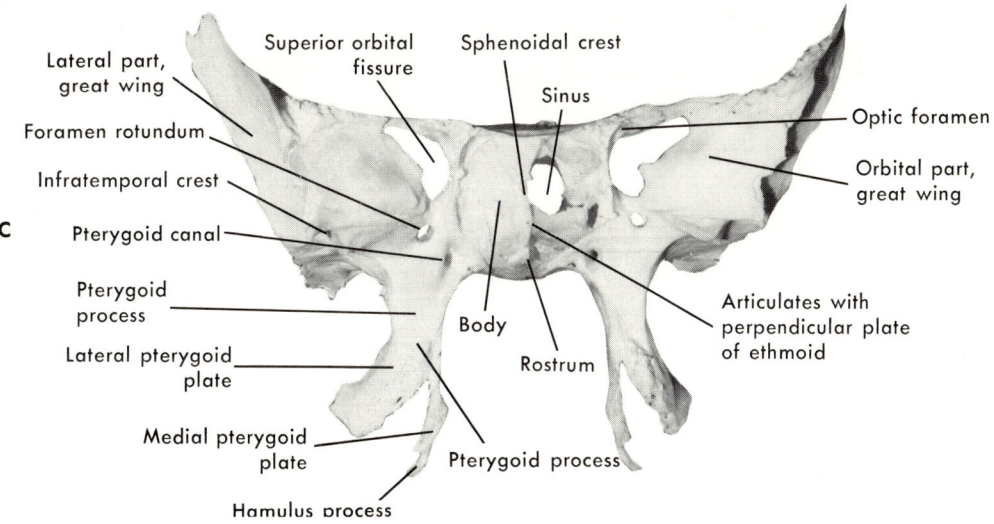

Fig. 4-52, cont'd. For legend see opposite page.

uated anteriorly and medially in each great wing; the maxillary nerve passes through this foramen. The *foramen ovale* is situated both posterior and lateral to the foramen rotundum; the mandibular nerve, accessory meningeal artery, and, in some skulls, the lesser superficial petrosal nerve pass through the foramen ovale. The *foramen spinosum* is situated close to and in front of the sphenoidal spine in the posterior angle of the great wing; the middle meningeal vessels and a recurrent branch from the mandibular nerve pass through this foramen. The medial half of the great wing margin forms the anterior boundary of the *foramen lacerum* and forms the posterior opening for the *pterygoid canal;* the pterygoid nerve and artery pass through this canal. The pterion is a point at the junction of the frontal, parietal, and temporal bones and the great wing of the sphenoid bone, about 3 cm. posterior to the external angular process of the orbit.

Radiographic demonstration of the sphenoid bone is achieved by use of the verticosubmental, submentovertex, and lateral skull positions. Visualization of the optic foramen is achieved by use of the anterior skull (Rhese) position.

The sphenoid bone articulates with both parietal bones, both zygomatic bones, both palatine bones, both temporal bones, the vomer, the ethmoid bone, the frontal bone, and the occipital bone. In some skulls the maxillary tuberosities articulate with the sphenoid bone.

Supernumerary bones. Small supernumerary bones may be found contained completely within a suture of the skull calvarium. These small bones are called *wormian* bones; if such bones occupy space in a suture in the region of and above the pterion, the bones are called *epipteric* bones.

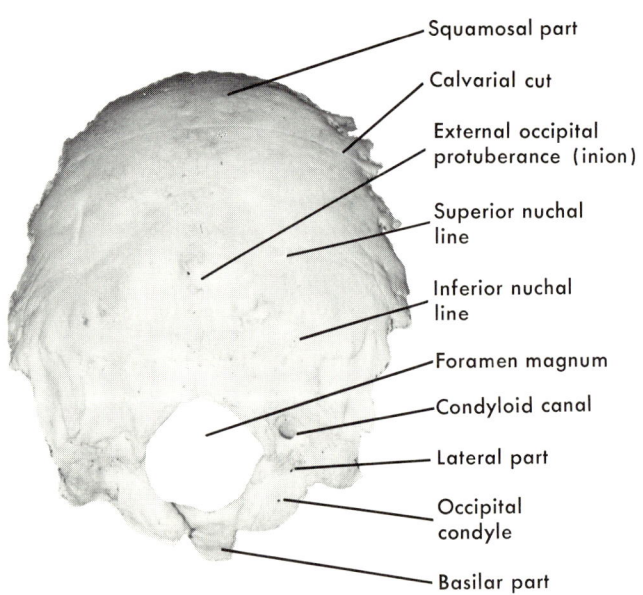

Fig. 4-53. Occipital bone, posterior surface.

Occipital bone. The occipital bone forms the major part of the *basal region* of the skull; this is the region of the skull involved in a *basal skull fracture*. The occipital bone includes three parts: the *basilar part* (pars basilaris), which articulates anteriorly with the basilar part of the sphenoid bone; the *lateral parts* (pars lateralis), which are bilateral to the *foramen magnum;* and the *squamosal part,* which is posterior and superior to the foramen magnum and contributes to the formation of the lambdoidal suture. The basilar and lateral parts and the adjacent part of the squama form the *posterior cranial fossa.*

The inferior surface of the occipital bone presents two articulating areas on the lateral parts; these are the *occipital condyles* for articulation with the superior articulating facets of the atlas. The squama of the occipital bone, like the squama of the frontal bone, curves toward its perimeter. The bony prominence in the median line and somewhat inferior to the center of the squama is the

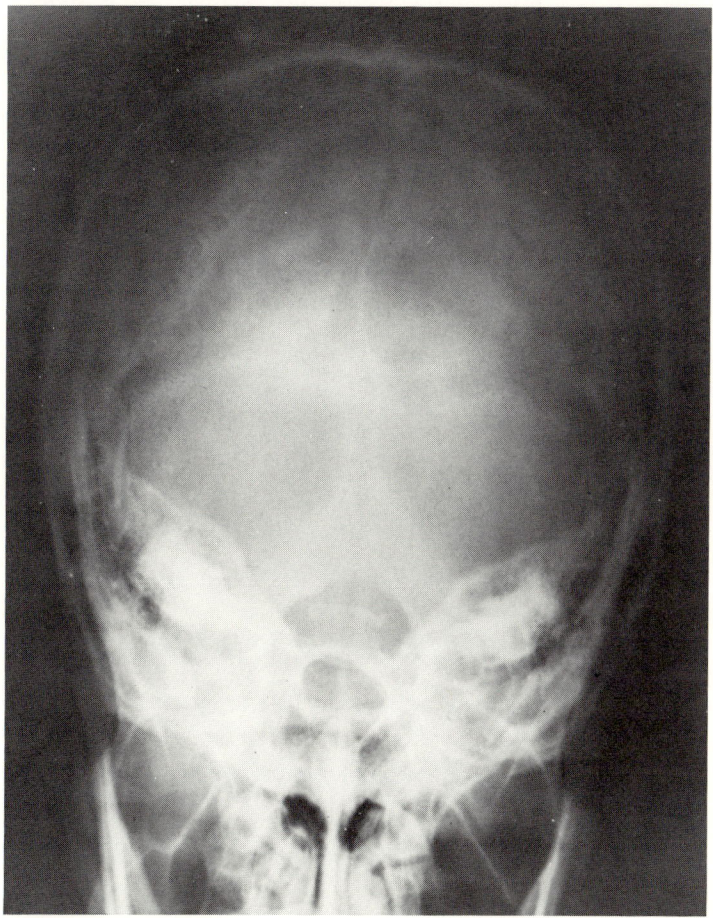

Fig. 4-54. Towne view of skull. (Courtesy Desert Springs Hospital.)

external occipital protuberance (E.O.P.). This raised area is quite easily palpated, although its size varies considerably between individuals. (The median nuchal line descends inferiorly from the E.O.P. and is the attachment of the ligamentum nuchae.) The cranial surface of the occipital bone faces the posterior part of the brain. The medulla oblongata of the brain is continuous with the spinal cord (medulla spinalis) through the foramen magnum (see Fig. 4-53). The foramen magnum is surrounded by the squama posteriorly and superiorly, the lateral parts bilaterally, and the basilar part anteriorly.

Radiographic demonstration of the occipital bone is achieved by use of the 35-degree caudad angle (Towne) and lateral skull positions (see Fig. 4-54).

The occipital bone articulates with the temporal bones, the parietal bones, the sphenoid bone, and the atlas.

Parietal bones. The paired parietal* bones articulate with each other in the midsagittal plane to form the sagittal suture. Each bone forms a major part of the side and vault of the calvarium. Each bone is quadrilateral in shape. Near the sagittal suture and anterior to the lambdoidal suture in each bone is the *parietal foramen;* either or both of these foramina may not be visualized. Each parietal bone articulates with the corresponding temporal squama to form the *squamosal suture.* The raised, smooth prominence in the approximate center of each parietal bone is the *parietal eminence,* the site of commencement of ossification. Each parietal bone articulates anteriorly with the frontal bone to form one-half of the coronal suture, and each parietal bone articulates posteriorly with the occipital bone to form one-half of the lambdoidal suture.

The parietal bone is demonstrated radiographically by use of the lateral skull and submentovertex positions.

Each parietal bone articulates inferiorly with the corresponding temporal bone, posteriorly with the occipital bone, anteriorly with the frontal bone, and medially with the opposite parietal bone (see Fig. 4-55).

Temporal bones. The temporal bone, a compound bone in the side of the skull, consists of three parts—*squamous, petrous,* and *tympanic*—and two processes—*mastoid* and *styloid.* The squamous part is large and scalelike and extends superiorly from the petrous part to form a part of the lateral wall of the cranial case and the *mandibular fossa* for articulation with the mandibular condyle. Also, the squama presents the *zygomatic process* anterolaterally to articulate with the temporal process of the zygoma and to form the *zygomatic arch.* Fractures of the zygomatic arch (cheek bone) are often incurred because of its exposed position. The fractured fragment is depressed and is reduced surgically. The squama of the temporal bone is surrounded anteriorly, superiorly, and posteriorly by the sphenoid bone and the parietal bones; it articulates with these bones to form the squamosal suture.

The petrous part of the temporal bone derives its name from its very hard structure; it contains most of the important auditory organs, among which are

---

*Parietal refers to wall, as of a cavity.

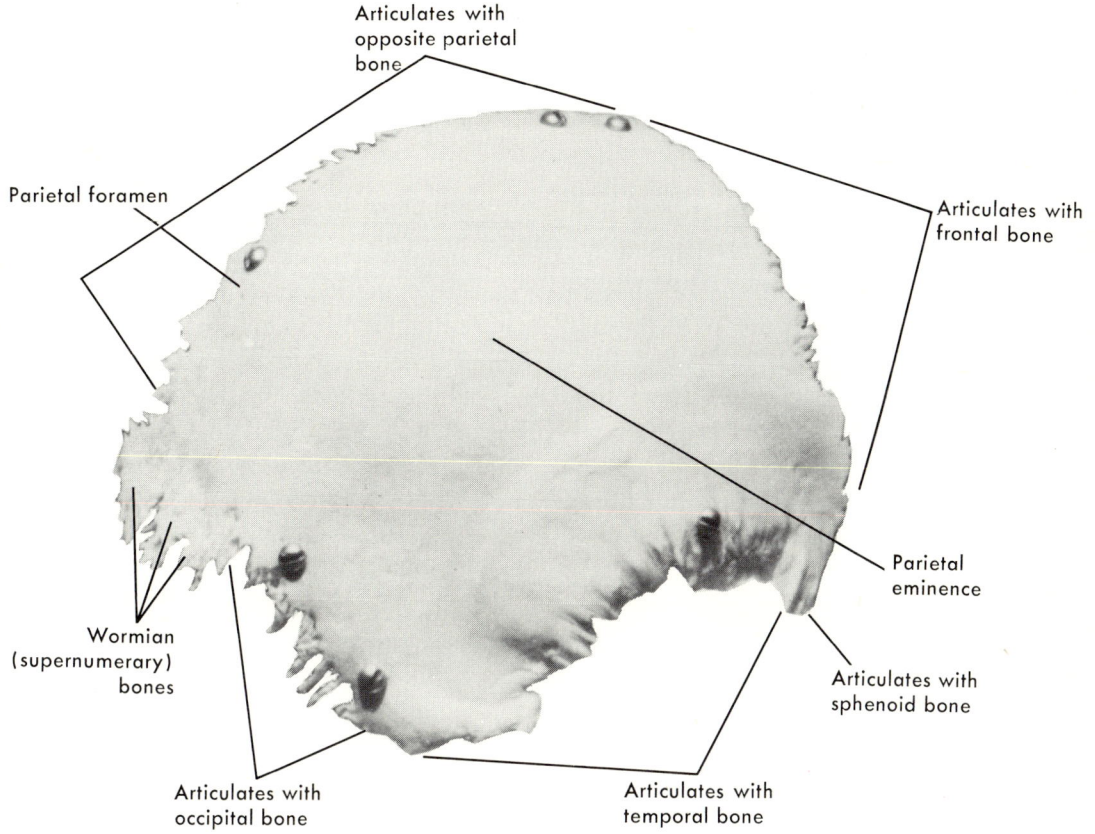

Fig. 4-55. Right parietal bone, supero-oblique aspect.

the *semicircular canals, membranous labyrinth,* and *auditory ossicles.* The petrous part separates the occipital bone from the sphenoid bone and extends superiorly to terminate in a series of sharp, continuous eminences, the *petrous ridge* (pyramids). The ridge is demonstrated radiographically in the *45-degree facial* (Stenvers) *position;* this position frequently visualizes one or more of the semicircular canals. The *cone-down occipital* (Towne) *position* is often used to demonstrate the petrous ridges.

The tympanic part of the temporal bone is a curved plate inferior to the squama and anteromedial to the mastoid process. The antero-inferior surface of the tympanic part is the posterior boundary of the mandibular fossa. The concave posterosuperior surface constitutes a major part of the osseous *external acoustic meatus* (E.A.M.). Internally the tympanic part fuses with the petrous part.

The lateral border of the tympanic part serves as the attachment site for the cartilaginous part of the external acoustic meatus. The lower border presents a medial and a lateral part; the root of the styloid process arises from the lateral part.

The external acoustic meatus extends medially and anteriorly for approximately 2 cm., forming a slight curve so that the canal floor presents a superior convexity.

The mastoid process contains the *mastoid cells* and extends posteriorly and inferiorly below the external auditory meatus. The mastoid portion is both posterior and inferior to the squama.

The *styloid process* of the temporal bone extends both inferiorly and anteriorly from beneath the tympanic part. The external auditory (acoustic) meatus opens from the *auditory canal* in the petrous part to the outside.

Other positions for radiographic demonstration of the temporal bone and/or its parts include the lateral skull position to demonstrate the squama, and the 15-degree–15-degree (Law) position to demonstrate the mastoid process. The tips of the mastoid processes are visualized in the posterior (A-P) position for the tips. The tangential for the facial bones (May) position demonstrates the zygomatic arch. Variations of the 15-degree–15-degree position employing 35-degree angulation (Mayer) or 50-degree angulation (Owens) are also used to demonstrate the mastoid cells.

The temporal bone articulates with the parietal bones, occipital bone, sphenoid bone, zygomatic bone, and mandible (see Fig. 4-56).

Cranial floor. Six bones form the floor of the cranial case; these are the sphenoid, ethmoid, two temporal, frontal, and occipital bones. The basilar part of the sphenoid bone articulates anteriorly and inferiorly with the vomer and posteriorly with the basilar part of the occipital bone.

The *anterior, middle,* and *posterior fossae* comprise the floor of the cranial case (see Fig. 4-57, A). The three fossae occupy different levels: the anterior

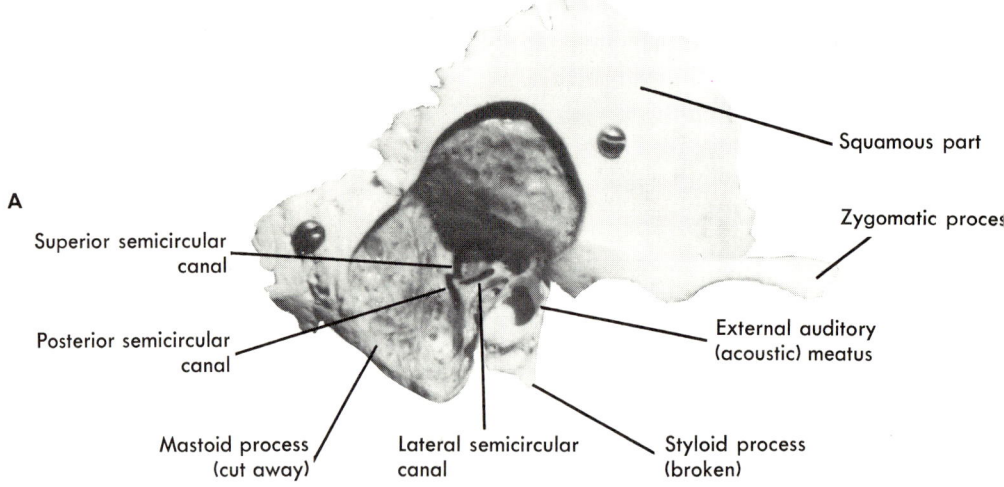

Fig. 4-56. **A,** Right temporal bone, lateral surface. **B,** Left temporal bone, medial surface. **C,** Left temporal bone, anterior half of medial aspect.

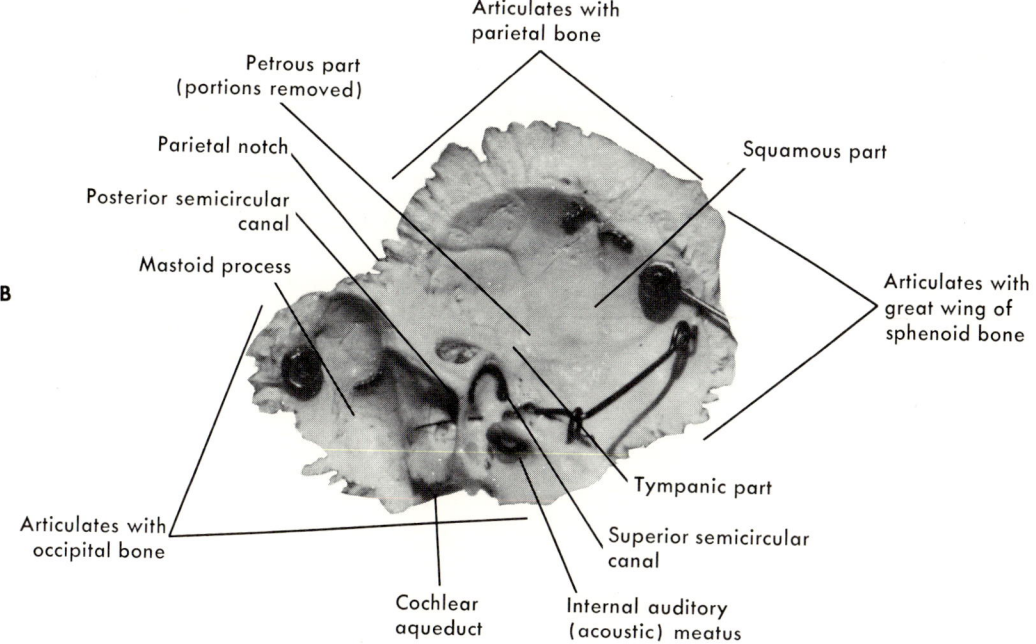

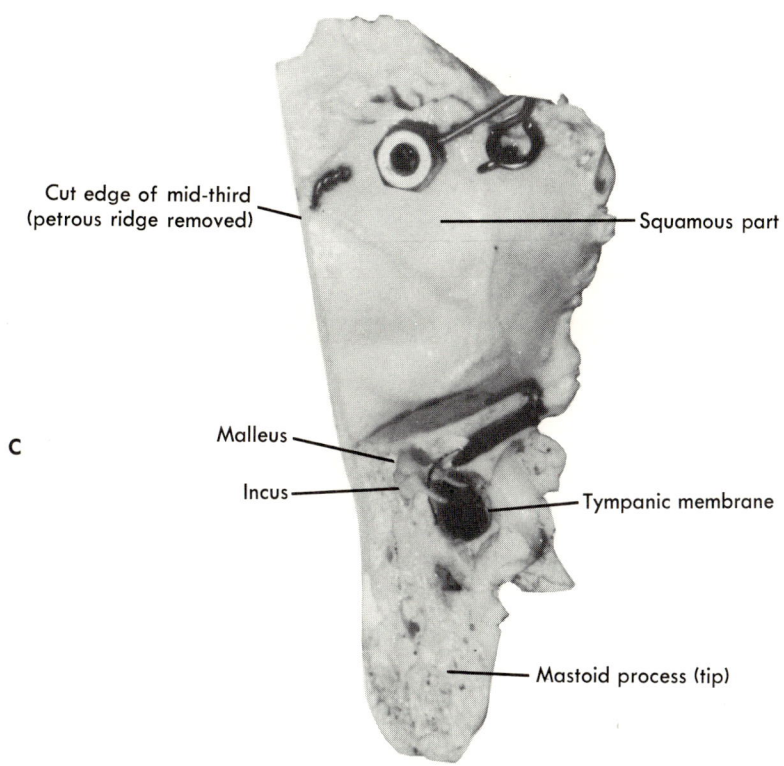

Fig. 4-56, cont'd. For legend see opposite page.

fossa is the highest, the middle fossa is somewhat lower, and the posterior fossa is considerably lower. Numerous foramina and other openings are spaced throughout the cranial floor; many of the important arteries, veins, nerves, and lymphatics pass through these openings. The radiologic technologist must have detailed knowledge of these openings and the relation of each to the entirety of the cranial floor. Many radiographic requests will include emphasis upon a specific part of the cranial floor and perhaps upon a specific foramen or other opening. Careful study and comparison of Figs. 4-57, A and 4-58 with Table 4-6 will acquaint the student technologist with these openings.

The occipital view demonstrates many of the structures and openings of the middle and posterior cranial fossae. The verticosubmental position is also useful to visualize these bony parts. Lateral views of the skull visualize the levels of the three cranial fossae.

FACIAL BONES. In general, the fourteen facial bones are responsible for the shape and appearance of the face. This group of bones includes two nasal bones, two lacrimal bones, two zygomatic bones, two maxillae, two palatine bones, two inferior conchae (nasal turbinates), one vomer, and one mandible.

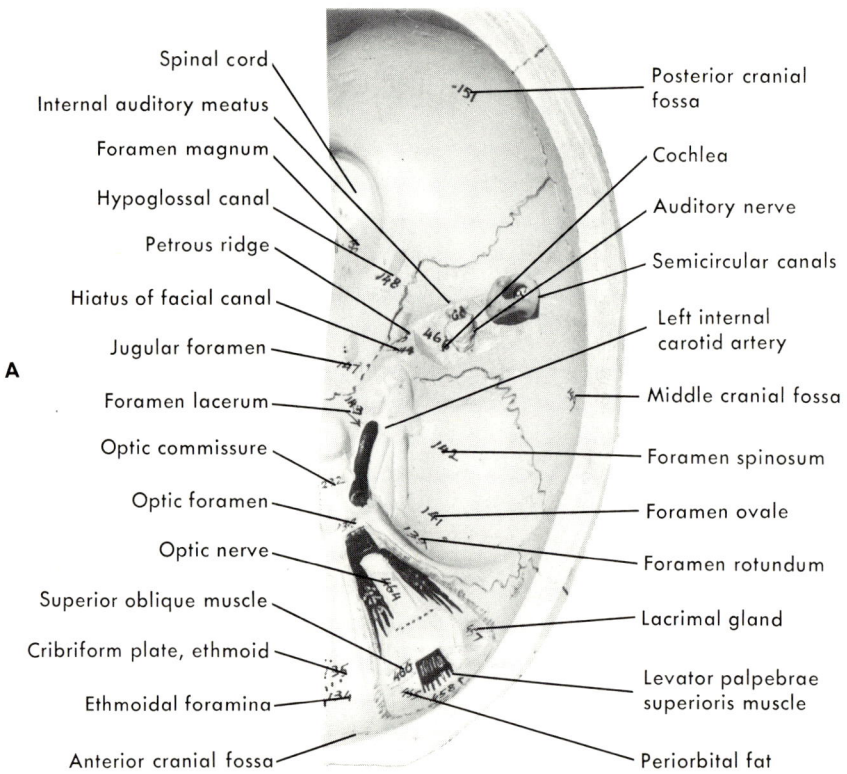

Fig. 4-57. **A,** Left half of cranial floor, superior aspect. **B,** Right labyrinth (model) removed from osseous enclosure, anterolateral aspect. **C,** Right labyrinth (model) removed from osseous enclosure, posteromedial aspect.

The skeletal system **145**

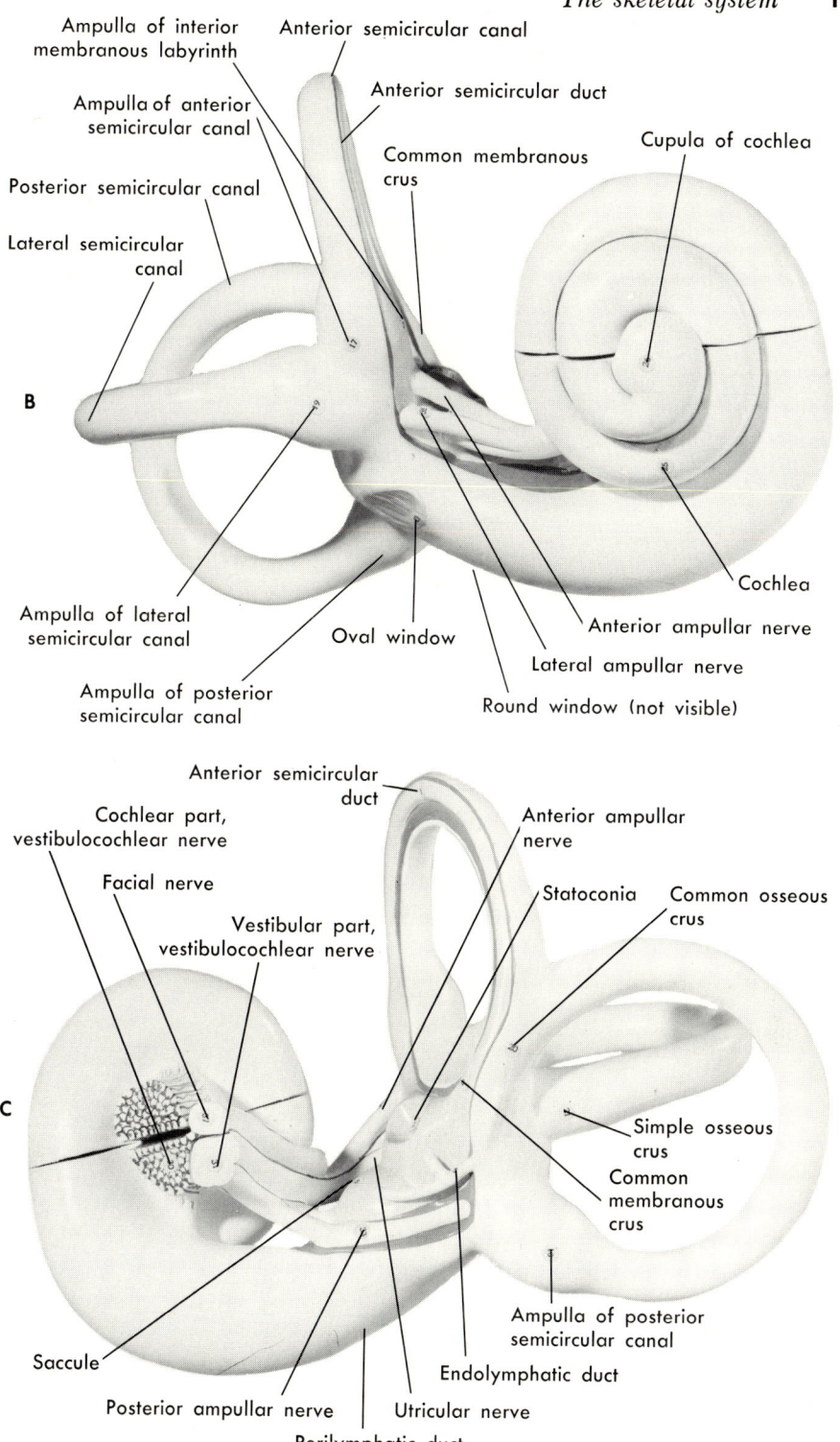

Fig. 4-57, cont'd. For legend see opposite page.

Table 4-6. Openings in the cranial floor

Fossa	Opening	Location	Description	Viscera passing through
Anterior	Foramen caecum	Frontal bone	Articulation between inferior part of frontal crest with ethmoid bone forms a small opening, the foramen caecum	Vein, from nose to superior sagittal sinus
Middle	Optic foramen	Sphenoid bone	Between lesser wing and middle clinoid process	Optic nerve, ophthalmic artery
	Superior orbital fissure	Sphenoid bone	Between lesser and greater wings	Oculomotor, trochlear, and ophthalmic divisi on of trigeminal and abducent nerves; filaments of the cavernous plexus of sympathetic and orbital branches of middle meningeal artery; ophthalmic vein; and recurrent branch of lacrimal artery
	Foramen rotundum	Sphenoid bone	In anteromedial part of great wing	Maxillary nerve
	Foramen ovale	Sphenoid bone	Posterolateral to foramen	Mandibular nerve, accessory meningeal artery; and (occasionally) lesser superficial petrosal nerve
	Foramen spinosum	Sphenoid bone	In posterior angle of great wing	Middle meningeal vessels and recurrent branch of the mandibular nerve
	Foramen lacerum	Bounded by great wing of sphenoid bone, apex of petrous portion of temporal bone, body of sphenoid bone, and basilar part of occipital bone	Medial to foramen ovale	Internal carotid artery, a plexus of sympathetic nerves, nerve of pterygoid canal, and a meningeal branch of ascending pharyngeal artery

The skeletal system

	Carotid groove (sometimes the caroticoclinoid foramen)	Sphenoid bone	Lateral to sella turcica	Cavernous sinus, internal carotid artery, and a plexus of sympathetic nerves
Posterior	Internal auditory meatus	Temporal bone, petrous part	Posterior to foramen lacerum	Facial and auditory nerves, nervus intermedius, and auditory branch of the basilar artery
	Jugular	Between petrous part of temporal bone and pars lateralis of occipital bone and anterior terminus of sigmoid sinus	Posterolateral to internal auditory meatus	Inferior petrosal sinus; glossopharyngeal, vagus, and accessory nerves; transverse sinus, meningeal branches of the occipital and ascending pharyngeal arteries
	Hypoglossal canal	Occipital bone	At base of occipital condyle in pars lateralis	Hypoglossal nerve and meningeal branch of ascending pharyngeal artery
	Condyloid canal	Occipital bone	In pars lateralis, posterolateral to hypoglossal canal	Emissary vein from the transverse sinus
	Foramen magnum	Occipital bone	Separates the pars basilaris, pars lateralis, and the squama	Medulla oblongata and its membranes; the accessory nerves; vertebral arteries; anterior and posterior spinal arteries; membrana tectoria; and alar ligaments

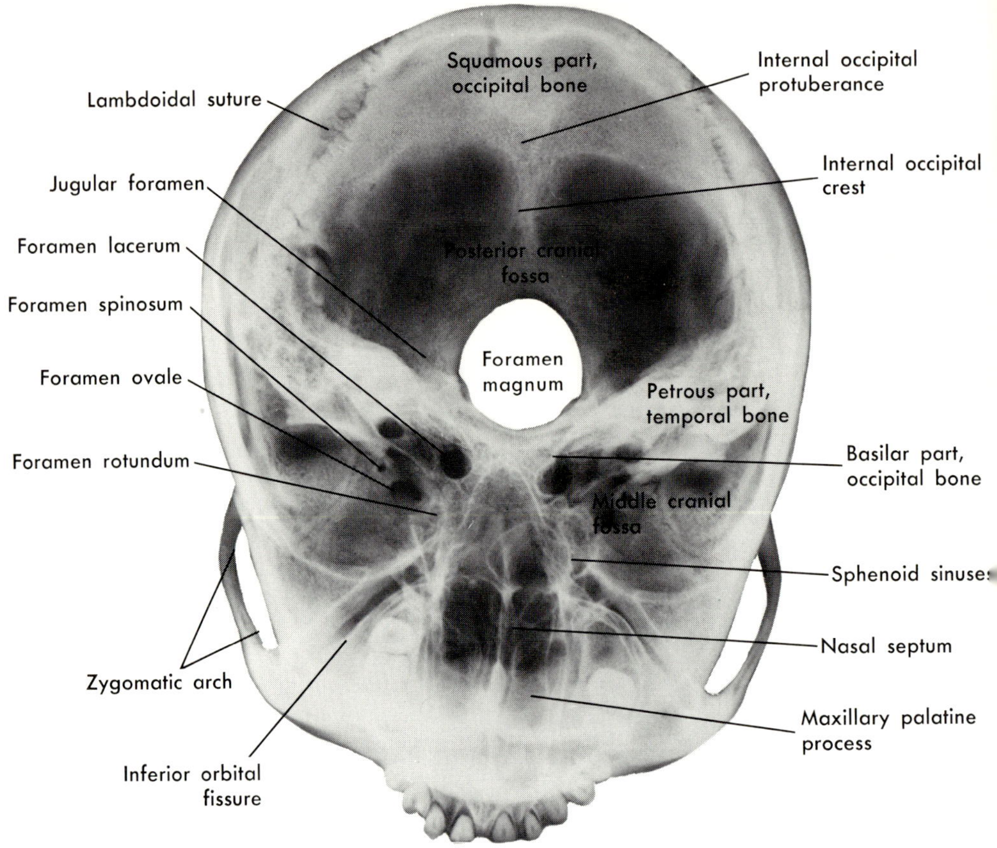

148 *Textbook of anatomy and physiology in radiologic technology*

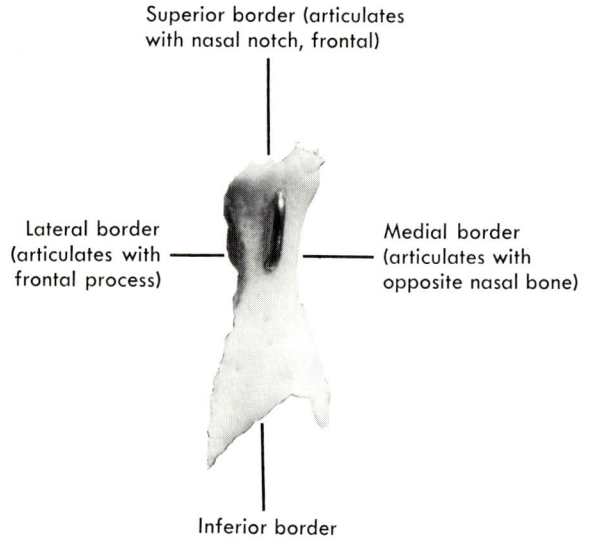

Fig. 4-58. Verticosubmental view of skull (calvarium and mandible removed).

Fig. 4-59. Right nasal bone, lateral surface.

Nasal bones. The nasal bones are quite small and articulate with each other in the midsagittal plane to form the *nasal suture*. Each nasal bone is thin and rectangularly shaped. The lateral cartilage of the nose is inferior to each nasal bone. Each nasal bone articulates inferiorly with the corresponding frontal process of the maxilla, thus forming the nasal bridge.

The nasal bones articulate superiorly with the frontal bone to form the nasofrontal suture. The *vertical crest* articulates with the frontal spine and the perpendicular plate of the ethmoid bone and extends below the sagittal border. Each nasal bone articulates with the frontal bone, maxilla, ethmoid bone, and opposite nasal bone (see Fig. 4-59).

Depressed fractures of the nasal bone are rather common. Such fractures are reduced while the patient is under anesthesia. Unreduced depressed fractures may cause a breathing obstruction, a nose deformity, or both.

The nasal bones are demonstrated radiographically in the axial and lateral skull positions.

Lacrimal bones. The smallest of the facial bones are the lacrimal bones. Each lacrimal bone occupies the space surrounded by the laminae papyraceae of the ethmoid bone, two parts of the maxilla, the orbital plate of the frontal bone, and the inferior concha. The lacrimal bone forms a part of the medial orbital wall (see Fig. 4-60).

Radiographic demonstration of the lacrimal bone is quite difficult because of its relation to the surrounding bones and because of its small size and thin structure. However, it is a frequent site of traumatic involvement and may require visualization in a radiograph. This is best achieved in the 45-degree (Rhese) facial position.

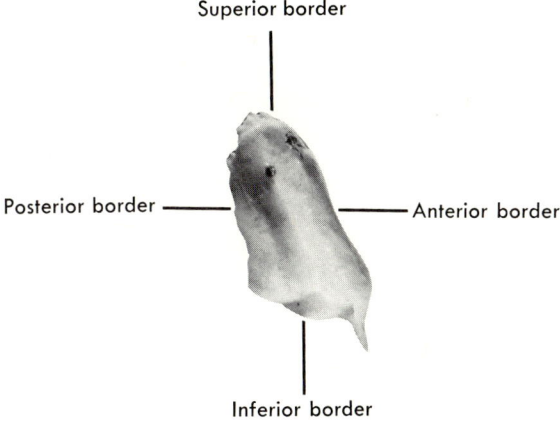

Fig. 4-60. Right lacrimal bone, lateral surface. The superior border articulates with the frontal bone. The anterior border articulates with the frontal process, maxilla. The posterior border articulates with the lamina papyracea. The posterior section of the inferior border articulates with the orbital plate; the anterior part is the descending process and articulates with the lacrimal process, inferior concha.

150 *Textbook of anatomy and physiology in radiologic technology*

Zygomatic (malar) bones. The zygomatic bone, commonly called the cheek bone, is both inferior and lateral to the outer canthus* of the eye. The zygomatic bone is surrounded by and, through its processes, articulates with the temporal bone, frontal bone, maxilla, and sphenoid bone (see Fig. 4-61). The zygomatic bone resembles a small quadrangle with four processes, one of which is quite prominent.

The large *frontosphenoidal process* articulates with the zygomatic process of the frontal bone to form the lateral border of the eye orbit. The *orbital process* projects both dorsally and medially from the orbital border of the zygomatic bone and forms a part of the orbital wall. The posterior margin of this process articulates with the great wing of the sphenoid bone and with the orbital surface of the maxilla. The *temporal process* extends dorsally on the lateral side to articulate with the zygomatic process of the temporal bone, thus completing the zygomatic arch. The *maxillary process* articulates with the zygomatic process of the corresponding maxilla.

Radiographic demonstration of the zygomatic bone is achieved by use of the 37-degree, facial-orbital (Waters) position and the oblique-facial position. The zygomatic arch is demonstrated in the tangential facial bone (May) position. When this structure is requested, both sides should be rayed for comparison.

Maxillae. The second largest bone of the facial group is the maxilla, which articulates with the opposite member in the midsagittal plane to form the upper

*The canthus is the angle at either end of the slit between the eyelids: the canthi are distinguished as outer, or temporal, and inner, or nasal.

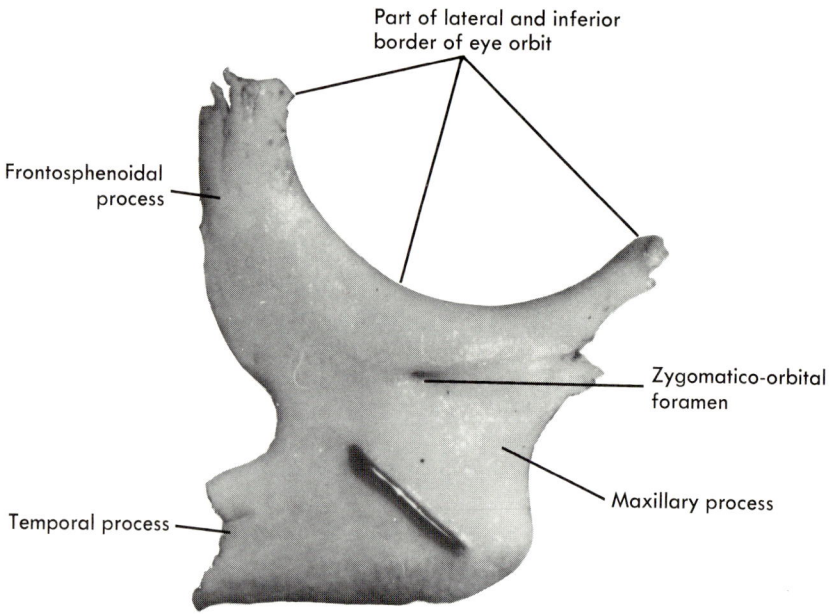

Fig. 4-61. Right zygoma, lateral surface.

jaw. The maxilla has an irregularly shaped body, most of which is hollow, and supports four processes. Each maxilla articulates with and occupies the space bordering on the nasal and lacrimal bones, zygomatic bone, inferior concha, vomer, palatine bone, ethmoid bone, frontal bone, and opposite maxilla.

The body of the maxilla contains the nasal antrum (maxillary sinus, or antrum of Highmore) lateral to the median plane and superior to the bicuspid tooth when visualized in an anterior radiograph. The *maxillary sinus* borders on the oral, nasal, and orbital cavities.

Each maxilla forms one-half of the *anterior nasal spine,* which is situated

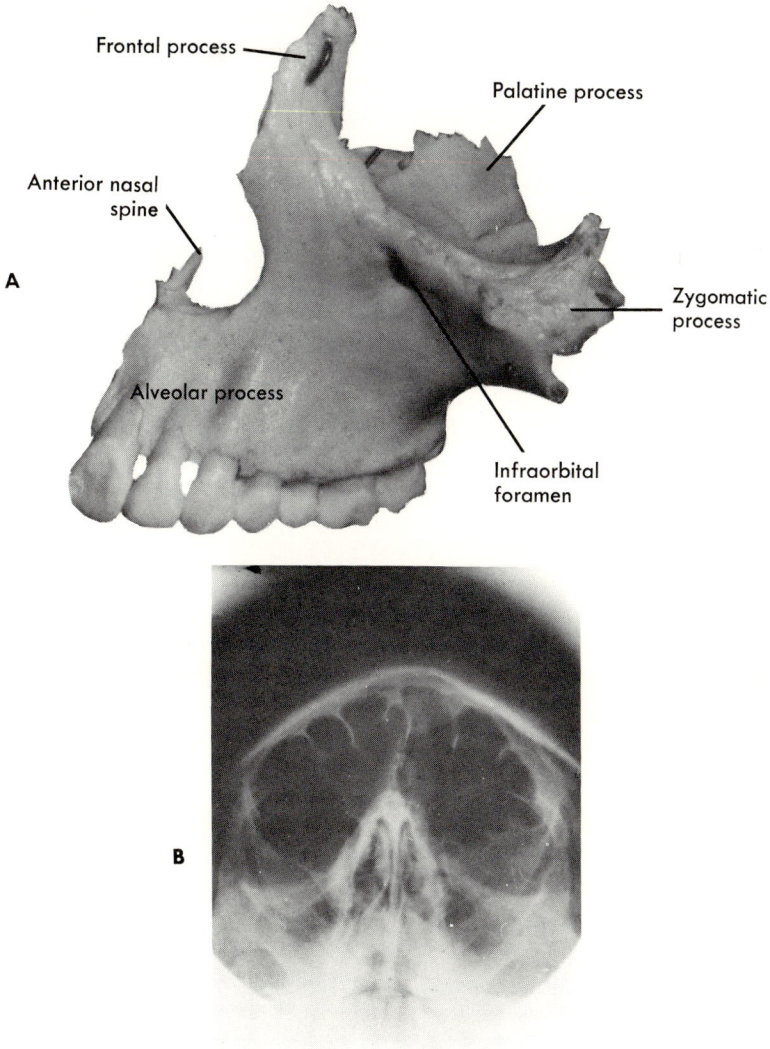

Fig. 4-62. A, Left maxilla, superoanterior aspect. **B,** Maxillae, Waters' view for orbital ridges.

anterior and inferior to the nasal septum. The *frontal process* extends superiorly between the nasal and lacrimal bones to articulate superiorly with the frontal bone. The *zygomatic process* faces and articulates posteriorly with the zygomatic bone. The strong *palatine process* is horizontal and extends medially to articulate with the palatine process of the opposite maxilla and form approximately the anterior two-thirds of the *hard palate*. Posteriorly each palatine process articulates with the horizontal part of the corresponding palatine bone; these articulations and bony parts form the remaining one-third of the bony hard palate. The hard palate separates the nasal and oral cavities. Failure of these bones and parts to form a union in the midline of the roof of the oral cavity results in a *cleft palate*. The *alveolar process* is the ridge from which the teeth erupt in the upper jaw. The *alveolar arch* is composed of the two alveolar processes (see Fig. 4-62, A). The upper teeth extend downward to mesh with the lower teeth; closing the jaws is called *occlusion*. The plane between the closed teeth is the plane of occlusion.

The maxilla is demonstrated radiographically by use of the oblique facial and lateral skull positions. The nasal antra are demonstrated by use of the 37-degree facial (Waters) positions (both open mouth and closed mouth), while the hard palate, palatine processes, and the alveolar arch are demonstrated by use of the axial skull position.

Palatine bones. Each palatine bone forms one-half of the posterior third of the hard palate and, for the most part, is superior to the hard palate. The palatine bone occupies the space bordering on the palatine process of the maxilla, the opposite palatine bone, the ethmoid bone, the vomer, and the inferior concha and articulates with these bones. The nasal and oral cavities and the eye orbit border on the palatine bone.

A palatine bone consists of five parts; the *pyramidal, orbital,* and *sphenoidal processes* and the *horizontal* and *vertical parts*. The horizontal part articulates midsagittally with the corresponding part of the opposite palatine bone and anteriorly with the palatine process of the maxilla. The horizontal part is inferior on the palatine bone and extends medially from the main structure. The vertical part of the palatine bone projects superiorly from the horizontal part and includes three processes (see Fig. 4-63).

Radiographic demonstration of the palatine bone is achieved by use of the axial skull position for the horizontal part and either the anterior or posterior skull position for the vertical part. The horizontal part is always included in radiographs of the hard palate.

Inferior conchae (nasal turbinates). The inferior nasal concha derives its name from its resemblance to a conch shell; it extends medially into the lower part of the nasal cavity, forming an incomplete shelf. This bone is completely osseous whereas the middle and superior conchae of the ethmoid bone are at least partially cartilaginous. The inferior conchae are lateral and inferior in the nasal cavity, one concha being situated in each lateral wall. The inferior concha articulates with the ethmoid, maxilla, lacrimal, and palatine bones (see Fig. 4-64).

The skeletal system   153

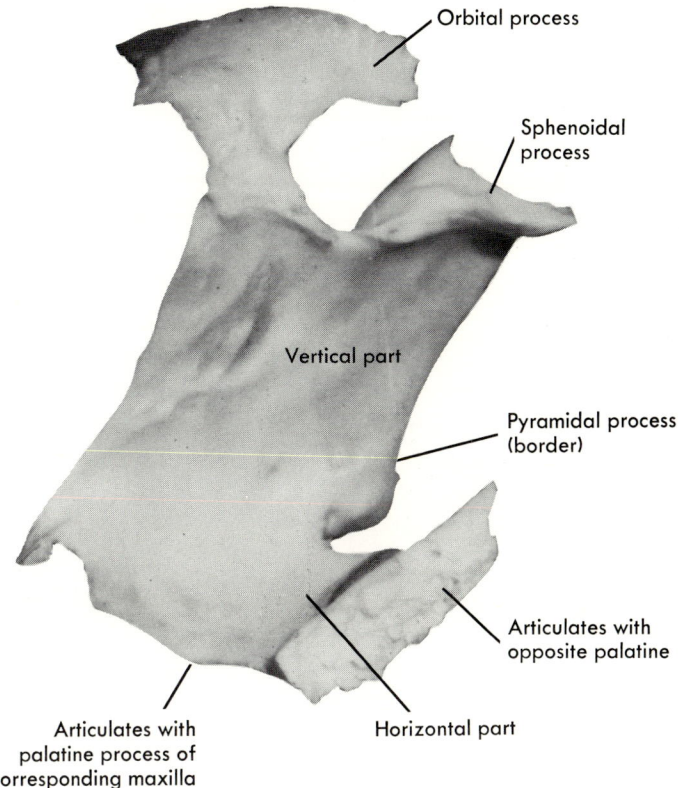

Fig. 4-63. Right palatine, anteromedial aspect.

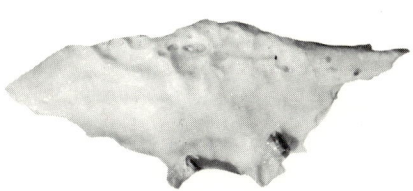

Fig. 4-64. Left inferior nasal concha, anteromedial surface.

The inferior concha is best demonstrated radiographically in the anterior skull position; it is also visualized in the 37-degree facial position.

Vomer. The vomer contributes to the formation of the nasal septum and is the most posterior part of the septum. The vomer is situated near the center of the base of the skull and occupies the space surrounded by the ethmoid bone, both maxillae, and both palatine bones and articulates with each of these bones.

Viewed laterally the vomer resembles a blunt arrowhead with the blunted end directed toward the *alveolar point,* a point on the alveolar process midway

154 *Textbook of anatomy and physiology in radiologic technology*

between the median upper incisor teeth. The *choanae* (posterior nares) are separated by the posterior margin of the vomer (see Fig. 4-65).

Although the vomer is difficult to visualize, radiographic demonstration of it is achieved by use of the lateral skull, submentovertex, and verticosubmental positions. An anterior skull position will visualize deviations in the nasal septum that may result from pathology, including trauma, in the vomer.

Mandible. The mandible is the largest of the facial bones and is the only moveable member of this group. This bone resembles a horseshoe in shape and comprises the entire lower jaw.

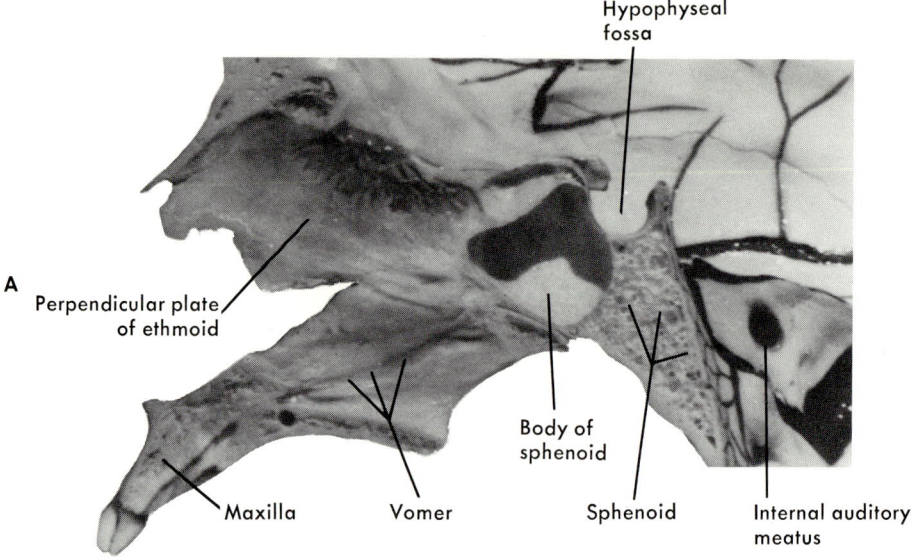

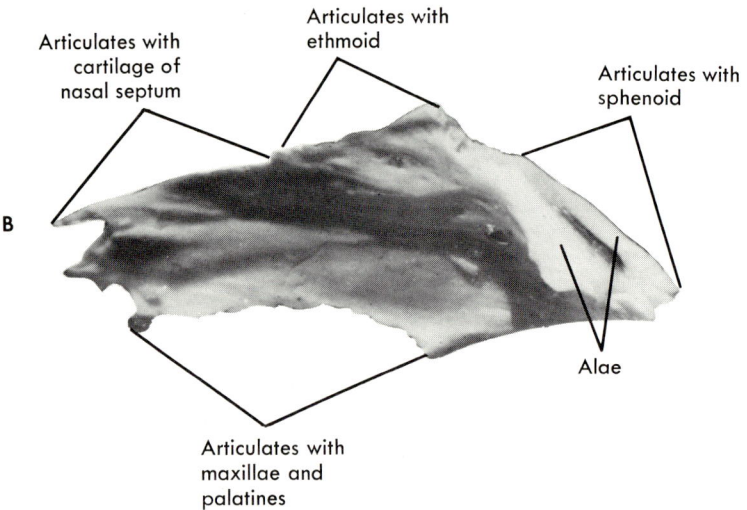

Fig. 4-65. A, Vomer in situ. **B,** Vomer, supero-oblique aspect.

Ossification of the mandible begins in two centers in intrauterine life, and the mandibular body is in two parts at birth. In the first few months of extrauterine life the two parts of the body fuse anteriorly in the midsagittal plane to form the *symphysis mandibulus*.

Superiorly in each side of the body is the *alveolar process* (ridge) from which the teeth of the lower jaw erupt. From the superior ends of the body the bone turns upward and slightly outward, forming the *ramus* on each side. The inferior and posterior margins of the body and ramus form the *mandibular angle*. The *condyloid* (condylar) *process* extends posteriorly and superiorly from

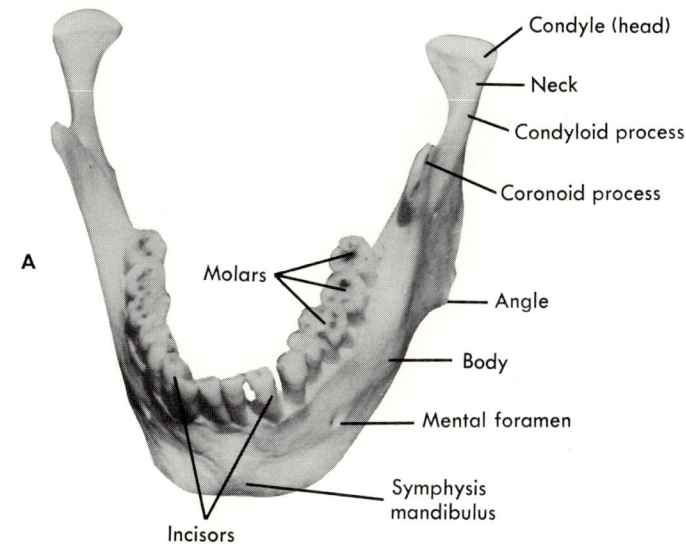

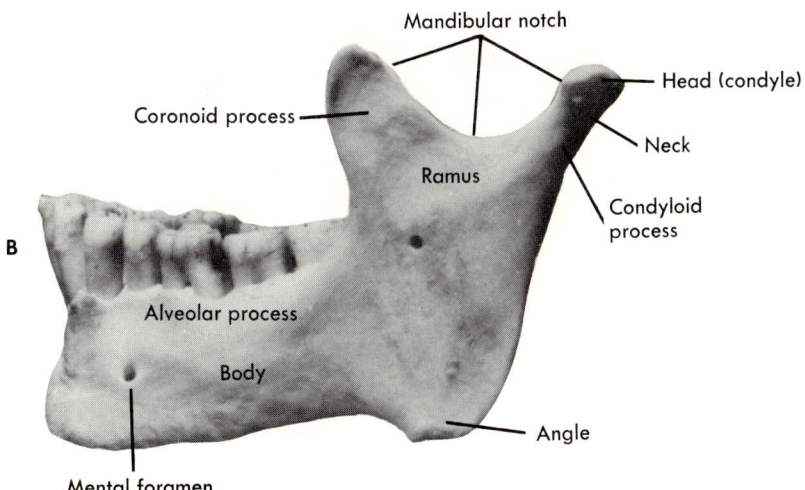

Fig. 4-66. Mandible. **A,** Supero-oblique aspect. **B,** Lateral surface.

the ramus and includes the *condylar neck,* which supports the *head* (condyle). The head articulates in the mandibular fossa of the temporal bone, forming the *temporomandibular joint* (TMJ). There is one head for each ramus. The *coronoid process* extends anteriorly and superiorly from the ramus and has no articulation. There is one coronoid process for each ramus. The deep, semilunar depression between the condyloid and coronoid processes is the *mandibular notch* (see Fig. 4-66).

The body of the mandible is demonstrated radiographically by use of the lateral facial bones position (see Fig. 4-67, A). In the body of the mandible, and on each side, the mental foramina are inferior to the second bicuspid tooth. Each mental foramen communicates with the corresponding mandibular canal.

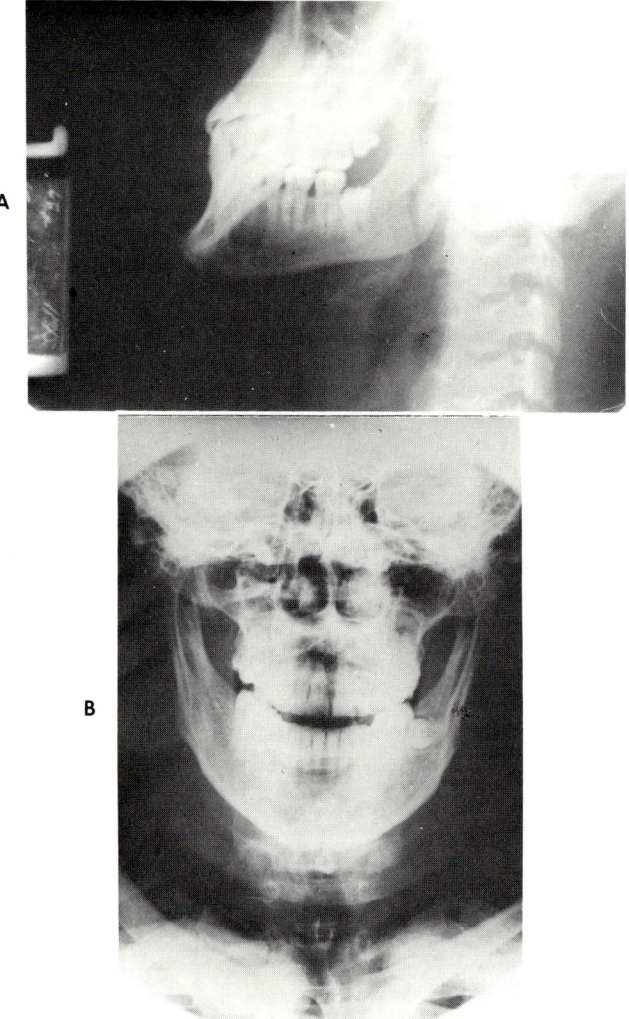

Fig. 4-67. A, Left mandible, lateral radiograph. **B,** Mandibular rami, posterior radiograph.

The mandibular canals extend downward in the ramus and forward in the body from the mandibular foramina. The inferior alveolar vessels and nerve (on each side) enter the mandible through the mandibular foramina, situated on the medial surfaces of the rami.

The symphysis mandibulus is demonstrated by use of the submentovertical position. The rami are demonstrated in the anterior position (see Fig. 4-67, B). The condyles are demonstrated in the anterior position if the tube is angled to direct the central ray 12 degrees cephalad. It is sometimes difficult to visualize pathology in the condylar neck and head from routine positions; various modifications of an antero-oblique position of the condylar neck and head will usually visualize pathology in this region (pathology that may be suspected but not demonstrated). The temporomandibular joints are visualized in the lateral skull position with the central ray directed 15 degrees caudad. These views are exposed routinely, one with the mouth open and one with the mouth closed (Fig. 4-68). The use of dental films, both intraorally and extraorally, is often useful.

HYOID BONE. Although the hyoid bone is not one of the twenty-two skull bones, it is best described in its relation to these bones in this section of the text.

The *horseshoe-shaped* hyoid bone is suspended (attached) by the stylohyoid ligaments from the tips of the temporal styloid processes. It is superior to the thyroid cartilage and both anterior and inferior to the base of the tongue. The extrinsic muscles of the tongue that are attached to the hyoid are the genioglossus, hypoglossus, and chondroglossus (see Fig. 4-69, A).

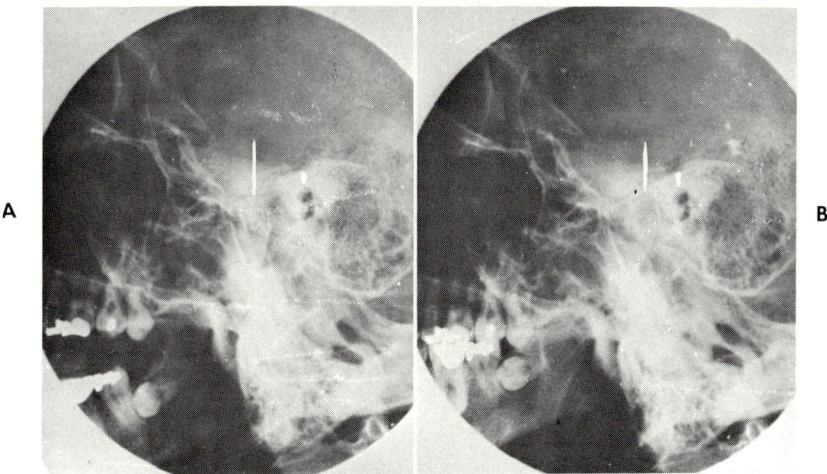

Fig. 4-68. Radiograph, temporomandibular joint. **A,** Mouth open. The white dot is superior to the external auditory meatus (EAM); the white line points to the mandibular condyle. The lower jaw is opened. Note the space between the condyle and the EAM and compare with the same structures in **B**. **B,** Mouth closed. The white dot is superior to the EAM; the white line points to the mandibular condyle. The lower jaw is closed.

158 *Textbook of anatomy and physiology in radiologic technology*

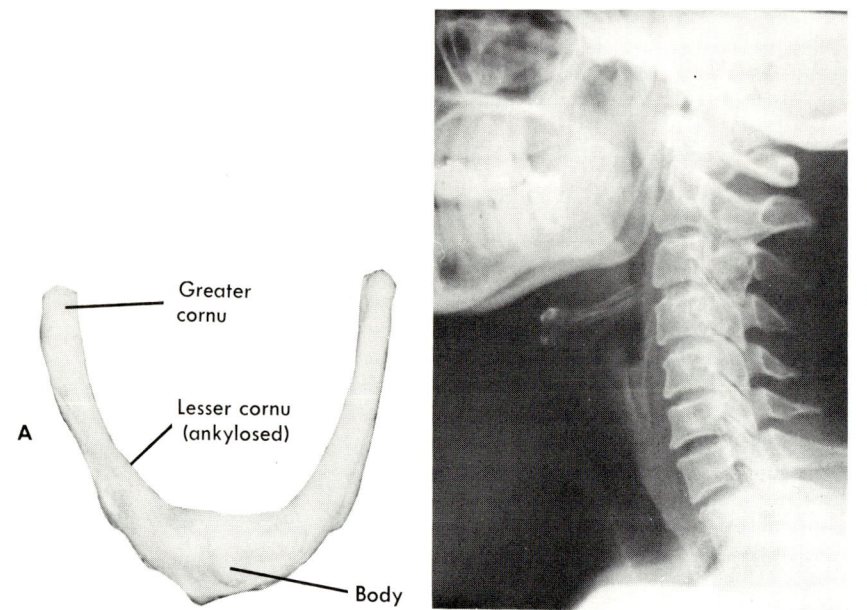

Fig. 4-69. Hyoid bone. **A**, Supero-oblique aspect. **B**, Lateral cervical spine view.

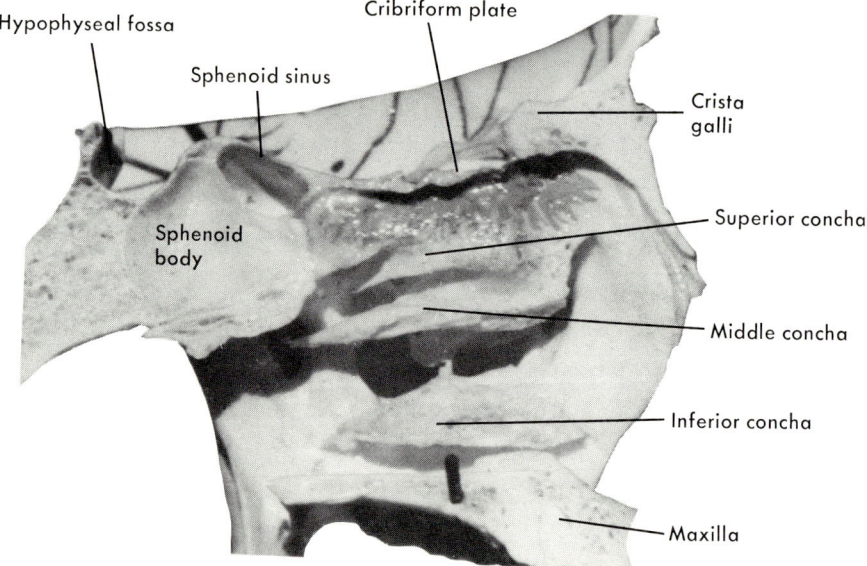

Fig. 4-70. Nasal cavity, median aspect.

The lateral cervical spine position is employed to visualize the hyoid bone, as demonstrated in Fig. 4-69, *B* (see also Fig. 5-2).

AIR SINUSES. The air sinuses of the skull include all the bony air spaces. These are the sinuses immediately adjacent to the nasal cavity (paranasal sinuses), the mastoid air cells of the temporal bones, and the nasal cavity. These sinuses and cavities may become infected and/or inflamed, and the infection may spread from one air space to another. The total function of the air spaces is not fully established; however, two very important contributions to homeostasis are lending quality and inflection to the voice and permitting less weight of the head.

The paranasal sinuses (frontal, ethmoidal, sphenoidal, and maxillary) form a border of air spaces around the nasal cavity. Each of these eight sinuses has direct communication with the nasal cavity. The paranasal sinuses are the subject of many radiographic requests.

Frontal sinuses. The frontal sinuses are in the frontal bone, posterior to the glabella and medial between the superciliary arches in the diploic layer. The two sinuses are often asymmetric, and frequently one sinus may be either malformed or missing entirely. A septum usually separates the rather large space into two sinuses; the septum is usually in the midsagittal plane. Each of the frontal sinuses communicates with the nasal cavity via the anterior part of the corresponding middle meatus of the nose (see Figs. 4-70 and 4-71).

The frontal-ethmoidal (modified Caldwell) position is best employed to demonstrate these sinuses; they are also demonstrated by use of the lateral skull position.

Ethmoidal sinuses. Each of the ethmoidal sinuses consists of anterior, middle,

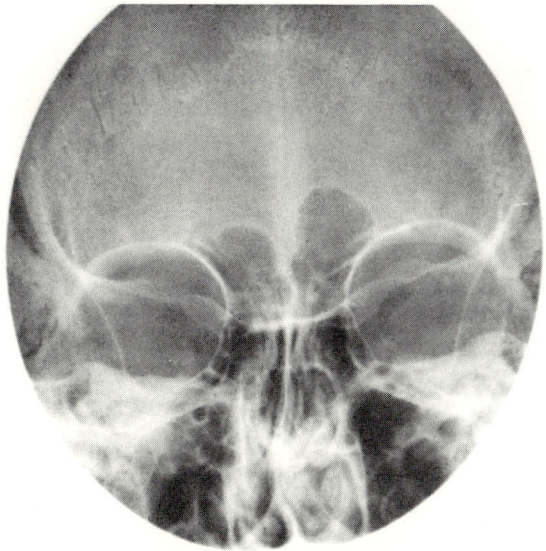

Fig. 4-71. Frontal and ethmoidal sinuses, frontal-ethmoidal (Caldwell) view.

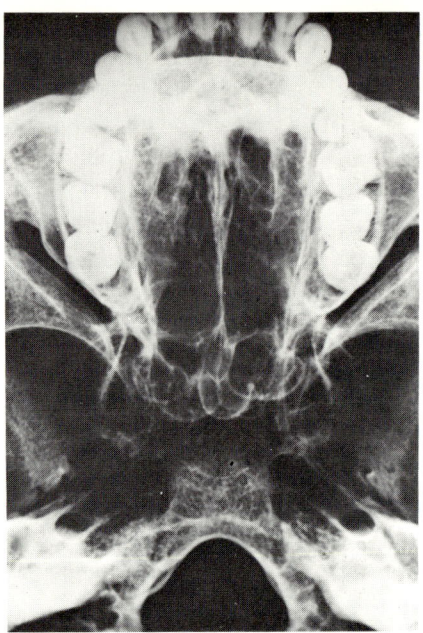

Fig. 4-72. Dry skull, ethmoidal and sphenoidal sinuses, verticosubmental view. (Compare with Fig. 4-58.)

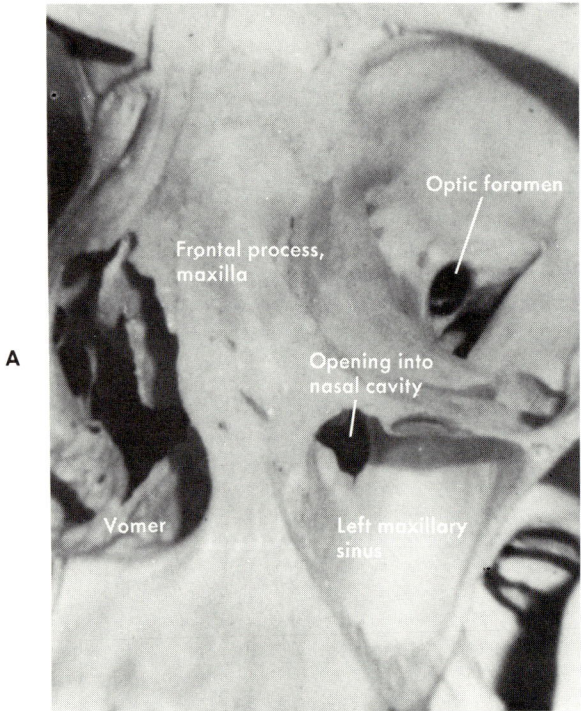

Fig. 4-73. A, Left maxillary sinus, cutaway aspect. **B,** Nasal antra, Waters' view. **C,** Paranasal sinuses, lateral view.

The skeletal system 161

and posterior parts. Each ethmoidal sinus is situated in the ethmoid bone, superior and lateral to the nasal septum and anterior and superior to the sphenoidal sinuses. The ethmoidal sinus communicates with the nasal cavity through the middle and superior meati of the nose. In many persons there are direct openings between the ethmoid cells and the sphenoidal sinus.

The 45-degree anterior facial bones (Rhese) and lateral skull positions are those most commonly employed to demonstrate the ethmoidal sinuses (air cells), although the submentovertex and verticosubmental positions are also used. The frontal-ethmoidal position is frequently employed to demonstrate certain aspects of these air cells (see Fig. 4-72).

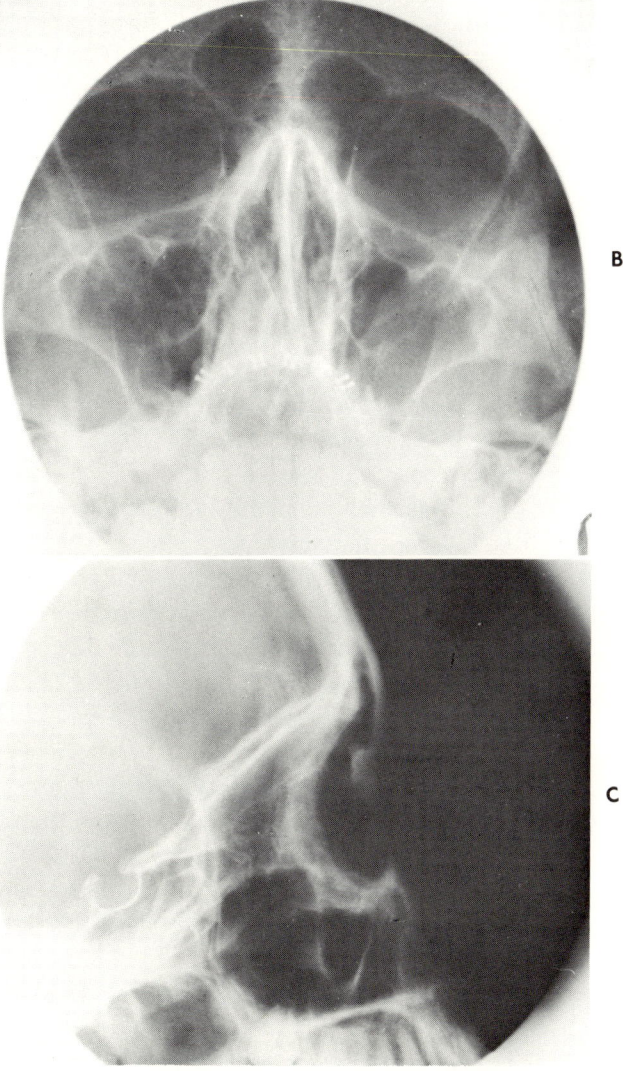

Fig. 4-73, cont'd. For legend see opposite page.

Sphenoidal sinuses. The sphenoidal sinuses are situated in the body of the sphenoid bone, anterior and inferior to the sella turcica. The two sinuses are separated in the midsagittal plane by a septum and are frequently asymmetric. Each sinus communicates with the nasal cavity through an aperture in the upper part of its anterior wall (see Fig. 4-72).

The 45-degree anterior facial bones and lateral skull positions are most commonly used to demonstrate these sinuses. The verticosubmental, submentovertex, and frontal-ethmoidal positions are used also to demonstrate the sphenoidal sinuses.

Maxillary sinuses. Each maxillary sinus (nasal antrum) is situated in the body of the corresponding maxilla. The paired sinuses are bilateral to the nasal cavity and inferior to the orbital floors of the eye cavities. These are the largest of the paranasal sinuses. Each sinus communicates with the nasal cavity through an opening (or openings) in the anterosuperior part of the nasal wall into the middle meatus; sometimes there are openings through this wall into the inferior meatus. One or more molar roots may be visualized in radiographs of these sinuses, and molar roots may even penetrate into the sinuses. A small part of the inferior portion of the maxillary sinus should be demonstrated in all dental radiographs of the upper jaw (see Fig. 4-73, *A*).

The 37-degree orbital facial bones (Waters), lateral skull, and oblique maxilla positions are used to demonstrate the maxillary sinuses (see Fig. 4-73, *B* and *C*). Routine positioning technique projects the petrous ridges of the temporal bones below the maxillary sinuses.

Mastoid cells. The mastoid sinuses (air cells) are contained in the mastoid processes of the temporal bones; they are *not* included in the group of paranasal sinuses. Each sinus consists of several individual cells scattered throughout this part of the temporal bone and communicates *indirectly* with the nasal cavity via the auditory tube into the pharynx.

These cells are demonstrated radiographically in several positions for this portion of the temporal bone. Among these positions are the posterior position of the skull for the mastoid tips, the 15-degree–15-degree lateral (Law) position, the antero-oblique (Stenvers) position, the occipital (Towne) position, and the axial (Mayer) position (and its variation, the Owens position) (see Fig. 4-74).

Nasal cavity. The nasal cavity is superior to the roof of the mouth between the maxillary sinuses and inferior to the ethmoid and sphenoid bones. The cavity is bounded anteriorly by the nose and posteriorly by the pharynx into which the cavity opens via the choanae.

The nasal septum divides the nasal cavity into two parts. The septum is situated between the oral cavity and the cranial floor. The following bones and parts contribute to the formation of the nasal septum: nasal crests and frontal spine, perpendicular plate of the ethmoid bone, sphenoid rostrum and vomer, and maxillary crests.

The nasal bones, frontal spine, cribriform plate of the ethmoid bone, body of the sphenoid bone, and palatine-sphenoidal processes form the roof of the

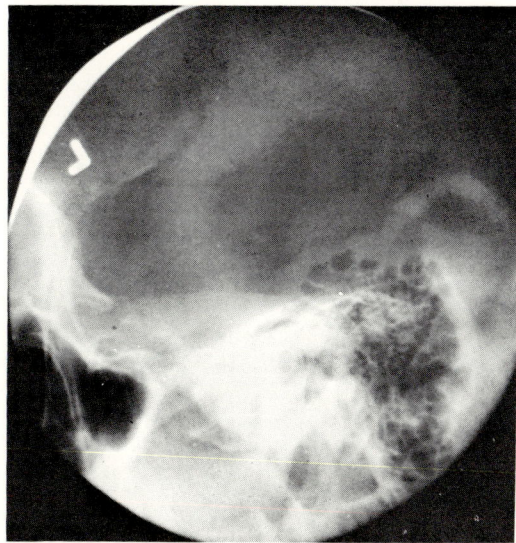

Fig. 4-74. Mastoid air cells, lateral (Law) view. (Courtesy Dr. J. M. Hilton.)

nasal cavity. The palatine processes of the maxillae and one-half of the horizontal parts of each palatine bone form the floor of the nasal cavity. The medial wall of the nasal cavity is the nasal septum. The frontal process of the maxilla, the lacrimal, the ethmoid bone, the body of the maxilla, the inferior (nasal) concha, and the medial pterygoid plate of the sphenoid bone form the lateral wall of the nasal cavity.

The nasal cavity opens to the outside through the *anterior nares* (bony parts of the nose) and into the pharynx via the choanae. The cavity communicates directly with each of the paranasal sinuses and indirectly with the mastoid air cells. It is demonstrated radiographically by use of the routine sinus position and the posterior skull position.

EYE ORBIT. Parts of seven bones form the orbit of the eye, and the orbital wall consists of six parts (divisions); within these parts are found nine openings, one of which is the orbital base.

The *orbital base* is quadrilaterally shaped: superiorly by the supraorbital arch of the frontal bone (containing the *supraorbital notch* or *foramen*); inferiorly by the zygomatic bone and maxilla; laterally by the zygomatic bone and zygomatic process of the frontal bone; medially by the frontal bone and the frontal process of the maxilla.

The orbital *apex* is posterior in the orbit and conically shaped. It is contained totally in the sphenoid and contains the *optic foramen.*

The orbital *floor* slopes upward and lateralward, being formed by the orbital surface of the maxilla, orbital process of the zygoma, and orbital process of the palatine. The openings of the floor include the *upper opening of the nasolacrimal canal* and the *infraorbital groove and canal.*

164 *Textbook of anatomy and physiology in radiologic technology*

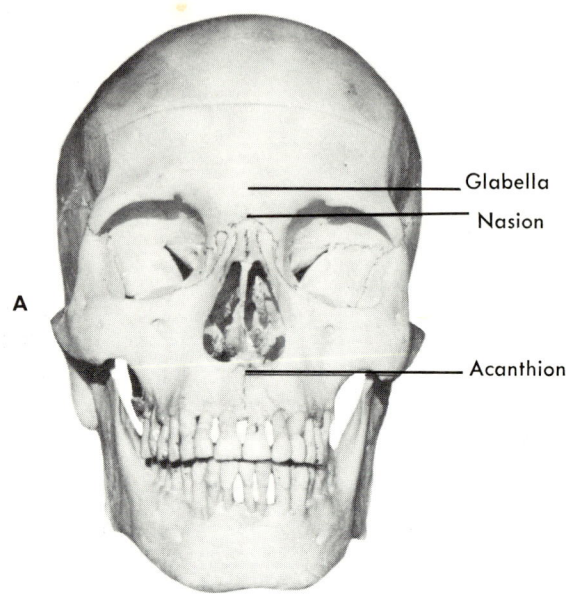

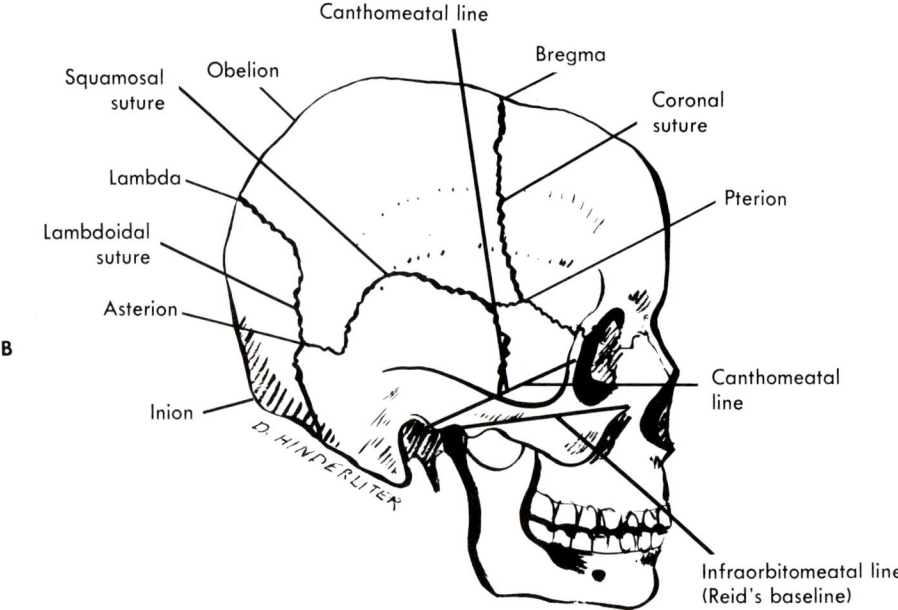

Fig. 4-75. A, Anterior landmarks of skull. **B,** Lateral landmarks of skull.

The *lateral wall* extends medialward and forward, being formed by the orbital process of the zygoma and the orbital surface of the great wing of the sphenoid. The *superior orbital fissure* is between the lateral wall and roof, and the *inferior orbital fissure* is between the lateral wall and floor.

The orbital *roof* is concave, and extends downward and forward, being formed by the orbital plate of the frontal and the small wing of the sphenoid.

The *medial wall* extends nearly vertically, being formed by the frontal process of the maxilla, the lacrimal, the lamina papyracea of the ethmoid, and the body of the sphenoid. The frontoethmoidal suture contains the *anterior* and *posterior ethmoidal foramina*.

Radiographic interest of the skull

Various points and landmarks on the skull are useful in positioning and/or localizing certain specific structures (see Fig. 4-75). These landmarks include the following structures:

acanthion A point at the base of the anterior nasal spine.
asterion A point located at the junction of the lambdoidal and squamosal sutures.
bregma The point of junction of the sagittal and coronal sutures.
canthomeatal line A line between the external auditory meatus (E.A.M.) and other outer canthus of the eye.
canthus Either inner or outer, and formed by the junction of the eyelids.
glabella A point midway between the superciliary arches.
infraorbitomeatal line (Reid's, anatomical, or anthropological baseline) A line between the infraorbital ridge and the E.A.M., and extending to the center of the occiput.
inion The midpoint of the external occipital protuberance.
lambda The point of junction of the sagittal and lambdoidal sutures.
nasion The median point of the nasofrontal suture.
obelion A point in the midsagittal suture formed where a line between the parietal foramina crosses the suture.

Table 4-7. Articulations

Class	Movement	Example	Location
Fibrous	None or limited	Suture	Calvarium
		Syndesmosis	Tibiofibular articulation
		Gomphosis	Tooth socket
Cartilaginous	Limited	Hyaline cartilage joint	Epiphysis
		Fibrocartilaginous joint	Vertebrae and symphysis pubis
Synovial	Freely moving	Plane joint	Wrist
		Hinge joint	Interphalangeal
		Pivot joint	Proximal radio-ulnar
		Condyloid joint	Radiocarpal
		Saddle joint	Carpometacarpal of thumb
		Ball-and-socket joint	Shoulder

166 *Textbook of anatomy and physiology in radiologic technology*

pterion A point at the junction of the coronal and squamosal sutures.

sella turcica In a lateral skull, at the midpoint of the infraorbitomeatal line.

Articulations

"Articulation" and "joint" are anatomic terms referring to a connection in the skeleton between two or more of its rigid components, either bony or cartilaginous. Specifically, *an articulation (joint) is the place of union or junction between two or more bones of the skeleton.* Joints are classified according to the nature of the tissue or tissues effecting the bony union. The three classifications are fibrous, cartilaginous, and synovial (see Table 4-7).

Fibrous joints. Fibrous tissue connects the bones that form fibrous joints. The shapes of the adjoining surfaces and the nature of the fibrous tissues permit very limited movement, if any movement does occur (see Fig. 4-76).

SUTURES. Sutures occur between the bones that form the calvarium of the skull. The toothlike margins of one bone fit closely into correspondingly shaped margins of adjacent bones. Several layers of fibrous tissue connect the bones across each suture. The sutures continue throughout life, and some are demonstrable radiographically: e.g., the sagittal, coronal, lambdoidal, and often the squamosal.

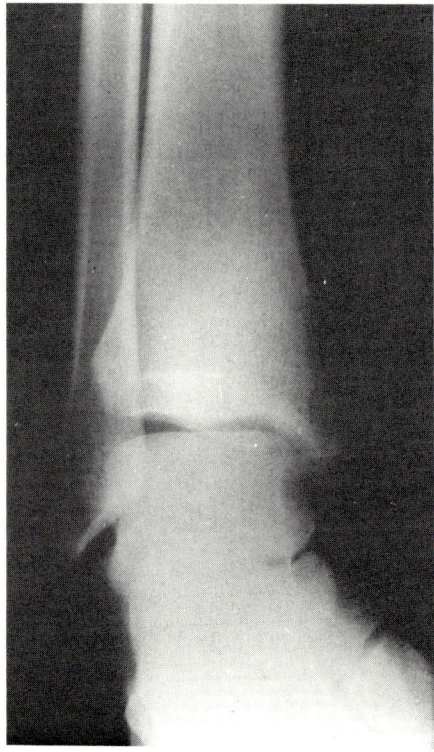

Fig. 4-76. Fibrous joint, right tibiofibular.

The sutures are not solidly united during intrauterine life. The mechanisms of growth of the calvarial bones at the sutures is not fully understood; however, increased cranial capacity does occur to accommodate increased brain mass.

SYNDESMOSES. The quantity of intervening fibrous connective tissue in syndesmoses is much greater than in sutures. Also, few, if any, of the bones that form these joints have serrated margins. The tympanostapedial and tibiofibular articulations are classed as syndesmoses.

GOMPHOSES. The teeth fit tightly in the alveolar sockets of the maxillae and the mandible and are held firmly in position by fibrous tissues.

Cartilaginous joints. The classification of cartilaginous joints includes two subclassifications: hyaline cartilage joints and fibrocartilage joints. There is limited motion in these joints (see Fig. 4-77).

HYALINE CARTILAGE JOINTS. Hyaline cartilage joints form a temporary union found, for the most part, as the growth area for long bones between the centers of ossification in the diaphysis and epiphyses. The cartilage is replaced by osseous tissue when growth ceases. The epiphyseal plates (metaphyses) exemplify hyaline cartilage joints.

FIBROCARTILAGINOUS JOINTS. Amphiarthroses and fibrocartilaginous joints are the same. The bony parts of such joints unite by means of fibrocartilage, often in the form of a disc. Thin layers of hyaline cartilage usually separate the bone ends from the fibrocartilage. Fibrocartilaginous joints are exemplified in the intervertebral spaces and in the symphysis pubis.

Synovial joints. The skeletal elements of synovial (diarthrodial) joints are bound together by capsular ligaments. In addition to the capsular ligaments,

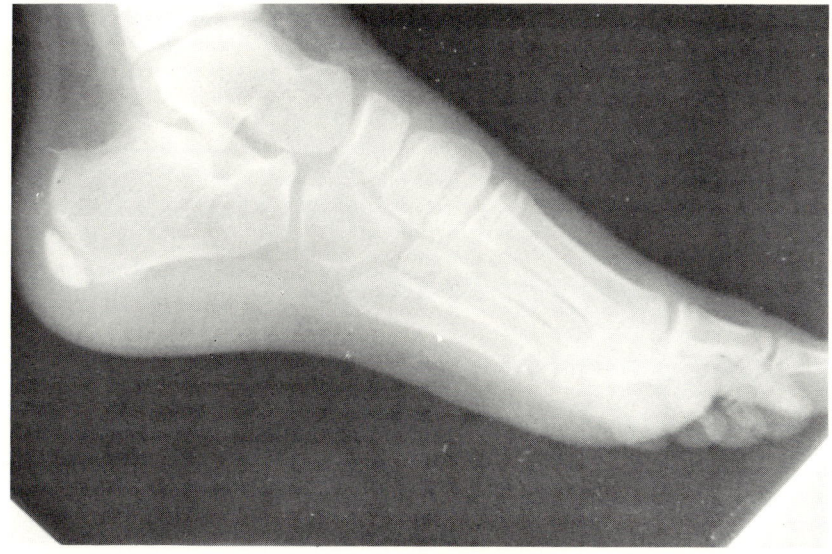

Fig. 4-77. Cartilaginous joint.

168 *Textbook of anatomy and physiology in radiologic technology*

specific ligaments extend from certain bony prominences, as in the knee; collateral ligaments are thickened extensions of and are continuous with the capsular ligament. However, collateral ligaments are much thicker and stronger than the capsular ligament. The capsular ligament completely encloses the joint cavity. An outer layer of fibrous connective tissue and an inner, specialized layer *(synovial membrane)* comprise the capsular ligament. The synovial membrane covers the bony parts within the capsule but not the articular cartilage. The ar-

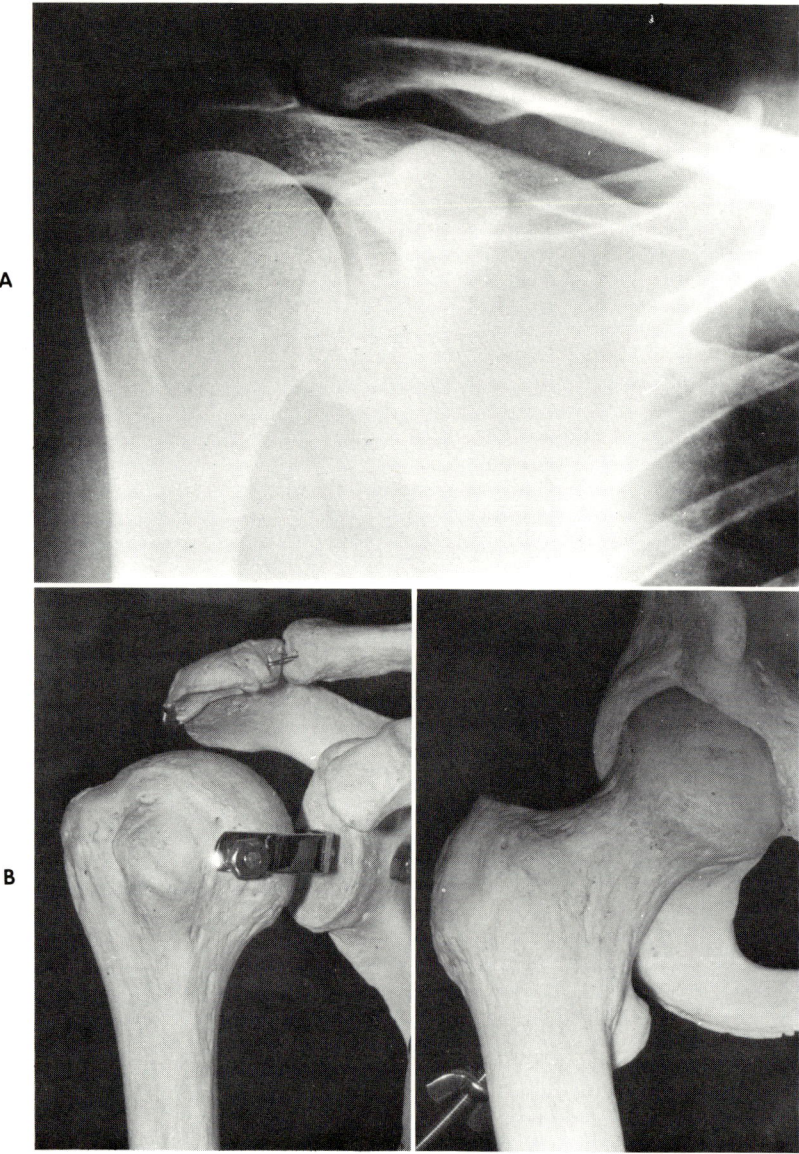

Fig. 4-78. **A,** Synovial joint. **B,** Right shoulder and hip, articulated skeleton.

ticular surfaces of the bones involved in the joint are covered with hyaline cartilage (see Fig. 4-78).

Synovia is a viscid fluid containing synovin, or mucin, and a small proportion of mineral salts. The fluid is transparent and alkaline, resembling the white of an egg. Synovia is secreted by the synovial membrane and constitutes the fluid substance of joint cavities, bursae, and tendon sheaths. A common misstatement is to speak of "losing the fluid from a joint." Such a statement is erroneous since a normal knee contains only a small quantity of fluid. Following injury or inflammation of a synovial joint, the synovial membrane secretes an excessive quantity of fluid, which is thin and more aqueous than usual. In cases of rheumatoid arthritis and gout the fluid develops quantities of crystals, fibrin, and other debris.

The radiographic examination of joints is named "pneumoarthrography." During this examination a part or all of the synovia is removed and replaced with air or oxygen. Oxygen is absorbed quickly into the surrounding tissues, whereas air is not. The bony parts of the joint are then contrasted against air; in cases of ruptured cartilage, air is found in places formerly occupied by cartilage. This procedure results in excellent outlining of the synovial lining, especially when it is not smooth.* Radiopaque contrast media are sometimes used instead of air or oxygen, although such procedure precludes the term "pneumoarthrography."

*Well-performed pneumoarthrography of the knee joint demonstrates oxygen both above and below the semilunar cartilage; however, diagnosis from the procedure is not 100 percent certain. When a foreign object moves into the hinge joint, the joint will lock and prevent extension of the knee. Such is the case with a torn semilunar cartilage when the cartilage slips back into the hinge.

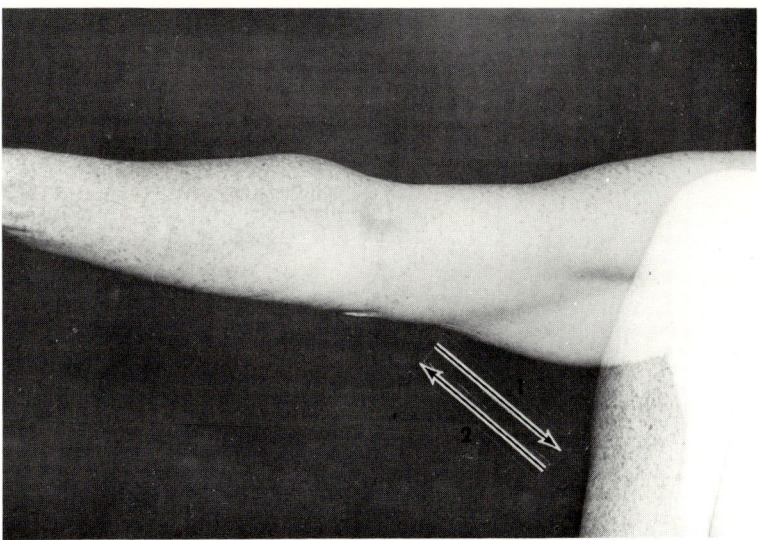

Fig. 4-79. Adduction-abduction. **1,** Adduction, movement of the arm toward the midsagittal plane; **2,** abduction, movement of the arm away from the midsagittal plane.

170 *Textbook of anatomy and physiology in radiologic technology*

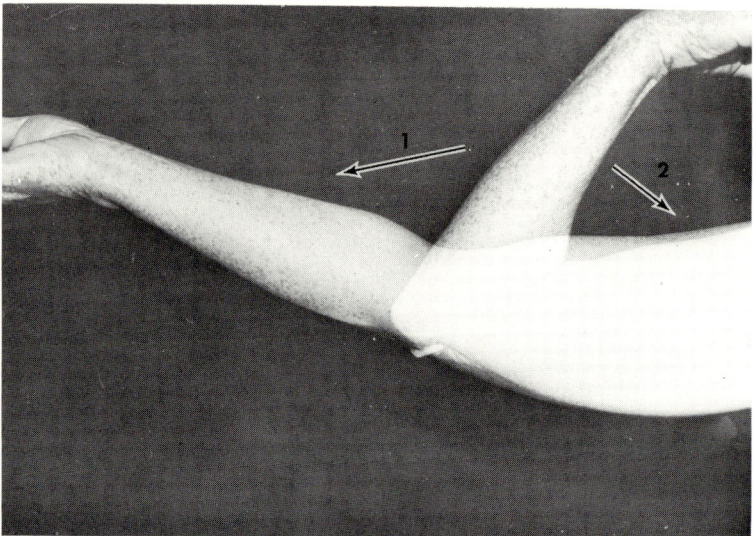

Fig. 4-80. Extension-flexion. **1,** Extension, increasing the angle of the joint (elbow); **2,** flexion, decreasing the angle of the joint (elbow).

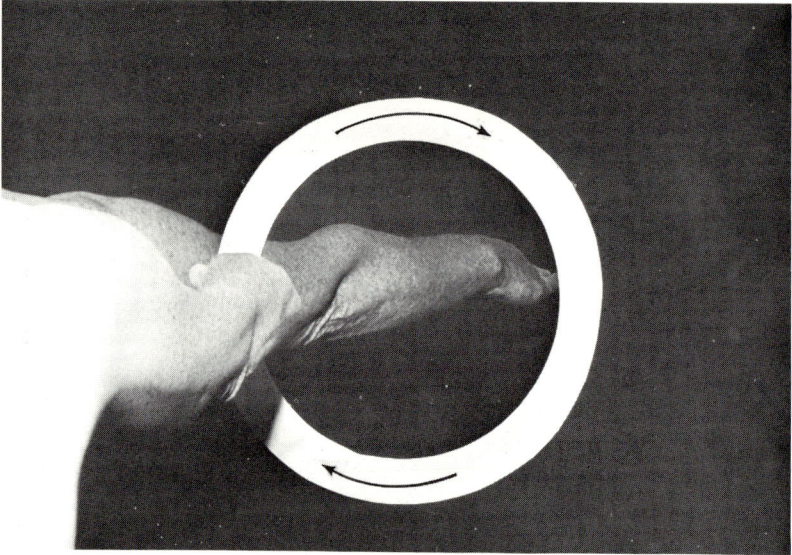

Fig. 4-81. Circumduction. Motion of the arm and hand in a circular (either clockwise or counterclockwise) direction constitutes the action of circumduction.

The skeletal system **171**

The physical nature of synovial joints permits far greater movement than in other joints. In synovial joints three types of movement occur:
1. Gliding (slipping)
2. Angular (about an axis in a transverse plane)
3. Rotary (about an axis in a vertical plane)

Particular articulations are described with each bone; however, articular movements are described in the following paragraphs.

Abduction and adduction are movements of body parts in coronal (frontal) planes. Abduction of a body part is the withdrawal of that part from the axis or midsagittal plane of the body. Adduction of a body part is the drawing of that part toward the midsagittal plane (see Fig. 4-79).

Extension and flexion are movements of body parts in sagittal planes. Extension of a body part is the movement by which the two ends of any part are pulled away from each other. Extension can be described as the movement that brings the members of a limb into or toward a straight line. Flexion is the act of bending or the condition of being bent. In the movement of flexion, the angle between two bones becomes more acute (see Fig. 4-80).

Circumduction is the active or passive circular movement of a limb; circumduction means to draw around (see Fig. 4-81).

Gliding (slipping) refers to a smooth motion and occurs in a very limited amount in any one site. The cumulative motion of gliding in several connected sites enables considerable movement.

Rotating (twisting) is the process of turning of the body or a limb about its

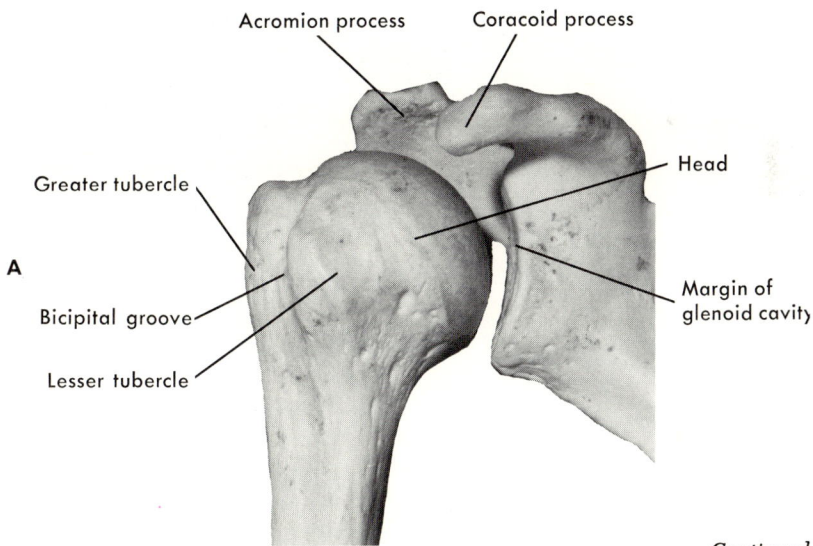

Continued.

Fig. 4-82. Right shoulder. **A,** Skeleton, external rotation. **B,** Skeleton, internal rotation. (Compare the positions of the greater and lesser tubercles in **A** with the same structures and positions in **B.**) **C,** External rotation view. **D,** Internal rotation view.

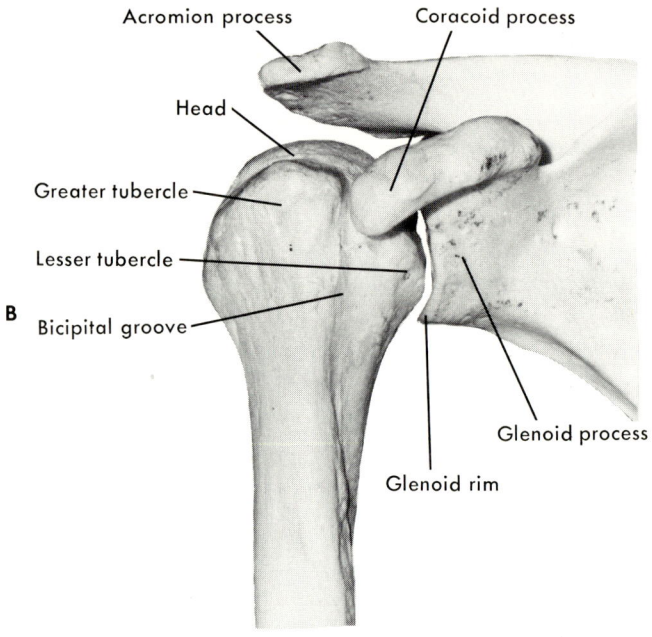

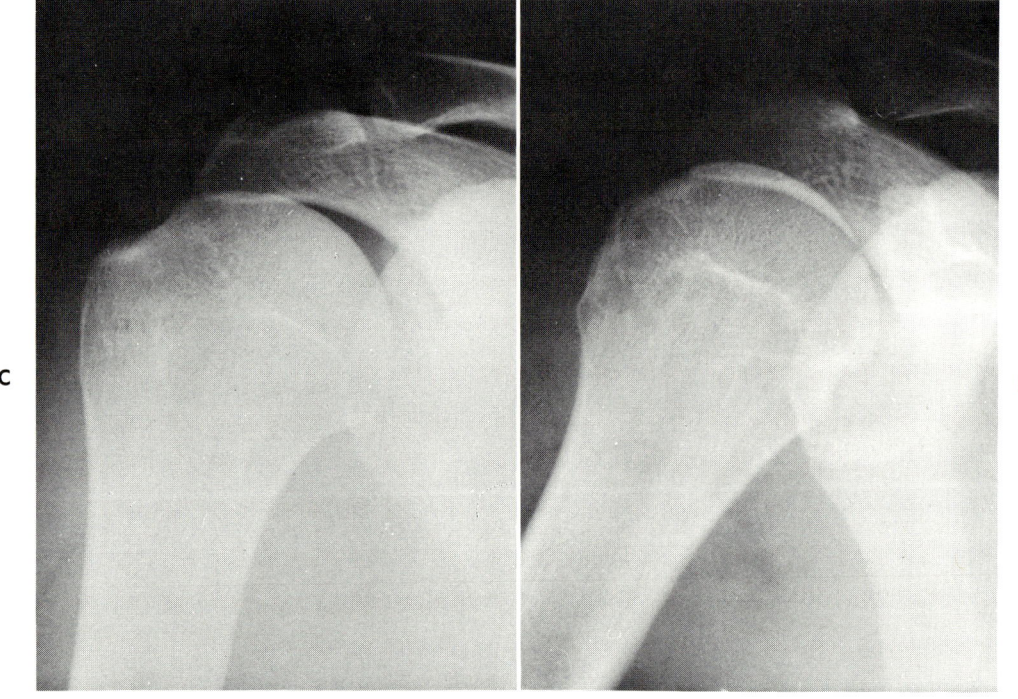

Fig. 4-82, cont'd. For legend see p. 171.

axis. Rotation occurs about the lengthwise axis; rotation of a limb may be either external or internal. External (lateral) rotation of the forearm places the forearm in supination; the volar surface turns away from the midsagittal plane. Internal (medial) rotation of the forearm places it in pronation; the volar surface turns toward the midsagittal plane (see Fig. 4-82).

PLANE JOINTS. The bones of plane joints have flat to slightly curved articular surfaces that permit gliding movements, as in the wrist.

HINGE JOINTS. Hinge (ginglymus) joints permit motion in one plane only. Examples of hinge joints are the joints of the elbows and knees and interphalangeal joints having the movements of extension and flexion.

PIVOT JOINTS. Pivot joints are rotary joints having a vertical axis. The articulation of the radial head in the radial notch of the ulna is exemplary of a pivot joint.

CONDYLOID JOINTS. A condyloid joint consists of an ovoid head of one bone that moves in an elliptical cavity of another bone, permitting all movements except axial rotation. The radiocarpal joint exemplifies condyloid joints.

SADDLE JOINTS. Saddle joints present two saddle-shaped surfaces at right angles to each other. The carpometacarpal articulation of the thumb is an example of a saddle joint.

BALL-AND-SOCKET JOINTS. Ball-and-socket joints are enarthroses. The recessed (cavity-shaped) socket of one bone receives the globular head of the other bone in the joint. Examples of the ball-and-socket joint are the shoulder and hip joints. The hip socket is quite deep, but the shoulder socket is rather shallow and approximately the diameter of a half-dollar coin; the remainder of the shoulder joint is a cartilaginous rim, the *glenoidal labrum.*

BONE PHYSIOLOGY
Tissues derived from the mesenchyme

The skeletal system consists of supporting tissues—cartilage and bone. Cartilage and bone derive from the mesenchyme, which is the embryonic connective tissue arising from the mesoderm.

Cartilage

Cartilage is a translucent, elastic tissue that comprises most of the skeleton of embryos and very young vertebrates. Cartilage appears during the fifth week of development in the human embryo.* Skeletal development evolves from the following four layers: *perichondrium, mesenchyme, precartilage,* and *cartilage.*

The perichondrium is the outer connective tissue sheath that covers the cartilage except on the articular surfaces of hyaline cartilage, *which covers the articulating ends of bones.* Blood vessels lie in this layer and supply the nutrition required for sustenance of the cartilage. Cartilage tissue is quite compact—so

---

*The term "fetus" applies to the new individual after the end of the second month of pregnancy in the first trimester.

compact as to prevent the presence of blood vessels. The avascular nature of cartilage explains why the semilunar cartilage in the knees of adults heals so poorly following fracture. Since there are no vessels, the cartilage heals poorly, if at all, and frequently the loosened piece of cartilage remains floating in the joint and can cause this hinge type of joint to *lock*. This is not true in juveniles in whom the cartilage grows rapidly. Most torn cartilages in children will heal; the reverse is true in adults, the dividing line being somewhere around 14 years of age. Very few injured knee cartilages require removal prior to this age.

The mesenchyme arises primarily from the primitive streak and secondarily from the *mesodermal somites*. The somites are the blocklike masses of mesodermal cells arranged along each side of the neural tube in the embryo. Each somite develops a cavity, the *myocele*. The medial and ventral walls of the somite form the sclerotome, which eventually gives rise to a vertebral body; the lateral walls form the dermatome; the dorsal wall forms the myotome (muscle plate). The cells of the mesenchyme transform into young cartilage cells, *chondrocytes*, and form the basic precartilage layer. The chondrocytes later deposit *matrix* and become buried by their own activity. *Matrix is the fibrous body of the cartilage, keeping the cartilage together as one tightly knit unit.*

Precartilage contains the young, immature cartilage cells that deposit the matrix, in which the encapsulated and mature chondrocytes later exist. As a result, cartilage contains these chondrocytes lying within the matrix.

Hyaline (glasslike) cartilage is the most widespread and the most typical of all types of cartilage. The other cartilage types are actually modifications of hyaline cartilage. In adult mammals, hyaline cartilage is found in the following three locations:

1. The respiratory passages
2. The ventral (sternal) ends of the ribs
3. The surfaces of bones within the joints (Here the average thickness of cartilage is one-eighth inch, and in radiographs the average spacing of all joints is approximately one-quarter of an inch. When this space is narrowed, the physician can deduct that the cartilage has been damaged. Narrowing occurs in arthritis, especially in degenerative arthritis.)

Hyaline cartilage is widespread in the embryo, where it comprises the majority of the temporary (embryonic and fetal) skeleton. Hyaline cartilage possesses the distinguishing properties of being very flexible and somewhat elastic. Hyaline cartilage is a semitransparent mass with an opalescent, bluish color that is seen in some types of glass. Calcification of hyaline cartilage, seldom seen in persons under 30 years of age, is found usually with advancing age. It is seen quite often in radiographs of the chest (lungs) as lines of continuing density between the sternal ends of the ribs and the sternum.

Elastic cartilage is frequently referred to as yellow elastic cartilage, or reticular cartilage. In mammals this variety of cartilage tissue is found in the following three locations:

1. The external ear in the pinna and the tragus

2. The walls of the external auditory meatus and the auditory tube
3. The epiglottis and parts of the cuneiform and corniculate cartilages of the larynx

Macroscopically, the yellowish color of elastic cartilage differentiates it from hyaline cartilage. Elastic cartilage is more opaque, flexible, and elastic than true hyaline cartilage. Calcification with advancing age is a phenomenon of elastic cartilage as well as hyaline cartilage.

Fibrocartilage occurs as indistinctly outlined cartilage in small accumulations in a few places in the bodies of mammals. Fibrocartilage is found in the following four locations:

1. The intervertebral discs (the chief constituent)
2. Certain articular cartilages of the symphysis pubis
3. The ligamentum teres femoris
4. The sites of attachment of certain tendons to bones, etc.

Fibrocartilage is a transitional form between cartilage and connective tissue; it is very strong and resilient. However, fibrocartilage tends to degenerate in adult life, and the so-called disc protrusion is merely a degeneration that takes place over a period of from 8 to 10 years. The radiograph will then show the vertebral bodies to be closer together, and when the degeneration is extreme, disc material extrudes. This condition is true particularly in the nucleus pulposus, which, when it protrudes, frequently causes pinching of the nerve at the foramen. Such a condition is the cause of the radiating pain down the leg and the numbness of the leg in the so-called intervertebral disc protrusion. It has been postulated that the most highly developed tissues are most sensitive to radiating pains and that the lining of blood vessels is involved early, thus causing a decrease in circulation to the area of involvement.

Bone

Bone is the hard tissue of which the adult skeleton of most vertebrates is largely composed. True bone appears during the seventh week of development in the human embryo. However, fetal bones (in utero) are seldom visible radiographically prior to the sixteenth week of pregnancy (the end of the first month of the second trimester). It is deemed unwise to radiograph the abdomens of pregnant women earlier than the eighth month of pregnancy because of the possibility of stimulating changes in the tender, growing tissues of the fetus or embryo. The ends of the long bones in the unborn and the young child should be given special protection (when possible) since damage would impair or even terminate the growth of the bones.

Numerous and diverse opinions and theories exist regarding the nature and quantity of the potential change that may result from exposing the unborn fetus to ionizing radiations. There is insufficient evidence to point positively to a particular type of change; however, the *radiologic technologist must observe all possible precautions in making radiographic exposures*. Precautionary measures are equally as necessary in making exposure of extrauterine individuals.

Protection of the gonads of persons of both sexes, especially prior to 30 years of age, is very important. To become transmissible, gene mutations and chromosome aberrations must occur in the gametes of the individual; the mutations and aberrations will be transmitted or evidenced in succeeding generations in other tissues as well as in bone tissue (see p. 49).

Bone structure. Bone is either *spongy* (cancellous) or *cortical* (compact). Since spongy bone (so-called because of its structure) is the first to develop, it is logical to discuss this class of bone first. In its embryonic development, bone is laid down originally in the spongy form. Spongy bone consists of a framework of connecting and intercrossing osseous bars of varying thicknesses and shapes. These irregular trabeculae* of the spongy substance consist of a varying number of closely adjoining bone plates, the bone *lamellae*, and subdivide incompletely in the bone marrow (medullary) cavity. The trabeculae are visible in radiographs of the cancellous parts of bones wherein the x-ray penetration has been correct and the condition of the bone is healthy. The lamellae branch, unite with one another, and partially surround the intercommunicating spaces, which are filled with bone marrow. *The points of contact and the direction of these crossbars are such that maximum rigidity and resistance to change in shape are imparted to all parts of the skeleton.* Examine the photograph (Fig. 4-83) of a coronal section through the upper end of a dry bone femur. Bridge builders and structural engineers have marveled at this logical configuration, which gives maximum strength to this area of the bone without adding weight. It also corroborates what has been said previously about Wolff's law, that nature places bone where the need is greatest (at a location of stress and strain) and removes bone from areas where need is least (a fractured bone in a cast).

Cellular and fibrous derivatives of the mesenchyme fill the open spaces of the spongy bone. The *red bone marrow* consists of this type of reticular tissue as do fat cells, sinusoids, and immature blood cells (both erythrocytes and leukocytes). Spongy bone is found in adults in the body of the sternum,† in the crests of the ilia, and in certain other regions of the body, such as the diploë of the flat bones of the cranium. The *epiphyses* (ends) of long bones continue to contain a considerable quantity of spongy bone throughout adult life. There is a single center of ossification in each epiphysis. The calcification of any bone or bone part begins with and radiates around the centers of ossification.

Cancellous bone, because of the included osteoblasts (many times more than in cortical bone), is the most osteogenic type of bone and, therefore, is in the greatest demand for bone grafting. Such bone tissue is usually taken from the iliac crest or from the proximal end of the tibia. However, when structural strength is desired, cortical bone is taken, usually from the medial face of the tibia. After the surgeon opens this *trough* some cancellous bone may then be *borrowed* from the wide, proximal end of the tibia.

---

*A trabecula is a small bar, rod, bundle of fibers, or septal membrane in the framework of an organ or part.
†The sternum is the usual site of puncture for extraction of embryonic blood for laboratory examination in bizarre diseases and in blood dyscrasias.

The skeletal system **177**

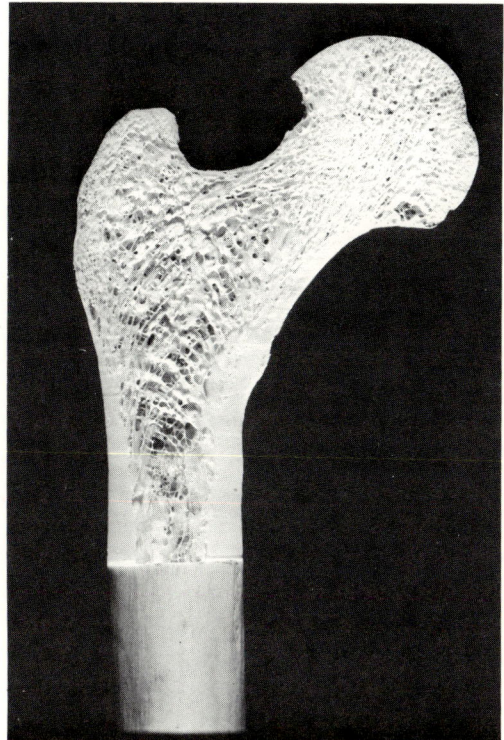

Fig. 4-83. Coronal longitudinal section, upper end of right femur.

The primary centers of ossification are principally in the *diaphyses* (shafts) of long bones and form in intrauterine life. The secondary centers of ossification are, for the most part, in the epiphyses and form in extrauterine life.

Very close connection exists between the regular arrangement of the lamellae in the *compact substance* and the distribution of the blood vessels that nourish the bone. Histologically, the compact substance of the diaphysis of any long bone is seen to be penetrated by a number of cylindric branching and anastomosing canals. These are *haversian canals,* named after the English anatomist, Havers. Haversian canals are, for the most part, parallel to the length of the bone so that a cross section of the bone reveals them as small, round openings; longitudinal sections of the bone reveal these canals as slender slits. Haversian canal shadows can be mistaken for tiny *crack* fractures. In living bone the haversian canals contain blood vessels with a small amount of accompanying connective tissue. The haversian systems[*] communicate with the external surfaces of the bone and with the bone medulla by means of the canals of Volkmann.[†] At the junction of the

---

[*]One haversian canal, its concentric lamellae, their canaliculi, and their lacunae comprise a single haversian system.
[†]A Volkmann canal is one of many nutrient canals transmitting blood vessels from the periosteum into the bone but not forming the center of an haversian system.

cortical and the spongy bones, the haversian canals of the former expand and pass directly over into the marrow spaces of the latter.

Haversian systems. Study of the haversian systems reveals the method by which bones grow. Each haversian system consists of a central canal, the *haversian canal,* surrounded by from 8 to 15 concentric rings, the *lamellae.* The lamellae are thin plates of osseous matrix—bone fibers cemented together with an intercellular matrix. The lamellar rings include some mature bone cells, *osteocytes.* If the bone is healing or growing following an injury, a certain number of *osteoblasts* will be present in the areas where new bone is being laid down. It is axiomatic among orthopedic surgeons that healing is directly proportional to the degree of comminution; a fracture producing many bony pieces releases many osteoblasts* into the blood clot, resulting in rapid callus formation. The reverse is true; when there is a linear fracture without displacement, little callus forms. For the same reason usually *no* callus is seen in compression fractures, such as in the vertebral bodies, where no osteoblasts *spill out* into the surrounding blood clot.

The haversian canals anastomose freely with each other by oblique and transverse communications. From the periosteal (outer) and endosteal (inner) surfaces of cortical bone, small channels arise to pierce the bone at right angles to the long axis of the bone and, therefore, to communicate with the haversian canals. The blood and nerve supplies of the bones are contained within these several canals, the canals of Volkmann. The *lacunae,* small cavities that contain the bone cells, are situated both between and within the lamellae. The lacunae usually are found to be flattened or oval-shaped cavities that are incompletely occupied with the osteocytes (one in each lacuna). The lacunae communicate with the haversian canals through the very small *canaliculi.* The canaliculi also communicate with other canaliculi (see Fig. 4-84).

Osteoblasts are the young bone cells that are in a state of continuous growth and formation in the trabeculae. The osteoblasts join together to form small spicules of bone. Upon the deposition of calcareous material, the bony spicules become walled off; at this time the osteoblasts mature into osteocytes. The process of growth causes the bony spicules to increase in size, which tends to occlude and/or obliterate the marrow spaces. However, *osteoclasts,* bone cells of special nature, function to digest or otherwise cause reabsorption of the innermost portions of the spicules and thereby keep the nutritive canals open. The process of reabsorption includes the removal of excess calcium salts; according to some au-

---

*The greater the degree of comminution or the more pieces into which the fracture is broken, the greater the surface area, from which large quantities of blood clot are released. Into this blood clot appear osteoblasts, which later form the callus by laying down new bone. The osteoblasts come from two sources: (1) the innermost layer of the periosteum in children and in young, growing bones (this layer causes the laying down of new bone on the surface) and (2) the bone itself from the lacunae. Both of these sources provide osteoblasts that are released into the clot and cause ossification. In the case of a fracture, the osteoblasts cause healing. There is evidence that osteoblasts also come from the endosteum.

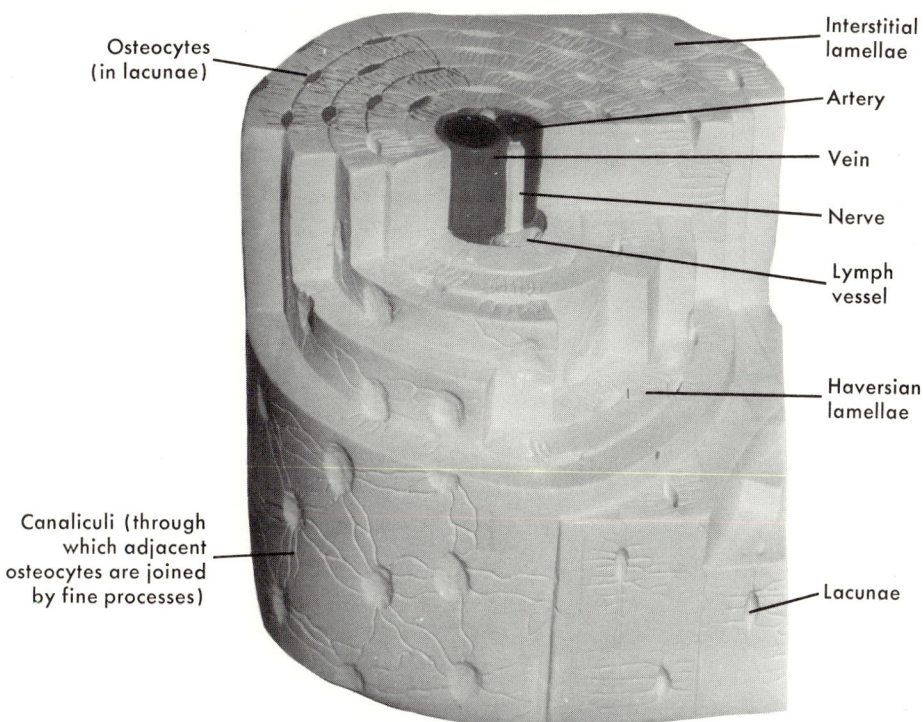

Fig. 4-84. Haversian system model. (Courtesy Dr. Justus F. Mueller and Ward's Natural Science Establishment, Inc., Rochester, N. Y.)

thorities this process is aided, at least, by the acidity* of the blood and by the carbonic acid, which is potentially present at all times in the blood and other body fluids. The osteoclasts phagocytize the osseous matrix. From the preceding remarks it is seen that osteoblasts are most numerous in the ends of long bones and least numerous in the center of the shafts of long bones. As a result, fractures in the middle third of a long bone heal quite slowly (and poorly), whereas fractures near the ends of long bones usually heal promptly.

Other than carbonic acid (H_2CO_3), any free acid present in the blood would cause severe hemolysis resulting in serious damage to, and probably death of, the patient if the acid were not neutralized instantaneously. Carbonic acid forms upon demand when a molecule of water (H_2O) joins with a molecule of carbon dioxide (CO_2):

$$H_2O + CO_2 \rightarrow H_2CO_3$$

When formed, carbonic acid either diffuses one hydrogen ion, thus being

*Other authorities state that there is no evidence of *any* change in the pH of the tissue fluid surrounding the area of bone reabsorption, thus precluding the possibility of increased acidity in the local region. It is further suggested that proteolytic activity of the osteoclasts removes certain organic components of the intercellular substances, thus initiating reabsorpion of bone.

changed, or separates into one molecule of water and one molecule of carbon dioxide, whichever is needed. As with other acids, H_2CO_3 *cannot* exist freely in the blood; if it did, it, too, would cause hemolysis. (For further discussion of blood pH see pp. 261 and 262.)

The pH of the blood clot surrounding a fracture quickly becomes strongly alkaline; this condition favors deposition of new calcium. The clot surrounding a fracture is most important for the formation of new callus and deposition of calcium. Thus, anything that will alter the natural alkalinity of this clot delays union of the fractured parts. (An example of this is seen in an injection of the fracture site between the bone fragments with procaine hydrochloride, which is strongly acid.)

Concurrent with bone fracture a variable amount of blood vessel rupture occurs along with a generally corresponding amount of soft tissue damage both within and around the bone. This condition results in an extravasation* (escape) of blood into the tissues with eventual clot formation. The fibrin (see p. 266) contracts, leaving a blood clot surrounding the fractured ends of the bone.

Blood clots usually become organized as a result of the ingrowth of young blood vessels and fibroblasts (formation of granulation tissue). In conjunction with this process, macrophages (wandering phagocytic cells) phagocytize extravasated erythrocytes and debris. Normally, the granulation tissue forms bone callus that is replaced by bone. Thus, ultimate healing of the fracture depends on the increased flow of blood into the fracture site.

Throughout each bone, growth occurs in single areas because of cellular activity either in organized haversian systems or in interstitial lamellae. The primary function of the haversian system in cortical bone is to assure nourishment to the bone cells. It is not possible to make a sharp line of demarcation between spongy bone and cortical bone since both classes of bone are merely different arrangements of the same histologic elements. Both spongy bone tissue and cortical bone tissue are found in most of the bones. Ossification of entire bones includes the growth achieved as described previously; however, ossification treats the development of whole bones. Ossification begins at one or more points (centers of ossification) for each future bone and is accomplished by mesoblastic cells (osteoblasts), which deposit the inorganic bone substance about themselves to become the *bone cells (corpuscles)* of completed bone.

Ossification. Bone invariably develops as a transformation of embryonic or adult connective tissue into calcifiable tissue, the *bone matrix*. When this transformation occurs in cartilage, the phenomenon is called *intracartilaginous*, or *endochondral, ossification* and is usually contrasted with simple *intramembranous* bone formation. The difference between the two methods of ossification is simple. *Endochondral ossification occurs within the cartilage, which must be removed before any ossification can occur;* however, some of the cartilage matrix may remain as a framework on which the bone is laid down. *Intramembranous bone ossifies (forms) in connective tissue instead of cartilage.* The actual process of

---

*See glossary.

bone formation is identical in each ossification; the first bone formed is always spongy, although some of it later becomes compact through internal reconstruction.

As bone forms after a fracture, it becomes more compact and stronger through the application of stress (Wolff's law). In the healing of a fracture, the bone always forms and heals most rapidly in those fragments that are subject to stress, whereas absorption or rounding off occurs in those fragments that carry no weight or strain. In young persons particularly the growth of bone will completely obliterate any sign of a fracture so that radiography after 1 or 2 years will usually fail to visualize differences in densities at the old fracture site (see Fig. 4-85, A and B).

Primitive connective tissue membrane, the mesenchyme, precedes the flat cranial bones; i.e., the connective tissue membrane exists in place of and prior to the formation of the cranial bones. Ossification begins at one or more internal points. A distinguishing characteristic of these centers of ossification is the appearance of osteoblasts, which promptly deposit bone matrix in the form of spicules (cell-like structures with spiny processes). The spicules unite into a meshwork of trabeculae; the meshwork spreads radially from each center. Since

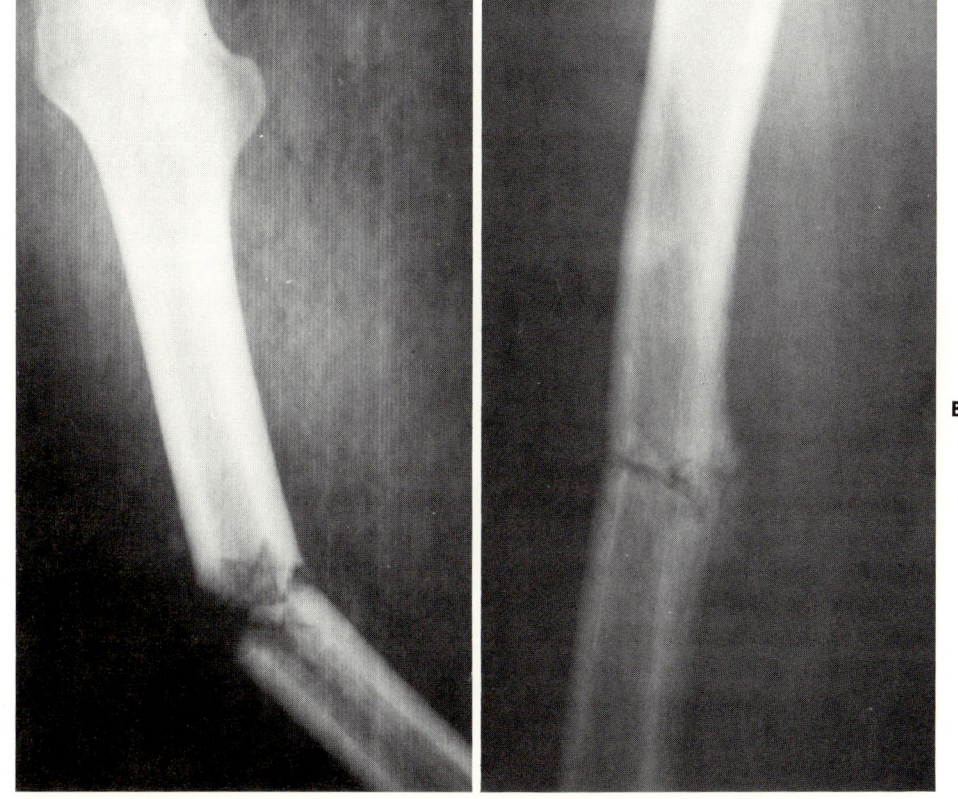

Fig. 4-85. A, New fracture of femur. **B,** Healing fracture of femur. (Courtesy Dr. Frank S. Cavallaro and Valley Hospital, Las Vegas, Nevada.)

Table 4-8. Approximate appearance time of bones of extremities, shoulder, and pelvis

Year	Carpals	Clavicle	Femur	Fibula	Humerus	Ilium	Ischium	Metacarpals
At birth	Capitate, hamate		Distal epiphysis		Head			
1			Head	Distal end	Capitulum, lateral part of trochlea			
2	Triquetrum, F							Base of 1
3	Triquetrum, M				Greater tubercle			Heads
4	Lunate, scaphoid, F		Greater trochanter, F	Proximal end				
5	Trapezium, F; scaphoid, M		Greater trochanter, M		Lesser tubercle; medial epicondyle, F			
6	Trapezium, M; trapezoid, F				Head unites with tubercles			
7	Trapezoid, M				Medial epicondyle, M			
8							Rami join, F	
9	Pisiform							
10					Medial part of trochlea			
11			Lesser trochanter					
12								
13					Lateral epicondyle			
14					Capitulum, trochlea, lateral and medial epicondyles	Crest, antero-inferior spine	Tuberosity	
15								
16		Medial end	All secondary epiphyses, F		Lateral epicondyle			All fuse, F

Metatarsals	Patella	Phalanges Manual	Phalanges Pedal	Pubis	Radius	Scapula	Tarsals	Tibia	Ulna
							Calcaneus, talus, cuboid	Proximal end	
					Distal end	Coracoid process	Lateral cuneiform		
Base of 1, heads of 2, 3, 4, and 5, F		Bases, F	Bases of proximal, F					Distal end	
Base of 1, heads of 2, 3, 4, and 5, M		Bases, M	Bases of proximal, M				Navicular, intermediate and medial cuneiform		
	Center								Head and styloid process
			Bases of distal		Head, F				
					Head, M				
				Rami join, F			Secondary calcaneal, F	Medial malleolus	Distal end
						Subcoracoid, superior ⅓ of glenoid	Secondary calcaneal, M	Tubercle	
									Tip of olecranon process
	Ossification completed				Tuberosity	Margins of glenoid and inferior angle			
						Margin of vertebral border, acromion process			
				Secondary symphysis	Head, F		Completed		
		All fuse, F						Completed, F	

Continued.

Table 4-8. Approximate appearance time of bones of extremities, shoulder, and pelvis—cont'd

Year	Carpals	Clavicle	Femur	Fibula	Humerus	Ilium	Ischium	Metacarpals
17						Triradiate epiphysis	Triradiate epiphysis	
18			All secondary epiphyses, M	Distal epiphysis	All proximal epiphyses, F			
19					All distal epiphyses			All fuse, M
20								
21					All proximal epiphyses, M	All secondary epiphyses	All secondary epiphyses	
22				Proximal epiphyses				
23								
24								
25		Medial end completed						

the osteoblasts are arranged in an epithelioid layer upon the surface of a spicule, growth of the spicule occurs as increased thickness of its shaft and at the tip. As the matrix is laid down progressively, some osteoblasts become trapped, remain imprisoned as osteocytes, and are lodged in the lacunae.

After the development of these primary internal centers is well underway, the entire *primordium* (an entire collection of cells) becomes enclosed within a periosteum. Osteoblasts differentiate upon the inner surface of the periosteum to form the osteogenetic layer and deposit parallel plates (lamellae) of cortical bone; this is the process of *periosteal* ossification.* The dense inner and outer

---

*Periosteal ossification is well visualized in radiographs during osteomyelitis, particularly the acute hematogenous osteomyelitis of childhood. In this process the periosteum becomes elevated from the surface of the bone, then lays down a new calcified layer in its new position. This is clearly seen as a white line, separate from the former shaft. The pressure of pus beneath the periosteum may move it out more, and it will again lay down another layer. If this process is continued several times, the layer will sometimes appear laminated, as an onion. Sometimes the suppuration destroys the entire former bony shaft and the shaft dies, leaving the very active periosteum that lays down enough new bone to support weight bearing. This new shell is called the involucrum. The surgeon may remove the entire old shaft, and the patient may soon walk on the involucrum. It then becomes thicker and stronger with weight bearing (Wolff's law). This periosteal bone will form an entire new scapula when the bone has to be removed because of osteomyelitis.

Metatarsals	Patella	Phalanges		Pubis	Radius	Scapula	Tarsals	Tibia	Ulna
		Manual	Pedal						
				Triradiate epiphyses					
					Head, M			Completed, M	
All fuse		All fuse, M	All fuse		Distal end, F	Two centers for acromion process, vertebral border			Distal end, F
						Margin of glenoid and inferior angle completed			
				All secondary epiphyses	Distal end, M				Distal end, M

tables of the skull (calvarium) bones develop by this method. The layer of cancellous bone joining these tables is the *diploë*.

In the process of intracartilaginous ossification the hyaline cartilage undergoes degenerative changes, disappears, and is replaced with bone that develops exactly as in intramembranous ossification. Either of two terms, "replacement bones" or "substitution bones," is frequently applied to these skeletal elements.

In endochondral ossification the cells in the center of the cartilage enlarge and become arranged radially; some calcium is deposited in their matrix. The cartilage cells and a part of the calcified matrix disintegrate and disappear in an unknown manner, thereby permitting the formation of the primordial marrow cavities. This destruction is apparently caused by the penetration of the cartilage by the blood vessels of the embryonic connective tissue.* As the embryonic marrow penetrates the cartilage capsules, the connective tissue cells adjacent to the cartilage matrix become a layer of osteoblasts.

From this point, ossification by the osteoblasts progresses in the same manner

---

*Loops of blood vessels and accompanying mesenchyme cells arise and spread toward the cartilage from the embryonic connective tissue. The vascular endothelium and the surrounding elements dissolve the *interstitial substance* in an unknown manner and open up the distended cartilage capsules, permitting the vascular, embryonic connective tissue to penetrate them.

as in intramembranous ossification. Similarly, the layers of the interstitial substance increase in size and surround isolated osteoblasts, which are thus transformed into osteocytes. At this time the spongy endochondral bone is composed of variously sized bony plates, covered by layers of osteoblasts and containing remnants of calcified cartilage in their interiors. For this reason endochondral bone is characteristically spongy, and replacement continues until the entire cartilage is eventually superseded by cancellous bone.

During the formation of endochondral bone (i.e., when the embryonic bone marrow penetrates the capsules), a thin layer of cells arises between the osteoblasts and the cartilage, thickens gradually, and surrounds the contours of the cartilage bars. This tissue is calcifiable when laid down and, under favorable conditions, begins to calcify as it is deposited; thus calcified osseous tissue (bone) is formed. The haversian systems, which derive from the osteogenetic layer of the periosteum, become established within this cortical bone. This is the method of perichondral and periosteal bone formation.

As bones grow in length a line of junction forms between the expanding centers of ossification. A layer of unossified cartilage, called the *epiphyseal plate,* lies between the epiphysis and the metaphysis.* The epiphyseal plate is easily damaged, much more so than the bony shaft. This plate continues as noncalcigerous cartilage until the bone has ceased to grow; conversely, bone growth ceases upon epiphyseal plate calcification.† New bone grows in length in the following manner. Cartilage grows on the epiphyseal side of the epiphyseal plate by the process of cell division (mitosis); the cartilage on the diaphyseal side of this plate is destroyed and replaced with osseous tissue. Thus, cartilage exists between the diaphysis and epiphysis of any long bone during the time of growth. Such areas of cartilage appear as dark areas between the bone shaft and bone end in radiographs of growing bones. In these areas the future bone now exists entirely as cartilage, and the ossification centers will not appear until later. The cartilage is growing rapidly and responds well at this time to corrective surgery (as in club foot). The time at which calcium is deposited in an ossification center in sufficient quantity to make a visible density in a radiograph varies considerably between individuals and between sexes. Table 4-8 lists the approximate (average) appearance time of the bones of the shoulder and pelvic girdles and the bones of both extremities. Appearance time is the average age at which the particular structure becomes visible radiographically in 50 percent of normal children.

Knowledge of the approximate appearance time of a bone, group of bones, or particular bony structure often enables the radiologic technologist to calculate

---

*Metaphysis is the line of junction of the epiphysis with the diaphysis of a long bone.
†In young persons severe contact often results in an injury (football casualty) in which the leg is fractured through the epiphyseal plate. Such a fracture reduces with a fair degree of ease, but there is always the risk of damage to the growth center, which could result later in a difference (from the other leg) in leg length. When one leg is short as a result of poliomyelitis or some other cause, the epiphyseal plate in the longer leg can be arrested in order to permit the shorter leg to grow faster than the longer leg and finally approach, if not equal, the length of the longer leg.

accurately the peak kilovoltage and milliampere-second requirements for a particular exposure of growing bones. Physicians often request radiographs of both extremities for comparison of bone development.

REFERENCES

Francis, C. C, and Martin, A.: Introduction to human anatomy, ed. 7, St. Louis, 1975, The C. V. Mosby Co.

Gardner, E., Gray, D. J., and O'Rahilly, R.: Anatomy, Philadelphia, 1960, W. B. Saunders Co.

Goss, C. M.: Gray's anatomy, ed. 27, Philadelphia, 1959, Lea & Febiger.

Jacobi, C. A., and Paris, D. Q: Textbook of radiologic technology, ed. 5, St. Louis, 1972, The C. V. Mosby Co.

Meschan, I.: Normal radiographic anatomy, ed. 2, Philadelphia, 1959, W. B. Saunders Co.

Weinmann, J. P., and Sicher, H.: Bone and bones, St. Louis, 1955, The C. V. Mosby Co.

CHAPTER 5

The digestive system

Extending between the mouth and the anus is a tube whose lumen varies in diameter according to function in each particular locus. Contained within the walls of this tube or pouring into the tubular lumen from the outside are numerous accessory glands and organs. The tube and its accessory glands and organs comprise the digestive system. The length of the tube (tract) may vary between individuals from 28 to 34 feet; it is approximately 9 meters long (see Fig. 5-1).

The digestive system (tube) lies outside the body; i.e., the tube forms the lining of the inside of the body, and the skin forms the lining of the outside of the body. A viscus must be situated between the digestive tube and the skin to be contained within the body.

FUNCTION

Specialized epithelial cells line each part of the digestive tract and form the principal substance of the digestive glands and organs. In a given locus the cells perform the specific function and/or secrete the specific juices peculiar to this region or structure.

Digestion includes the following steps:
1. Preparation of food in the oral cavity
2. Chemical hydrolysis in the stomach and small intestine including the enzymatic actions
3. Assimilation (absorption) of the reduced food elements through the linings of the small and large intestines

Limited absorption of certain substances (alcohol) occurs through the stomach wall; water is reabsorbed into the circulating blood through the large intestinal wall.

Each part of the digestive system functions in its specific way to aid in maintaining homeostasis. The digestive system as a whole functions to convert foods into their elemental (or smaller and more suitable) forms for assimilation into the bloodstream, all for the general purpose of nutrition; the system also functions to eliminate solid waste by-products of digestion.

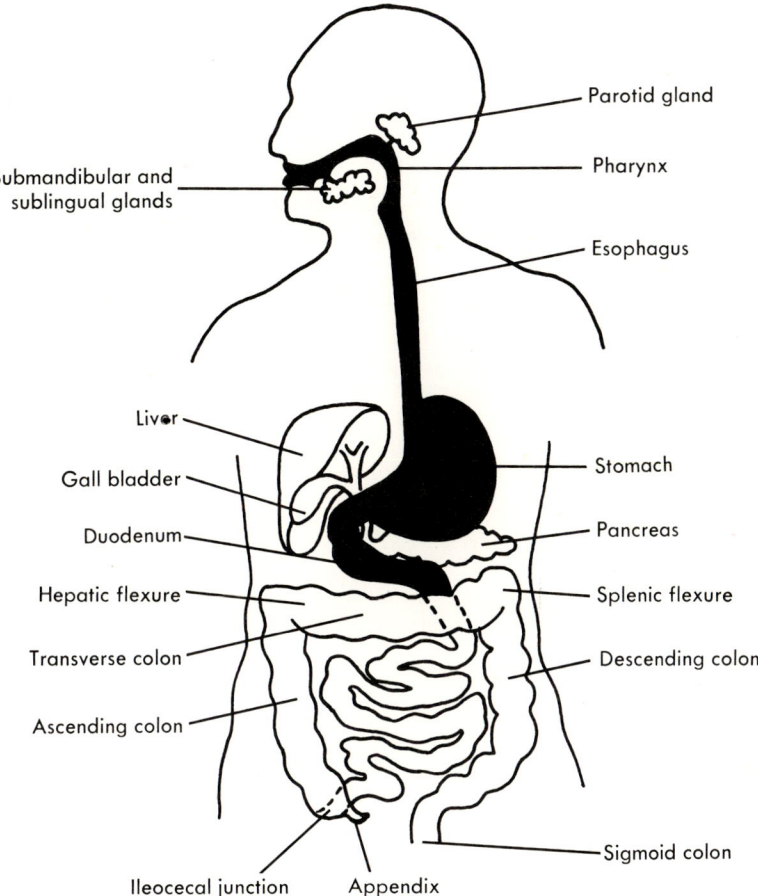

Fig. 5-1. Digestive viscera.

ANATOMY

The digestive system structures are situated in the buccal, oral, thoracic, and abdominal cavities. The principal structures of the digestive system are the salivary glands, tongue, teeth, fauces, pharynx, esophagus, stomach, small and large intestines, liver, gallbladder, and pancreas.

Buccal cavity

The buccal cavity is the vestibule of the oral cavity; it lies between the teeth and the cheeks. The buccinator muscles form the principal tissue of the cheeks; other types of cheek tissue include the skin and masseter muscles. The cheeks are the outer boundary of the buccal cavity. The opening into the buccal and oral cavities is the mouth (lips). The orbicularis oris muscles comprise the principal structure of the lips.

Parotid glands

The parotid glands are situated in the cheeks immediately inferior and anterior to the external ears. Each parotid gland is situated on the corresponding masseter muscle beneath the skin. The parotid gland is drained of its secretions by the parotid (Stensen's) duct, which opens into the buccal cavity through a small papilla in the cheek opposite the upper second molar (see Fig. 5-2).

In general, the parotid gland is quadrilaterally shaped and has a tonguelike extension between the temporomandibular joint and the mastoid process. Mumps is a contagious, febrile, virus-caused disease in which the parotid glandular tissue swells. The action of opening the mouth of the person suffering from the mumps compresses the tonguelike extension of the parotid gland and causes rather severe pain. Mumps may also affect the other salivary glands and the pancreas and/or the gonads.

The parotid gland receives blood from the external carotid artery and its branches; it is drained of blood by the posterior facial vein. This gland has sensory innervation from the facial nerve and branches; its secretory innervation is from the ninth cranial nerve autonomic outflow and the thoracolumbar outflow along the blood vessels (see Fig. 5-2).

Oral cavity

The oral cavity contains the submandibular* (submaxillary) and sublingual salivary glands, the tongue, and the teeth. The hard and soft palates form the roof of the oral cavity. The *uvula,* an extension of the soft palate, partially closes the oropharyngeal opening except during deglutition. The teeth form the sides and front of the oral cavity, and the tongue muscles form the floor of the cavity. The oral cavity opens posteriorly into the pharynx through the fauces (see Fig. 5-3).

Salivary glands

The salivary glands are the following three pairs: parotid (discussed under the buccal cavity), submandibular, and sublingual. The parotid glands are the largest of the salivary glands.

Submandibular glands. The submandibular glands are slightly smaller than the parotid glands; they are situated, one on each side, medial to the mandible. Each submandibular gland lies about halfway between the angle of the mandible and the point of the lower jaw and somewhat inferior to the mandibular body; the gland rests inferiorly on the mylohyoid muscle.

Each submandibular gland is drained of its secretions by the corresponding submandibular (Wharton's) duct, which opens onto the floor of the oral cavity lateral to the *frenum* of the tongue and below the free end of the tongue.

---

*Early anatomic literature described two superior and one inferior maxillae. With this nomenclature, submaxillary fitted these glands. Current anatomic literature describes two maxillae and one mandible: hence, the name, submandibular glands.

The digestive system 191

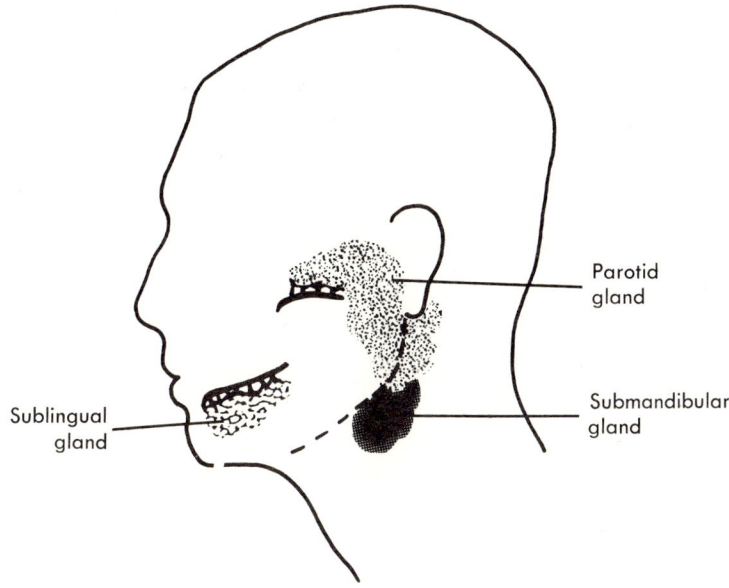

Fig. 5-2. Left salivary glands.

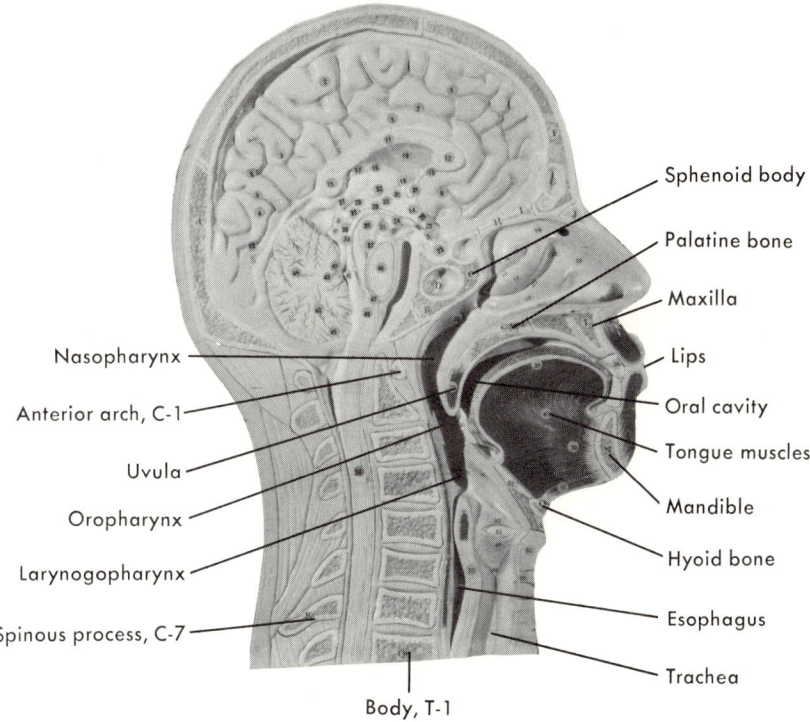

Fig. 5-3. Median section through the head and neck showing the oral cavity, the pharynx, and the esophagus and related structures.

The submandibular gland receives blood from branches of the external maxillary and lingual arteries. It is drained of blood by veins that follow the corresponding arterial courses. The submandibular gland has secretory innervation from the chorda tympani branch of the seventh cranial nerve autonomic outflow and the thoracolumbar outflow along the blood vessels.

Sublingual glands. The sublingual glands, the smallest of the three pairs of salivary glands, are situated in the floor of the oral cavity beneath the free end of the tongue and bilateral to the midline of the skull. These glands empty their secretions onto the floor of the buccal cavity through from 8 to 20 individual papillae (from 4 to 10 papillae for each gland).

The sublingual gland receives blood from the sublingual and submental arteries and is drained of blood by veins that follow the corresponding arterial courses. The sublingual gland is innervated for secretory function by branches of the same nerves that supply the submandibular gland.

In addition to the three pairs of glands having names, numerous small unnamed glands, having few, if any, duct systems, assist in the production of saliva. These are the branched tubular and tubuloacinous glands found in the cheeks, palate, tongue, and lips.

Each of the three named salivary glands is composed of an organized duct system, tubules, and secreting acini.* These structures collectively comprise the parenchyma of the gland. The stroma consists of a supporting framework of fibrous connective tissue. The parotid and submandibular glands are contained within a distinct fibrous connective tissue capsule; the sublingual gland lacks this capsule.

The ducts of the named salivary glands are secretory, excretory, and intercalated. The secretory ducts are formed by simple columnar epithelium; the excretory ducts are formed by squamous and columnar epithelium; the intercalated ducts are formed by low (or flattened) columnar epithelial cells.

Radiography of the salivary glands and associated ducts is called *sialography,* an examination performed to determine the presence (or absence) of calculi and other pathology, principally of the ducts.

Tongue

The entire tongue consists of muscles encased within a specialized covering containing numerous sensory nerve endings within the taste buds. The nerves convey the sense of taste to the corresponding receptive areas of the brain (see p. 353. The motor innervation of the tongue derives from the hypoglossal nerve.

Scattered on the tongue surface (and occasionally on adjacent parts) are the *taste buds,* which are spherical- to oval-shaped collections of cells in the stratified squamous epithelium of the tongue; these form the *papillae.* Approximately 250

---

*Acini is the pleural of acinus. "Acinus" is a general term designating a small saclike dilation, particularly one found in various glands; it is commonly used synonymously with alveolus.

individual taste buds form a single papilla, which opens onto the upper surface of the tongue through the small openings called *pores*. The taste impulse travels along a nerve pathway to the taste center. Four primitive taste areas are located on and, sometimes, adjacent to the tongue. Bitter is received at the back of the tongue. Sour is received at the sides of the tongue. Salt is received at the sides and tip of the tongue. Sweet is received at the tip of the tongue. Several stimuli of differing tastes may enter the pores simultaneously and cause blending of the tastes. The general stimulus travels along one branch of the trigeminal nerve.

A taste bud is situated in an ovoid pocket filling the thickness of the epithelial layer. Both supporting and gustatory cells comprise the taste buds. The peripheral end of a gustatory cell extends through the surface opening, which is called the *gustatory pore*. The terminal part of this cell is hairlike and is called the *gustatory hair*.

Two sets of muscles, the intrinsic and extrinsic, comprise the tongue substance. The *extrinsic* muscles, those with the origin outside and the insertion inside the tongue, are the palatoglossus, styloglossus, genioglossus, hypoglossus, and chondroglossus. The *intrinsic* muscles, those with both origin and insertion inside the tongue, are the longitudinalis superior, longitudinalis inferior, transversus, and verticalis. Each muscle of a pair has its corresponding member in the opposite side of the tongue; each tongue muscle is a member of a pair.

The tongue serves to move the food from one side of the mouth to the other so that the food may mix thoroughly with the juices of the salivary glands to initiate digestion and to form the food bolus. The tongue also functions importantly in deglutition (see p. 220).

The lingual branch of the external carotid, the external maxillary, and the ascending pharyngeal arteries supply blood to the tongue; blood flows from the tongue to the internal jugular vein.

Teeth

In the normal adult mouth the teeth number thirty-two. There are four central incisors, four lateral incisors, four canines, eight premolars, and twelve molars. The adult teeth are permanent; teeth that erupt early in childhood and are later shed are deciduous.

Specialized epithelium forms each tooth. The part of a tooth that is visible above the gum is the *crown*. *Enamel* covers the crown and *neck* (the narrowed part immediately beneath the crown). The layer beneath the enamel is the *dentin*, which extends below the gum line into the socket. *Dental pulp*, containing the nerves and vessels of the tooth, fills the central cavity of the tooth. At the gum line, cementum replaces enamel; the cementum grows to the *periodontal membrane*, a specialized form of periosteum (the lining of bones). In the mandible and maxillae the *alveolar* processes contain the sockets from which the teeth erupt (see Fig. 5-4).

The incisors and canines tear and rend the food; the premolars and molars

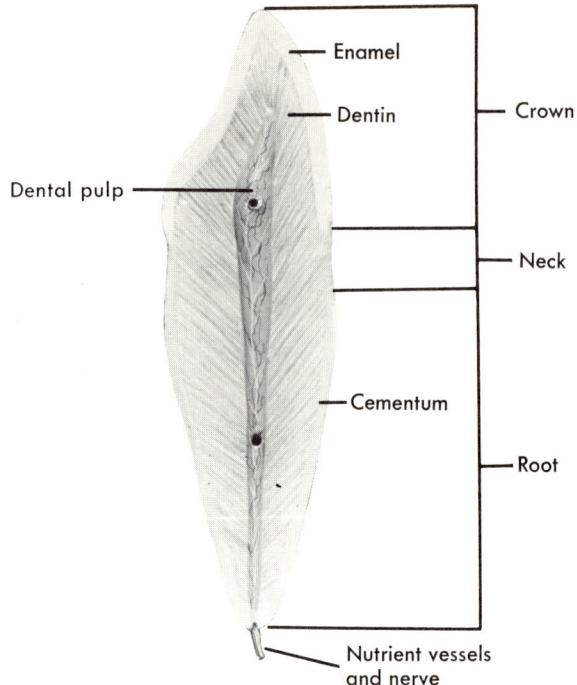

Fig. 5-4. Tooth, showing inner structure.

masticate (grind) the food after it is taken into the mouth. The teeth of the mandible and maxillae are supplied with blood from branches of the internal and external maxillary arteries. Blood drains from the teeth through veins that empty into the jugular vein. The sensory innervation of the teeth is from branches of the trigeminal nerve.

Radiography of the teeth is performed to determine the presence (or absence) of cavities, caries, and inflammation or disease of the periodontal membrane. With both dental and medical radiographic equipment, current practice emphasizes the use of the *long-cone technique** to avoid exposing patients to excessive quantities of radiation.

Fauces†

The *mouth* (oral cavity) opens posteriorly into the pharynx. The opening for communication is the *isthmus faucium*. The soft palate forms the superior boundary of the isthmus faucium; the glossopalatine arch forms the sides of the isthmus faucium; and the dorsum of the tongue forms the inferior part (floor) of the isthmus faucium. On either side between the glossopalatine arch and the adjacent pharyngopalatine arch is the site of the palatine tonsil.

---

*Long-cone technique employs a target-to-skin distance of not less than 16 inches. Twenty-inch cones are also available.
†Fauces is the plural of faux: a narrow pass or gorge.

The structures forming the oral cavity receive blood from branches of the internal maxillary, external maxillary, and ascending pharyngeal arteries. The blood from this region passes through veins that terminate in the pterygoid and tonsillar plexuses. Lymph from this region flows through lymphatics into the cervical nodes; sensory innervation is from the palatine, nasopalatine, and glossopharyngeal nerves.

Pharynx

"Pharynx" is the correct anatomic term for the common word, throat. It is a muscular tube common to the digestive and respiratory systems. The pharynx begins immediately beneath the brain and terminates with the opening into the larynx. The pharynx is approximately 12.5 cm. long (see Fig. 5-3).

Seven openings are found in the pharynx, each of which has significance physiologically and some of which have significance radiographically. The openings are as follows:
1. Two from the auditory (eustachian) tubes bilaterally and near the superior margin of the pharynx
2. Two from the posterior nares (choanae) anteriorly into the upper one-third of the pharynx
3. One from the oral cavity into the middle one-third of the pharynx
4. One into the esophagus posteriorly and inferiorly in the lower one-third of the pharynx
5. One into the larynx anteriorly and inferiorly in the lower one-third of the pharynx

The pharynx consists of three parts, and each part contains certain of the aforementioned openings: items 1 and 2 in the *nasopharynx,* item 3 in the *oropharynx,* and items 4 and 5 in the *laryngopharynx.*

The pharynx functions as the common passageway for the food bolus between the oral cavity and the esophagus and for air between the nasal cavity and the trachea. Infections in the pharynx may spread directly into the nasal cavity or into the auditory tubes. From these loci, infections may spread into the paranasal sinuses, the middle ear, or the mastoid air cells.

Branches of the external carotid arteries supply the pharynx with blood, which drains from this structure into the internal jugular vein. The six muscles of the pharyngeal walls are innervated from branches of the pharyngeal plexus, the external laryngeal and recurrent nerves, and the glossopharyngeal nerve.

Esophagus

The pharynx opens from above into the esophagus approximately on the level of the cricoid cartilage of the larynx, about even with the sixth cervical vertebra. The esophagus is a muscular tube approximately 24 cm. long, although it may vary between individuals from as short as 23 cm. to as long as 25 cm. It is separated from the cervical and thoracic vertebrae by connective tissues in the upper thorax and by the thoracic aorta and the azygos vein in the lower thorax. The esophagus pierces the diaphragm in the *esophageal hiatus,* a natural opening somewhat anterior and slightly to the left of the aortic hiatus, and terminates about 1 or 1½ cm. below the diaphragm on the level of the eleventh thoracic vertebra.

The esophagus has three parts: cervical, thoracic, and abdominal. The esophagus, in its thoracic part, passes first through the superior mediastinum and second through the posterior mediastinum. In the mediastinum the esophagus is in close relation to the other mediastinal viscera and is posterior to the trachea and bronchi. The esophagus, in its abdominal part (the *cardiac antrum*), terminates in the junction with the stomach in the cardiac orifice, which is guarded by the *cardiac sphincter.* The acute angle formed between the stomach and esophageal walls is named the *cardiac incisura.* In some individuals the abdominal esophagus or a part of the stomach or other viscera may extend upward through the hiatus if it is weakened or enlarged. This condition, named a *hiatus hernia,* can be demonstrated radiographically; approximately 1 per cent of patients having E.S.D. studies* are found to have a hiatus hernia.

As the esophagus descends through the thoracic cavity, it changes direction slightly at different levels; these deviations are often important to demonstrate radiographically. Shortly after the origination of the esophagus in the neck, the esophagus curves slightly to the left as it passes through the base of the neck; it then descends vertically for a short distance and curves left again near the level of the fifth, sixth, or seventh thoracic vertebra; below the seventh thoracic vertebra the esophagus *descends vertically* until it approaches the diaphragm, at which point the esophagus again turns left, pierces the diaphragm, and joins the stomach. The esophagus is best demonstrated radiographically with the patient in the R-A-O position, although the R-P-O position of the patient enables excellent visualization of the esophagus as seen in Fig. 5-17, A.

Striated muscle tissue comprises the wall of the upper two-thirds of the esophagus; nonstriated muscle tissue comprises the wall of the lower one-third of the esophagus and continues throughout the walls of the remainder of the digestive tube. The nonstriated muscles of the small and large intestinal walls maintain a continuous series of wavelike contractions that begin in the walls of the stomach. These contractions continue throughout the abdominopelvic parts of the digestive tube and are called *peristaltic waves.* This wavelike motion continues throughout day and night; the rate of contractions increases during digestion and decreases during quiescence; this is *true peristalsis.* True peristalsis does not occur in the esophagus; the food bolus passes through the esophagus with the aid of gravity in the upper one-third and by means of *pseudoperistalsis* in the lower two-thirds. Pseudoperistalsis (false peristalsis) occurs only during deglutition. The mucosal pattern of the abdominal esophagus and, sometimes, of the lower thoracic esophagus is usually parallel in direction with the length of the tube. These mucosal folds extend into the stomach and are continuous with the gastric mucosa.

*An E.S.D. study is a radiologic examination of the esophagus, stomach, and duodenum.

Radiologists perform fluoroscopic examinations of the esophagus with the patient in the erect position and in the Trendelenburg* position. Depending upon the type of the examination, the barium sulfate mixture, used as a contrast medium, will be varied from thin and watery to a thick paste. In addition to the radiologist's fluoroscopic report, spot films for permanent graphic record of a certain part of the esophagus may be required.

Among the pathologic conditions found in the esophagus are types of diverticula. *Diverticula* may be either of the *pulsion*† type, which occurs in the pharyngo-esophageal region, or the *traction*‡ type, which occurs more commonly in the middle one-third of the esophagus.

The esophageal artery and branches from the inferior thyroid, bronchial, left gastric, and left inferior phrenic arteries supply the esophagus with blood, which returns to the venous circulation via collateral branches connecting the systemic and portal systems. The veins receiving the esophageal blood are the inferior thyroid, azygos, hemiazygos, and gastric. Collateral communication usually exists between the gastric and esophageal veins. This and other collateral routes serve to relieve portal obstruction in the liver. The esophageal walls are innervated from the vagus and sympathetic nerve trunks.

Abdominopelvic cavity

The abdominal cavity includes both the abdominal and pelvic parts; it often is named the abdominopelvic cavity, since there is no dividing membrane between the parts. The abdominal part of this cavity contains the stomach and small and large intestines (except for the rectum and anal canal) and is bounded superiorly by the diaphragm; bilaterally and anteriorly by the abdominal muscles; posteriorly by the lower part of the thoracic spine, the entire lumbar spine, and the psoas and quadratus lumborum muscles; and inferiorly by the superior pelvic strait and the plane of the pelvic inlet. The pelvic part of the abdominal cavity is extraperitoneal and is inferior to the abdomen, lying, for the most part, in the hypogastric region.

The esophagus descends through the thoracic cavity (one of the two large divisions of the ventral cavity), pierces the diaphragm, and enters the abdominal cavity (the other large division of the ventral cavity of the body). The serous membrane of two layers lies immediately inside the thoracic cavity and covers the viscera. This is also true in the abdominal cavity. In this latter cavity the serous membrane is called the *peritoneum,* which consists of the parietal (outer) layer and the visceral (inner) layer. The potential space between these two layers is the *peritoneal cavity.*

Within the abdominopelvic cavity are viscera that are *extraperitoneal* and other viscera that are nearly or completely enclosed by the peritoneum. The

---
*Trendelenburg's position is a supine recumbent position of the patient on the radiographic table—the coronal plane angled, the pelvis from 30 to 40 degrees above the head, and the feet and legs hanging over the table end.
†Pulsion diverticula result from intraluminal pressure.
‡Traction diverticula result from an extrinsic pull.

extraperitoneal viscera lie against the walls of the abdomen or pelvis or in any position that permits peritoneal covering but not envelopment. These viscera are the duodenum, pancreas, kidneys, ureters, urinary bladder, and a considerable part of the large intestine (and the uterus in the female).

The nearly or completely enclosed viscera are the liver, stomach, small intestine (except the duodenum), transverse and pelvic colons, and spleen (and uterine tubes in the female).

Arising from the posterior abdominal wall is a loose fold of the peritoneum called the *mesentery,* which suspends the small intestine. Arising in the same manner as the mesentery are the *transverse mesocolon,* which suspends the transverse part of the colon, and the *mesocolon,* which suspends the pelvic part of the colon (see Fig. 5-5).

Description of the complete peritoneum and its extensions best begins with the region of the umbilicus. From this level the peritoneum spreads lateralward and upward to cover the inferior surface of the diaphragm, after which the peritoneum extends downward onto the superior surface of the right lobe of the liver and invests this organ. The peritoneum, in its investment of the liver, ascends on the inferior surface of the liver in the *great transverse fissure* (porta hepatis) and folds downward upon itself to meet that part of the peritoneum extending from the left side to form a double-layered part, the *lesser omentum.* The lesser omen-

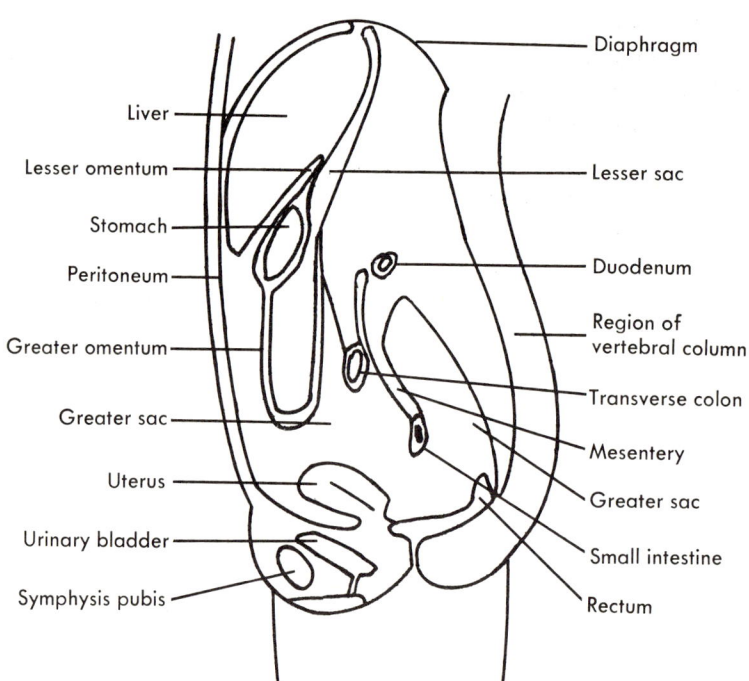

Fig. 5-5. Peritoneum, omenta, and mesentery in the median section of the female abdominopelvic cavity.

tum descends to the lesser curvature of the stomach, where the two layers separate to envelope the stomach and rejoin inferiorly on the greater curvature. The lesser omentum descends for a short space and then curves upward upon itself, forming the four-layered *greater omentum*. The greater omentum extends posteriorly to enclose the transverse colon and beyond the transverse colon to the posterior abdominal wall, where the layers again separate. One of these layers extends upward over the pancreas, aorta, inferior vena cava, and duodenum to the inferior surface of the diaphragm to cover the left part of the diaphragm. This layer next turns downward over the left lobe of the liver and enters the portal fissure to meet the layer of peritoneum from the right side and to form the lesser omentum.

The omentum, in its extensions throughout the abdominal cavity, separates the cavity into two sacs: the *greater* and *lesser sacs*. The principal part of the abdominal cavity is the greater sac; that part of the abdominal cavity lying posterior to the stomach is the lesser sac.

The anterior boundary of the lesser sac is formed by the anterior two layers of the greater omentum, the stomach, the lesser omentum, and the left part of the liver. The posterior boundary of this sac is formed by the posterior two layers of the greater omentum, transverse colon and mesocolon, and the peritoneum, which covers the upper part of the posterior abdominal wall.

Both the lesser and greater sacs are potential cavities separated from each other by the omenta. In the upper part of the lesser omentum and near its right border is the *epiploic foramen*, which serves as a pathway of communication between the two sacs.

The greater omentum possesses two most important protective functions. It has the ability to move toward and surround inflamed or infected parts, such as a ruptured appendix, thereby preventing the spread of infection and infectious material. The omentum also contains many macrophages in the areolar tissue that act as phagocytes upon many invading organisms and upon other debris which may enter the abdominal cavity.

Extending inferiorly along the posterior abdominal wall from the site of origin of the mesentery, the peritoneum forms the lining of the pelvic cavity to a level approximately 7.5 cm. above the anus. The pelvic part of the peritoneum covers the upper part of the uterus in the female, the upper part of the urinary bladder, the seminal vesicles in the male, and the rectum. The peritoneum forms a complete seal from the exterior in the male, but not in the female, since the uterine tubes open into the uterine lumen exteriorly and into the abdominal cavity interiorly.

Stomach

The stomach, a pouchlike enlargement of the digestive tube, lies immediately inferior to the diaphragm, principally in the epigastric and left hypochondriac abdominal regions. The stomach may sometimes extend into the right hypochondriac and/or umbilical regions. Two sagittal and two transverse lines (planes)

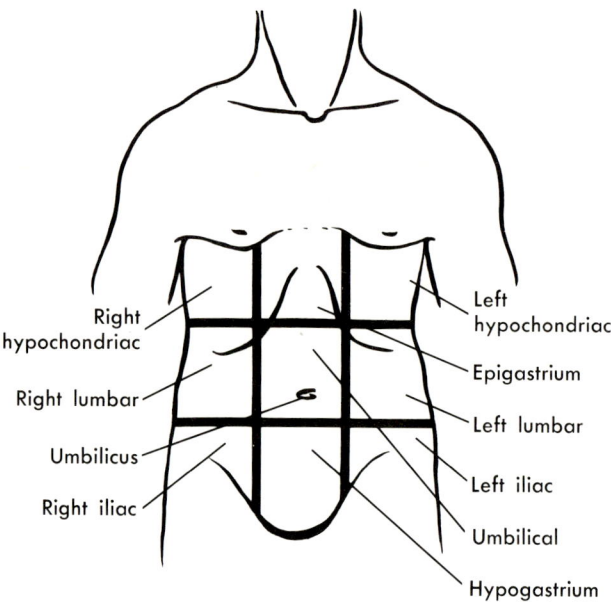

Fig. 5-6. The nine abdominopelvic regions.

Table 5-1. Major viscera of the nine regions of the abdomen

Right hypochondrium	Epigastrium	Left hypochondrium
Most of right lobe of liver Hepatic flexure of colon Part of right renal body	Most of left lobe and remainder of right lobe of liver Gallbladder Most of stomach, duodenum, and pancreas Part of spleen Suprarenals Parts of renals	Greater curvature of stomach Remainder of spleen Tail of pancreas Splenic flexure of colon Part of left renal body
Right lumbar	*Umbilical*	*Left lumbar*
Ascending colon Most of right renal body	Most of transverse colon Parts of duodenum, jejunum, and ileum Parts of renal bodies Most of ureters	Descending colon Part of jejunum Most of left renal body
Right iliac	*Hypogastrium*	*Left iliac*
Cecum Appendix Terminal end of ileum Ileocecal valve	Ileum Flexure of sigmoid colon	Sigmoid colon Jejunum Ileum Iliac colon

From Jacobi, Charles A., and Paris, Don Q: Textbook of radiologic technology, ed. 5, St. Louis, 1972, The C. V. Mosby Co.

divide the abdomen into nine regions. The sagittal lines are the right and left lateral lines. A lateral line extends vertically through a point halfway between the midline and an anterior superior iliac spine. The superior transverse line is the transpyloric line. The transpyloric line extends transversely through a point halfway between the sternal jugular notch and the superior border of the symphysis pubis. The inferior transverse line is the transtubercular line. The transtubercular line extends transversely between the iliac tubercles and across the body of the fifth lumbar vertebra (see Fig. 5-6 and Table 5-1).

Four layers of tissue form the stomach walls; they are, from inside to outside, the *mucous, areolar, muscular,* and *serous layers.* The mucous layer is quite smooth, and its thickness increases from near the cardiac end as it approaches the pylorus. Specialized glands that manufacture the juices for gastric digestion form parts of the mucous layer, which is arranged in folds called *rugae* and which extends longitudinally with the long axis of the stomach. The rugal folds extend into the lumen of the stomach and form crypts* between each two folds; thus, the digestive surface of the stomach enlarges. The gastric glands pour their secretions into the crypts, and the churning motion of the stomach mixes the gastric juices with the food bolus during gastric digestion. The rugae and crypts are visualized radiographically when the stomach fills with barium sulfate suspension; complete distension of the stomach wall will obviate the rugal pattern. Because the rugal pattern and its visibility have clinical significance, it is necessary to demonstrate the rugae for radiologic diagnosis.

The mucosa (mucous layer) of the stomach is a loose arrangement of tissue that changes constantly under the varying conditions of diet, nutrition, and health. The changes evidenced in the mucosal pattern indicate corresponding contractions in the muscularis mucosa.

The areolar layer is very thin and consists of areolar tissue.

The muscular layer contains three sublayers of nonstriated muscle tissue. The fibers of the innermost sublayer are arranged in an oblique fashion extending from the pylorus to the esophagus. The fibers of the middle sublayer extend in a circular fashion, completely envelop the oblique layer, and encompass the pyloric valve to form the *pyloric sphincter*—the outlet of the stomach. The fibers of the outermost sublayer extend longitudinally from the esophageal wall to the pyloric (lower) end of the stomach.

Muscular tone divides the stomach into four types, which are *orthotonic, hypertonic, hypotonic,* and *atonic.* These types of stomach are associated with the corresponding body types: *sthenic, hypersthenic, hyposthenic,* and *asthenic.*

The sthenic (normal) type of individual has his stomach in a normal position, which is usually even with the umbilicus. Sthenic persons also have normal emptying time of the stomach and a normal position of the colon, i.e., no appreciable ptosis.

The hypersthenic type of individual is heavy and has a large body, a short

---

*A crypt is a minute tubular depression opening on a free surface.

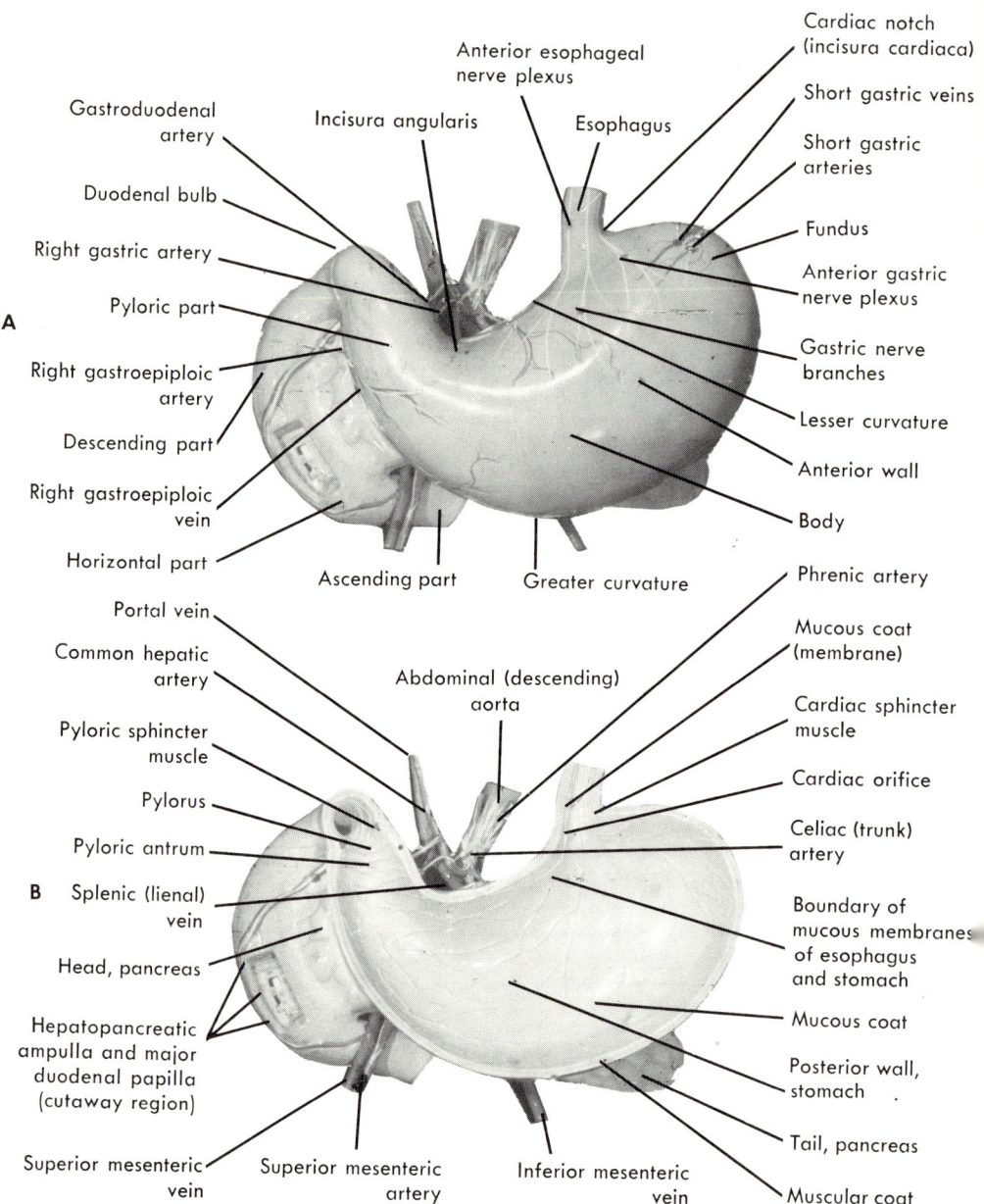

Fig. 5-7. **A**, Stomach and duodenum, anterior surfaces. **B**, Stomach, anterior wall removed showing internal surface of posterior wall. **C**, Stomach, posterior view.

thorax, wide intercostal angles, and a broad abdomen. Such a person has his stomach high and on the left side. The stomach is small, is transverse in position, is shaped somewhat like a cow's horn, and empties rapidly. The hepatic and splenic flexures of the colon are high, so the transverse colon is usually above the umbilicus. The sigmoid colon is usually quite short in the hypersthenic type of individual.

The hyposthenic type of individual is thin and has a long, narrow thorax and a wide pelvis. Such a person has his stomach, which is elongated and fishhook shaped, below the umbilicus. The stomach empties sluggishly, but emptying is complete within 5 hours. Both the cecum and transverse colon are somewhat ptosed.

The asthenic type of individual is a strongly exaggerated hyposthenic type. Such a person has a ptosed stomach, which is often situated on the pelvic floor. The stomach is atonic (without tone) and empties quite sluggishly, but usually within 6 hours. Both hyposthenic and asthenic persons may suffer from *visceroptosis*, the prolapse or falling down of the abdominal viscera. This condition is known also as abdominal ptosis, *Glenard's disease*.

The serous layer (coat) is a part of the peritoneum that covers the stomach wall, except for the following places:
1. A small portion on the posterior surface of the fundus that is in contact with the diaphragm
2. Along both curvatures (the lesser and greater) where the stomach attaches to the greater and lesser omenta

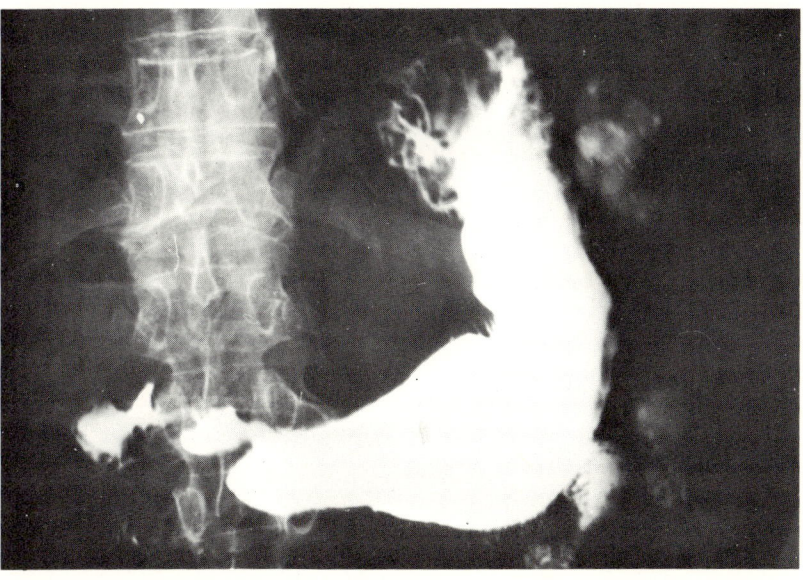

Fig. 5-7, cont'd. For legend see opposite page.

The stomach presents two *curvatures*—the *lesser* curvature in the superior surface and the *greater* curvature in the inferior surface—and three parts—the *fundus*, the *body*, and the *pyloric part*. The body, the largest of the three parts of the stomach, is between the fundus and pyloric part. The fundus is the superior end of the body of the stomach and lies above the opening of the esophagus. The upper part of the body and the fundus are often included in a general term "the cardia." The *incisura angularis* (gastric notch), the deeply indented notch in the lesser curvature, is nearer to the pylorus than the cardia. An oblique plane extending from the incisura through the greater curvature, generally toward the left greater trochanter, separates the body and pyloric parts. The incisura is demonstrated radiographically in normal stomach radiographs (see Fig. 5-7).

A small sphincter muscle, the *cardiac sphincter,* surrounds the cardiac orifice of the stomach, and the very strong *pyloric sphincter* surrounds the pyloric outlet into the duodenum. The pyloric sphincter opens to permit passage of the *chyme,* the gastric contents, from the stomach into the duodenum. The pyloric antrum, pylorus, and duodenal bulb act as a unit to control (open and close) the pyloric sphincter. A current theory with wide acceptance is that a pressure gradient exists between the pyloric antrum and the duodenal bulb. The difference in pressure between these spaces equals from 3 to 4 cm. of water. This pressure difference is explained on p. 223.

During digestion, the three layers of the muscularis mucosa keep the food bolus in constant motion, which thoroughly mixes the bolus with the gastric juices; the net result of this action is the breakup of the bolus into the more fluid chyme. Peristaltic waves begin in the cardia of the stomach and extend throughout the remainder of the digestive tube. Visualization of peristaltic waves is clinically significant in fluoroscopic and radiologic diagnoses. Routine radiographic procedure usually requests several exposures of the pylorus and duodenal bulb. Serial exposures are obtained by use of either a spot-film device or a tunnel.

Numerous gastric glands are situated in the bottom of the *gastric pits,** minute depressions found throughout the gastric mucosa, and empty their secretions directly into the pits. The location of the gastric glands classifies them as follows:

1. *Fundic* (peptic) glands, found throughout the greater part of the gastric mucosa
2. *Pyloric* glands, found only in the mucosa of the pyloric region
3. *Cardiac* glands, found only in the cardiac third near the orifice

The use of the terms "fundic" or "peptic," as applied to certain of these glands, is somewhat misleading, since both terms indicate that the glands are located only in the fundus or that they manufacture the precursors of pepsin. Mucus is the principal product of the pyloric and cardiac glands; the essential digestive substances of the gastric juice are secreted by the fundic glands.

---

*A pit is a hollow fovea or depression.

All three types of gastric glands are simple or branched tubular glands, although the cardiac glands are less typical and frequently resemble the glands of the tunica propria in the distal end of the esophagus.

Small intestine

The small intestine consists of three subdivisions that do not possess readily distinguishing characteristics from each other. The subdivisions of the small intestine, beginning with the gastric end, are the duodenum, the jejunum, and the ileum. The characteristics of each will be discussed separately.

Duodenum. The duodenum, the first part of the small intestine, is located, for the most part, in the epigastric region, although it may extend into the right hypochondriac and/or umbilical regions. The duodenum is slightly less than 1 foot in length, the average being from 22.5 to 29 cm. in length. A considerable part of food absorption occurs through the villi* of the walls of the duodenum, although most of the products of digestion are absorbed through the walls of the combined jejunum and ileum.

Directionally, the duodenum is divided into four parts as follows:
1. The *bulb,* or *cap,* which is immediately adjacent to the pylorus
2. The *descending part,* which is the longest single part and into which the pancreatic juices and the bile empty
3. The *transverse part,* which makes a rather sharp turn to the left and encompasses, along with the bulb and descending part, the head of the pancreas
4. The *ascending part,* which arises from the transverse part approximately at the midsagittal plane to join the jejunum in the *duodenojejunal flexure* (The flexure forms by the abrupt turn of the duodenum in an anterior and right lateral direction) (see Fig. 5-8).

Radiologic technologists attach little significance to the lack of histologic differentiation of the four parts of the duodenum; however, both radiologists and radiographers recognize and appreciate the radiographic differentiation of these parts. Diverticula occur only at the junctions of the transverse with the descending portions of the duodenum and of the ascending with the transverse portions of the duodenum. *A diverticulum is a blind pouch or pocket leading off from a main cavity or tube.* In certain pathologic conditions a false diverticulum may be visualized in other parts of the intestines. A false diverticulum results when the mucous membrane protrudes through a tear in the muscular coat.

The duodenal wall layers are continuations of the stomach wall layers and are named, as is the stomach, from inside to outside as follows: the mucosa, submucosa, muscularis mucosa, and serosa. Numerous folds of the mucosa, visible radiographically by the aid of contrast media, extend into the lumen of the

---

*A villus (singular of villi) is a small vascular process or protrusion, especially such a protrusion from the free surface of a membrane.

small intestine. These folds, arranged in an annular pattern, have several names, among which are the *valvulae conniventes, plicae circularis,* and *valves of Kerckring*. The valvulae conniventes usually are absent in the upper duodenum, quite pronounced in the jejunum, and decreasingly prominent in the ileum as it progresses toward the colon. In addition to the valvulae conniventes, the mucosa extends into the lumen in numerous fingerlike projections, the villi.

The villi are leaf shaped in the duodenum, round in the jejunum, and club shaped in the ileum. The increased surface afforded by the villi enormously increases the digestive and absorptive surface of the entire small intestine. The *crypts of Lieberkühn* appear as simple glandular pits between the villi. The crypts of Lieberkühn are minute tubes or depressions in the mucosa of the small intestine; they have secretory function. Action of the muscularis mucosa causes the villi to be in a state of continuous motion, similar to the motion of the kneading of bread dough. This motion forces the fluid chyme between the villi for further hydrolysis and eventual absorption through the villi walls into the venous tributaries of the portal circulation.

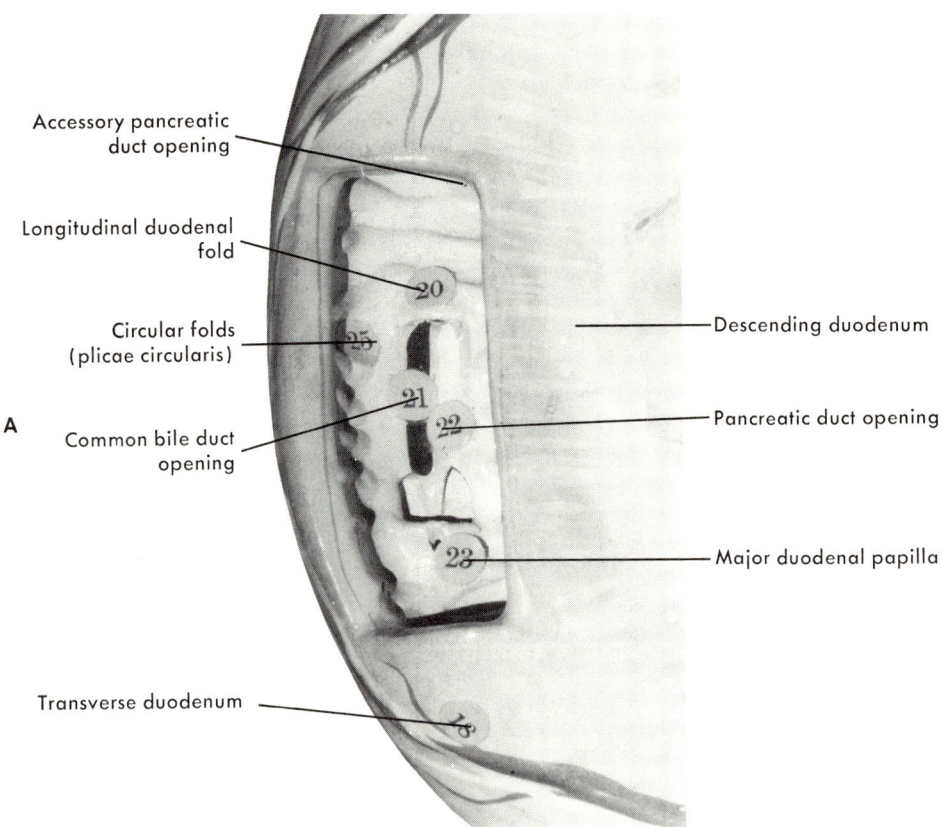

Fig. 5-8. **A,** Region of the hepatopancreatic ampulla. **B,** Section through duodenal wall.

Each villus has its individual arterial blood supply from the right gastric and pancreaticoduodenal branches of the superior mesenteric artery. Blood drains from each villus in a venous branch terminating either in the splenic (lienal) or the superior mesenteric veins, which join between the pancreatic head and the inferior vena cava to form the portal vein at about the level of the second lumbar vertebra. The duodenal lymphatics consist of both anterior and posterior sets opening into the pancreaticoduodenal nodes. There is a single lymph vessel in each villus adjacent to the vascular supply.

Within the crypts of Lieberkühn are numerous coarsely granular cells, the *cells of Paneth*, which probably are serozymogenic.* These cells are known to pour a special secretion into the small intestine. This secretion is recognized as an integral part of the digestive juices, but its nature is not completely established. The crypts of Lieberkühn also contain the *argentaffine cells*, whose function is not completely established.

---

*A serozymogenic cell is one that produces a watery secretion containing an enzyme.

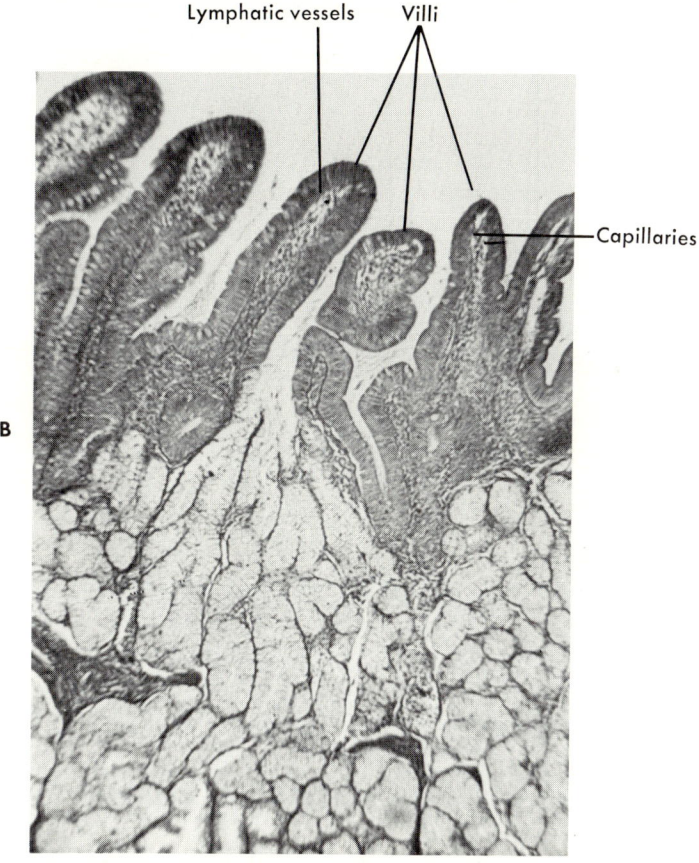

Fig. 5-8, cont'd. For legend see opposite page.

Within the wall of the duodenum and approximately 7.5 cm. inferior to the pylorus in the posteromedial margin is a small dilation formed by the junction of the pancreatic duct (duct of Wirsung) with the common bile duct. The dilation is the *hepatopancreatic ampulla* (ampulla of Vater), which opens into the duodenal lumen through the *major duodenal papilla*. The small *choledochal sphincter* (sphincter of Oddi) closes the terminal end of the common bile duct.

Jejunum. The jejunum joins the duodenum in the duodenojejunal flexure and extends for approximately 275 cm. (9 feet) in the umbilical, left lumbar, and left iliac regions to terminate in its junction with the ileum. The numerous circular folds of mucosa in the jejunal walls are larger than the circular folds in the duodenal walls. The larger folds present less surface area for absorption of food elements into the portal vein tributaries (see Fig. 5-9).

Ileum. The ileum is narrower than either the duodenum or jejunum; its wall is thinner and contains fewer blood vessels than the jejunal wall. The circular folds of the ileal mucosa diminish rapidly in number and size toward

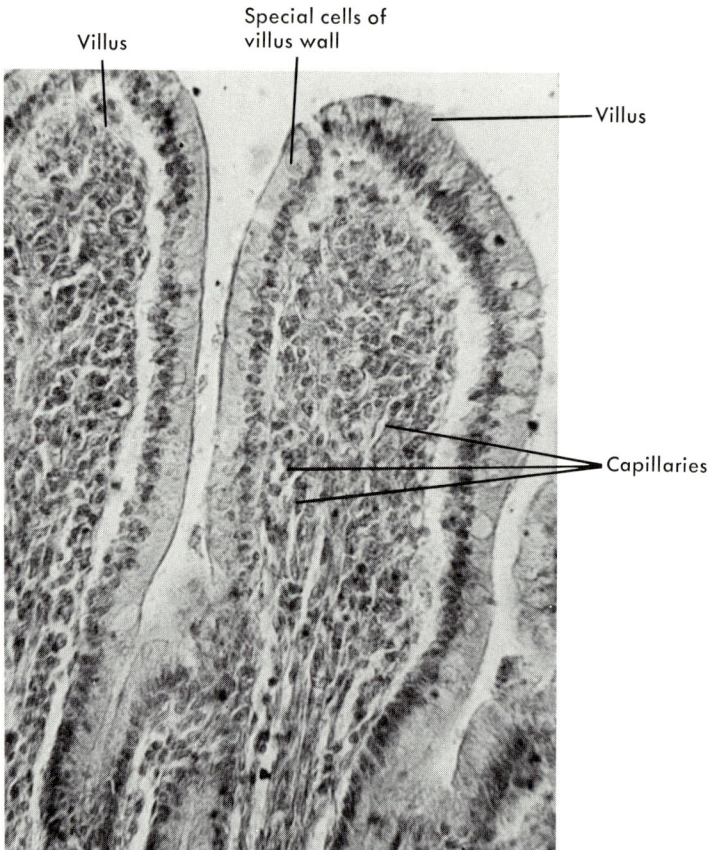

Fig. 5-9. Photomicrograph of villi of the jejunum. (No lymphatics appear in this upper section of the jejunum.)

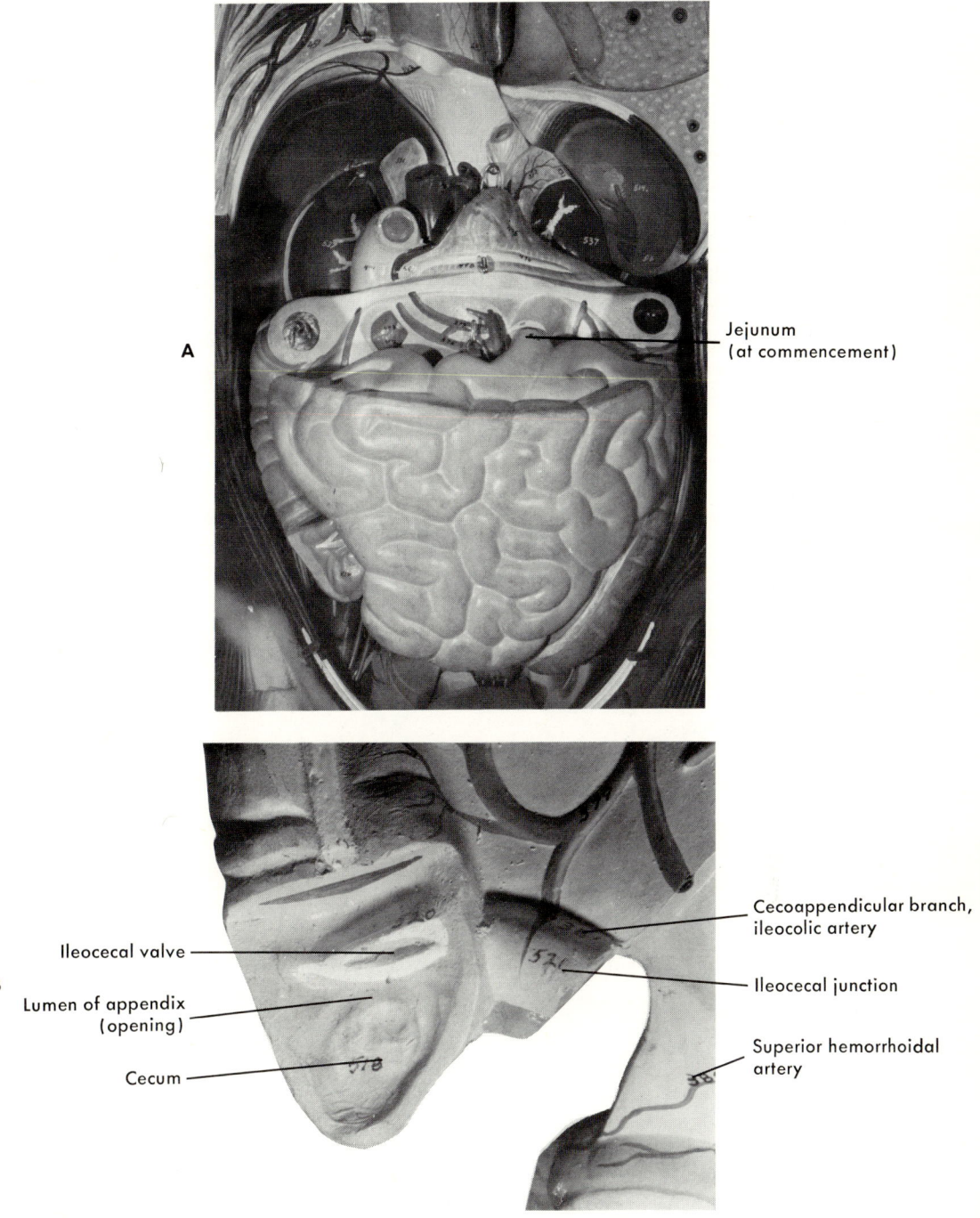

Fig. 5-10. A, Jejunum and ileum. The transverse colon is removed; jejunum joins ileum after extending approximately 275 cm. beyond the duodenojejunal flexure. **B,** Junction of small with large intestines.

its terminus. The ileum extends approximately 400 cm. (13 feet) through the umbilical, hypogastric, left iliac, and right iliac abdominal regions (see Fig. 5-10, A). The ileum terminates in the right iliac region in its junction with the cecum (see Fig. 5-10, B). This junction contains the *ileocecal valve,* which consists of two folds of mucosa that prevent, in normal conditions, a reverse flow of fecal material into the ileum. Barium enema radiography of the colon reveals patency of the ileocecal valve in a fairly high percentage of patients examined.

The blood supply to the stomach derives from the celiac trunk; the blood supply of the small intestine derives from the celiac trunk and the superior mesenteric arteries. Blood drains from the stomach into tributaries of the portal vein or directly into the portal vein; blood drains from the small intestine into the superior mesenteric vein. Stomach innervation derives from both the left phrenic nerve and gastric branches of the vagus nerve; sensory and autonomic nerves supply the stomach. Sensory and autonomic nerves supply the small intestine from the celiac and superior mesenteric plexuses.

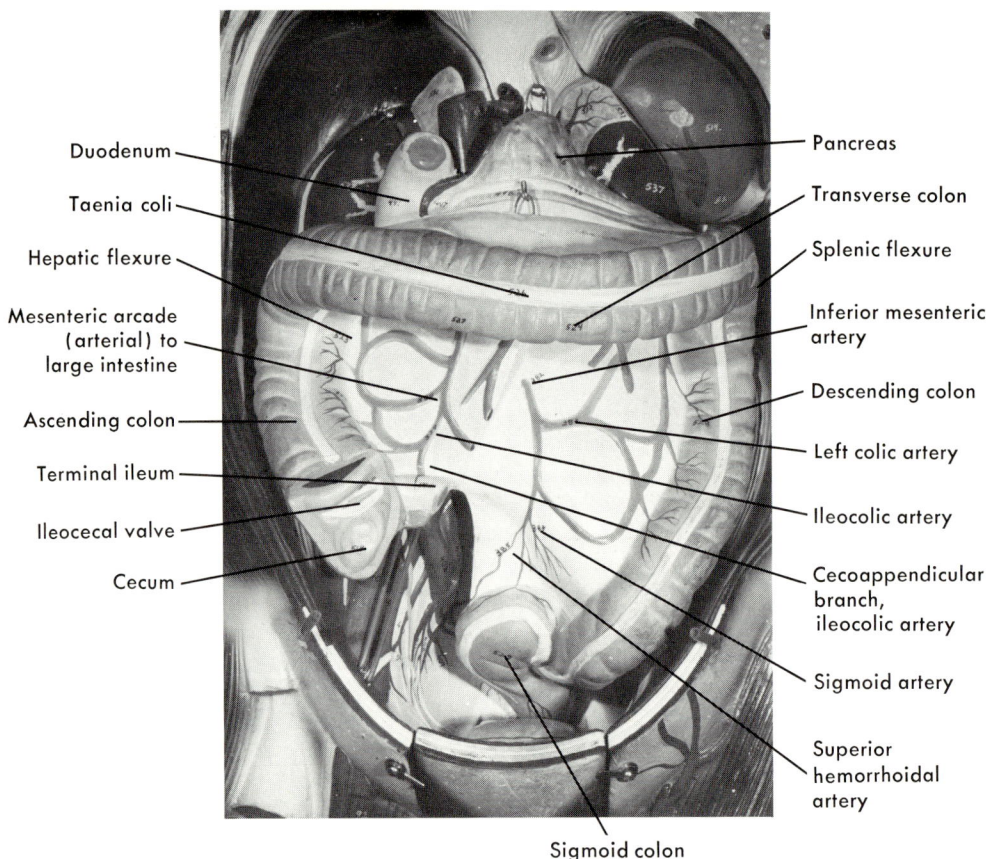

Fig. 5-11. Large intestine and associated structures.

Large intestine

The large intestine varies in length between individuals; it may be from 152 to 168 cm. (5 to 5.5 feet) long. It includes the cecum, colon, rectum, and anal canal and terminates with the anus. The large intestine normally extends throughout each of the nine abdominal regions; the transverse colon usually passes across the abdomen in both the epigastric and umbilical regions (see Fig. 5-11).

The layers of the large intestine are continuations of the layers of the small intestine with some notable differences. In the large intestine the mucosa supports neither villi nor valvulae conniventes; consequently, the surface area is greatly reduced. The reduced surface area appears to have no appreciable effect upon the volume of water absorbed through the walls of the large intestine. Approximately 3 liters of water are absorbed in each 24-hour period through the walls of the large intestine compared with approximately 5 liters in each 24-hour period absorbed through the walls of the small intestine.

The large intestine mucosa contains numerous intestinal glands, which are considerably longer than the intestinal glands of the small intestine and contain many goblet cells and fewer argentaffine cells. The mucosa secretes large quantities of mucus to lubricate the inner intestinal wall to permit easy passage of the fecal matter.

A well-defined inner, circular layer and a partly deficient outer, longitudinal layer of nonstriated muscle tissue forms the muscularis mucosa. The outer layer consists of three strong bands, the *taeniae coli,* equally spaced in the layer. The taeniae coli contract to pull the intestinal wall into a series of sacculations, the *haustral markings* seen in radiographs of healthy colons filled with contrast media. The haustral churnings complement the peristaltic movement, which propels the fecal contents toward the anus.

Cecum. The cecum lies below its junction with the ileum in the right iliac region and joins superiorly with the ascending colon. The cecum is the first part of the large intestine. The appendix usually arises near the inferior margin of the fundus of the cecum and extends both inferiorly and medially from its posterior surface. A healthy appendix may fill with barium sulfate suspension during a barium enema examination and thus become visible in fluoroscopy or cast a distinguishable shadow in radiography.

Ascending colon. The ascending colon begins at the superior margin of the cecum and is slightly smaller in diameter than the cecum. The ascending portion of the colon rises to the inferior margin of the right lobe of the liver, slightly to the right of the gallbladder. Immediately superior to the gallbladder the colon turns abruptly forward and medialward to form the transverse colon. The angle so formed is the *hepatic flexure* of the colon; the hepatic flexure is usually on a level with the ninth and/or tenth ribs posteriorly.

Transverse colon. The transverse colon usually extends transversely across the abdomen along the imaginary line between the epigastric and umbilical abdominal regions. This segment of the colon turns sharply upward, posteriorly and to the left, then forward and downward in the region of the spleen. The angle so formed, the *splenic flexure* of the colon, is somewhat higher than the hepatic flexure, usually on the level of the eighth left rib in the midaxillary line.

Descending colon. The descending colon extends downward through the left side of the abdomen, adjacent to the lateral margin of the left kidney. At the inferior surface of the left kidney the descending colon turns toward the midline of the body until the colon reaches the space between the iliopsoas and quadratus lumborum muscles. At this point the colon again turns downward toward the iliac crest. In this region of the left ilium, the colon is often named the iliac colon.*

Iliac colon. The iliac colon extends for approximately 15 cm. between the left iliac crest and the sigmoid portion of the colon at the superior aperture of the lesser pelvis.

Sigmoid colon. The sigmoid colon, the fourth and terminal part of the colon, derives its name from its S-shape. The sigmoid colon usually extends for approximately 40 cm. in the pelvic portion of the abdominal cavity; however, part of the sigmoid colon may be in the lower portion of the umbilical region. After a somewhat tortuous path the sigmoid colon terminates in the rectum immediately inferior to a level even with the third sacral segment.

Rectum. The rectum lies between the sigmoid colon and the anal canal. In man the rectum is quite tortuous and extends forward to the apex of the prostate gland. The rectum is about 12 cm. long and terminates in the rectal ampulla at the proximal end of the anal canal.

Anal canal. The anal canal is the terminal part of the large intestine and may vary between individuals, from 2.5 to 4 cm. in length. The anal canal is entirely inferior to the peritoneum and terminates in the levatores, sphincter ani internus, and sphincter ani externus muscles.

The cecum and colon receive arterial blood from the colic and sigmoid branches of the mesenteric arteries. The rectum and anal canal receive arterial blood from the inferior mesenteric, hypogastric, and internal pudendal arteries and from lesser branches of these arteries. Blood from the rectum drains into the hemorrhoidal plexus, which has indirect communication with the systemic and portal circulations. Blood from the remainder of the large intestine drains either into the superior mesenteric or the inferior mesenteric veins, which are a part of the portal circulation.

The lymphatics of both the small and large intestines are described on pp. 257 to 259.

Sensory and autonomic nerves supply the large intestine from the celiac, superior mesenteric, inferior mesenteric, and hypogastric plexuses.

Accessory organs

The accessory organs of the digestive system found in the abdominal cavity are the liver, the pancreas, and the several structures that comprise the biliary

---

*The iliac colon is considered by some authorities as a separate part of the descending colon.

system, the gallbladder, and the biliary ducts. Radiography occupies a most important place in the diagnosis of pathology in these structures.

Liver

The liver is the largest gland in the body; it weighs from 2.5 to nearly 4 kilograms* (3 to 4 pounds) and contains about seven times more tissue than is necessary to maintain life (see Fig. 5-12).

The liver consists of two major lobes: the left lobe and the right lobe. The right lobe is further subdivided into the caudate and quadrate lobes; thus, four

*1 kilogram (kg.) = 1,000 grams = 2.204622341 pounds avoirdupois.

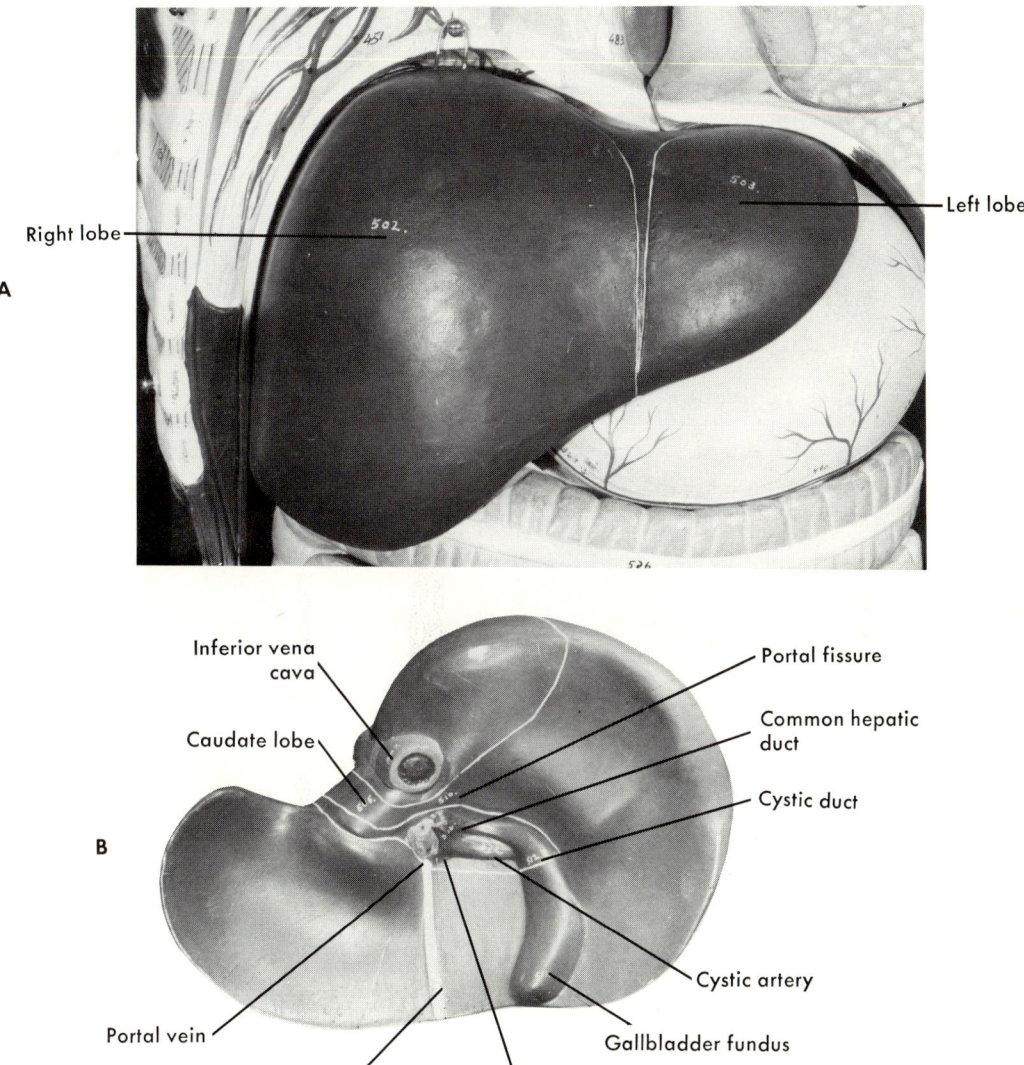

Fig. 5-12. Liver. **A,** Ventral surface. **B,** Inferodorsal surface (inverted).

lobes comprise the liver. The right lobe lies directly beneath the right dome of the diaphragm and is the largest of the four lobes. The left lobe lies beneath the left dome of the diaphragm. The quadrate lobe is the quadrilateral area situated both inferiorly and posteriorly on the right lobe and anterior to the portal fissure between the gallbladder and the fossa for the round ligament. The caudate lobe is situated superiorly and posteriorly above and to the left of the quadrate lobe and between the inferior vena cava and the fossa for the ligamentum venosum (the remains of the ductus venosus). The entire right lobe lies in the right hypochondrium, with the exception of a part that extends into the right lumbar region. For the most part, the left lobe lies in the epigastrium, with a small part extending into the left hypochondrium.

The superior surface of the liver is in contact with the diaphragm and presents a small *bare area* not covered with the peritoneum. A fossa on the inferior surface of the right lobe of the liver lodges the gallbladder. A connective (areolar) tissue sheath, *Glisson's capsule,* encases the entire liver and passes into the liver through the portal fissure as the enveloping membrane of the hepatic artery, portal vein, and bile ducts.

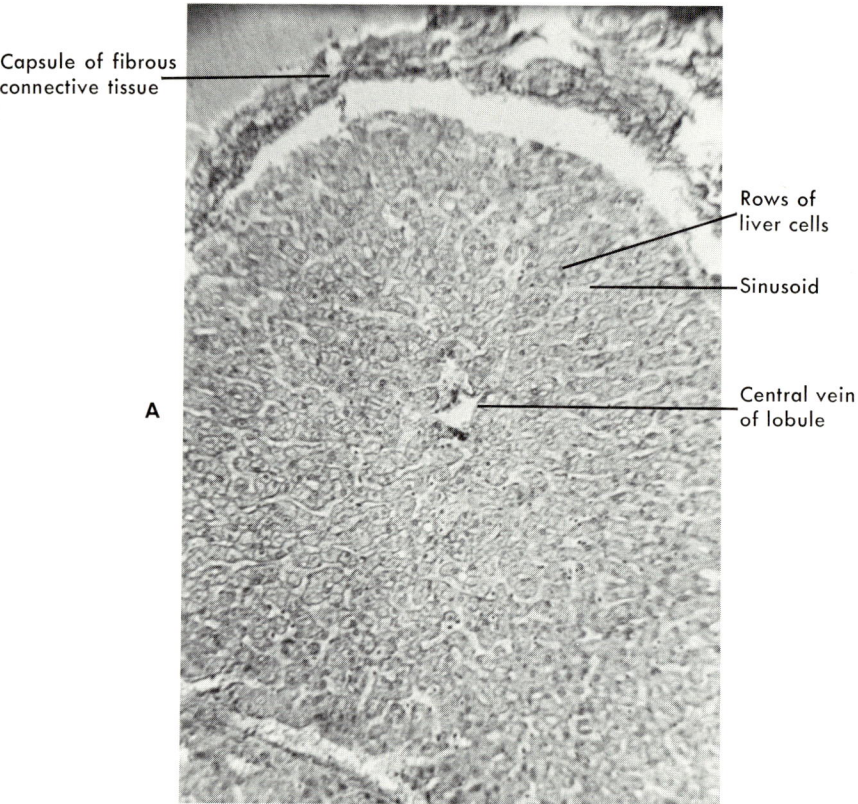

Fig. 5-13. **A,** Cross section of a liver lobule. **B,** Section of a liver lobule showing the central vein.

Five ligaments suspend the liver from the diaphragm and the anterior wall of the abdomen. One of these, the *round ligament,* is a fibrous cord that is the obliterated umbilical vein. The four remaining ligaments are the *falciform ligament,* the *coronary ligament,* and the two *lateral ligaments.* The visceral layer of the peritoneum reflects from the liver in various loci to the parietal layer of the peritoneum and forms these ligaments. The falciform ligament extends anteriorly along the anterior abdominal wall to the umbilicus. The right and left lateral (triangular) ligaments and the coronary ligament extend superiorly from the liver to attach it to the diaphragm. The round ligament extends between the umbilicus and the portal fissure and is embedded in the free border of the falciform ligament. The *portal fissure* (porta hepatis) is situated in the inferior surface, centrally between the right and left lobes. The hepatic artery, portal vein, lymphatics, bile (hepatic) ducts, and nerves pass into and/or from the liver through the portal fissure.

Fibrous bands from Glisson's capsule and branches of the blood vessels pass

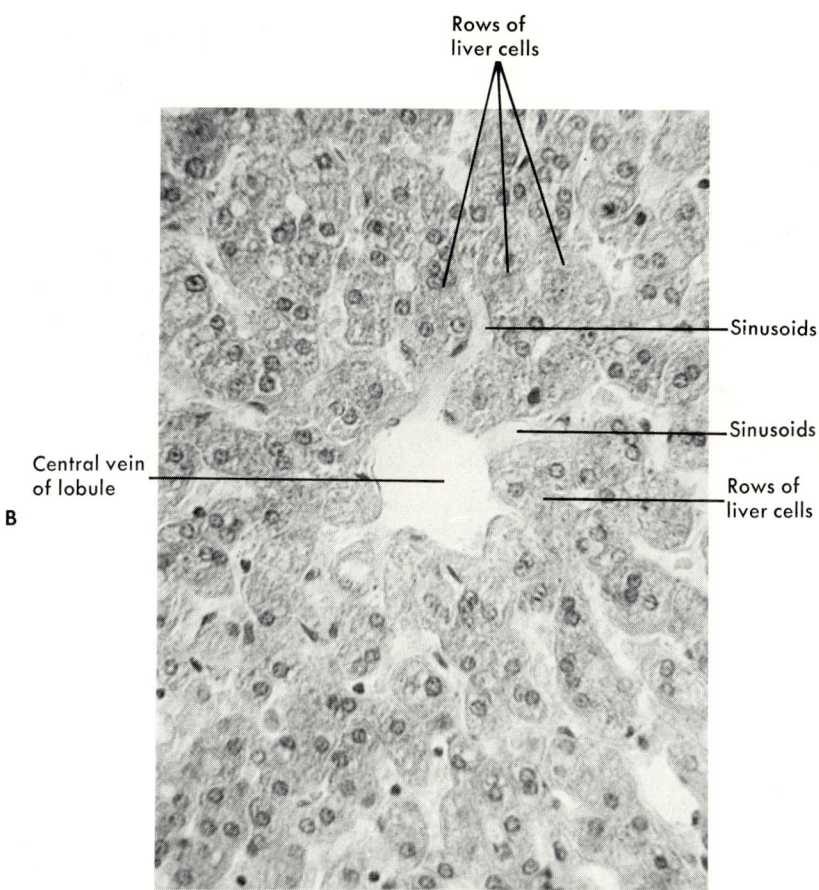

Fig. 5-13, cont'd. For legend see opposite page.

into the substance of the liver and further divide it into its functional units, the *liver lobules.* Each lobule measures about 2 mm. × 1 mm. × 1 mm. The lobules are incompletely separated from each other in man. Each lobule consists of a *central intralobular vein* surrounded by rows of hepatic cells arranged radially about the vein in the center. The several intralobular veins finally drain into the hepatic vein. Between each two rows of hepatic cells is an irregular space, the *liver sinusoid.* The liver sinusoids are the terminal areas for the hepatic artery and portal vein branches. The branches of the portal vein, as interlobular veins, connect with the central intralobular vein, and the interlobular bile ducts connect with the hepatic duct. Macrophages, named Kupffer's cells, are found in the linings of the liver sinusoids (see Fig. 5-13).

The innervation for the liver derives from the vagus nerve and the celiac plexus.

Although many of the functions of the liver do not relate directly to the processes of either digestion or food assimilation, liver functions are grouped and discussed on pp. 224 and 225.

Gallbladder

The relation of the gallbladder to the liver is referred to in discussion of the anatomy of the liver. The gallbladder, the terminal extension of the cystic duct, is a pear-shaped sac lodged in its fossa inferiorly on the right lobe of the liver. The gallbladder consists of a neck, body, and fundus; it extends from 7 to 10 cm. between the portal fissure and the anterior margin of the liver. The fundus of the gallbladder usually extends somewhat inferior to the inferior margin of the right lobe of the liver. *Radiographic location of the gallbladder places it 3 inches to the right of and at the level of the second lumbar vertebra;* however, it may be ptosed with the fundus as low as the pelvic cavity. To demonstrate the gallbladder radiographically, it is often necessary to employ the cone-down, erect, and decubitus* techniques and positions; sometimes with the breath in, and others with the breath out (see Fig. 5-14).

The gallbladder and associated duct walls consist of an inner mucous membrane layer, a middle nonstriated muscle layer, and the outer peritoneal layer. On the average, the gallbladder contains between 30 and 50 ml. of fluid and is capable of concentrating the fluid to approximately one-twelfth of its original volume.

The arterial blood supply of the gallbladder derives from the cystic artery; the blood drains from the gallbladder walls into the liver capillaries and into the portal vein. Innervation of the gallbladder derives from vagus nerve branches in the celiac plexus.

---

*A decubitus position refers to a situation wherein the skin of a certain surface of the body is in contact with the table top. The patient is recumbent (horizontal) and the central ray is horizontal. In a left lateral decubitus position the patient's left side is on the table top and the central ray (usually) passes from anterior to posterior through the body.

The digestive system

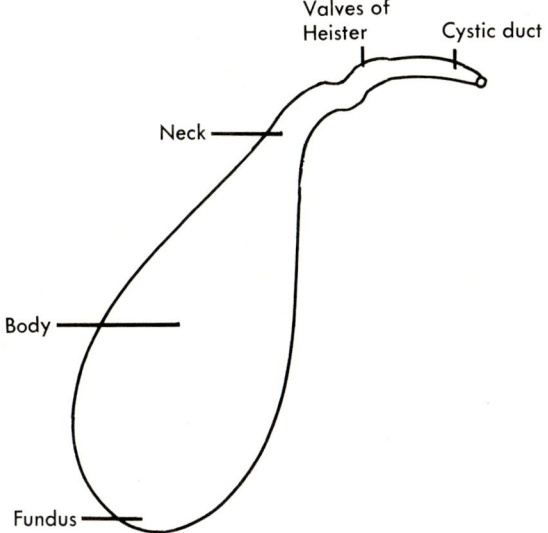

Fig. 5-14. Gallbladder, anterior aspect.

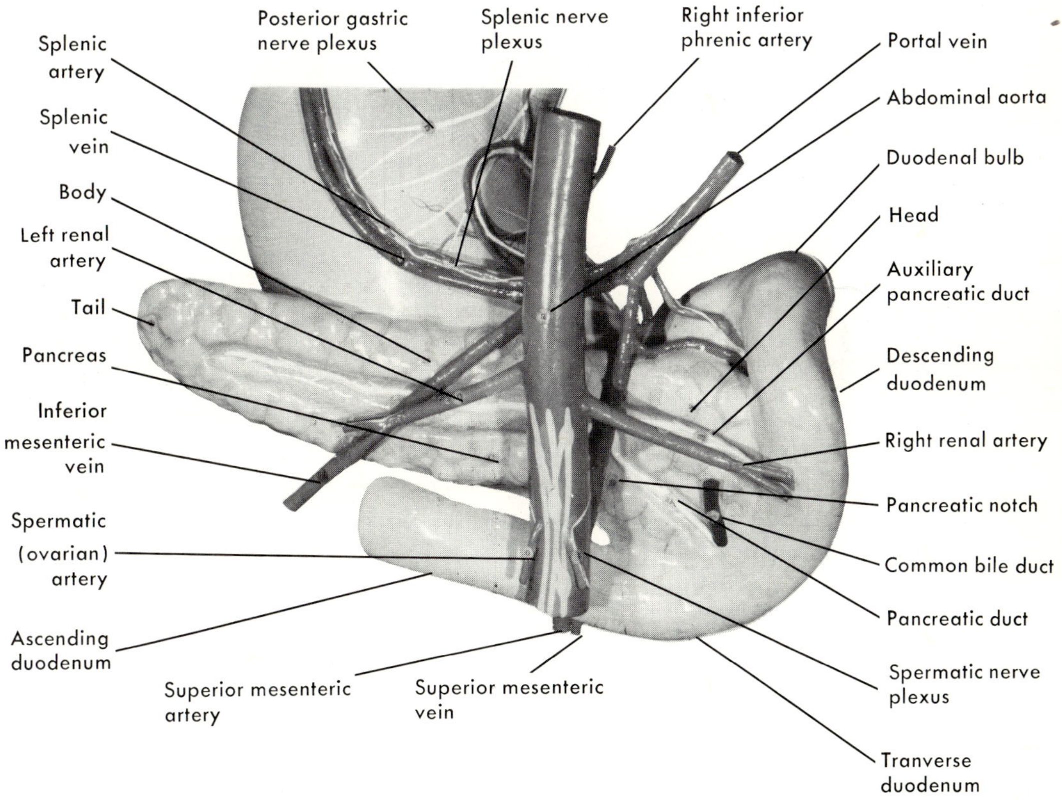

Fig. 5-15. Pancreas, duodenum, and associated structures. Posterior aspect.

Pancreas

For the most part the pancreas lies in the posterior part of the abdominal cavity behind the epigastric and left hypochondriac regions. The pancreas is a compound racemose gland, which secretes *insulin* from the *islands of Langerhans* in addition to secreting enzymes—lipase, amylase, and the precursor to trypsin—and an alkaline substrate through the *pancreatic duct* (duct of Wirsung) (see Fig. 5-15).

The *neck* of the pancreas, a slightly narrowed transverse portion, separates the *head* from the *body* of the pancreas, which extends toward the left side of the abdominal cavity. The gland continues to extend posteriorly, terminating in the *tail*, which is enveloped by the peritoneum, the remainder of the pancreas being retroperitoneal.

The duodenum entirely encompasses the head of the pancreas. The *auxiliary pancreatic duct* (duct of Santorini) arises from the pancreatic duct in the region of the head and neck, extends to the duodenum, and empties into it through the *minor duodenal papilla,* approximately 2.5 cm. superior to the major duodenal papilla. In many persons the common bile duct and the pancreatic duct open separately into the duodenum. In the latter instance there is no apparent hepatopancreatic ampulla, and the major duodenal papilla may be the outlet of either or both the common bile duct and the pancreatic duct. In other instances there may be two hepatic ducts extending between the liver and the duodenum: one receiving the cystic duct and joining with the duodenum at the site of the minor duodenal papilla and the other joining with the pancreatic duct to join the duodenum at the site of the major duodenal papilla. In either instance the pancreatic enzymes flow into the duodenum to effect their part of the hydrolytic actions of digestion.

The blood supply of the pancreas derives from the splenic and pancreaticoduodenal branches of the hepatic and superior mesenteric arteries. Innervation of the pancreas derives from the splenic and celiac plexuses.

Biliary system

The biliary system includes the hepatic duct and branches from the liver, the cystic duct and gallbladder, the common bile duct, and the choledochal sphincter. This system of ducts is closely related to the pancreas and its ducts and to the duodenum (see Fig. 5-16).

The several lesser hepatic duct branches arise in the liver and leave the liver by two main branches, one each from the left and right lobes. These two branches immediately join inferior to the liver to form the hepatic duct, which descends for approximately 4 cm. to join with the cystic duct and form the common bile duct.

The cystic duct extends upward and to the right for approximately 4 cm. to terminate in a blind pouch, the gallbladder. The inner lining of the cystic duct presents several crescent-shaped folds, the *valves of Heister,* which function to prevent collapse or distention of the cystic duct as a result of changing pressures in either the gallbladder or the common bile duct.

The digestive system   219

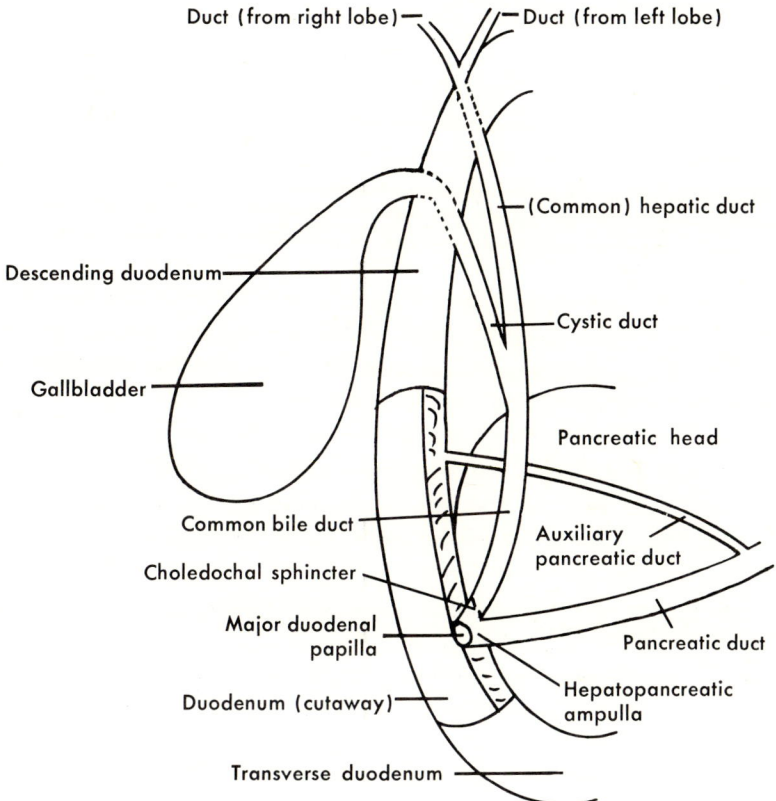

Fig. 5-16. Biliary system and related structures.

The common bile duct continues in a downward direction for approximately 7.5 cm. to join with the pancreatic duct in the hepatopancreatic ampulla. The common bile duct descends between the descending duodenum and the pancreas and is considered to lie in the epigastric region.

The blood supply, the venous return, and the innervation of this system of ducts is closely related to the corresponding items of the liver, gallbladder, and pancreas.

DIGESTION AND ABSORPTION OF FOOD

Digestion begins in the mouth (oral cavity) with chewing of the food and mixing of it with the salivary secretions.

Oral cavity

The salivary glands secrete a mixture of substances: some are digestants, others are alkalizers, another destroys bacteria, and still another acts to form the food into a sticky bolus. These substances are amylase, sodium bicarbonate, sodium chloride, lysozyme, and mucin.

The parotid glands secrete the serous exudate, amylase, possibly some maltase, and very little (if any) mucin. Amylase hydrolyzes polysaccharides to maltose, a disaccharide. Maltase hydrolyzes maltose to monosaccharides. Mucin is a protein substance.

The submandibular glands secrete both amylase and mucin, and in man the juice of the submandibular glands is preponderantly serous, containing amylase as its principal product of secretion.

The sublingual glands mainly secrete mucin.

The salivary juice contains approximately 99 percent water and 1 percent solid material, and is usually alkaline because of the presence of both sodium bicarbonate and sodium chloride. Alkalinity of the salivary juice is necessary for amylatic hydrolysis. Mucin functions to form the masticated food into a sticky bolus containing the large quantity of amylase.

Considerable starch hydrolysis occurs in the mouth and continues for approximately 30 minutes in the stomach because the bolus does not immediately break up as it enters the stomach. When the gastric juices penetrate into the center of the bolus, the acidity of the gastric juice is such that the alkalinity of the salivary juice in the food bolus is neutralized, and amylatic hydrolysis ceases.

Lysozyme, a bacteriostatic enzyme, is manufactured in the salivary juice; the enzyme functions to destroy harmful bacteria that would attack the teeth.

After the proper length of time for mastication of the food and for mixing of the food with the salivary juice, the food bolus is swallowed.

Deglutition

Deglutition is the process of swallowing the food bolus. The process is achieved in three stages; the first stage is voluntary, and the second and third stages are involuntary. The mechanics of deglutition are explained in the following three steps:

1. The larynx rises and temporarily closes the pharyngeal space.
2. Simultaneously, the bolus is conveyed toward the pharynx and over the tongue.
3. Finally, the pharyngeal space opens to receive the bolus, which is forced downward from the beginning of the middle one-third of the esophagus by *pseudoperistalsis.* The nasal, oral, and laryngeal openings of the pharynx are completely closed after the bolus "shoots" back over the tongue and the second and third stages of deglutition commence.

Deglutition is usually very closely observed by the radiologist during E.S.D. examinations. The esophagus functions in the digestive tract as a tube or passageway between the mouth and pharynx at its upper end and the stomach at its lower end.

Stomach

Upon entering the stomach the food bolus undergoes a strong churning action caused by the contractions of the layers of nonstriated stomach muscles. The churning action forces the bolus into contact with the gastric juices; this results in the slow breakup of the bolus so that the food is suspended in a rather thick liquid called the fluid chyme.*

The gastric juice is secreted by the gastric glands and consists of a mixture of substances including the strong inorganic hydrochloric acid, enzymes, and at least one precursor† to other enzymes and mucus.

The glands of the stomach wall are classified as fundic, pyloric, or cardiac (see p. 204). The gland names serve, for the most part, to locate the gland in the stomach; the secretions of the glands appear to be controlled by the kinds of cells within the glands. There are approximately 35,000,000 gastric glands in the stomach walls, and each gland consists of a *mouth*, behind which is the *neck*, the *body*, and the blind *fundus*.

The four types of epithelial cells lining each gland lumen are the *chief, parietal (oxyntic), neck chief (mucous neck),* and *surface epithelial cells.* Mucus-secreting cells are situated in the glandular wall near the mouth of the gland, parietal cells in the glandular wall in the body of the gland, and chief cells more deeply within the gland wall. Within the stomach wall, the mucus-secreting, parietal and chief cells are found in the body of the stomach; the parietal cells are absent in the glands of the walls of the pyloric part of the stomach.

The essential substance secreted by the chief cells appears to be *pepsinogen,* an inactive precursor of *pepsin.* Pepsin is the principal gastric proteolytic enzyme; it results from the conversion of pepsinogen in the presence of hydrochloric acid (or active pepsin).

Experimental evidence establishes that the parietal cells are the exclusive producers of hydrochloric acid. Formation of hydrochloric acid by living animal cells is a process that is not completely understood. From previous studies we have learned that water and carbon dioxide will combine in the body to form carbonic acid and that carbonic acid added to sodium chloride will result in the formation of hydrochloric acid and sodium bicarbonate. The foregoing is written in the following equations:

$$H_2O + CO_2 \longrightarrow H_2CO_3$$
$$H_2CO_3 + NaCl \longrightarrow HCl + NaHCO_3$$

The two preceding equations could be the source of hydrochloric acid, but this has not been proven. If this process of hydrochloric acid formation is valid, mystery remains regarding the direction of movement of the end products in the second equation. In other words, why sodium carbonate *always enters* the blood and why hydrochloric acid *always collects* in the lumen of the gland is not explained. Hydrochloric acid, regardless of the method of formation, is formed or secreted by the parietal cells, and it performs very importantly in the digestive process.

---

*Chyme is the semifluid mass of partly digested food resulting from the action of the gastric juice; it is expelled by the stomach into the duodenum.
†A precursor is generally called a zymogen.

Hydrochloric acid renders the gastric juice, without food, to a pH range of from 0.9 to 1.5; gastric juice with food has a pH range of from 2.0 to 3.0. The volume of gastric juice secreted in a 24-hour period ranges from 1 to 2 liters (1,000 to 2,000 ml.). In addition to making the gastric juice acid, which is necessary to activate pepsin and for pepsin to act, hydrochloric acid has the following four additional functions.
1. It destroys or makes impossible the growth of many of the bacteria taken into the digestive system through the mouth and nose.
2. It splits certain proteins (albumins and globulins) and forms acid albuminates.
3. It splits disaccharides into monosaccharides.
4. It is a stimulus to initiate or increase the flow of both bile and pancreatic juice.

The chief cells of the neck and the mucous cells located throughout the gastric mucosa, especially those that are prevalent in the pyloric end of the stomach, manufacture *mucus*. Mucus is a viscid, slippery secretion produced by mucous membranes, which it serves to moisten and to protect. The character of mucus is due largely to mucins. A mucin is any of the glycoproteins found in various secretions of man and other animals. The pH of the secretion of the pyloric mucous cells ranges from neutral to slightly alkaline.

Rennin appears to be a component of pepsin and is manufactured by the chief cells. Rennin is a limited proteolytic enzyme found only in infants; it hydrolyzes* caseinogen to paracasein.

Evidence points to the presence of an *intrinsic factor* in the gastric juice. There is little specific information regarding this factor; however, it is known to be essential to hemopoiesis.

Lipase, the lipolytic enzyme in the gastric juice, is relatively weak and is apparently capable of hydrolyzing only emulsified milk fats. Some authors state that lipase is manufactured only in the pancreas and in the small intestine; others believe that lipase is manufactured also in the stomach in certain specialized cells. The most recent evidence points to the latter condition as being correct, although the site of the manufacture of lipase remains unknown.

A brief summation of the results of gastric digestion is as follows:
1. There is some additional hydrolysis of polysaccharides to disaccharides due to the delayed breakup of the food bolus and continuing amylatic hydrolysis.
2. There is splitting of some disaccharides into monosaccharides by hydrochloric acid.
3. In the presence of hydrochloric acid, the pepsin hydrolyzes proteins to proteoses and peptones.
4. In the presence of hydrochloric acid, rennin hydrolyzes caseinogen to paracasein (only in infants).

---

*Hydrolysis is a chemical process in which a substance is split by the addition of water.

5. Soft fats are hydrolyzed to glycerol and stearic, palmitic, and oleic acids. The undigested starches and maltoses in the chyme pass into the duodenum, where further hydrolysis occurs. The proteoses and peptones undergo further hydrolysis by actions of the pancreatic and intestinal proteolytic enzymes. The more complex fats pass, unattacked by gastric lipase, into the duodenum and beyond, where bile emulsification occurs prior to further lipolytic hydrolysis. *Absorption of the food products through the stomach wall is quite limited.*

Emptying of the stomach

Two very important functions are performed by the stomach: (1) food storage and (2) subsequent slow release of chyme into the duodenum to prevent overloading of the stomach. The stomach can contain from 1,420 to 2,360 ml. (3 to 5 pints), the quantity varying between individuals and within one individual according to the state of health and dietary habits. In normal adults the emptying time of the stomach may vary between individuals from as short as 2 hours to as long as 5 hours. Normal extremes appear to be as short as 1.5 hours to as long as 5.5 hours. The stomach of an average normal young adult will probably empty in slightly more than 2 hours, although it is both normal and usual for the stomach to commence emptying barium sulfate mixture within a very few minutes after ingestion.

Since the pyloric antrum, pyloric sphincter, and duodenal bulb function as a single unit in the process of the stomach emptying into the duodenum, the sphincter does not act in opposition to the antrum and bulb. By action of the stomach muscles, the gastric contents are forced into the antrum and against the closed sphincter until a positive pressure builds up. At the same time and on the opposite (duodenal bulb) side of the sphincter, a negative pressure develops. When the pressure gradient between the two sides rises to equal from 3 to 4 cm. of water (a relatively low pressure), the sphincter relaxes until the pressure gradient ceases; it is during this short time interval that the chyme can pass (normally) through the pyloric sphincter.

Experimental evidence supports a chemical effect upon stomach emptying in addition to the pressure gradient theory. Gastric contents with a pH of 1.0 tend to increase the rate of emptying over similar contents with pH ranges from 2.0 toward neutrality.

At least two factors regulate the quantity of chyme that passes from the stomach during a given time interval: (1) the enterogastric reflex and (2) the chemical regulation effected by the hormone, enterogastrone. The continuous muscular rhythm evidenced by the stomach is subject to controlling influences from other parts of the body. This is the enterogastric reflex, which commences in the duodenum with the entry of the chyme; at the same time there is a pronounced decrease in the gastric peristaltic force. The chemical regulation effects the same end result as the enterogastric reflex. Enterogastrone in the duodenum mediates the humoral inhibition of gastric secretion and motility produced by the ingestion of fats.

Small intestine

The chyme, upon entering the bulb of the duodenum, stimulates a series of events that lead to the ultimate breakdown of food particles and the absorption of the food particles into the bloodstream. The processes occurring in the small intestine are the final digestive processes. *Digestion is the total process of converting food into lesser chemical products whose size permits their absorption into the body.* Digestion occurs in two stages: (1) chemical breakdown and (2) absorption by osmosis. Enzymatic hydrolysis breaks down food particles into molecules that are sufficiently small to pass through the thin walls of the villi into the venules, both blood and lymphatic, in the small intestine. The molecules pass through the walls as a result of differences in osmotic pressures between the lumina of the small intestine and the venules in the villi walls.

In addition to the numerous arterial capillaries and small venules in the villi walls are large numbers of lymphatic venules, which receive large quantities of hydrolyzed fats—glycerol and palmitic, stearic, and oleic acids. The hydrolyzed fats mix with the lymph to form the milky-white fluid, *chyle* (see p. 229).

The mucosa of the small intestine, especially in its upper parts, secretes many enzymes that contribute materially to the total hydrolysis of food. These enzymes are *erepsin, maltase, lactase, sucrase,* and *lipase.* Erepsin is proteolytic and carries the hydrolysis of amino acids (after trypsin) to the final form. Maltase hydrolyzes maltose to two molecules of glucose. Lactase hydrolyzes lactose to two hexose molecules—galactose and glucose. Sucrase hydrolyzes sucrose to one molecule of fructose and one molecule of glucose. Lipase hydrolyzes the molecules of fat that are emulsified with bile to glycerol and the three fatty acids (see Table 5-2).

Accessory organs

The accessory organs of digestion, located in the abdominal cavity, pour their secretions of digestive juices into the duodenum, usually through the major duodenal papilla. The digestive secretions of these organs include bile (from the liver and gallbladder), lipase, amylase, trypsin, and chymotrypsin (from the pancreas), and an alkaline substrate (from the pancreas).

Liver and gallbladder

The liver, because of its many chemical activities, is often called the chemical storehouse and laboratory of the body. The functions of the liver are numerous and diversified. The liver manufactures erythrocytes in the embryo (and in adults, in certain blood dyscrasias); stores glucose as the polysaccharide, glycogen; manufactures heparin, an anticoagulant; synthesizes prothrombin, aided by action of vitamin K; manufactures fibrinogen, necessary in clot formation, and (perhaps) other plasma proteins; synthesizes amino acids from less complex compounds; probably manufactures an intrinsic factor, which is necessary as an antianemic factor; is concerned in the internal excretion of urea; detoxifies nitrogenous wastes to less toxic urea; is responsible for the deamination of amino acids; stores vitamin B_{12}, an antianemic factor; stores iron and copper; synthesizes uric acid; destroys uric acid and hippuric acid; is the center of fat metabolism; is the center of carbohydrate metabolism; and stores vitamins A and D.

As radiologic technologists we are more concerned with the following activities of the liver, in that they have more direct application to radiography and radiology. By means of the Kupffer* cells within the liver sinusoids, old erythrocytes are destroyed, and a considerable quantity of the products of destruction is retained to be used again. Bile is formed in this fashion.

Bile formation. Blood from the portal vein and the hepatic artery flows through very small vessels lying adjacent to a row of hepatic cells. On the opposite side of the row of hepatic cells and extending parallel with them is a bile capillary. The bile pigments, bilirubin and biliverdin, derive from the hemoglobin of the destroyed erythrocytes and, with other materials obtained in similar fashion, are taken by the hepatic cells and emptied (secreted) into the bile capillaries (see Fig. 5-13, *A* and *B*). In the bile capillaries the bile flows to the periphery of the lobule, where the capillaries join with branches of the hepatic duct.

Bile is secreted in the liver at a rate of from 400 to 800 ml. in a 24-hour period. When the bile is secreted in the liver, it is thin and watery and has a specific gravity ranging from 1.01 to 1.05. Bile consists of water; inorganic salts (of which sodium bicarbonate is the principal one); mucin; lecithin; cholesterol; the bile salts, sodium taurocholate and sodium glycocholate; and the bile pigments, bilirubin and biliverdin.

Sodium bicarbonate maintains the pH of the bile from slightly above 7.0 to as high as 8.6. Mucin is added to the bile by the gallbladder walls. Lecithin and cholesterol are added in the liver.

Sodium taurocholate and sodium glycocholate function in the emulsification of fats by reducing the surface tension of the fat molecules. This permits an increase in the solubility of both fats and fatty acids in water, thereby enabling lipolytic hydrolysis. The enzyme, lipase, which hydrolyzes emulsified fats, is water soluble.

The bile pigments, bilirubin and biliverdin, are waste products and are excreted in the bile. Bilirubin is found in yellow bile, and biliverdin is found in green bile.

Bile is manufactured in the liver and transported from the liver via the right and left hepatic duct branches to their junction in the main hepatic duct. The hepatic duct conveys the bile to the junction with the cystic duct; the junction is the commencement of the common bile duct. The common bile duct usually terminates in the hepatopancreatic ampulla with the choledochal sphincter. When bile from the liver reaches the closed choledochal sphincter, the volume builds up until bile overflows via the cystic duct into the gallbladder. In the gallbladder,

---

*The Kupffer cells are a part of the reticuloendothelial system.

water is reabsorbed from the thin bile to effect a concentration of the volume to approximately one-twelfth of its original quantity. This process is continuous, except during digestion when all the bile flows into the duodenum. Mucin is added to the bile in the gallbladder, and therefore, the bile excreted from the gallbladder is very viscid and dark green; this bile has a relatively high specific gravity.

Bile excretion. Emulsified milk fats undergo hydrolysis in the stomach. More complex fats pass from the stomach (as a part of the chyme) in their original form into the duodenum. As these fats (and chyme) enter the duodenum, they cause the small intestine mucosa to secrete a hormone, *cholecystokinin*. The presence of cholecystokinin triggers a sensory nerve impulse to the central nervous system to relay a motor impulse to perform two functions simultaneously; the gallbladder contracts and expels the concentrated bile into the cystic duct, and the choledochal sphincter relaxes to permit the bile to flow into the duodenum. The concentrated bile emulsifies the nonhydrolyzed fats and permits hydrolysis by lipase to commence.

Both proteins and fats stimulate the secretion of cholecystokinin. Also, eating both of these types of food stimulates the liver to secrete bile at an increased rate. Other intestinal hormones are caused to be secreted by the chyme passing into the duodenum, especially *secretin* and *pancreozymin*. Secretin (probably) stimulates the pancreas to secrete sodium bicarbonate and other inorganic salts and water; pancreozymin is believed to stimulate the pancreas to secrete the other pancreatic enzymes. Pancreatic lipase is the enzyme important in this particular act.

Appreciation of this entire process will aid the radiologic technologist in an understanding of the principles of physiology involved in the radiographic examination of the structures in the biliary system: cholecystography for the gallbladder and cholangiography for the bile ducts. The contrast media used in cholecystography is taken orally and must reach the gallbladder by normal digestive processes. Iodine in the contrast medium is released from the liver with the bile salts and is stored and concentrated in the gallbladder. Ingestion of concentrated fat substance implements the mechanism of release. Intravenous contrast medium is most frequently used for cholangiography. This medium is in high concentration in the bile as it is excreted from the liver and is, therefore, well visualized in the ducts of the biliary system.

Pancreas

The pancreas secretes some of the principal digestive enzymes, water, and an alkaline substrate, all of which pass from the pancreas via the main pancreatic duct usually through the hepatopancreatic ampulla and the major duodenal papilla into the duodenal lumen. The accessory pancreatic duct may always be active, or it may function simply to convey the pancreatic juices in the event of a block in the hepatopancreatic ampulla or in the mouth of the main pancreatic duct. The accessory pancreatic duct empties into the duodenal lumen through the minor duodenal papilla.

The pancreas secretes the precursor, *trypsinogen,* to the strong proteolytic enzyme, trypsin. In the presence of *enterokinase,* an enzyme of the succus entericus (intestinal juice), trypsinogen is converted to the active enzyme, *trypsin.* Trypsinogen contains a second precursor, *chymotrypsinogen,* which is converted into the active enzyme, *chymotrypsin,* in the presence of trypsin. Trypsin hydrolyzes proteoses and peptones to polypeptides and amino acids. Chymotrypsin achieves the same end results. The hydrolysis accomplished by the pancreatic enzymes occurs in the small intestine. Following trypsin hydrolysis the peptides may consist of as few as two or three amino acids.

Pancreatic amylase is many times stronger than salivary amylase; however, the actions of the two amylases are similar. Pancreatic amylase, in conjunction with small traces of maltase, hydrolyzes a considerable part of the carbohydrate molecules to maltose and the remainder into two molecules of glucose. The remaining disaccharide molecules are reduced in the small intestine by specific enzymes—sucrase, maltase, and lactase.

When the fats are emulsified with concentrated bile, pancreatic lipase hydrolyzes the fatty particles to oleic, stearic, and palmitic acids and glycerol.

Small collections of specialized cells, the *islands of Langerhans,* are intermittently dispersed over the surface of the pancreas. The islands of Langerhans secrete the hormone, *insulin,* which exerts much of the control of carbohydrate metabolism.

The blood sugar is maintained at a normal level by the combined activities of several hormones in exact balance. The principal hormone in this group is insulin. The general effect of insulin is to burn excesses of glucose, which occur in a variety of ways, principally overingestion. The burning effect is that of lowering the blood sugar level. Excesses of sugar in the blood may result from deficient secretion of insulin by the pancreas. This is a pathologic condition called *diabetes mellitus.* In persons who are suffering from diabetes mellitus, the conditions of glucosuria and hyperglycemia are problems that require treatment. The opposite condition of diabetes mellitus is hypoglycemia, wherein there is insufficient sugar in the blood. When a diabetic patient receives an amount of insulin beyond his bodily requirements, he may suffer from *insulin shock.*

In the small intestine the succus entericus contains erepsin, maltase, sucrase, and lactase. Erepsin, a proteolytic enzyme, hydrolyzes polypeptides to amino acids. Maltase hydrolyzes maltose to two molecules of glucose. Sucrase hydrolyzes sucrose to one molecule of glucose and one molecule of fructose. Lactase hydrolyzes lactose to one molecule of glucose and one molecule of galactose.

Recent evidence indicates that other enzymes and hormones are secreted in parts of the digestive system; however, it is the purpose of this text to list only the more common digestants and hormones involved in digestion that are useful to the better understanding of assimilation and excretion of contrast media and of the general digestive processes. Table 5-2 lists these enzymes and hormones in a simplified form.

Table 5-2. Digestive juices and actions

Region, gland, organ, or tissue	Secretion	Substances acted upon	End product at particular stage
Mouth	Amylase Mucin (forms sticky bolus) Lysozyme (destroys decay-causing bacteria) Alkaline substrate (alkalizes)	Starches (polysaccharides)	Maltose
Stomach	Lipase Pepsinogen + Hydrochloric acid (acidifies) ↓ Pepsin ↓ Rennin (only in infants)	Emulsified milk fats Proteins Caseinogen	Glycerol and palmitic, oleic, and stearic acids Proteoses and peptones Paracasein

The liquified, acid contents including food in the stomach and passing into the small intestine constitute the chyme.

Region, gland, organ, or tissue	Secretion	Substances acted upon	End product at particular stage
Pancreas	Amylase Lipase (from liver, bile) Trypsinogen (containing chymotrypsinogen) Alkaline substrate (alkalizes)	Starches Emulsified fats (both in the small intestine)	Maltose Glycerol and palmitic, oleic, and stearic acids
Liver	Bile (emulsifies fats in small intestine)		
Small intestine mucosa	Cholecystokinin (causes gallbladder to contract and choledochal sphincter to relax, thus, expelling concentrated bile into duodenal lumen) Secretin and pancreozymin (stimulate both liver and pancreas to secrete) Trypsinogen (from pancreas) + Enterokinase ↓ Trypsin + Chymotrypsinogen (from pancreas) ↓ Chymotrypsin Erepsin Maltase Sucrase Lactase	 Proteoses and peptones Proteoses and peptones Polypeptides Maltose Sucrose Lactose	 Polypeptides and amino acids Polypeptides and amino acids Amino acids Glucose Glucose and fructose Glucose and galactose

Large intestine

After the multitude of intricate digestive processes, the remaining substance from the ingested food passes from the ileum into the cecum, the first part of the large intestine. If the person is healthy and if the ingested food was such that all vegetables were either well cooked or well masticated to break up the cellulose, very little, if any, nutritious material will pass from the small intestine into the large intestine. The walls of the colon secrete an alkaline mucinous fluid to expedite passage of the waste material. During its passage through the large intestine, fecal matter loses much of its original bulk through reabsorption of water from the fecal matter into venous tributaries of the inferior mesenteric vein.

Portal circulation

The end products of digestion ultimately reach the blood for distribution to the cells of the body so that homeostasis may continue. Different pathways convey the monosaccharides, amino acids, fatty acids and glycerol, and reabsorbed water in the return to peripheral blood.

Monosaccharides and amino acids are absorbed through the villi of the small intestine into tributaries of the superior mesenteric vein. The majority of hydrolyzed fats are absorbed through the lacteals. The fats with twelve or more carbon atoms in the molecule are absorbed via the thoracic duct; this is called *indirect absorption* of fats into the peripheral blood. The fats with less than twelve carbon atoms in the molecule are small enough to pass through the villi walls along with the monosaccharides and amino acids; this is called *direct absorption* of fats into the blood.

The short-chain (less than twelve carbon atoms) fats, monosaccharides, and amino acids enter tributaries of the superior mesenteric vein. This vein joins with the splenic vein to form the portal vein, which enters the liver through the portal fissure. The splenic vein receives the inferior mesenteric vein prior to joining the superior mesenteric vein. The inferior mesenteric vein receives tributaries from the large intestine and from the pelvic parts of the small intestine (see Figs. 6-13, *B* and *C*, and 5-15). Thus, the water reabsorbed from the fecal matter in the large intestine returns to the peripheral blood along with the end products of digestion.

Metabolism

The end products of digestion, upon entry into the peripheral blood, are available for use by the body cells; in other words, these products are ready for metabolism. *Metabolism is the sum of the processes concerned in the building up of protoplasm and its destruction incidental to life.* Metabolism includes the chemical changes in living cells by which the energy is provided for the vital processes and activities and new material is assimilated to repair the waste. Metabolism includes two opposing aspects: *constructive,* or *anabolism* (assimilation), and *destructive,* or *catabolism* (dissimilation). A few very brief statements con-

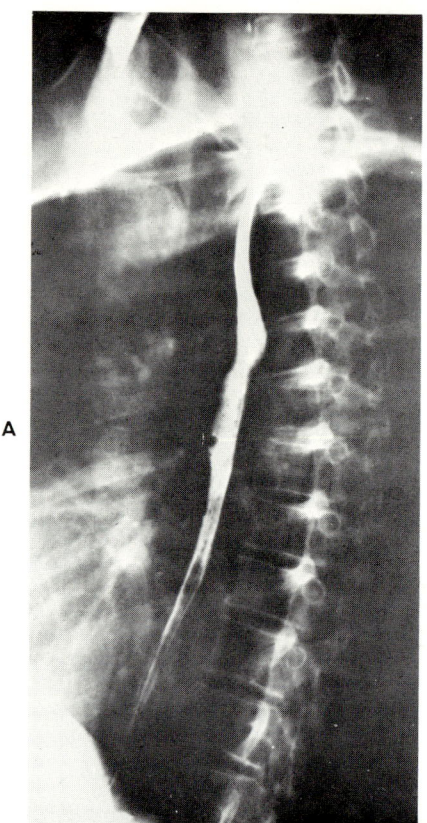
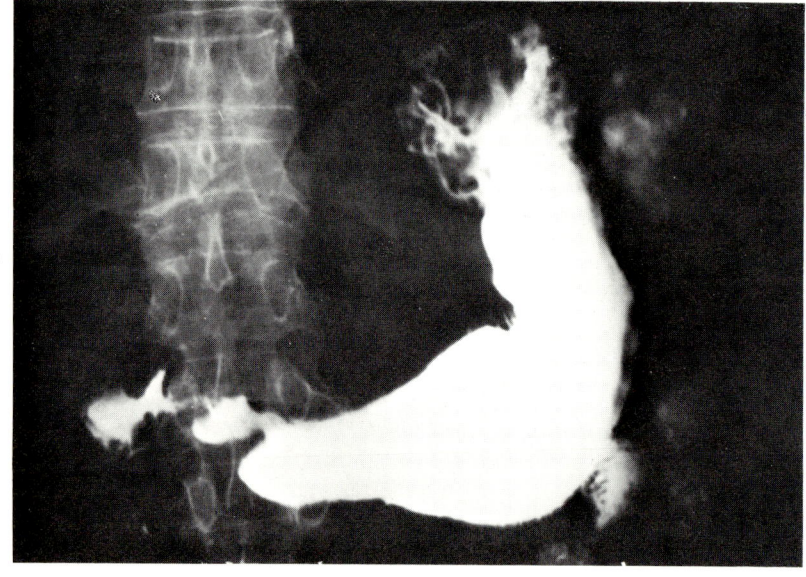

Fig. 5-17. A, Esophagus, R-P-O view. **B,** Stomach, barium filled.

cerning metabolism are presented to complete the overall survey of the digestive process.

Amino acids are the end products of protein hydrolysis. Amino acids may be oxidized to furnish energy, used to build new protein molecules of the body, converted to carbohydrates when the blood sugar level demands it, transformed into fats for storage in various parts of the body, stored in the liver for later use, and excreted. Through a process of deamination in the liver, ammonia is removed from the amino acid molecules; this leaves a simple carbon-hydrogen-oxygen chain, which can be oxidized to release energy, carbon dioxide, and water.

Monosaccharides are the end products of carbohydrate hydrolysis. Monosaccharides may be combined to form glycogen, a polysaccharide; stored in the liver and in striated muscles for oxidation upon demand; transformed to fats for storage in various parts of the body; circulated as blood sugar; oxidized immediately to release energy; and excreted.

Glycerol and fatty acids are the end products of lipoid hydrolysis. Fats, in their reduced forms, may be stored in various parts of the body or oxidized for immediate release of energy. It is possible, but extremely rare, that fats are converted to carbohydrates.

RADIOGRAPHIC INTEREST

Radiologists employ several methods to relate certain anatomic parts to a radologic examination by a certain name. Any of these combinations are satisfactory, but some are confusing to radiologic technologists and other allied health personnel. The *upper G.I. series* usually refers to radiologic examination by flu-

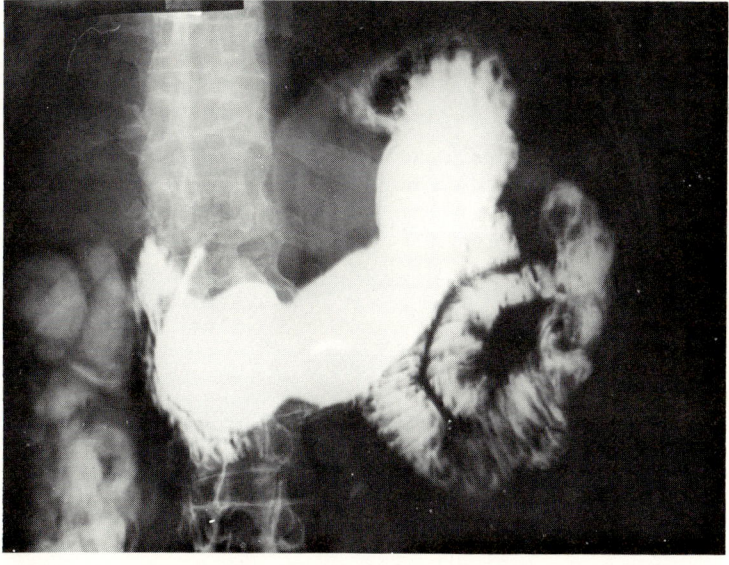

Fig. 5-18. Stomach and part of small intestine, barium filled.

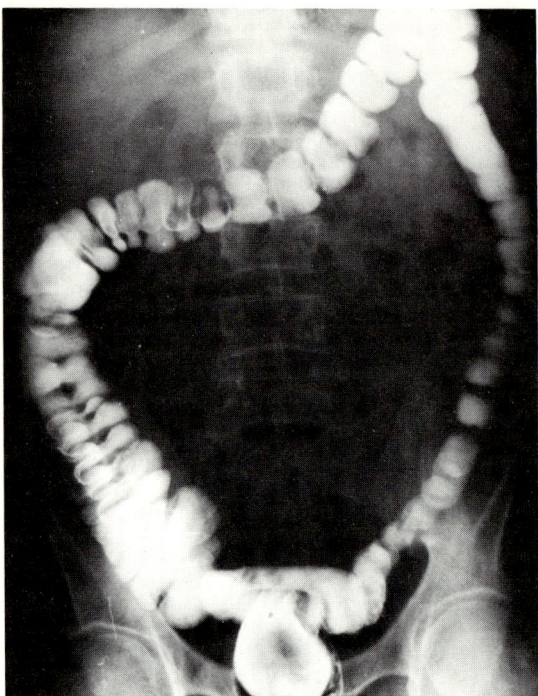

Fig. 5-19. Barium enema. (Courtesy Dr. J. M. Hilton.)

Fig. 5-20. Kidneys, ureters, and urinary bladder (K.U.B.). (Courtesy Dr. J. M. Hilton.)

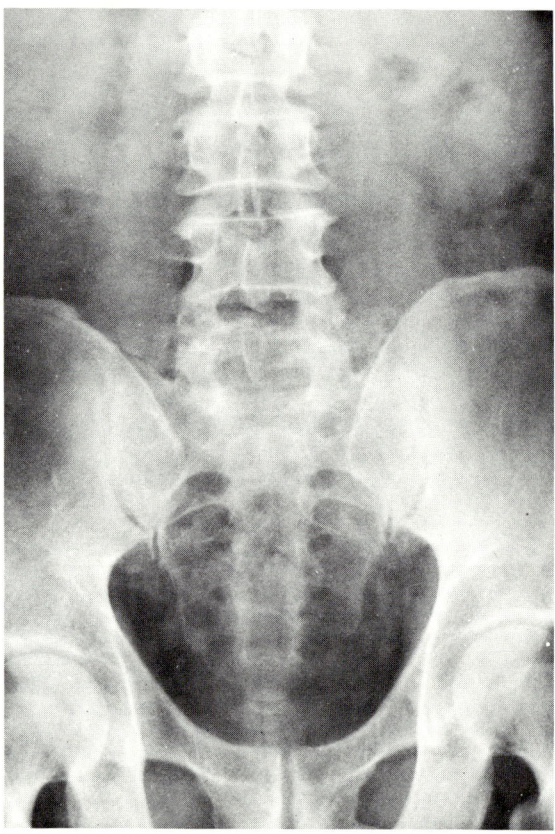

oroscopy and radiography of the esophagus, stomach, and duodenum; many radiologists term this examination by either of two abbreviations, E.S.D. or U.G.I. Radiologic examination of the small intestine other than the duodenum is usually called the *small intestine beyond the duodenum*. Both of these examinations are performed during and after oral ingestion of a suspension of barium sulfate in water (see Figs. 5-17 and 5-18).

Examinations of the large intestine that employ suspensions of barium sulfate in water are called B.E. (barium enema). The barium suspensions may be of several different viscosities, depending upon department routines and the pathology suspected. Other methods of large intestine examination employ air in conjunction with barium sulfate in *double air contrast studies* of the large intestine. This examination is performed during and after routine barium enema examinations. Among the conditions visualized are polyps and diverticula (see Fig. 5-19).

Other studies of the abdomen included in many department routines are as follows:
1. Flat abdomen, either erect or prone position
2. K.U.B. (kidneys, ureters, and urinary bladder), usually made with the patient in supine position (see Fig. 5-20)

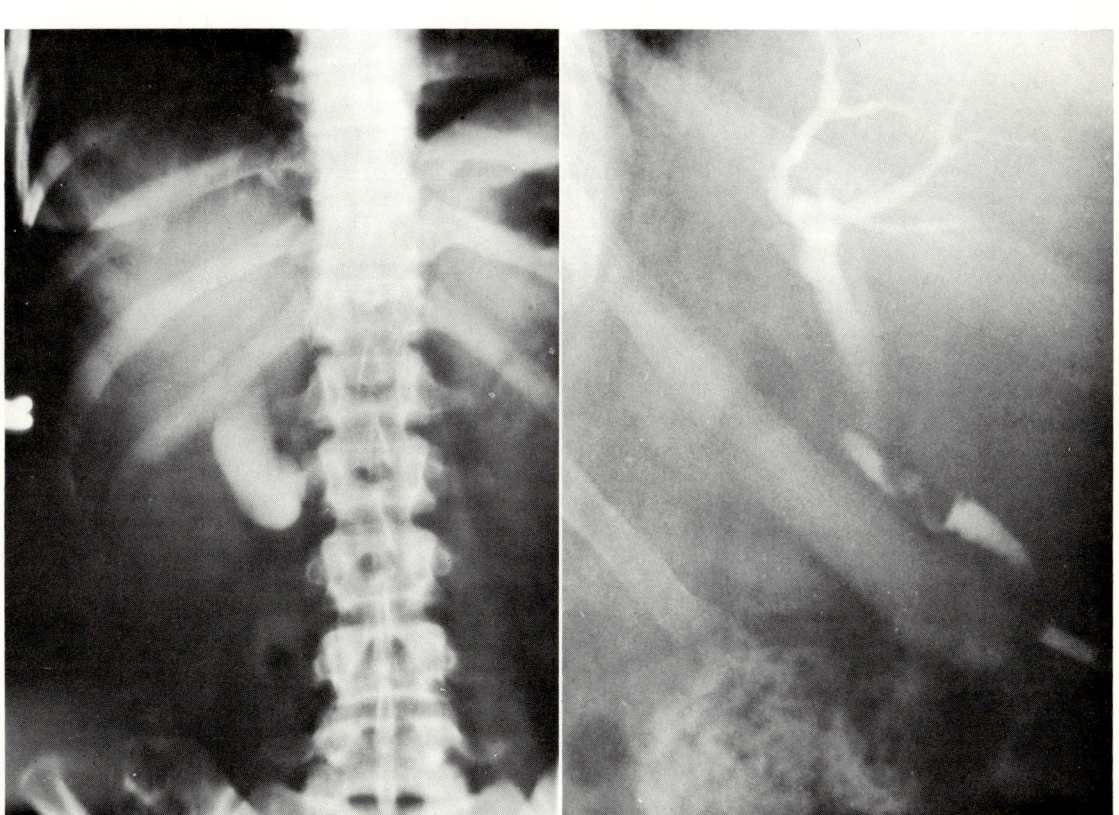

Fig. 5-21. A, Filled gallbladder. **B**, Filled bile ducts. (**B**, Courtesy Dr. J. M. Hilton.)

3. Cholecystography, in which the early exposures are made with the patient in prone position and subsequent exposures made with the patient in any of the decubitus positions (In these studies it is often necessary to position the patient so that the fundus of the gallbladder moves from its normal position near the vertebral column to a position with the long axis of the gallbladder at right angles to the vertebral column to permit visualization of possible stones.) (see Fig. 5-21, *A*)
4. Cholangiography, in which the patient is usually in supine position, although lateral, oblique, and prone positions are often employed (see Fig. 5-21, *B*)
5. Flat abdomen for *free air,* in which case the patient may be positioned in a variety of modifications of standard positions to demonstrate a gas bubble, as follows:
   a. A left lateral decubitus position places the left side of the patient against the table; the body is recumbent, and the roentgen rays (x rays) penetrate the body part from ventral to dorsal. (This position is used to demonstrate a gas bubble against the liver.)
   b. The patient may be positioned erect against the upright table; the exposure is made on a 14 × 17 inch film in a cassette with the upper margin of the cassette even with the nipple. (This position will demonstrate a gas bubble immediately beneath the diaphragm on the left side; the liver usually prevents gas from collecting on the right side. On the left side the gas bubble may be superimposed over the *normal gas bubble* usually found in the fundus of the stomach.)

REFERENCES

Feldman, M.: Clinical roentgenology of the digestive tract, ed. 2, Baltimore, 1945, Williams & Wilkins Co.

Gardner, E., Gray, D. J., and O'Rahilly, R.: Anatomy, Philadelphia, 1960, W. B. Saunders Co.

Goss, C. M.: Gray's anatomy, ed. 27, Philadelphia, 1959, Lea & Febiger.

Jacobi, C. A., and Paris, D. Q: Textbook of radiologic technology, ed. 5, St. Louis, 1972, The C. V. Mosby Co.

Meschan, I.: Normal radiographic anatomy, ed. 2, Philadelphia, 1957, W. B. Saunders Co.

Mountcastle, V. B.: Medical physiology, ed. 13, St. Louis, 1974, The C. V. Mosby Co.

Schottelius, B. A., and Schottelius, D. D.: Textbook of physiology, ed. 17, St. Louis, 1973, The C. V. Mosby Co.

CHAPTER 6

The circulatory system

The heart, arteries, aterioles, capillaries, venules, veins, lymphatics, blood, and lymph comprise the circulatory system. Numerous other organs and tissues contribute to the formation of the different cells of this system. Among the organs making specific contributions to blood cell formation are the liver and spleen. The lymph nodes and the thymus gland manufacture certain additional blood cells. The red bone marrow, found principally in the diploë of the flat bones and in the ends of some long bones, contributes a major part of the red cells (erythrocytes) to the blood (see Fig. 6-1).

The heart action forces the blood by way of the arteries and arterioles to the capillaries to supply nourishment to the cells and tissues of the body. The blood receives waste products from these cells and tissues and returns to the heart via venules and veins as a result of muscle contractions and replacement of each volume of blood with another volume of blood from the lesser venules and veins.

All of the fluid portion of the blood does not return immediately to the heart; some circulating blood leaves the bloodstream to become interstitial fluid. Interstitial fluid returns to the heart, either via the venules and veins as blood plasma or via the lymph vessels as lymph.

Some of the fluid part of the blood may replace, for a time, some of the intracellular fluid. The replaced volume of intracellular fluid may then pass through the cell membranes into the spaces between the cells and tissues, thus becoming the interstitial fluid.

GENERAL FUNCTIONS

The circulatory system, via the blood, performs three vital functions: transportation, water balance maintenance, and heat equilibration.

Transportation

As a transportation medium the blood moves food nutriments from the digestive viscera to the liver, then to the heart for eventual distribution to all body

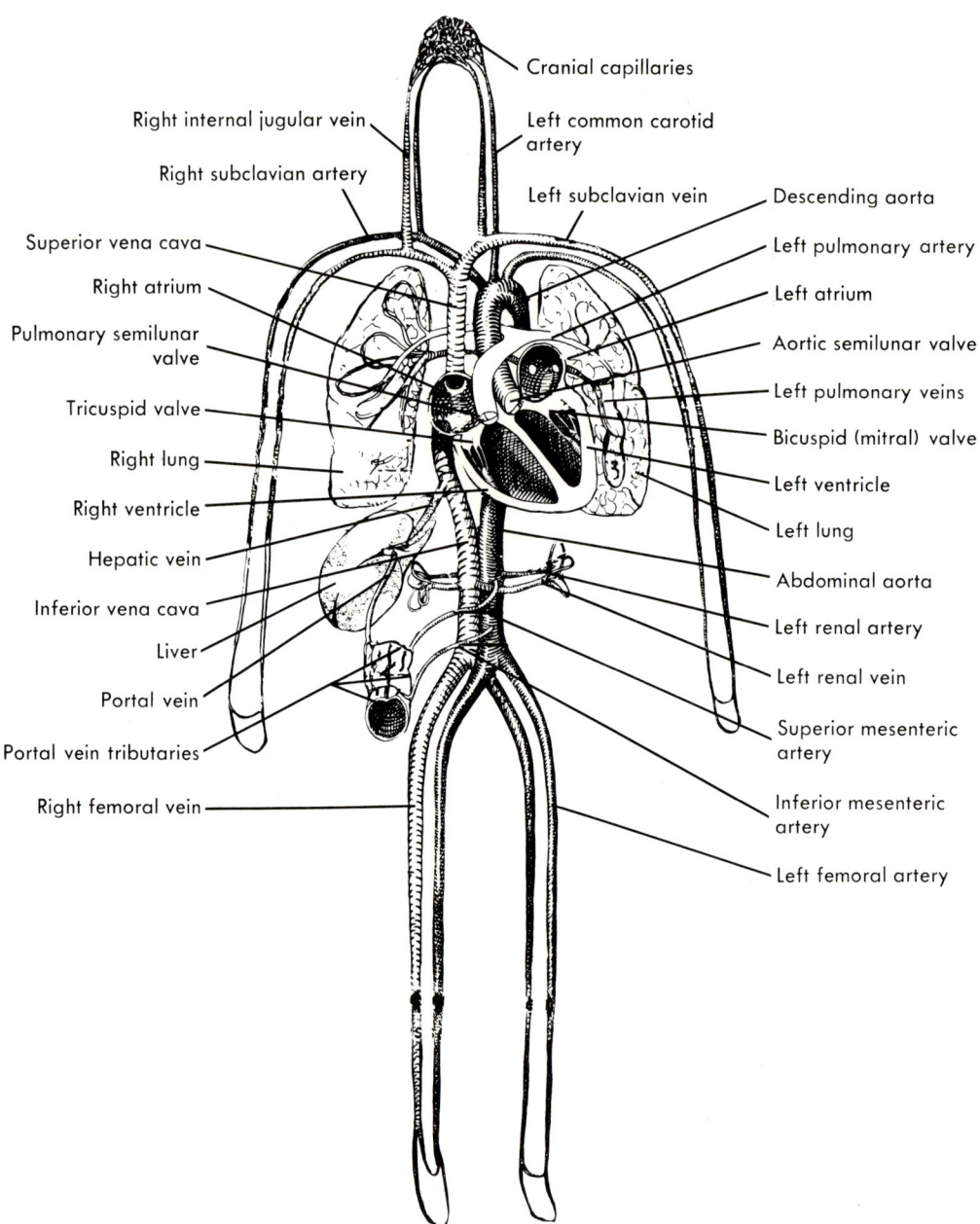

Fig. 6-1. Circulatory system (schematic).

cells. Blood moves oxygen from the lungs to the body cells and carbon dioxide from the body cells and tissues to the lungs to be exhaled. Blood conveys hormones from the various sites of secretion to all body tissues to aid in maintaining homeostasis. Blood conveys some of the enzymes to various sites to act as catalysts. Blood transports immune bodies throughout the body to combat infection, and fibrinogen, along with other substances, to hemorrhage sites to form clots.

In radiology another use of the transportation function of the blood is made; i.e., blood is used as a transportation medium to convey contrast media to various viscera for radiographic and fluoroscopic studies.

Water balance maintenance

In order to perform its part in maintaining the water balance of the body, the blood moves through the kidneys, where a small proportion of the fluid, including waste products, is extracted from the blood and excreted as urine. Blood circulates through the vessels in the skin, where small quantities of water may be lost. Finally, blood carries carbon dioxide to the lungs; in passing through the capillaries of the lung alveoli, about 500 ml. of fluid are exhaled each 24-hour period. If the water level of the body exceeds 8 percent of the body weight, the tissues become water-logged. Conversely, water losses exceeding from 6 to 10 percent of the body weight constitute the condition of dehydration.

Heat equilibration

As an aid in regulating body heat the blood moves the heat resulting from muscle activity to all parts of the body and the heat of oxidation from food hydrolysis to all parts of the body. When heat quantities increase, the peripheral vessels distend to permit increased losses of heat through the skin.

ANATOMY
Heart

The heart, a strong, mostly hollow, muscular organ, is situated obliquely in the median-left part of the lower two-thirds of the thoracic cavity; about one-third of the heart mass is to the right of the midsagittal plane, and the other two-thirds are to the left of this plane. The base of the heart faces upward and to the right; the apex points downward on a level with the interspace between the fifth and sixth left anterior ribs, approximately 2.5 cm. inferior to the left nipple (see Fig. 6-2).

The heart is one of the largest organs of the body; it varies in weight from 230 to 280 grams in the female and from 280 to 340 grams in the male. It is approximately 12 cm. from base to apex, 8.5 cm. in breadth, and 6 cm. in thickness. A loose-fitting sac, the pericardium, covers the heart. The pericardium consists of two layers; the parietal (outer) layer is fibrous, and the visceral (inner) layer is serous. The visceral layer is the *epicardium*. The very slight space between the parietal and visceral layers is the *pericardial cavity*.

The very delicate *endocardium* forms the inner lining of the heart chambers.

238 *Textbook of anatomy and physiology in radiologic technology*

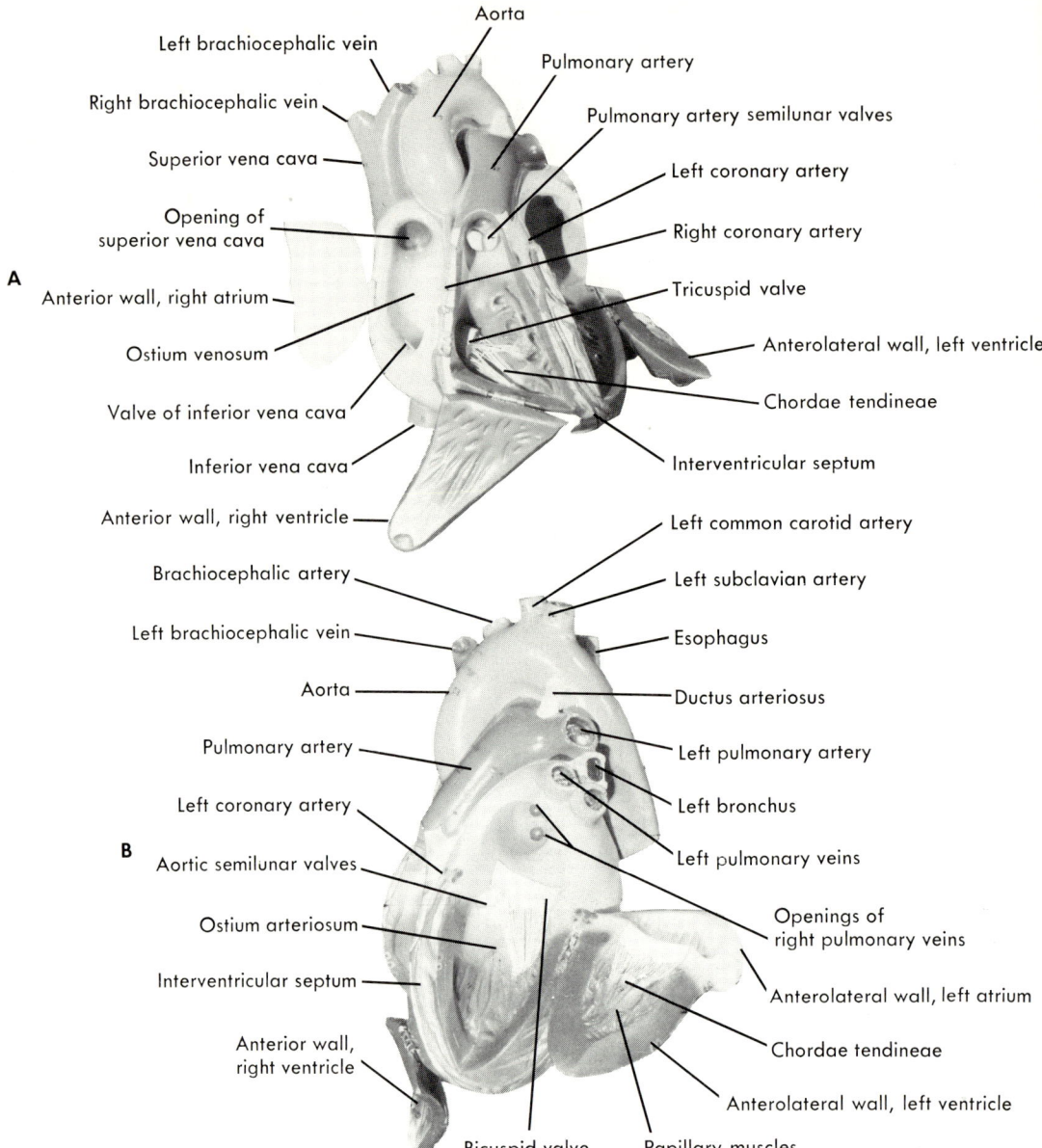

Fig. 6-2. Heart, internal aspect and associated structures. **A,** Right side. **B,** Left side.

This tissue is a form of endothelium and arises from the mesoderm in the embryo; the epicardium is composed of mesothelial tissue and arises from the mesoderm.

Structure

The myocardium (heart muscle) for the most part consists of branching, cardiac muscle cells. Some connective tissue cells function as basement membrane for the cardiac cells. A tough membrane, varying in thickness and consisting chiefly of cardiac muscle tissue, separates the right and left halves of the heart from each other. This membrane is the *septum* (interatrial and interventricular), which extends from the base of the heart to the apex. Large valves separate the upper chambers from the lower chambers on each side. Thus, there are four chambers (divisions) of the heart: the right and left atria (auricles) above and the right and left ventricles below.

Chambers

The right atrium includes two regions: the enlarged *sinus (ostium) venosum* to receive the blood and the smaller auricula. By far the larger percentage of venous blood flows into the right atrium via the superior and inferior venae cavae from the body, although some blood arrives via the coronary sinus from the myocardium. A very small quantity of venous blood flows directly into the right ventricle from the myocardium. A small, oval-shaped depression in the interatrial wall, the fossa ovalis cordis, is the site of the *foramen ovale* in the fetal heart. Blood is forced from the right atrium via the tricuspid valve into the right ventricle.

The right ventricle is much larger than the right atrium. The right ventricle receives blood from the right atrium and from a few small coronary veins. Blood is forced from the right ventricle via the pulmonary artery to the lungs for the gaseous exchange.

The inner walls of the right ventricle (and left ventricle) form the muscular papillae, to which attach the *chordae tendineae;* the opposite ends of the chordae tendineae attach to the edges of the tricuspid valve in the right ventricle and to the edges of the bicuspid valve in the left ventricle.

The left atrium receives blood from the four pulmonary veins (two from each lung). Blood is forced from the left atrium via the bicuspid (mitral) valve into the left ventricle. The left atrium resembles the right atrium in shape and size.

The left ventricle is the largest of the four chambers of the heart, and its walls are much thicker than the walls of the three other chambers. The left ventricle receives blood from the left atrium and forces the blood via the aorta to the body and via the coronary arteries to the myocardium. The *ostium arteriosum* in the left ventricle compares functionally with the ostium venosum in the right atrium. The muscular papillae in the left ventricle act similarly to the papillae in the right ventricle.

Valves

In addition to the tricuspid and mitral valves, two semilunar valves function importantly in heart actions. One of these, the *pulmonary semilunar valve*, is situated in the lumen of the pulmonary artery in the right ventricle. The other of these two valves, the *aortic semilunar valve*, is situated correspondingly in the lumen of the aorta in the left ventricle. Another valve, the inferior vena cava (eustachian valve), guards the entrance of the inferior vena cava but is generally functionless in the adult. The cuspid valves function to prevent backflow of blood from the ventricles into the corresponding atria when the ventricles are in systole. The semilunar valves function to prevent backflow of blood from the pulmonary artery and aorta into the corresponding ventricles when they are between systole and diastole or when they are in diastole.

The cords (chordae tendineae) pull downward on the cuspid valve leaves during ventricular systole to prevent the backflow of blood into the atria and to force the blood to flow into the proper arteries. Blood flowing through the semilunar valves from below forces the valves open. When there is no force from below, the semilunar valves close (the parts touch each other) and prevent the backflow of blood into the ventricles. Specialized structures in the right atrium and the interatrial septum regulate the heartbeat and the stages of systole and diastole.

Special heart structures

Systole and/or diastole occur simultaneously in the two atria or in the two ventricles because of the sinoatrial node, the atrioventricular node, the bundle of His, and the Purkinje system of fibers.

Situated in the right posterior atrial wall and to the right of the orifice of the superior vena cava is a small mass of specialized tissue (primitive cardiac muscle cells with numerous nerve fibers), the *sinoatrial (SA) node*. This tissue mass functions independently of nerve impulses but is influenced by nerve impulses and hormones. The structure transmits impulses (much like electrical impulses) about seventy-two times each minute over the atrial walls and to the *atrioventricular (AV) node,* situated posteriorly in the right atrium in the septal and annular fibers near the orifice of the coronary sinus.

The *bundle of His (AV bundle)* commences with the AV node and continues as the main bundle (of special fibers) to the tops of the ventricles, where it divides into the right and left septal divisions; the septal divisions of the bundle of His end in the terminal divisions, which supply the papillary muscles and the ventricular muscle generally. Impulses are received by the AV node and are transmitted over the bundle of His via the *Purkinje fibers* (the terminal network of the bundle of His) to the ventricles. These impulses keep the atria and ventricles beating in normal, regular rhythm. If some pathologic condition interrupts the transmission of impulses from the SA node to the AV node, the AV node is then cut off completely from control of the SA node. Such a condition is called *heart-block.* If disease has caused the interruption, the condition has occurred gradually and the AV node has developed a rhythmic beat in the ventricles that is independent of the atrial beat. The ventricular beat rate is usually less than the atrial beat rate. Incomplete systolic impulses of the atria result from arrhythmic contractions of individual fibers, which is called *atrial fibrillation*.

The impulses arising in the SA node properly identify it as the *pacemaker* of the heart. This system of special nodes and fibers transmits cardiac impulses at a rate four times faster than regular muscle tissue. The AV node is the *regulator* of the heartbeat.

Cardiac cycle

The cardiac cycle is the series of different and successive actions performed by the heart. A complete cycle occurs each successive time that a specific action is repeated in the same part of the heart. The SA node initiates the rhythm of the heartbeat, and the AV node and the Purkinje system conduct the rhythmic impulses. The cycle is completed as follows:
1. The two atria contract (are in systole) for a brief period of time.
2. A slight pause follows; then the two ventricles contract for a slightly longer period of time.
3. The entire heart relaxes (is in diastole) for a short period of time.

In a given cardiac cycle the atria contract for 0.1 second of the total 0.8 second required for one complete cycle. The ventricles contract for 0.3 second to pump the same quantity of blood. The cuspid valves are open for 0.4 second prior to atrial systole. The ventricles rest 0.5 second during each cycle. The atria require 0.1 second to contract, this interval being the final 0.1 second of the ventricular rest period. As a result, and while the atria fill, the ventricles complete systole and then rest; the atria are next in systole, and the ventricles are in diastole. During ventricular systole the atria are in diastole.

Blood flow through the heart

Blood from the body enters the right atrium via the venae cavae; blood from the heart enters the right atrium via three routes as follows:

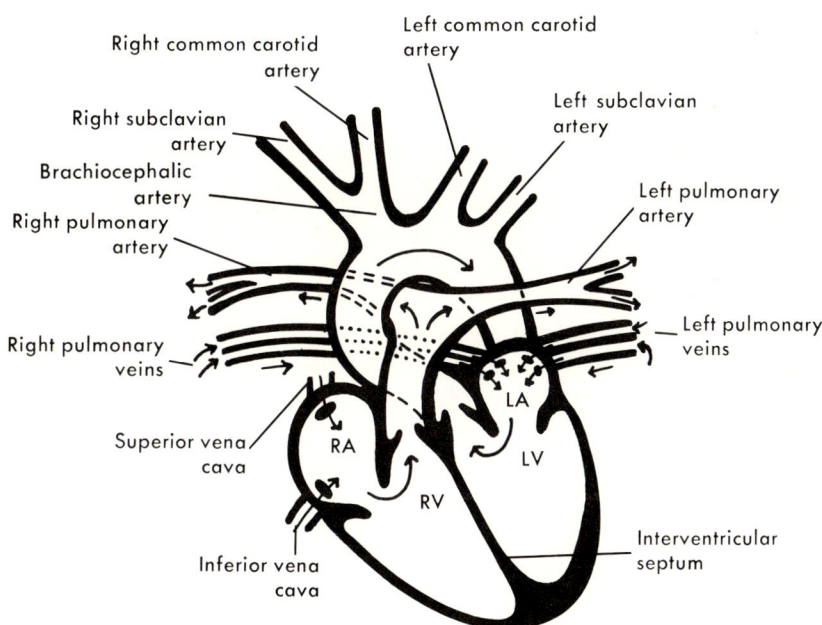

Fig. 6-3. Heart, anterior aspect showing flow of blood.

1. Most of the left ventricular venous blood flows into the coronary sinus and thence into the right atrium.
2. The right ventricular blood usually empties into the right atrium independently of the coronary sinus.
3. Thebesian veins,* important on the right side only, drain a small amount of the myocardial venous blood into the right atrium and the right ventricle.

At the same time that the right atrium fills, blood from the lungs enters the left atrium via the four pulmonary veins. When the atria are filled and enter the stage of systole, the tricuspid valve opens to permit the flow of blood from the right atrium into the right ventricle, and the mitral (bicuspid) valve opens to permit the flow of blood from the left atrium into the left ventricle. When the ventricles are filled and enter the stage of systole, blood is forced from the right ventricle through the pulmonary semilunar valve into the pulmonary artery, thence to the lungs; simultaneously, blood is forced from the left ventricle through the aortic semilunar valve into the aorta, thence to the coronary arteries, to the myocardium, and through the aorta to the remainder of the body (see Fig. 6-3).

Coronary vessels

Three small dilations, the three *aortic sinuses,* lie opposite the corresponding segments of the aortic semilunar valve in the commencement of the aorta, as shown in Fig. 6-11. The two coronary arteries arise from the aortic sinuses. The left coronary artery, the larger of the coronary arteries, usually arises from the left posterior aortic sinus. A short distance beyond its origin, the artery divides into two branches—the anterior descending branch and the circumflex branch. The right coronary artery frequently arises both from the anterior aortic sinus and from a second, very small adjacent sinus; otherwise, this artery arises from the anterior aortic sinus. The right coronary artery and the two branches of the left coronary artery extend along the surface of the myocardium and send smaller branches into it.

The left coronary artery bifurcates upon reaching the junction of the anterior longitudinal sulcus with the left transverse sulcus. One branch, the anterior descending ramus, extends along the anterior longitudinal sulcus to the apex of the heart. The other branch, the circumflex ramus, extends along the left transverse sulcus to approach the right coronary artery. This branch of the left coronary artery (circumflex ramus) and the right coronary artery anastomose frequently.

The right coronary artery extends along the transverse sulcus to the posterior longitudinal sinus. The artery extends along the latter sulcus to the apex of the heart. In the posterior longitudinal sulcus, the artery is called the posterior descending branch.

Throughout the myocardium small venules collect into larger veins, which

---

*Thebesian veins are the numerous, small veins that arise in the muscular walls and open independently into the cavities of the heart; most of them open into the right atrium.

collect eventually into two large coronary veins—the right and left coronary veins. In general, the veins follow the sulci, and most of the veins drain into the *great coronary vein* and *coronary sinus*. Along with numerous other veins, the coronary vein and coronary sinus drain the great, middle, and small cardiac veins, the posterior vein of the left ventricle, and the oblique vein of the left atrium.

Arteries

An artery is one of the tubular branching vessels that conveys the blood from the heart to the numerous organs, glands, tissues, systems, and other structures of the body. With the exception of the pulmonary arteries between the right ventricle and the lungs, the arteries carry blood that is quite high in oxygen content.

Structure

The tough, thick wall of an artery consists of three layers, which are, from inside to outside, the *tunica intima, tunica media,* and *tunica externa* (adventitia). The tunica intima is quite smooth and forms the lining of all blood vessels. The tunica media contains, in different arteries, varying amounts of both elastic tissue and nonstriated muscle fibers. The tunica media of the aorta contains essentially no muscle fibers and quite large quantities of elastic tissue fibers. The tunica externa consists of connective tissue. Some of the larger arteries contain within their walls smaller arteries, the *vasa vasorum*, to supply nourishment for the relatively thick arterial walls.

The elastic tissue in the arterial walls permits some elasticity during systole. The nonstriated muscle fibers are innervated from the autonomic nervous system and contribute to the control of blood pressure. In the trunk of the body, the arteries are protected by muscles and bones. In the extremities, the arteries are situated quite deeply on the side of the flexor surface for the dual purposes of protection against injury and provision for the least amount of stretching.

Names

In many instances the name of an artery derives from its location near a bone; in other instances arterial names derive from the structure supplied. No attempt is made in this text to name or to describe all of the arteries. Radiologic technologists need to know the names of the principal arteries and veins and the names of the structures involved with them (see Fig. 6-4).

Aorta. The aorta arises from the upper part of the base of the left ventricle, extends upward for about 5 cm. as the ascending aorta, and terminates in the upper limits of the pericardium in a continuation into the aortic arch.

The aortic arch begins at a level even with the superior border of the second, right sternocostal articulation. The superior margin of the aortic arch lies about 2.5 cm. inferior to the superior border of the manubrium of the sternum. The aorta, in its arch, curves posteriorly and slightly to the left to pass inferiorly on

244 *Textbook of anatomy and physiology in radiologic technology*

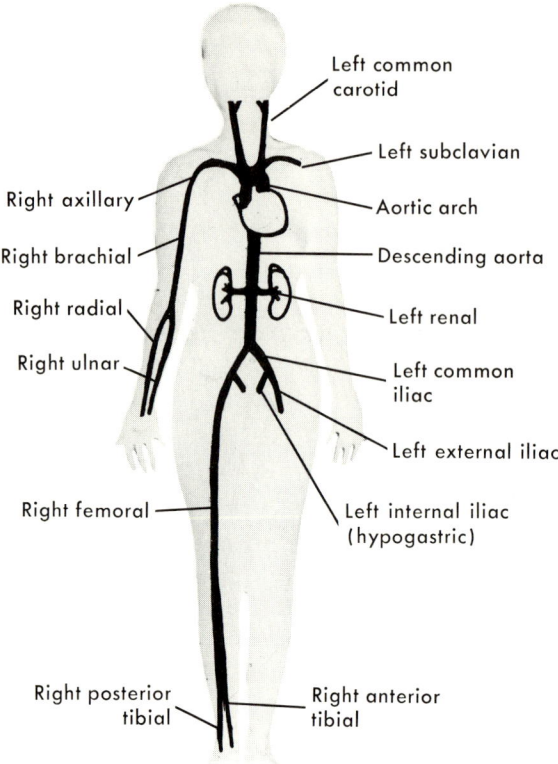

Fig. 6-4. Some major arteries (schematic).

the left side of the fourth thoracic vertebra, where the aorta continues as the descending portion.

In the descent of the aorta in the body, it commonly bears regional names—thoracic aorta in the thorax and abdominal aorta in the abdomen. The aorta descends posteriorly in the midline through the thorax to the diaphragm and pierces this muscle in the *aortic hiatus*. The abdominal part of the aorta commences immediately below the diaphragm. The abdominal aorta continues downward in the midline to terminate in its bifurcation on a level with the inferior margin of the fourth lumbar vertebra. The arteries resulting from the bifurcation of the aorta are the right and left common iliac arteries.

Arteries arising from the ascending aorta. The right and left coronary arteries arise from the aortic sinuses of the ascending aorta (see pp. 242 and 243).

Arteries arising from the aortic arch. From right to left (anterior to posterior), the arteries arising from the aortic arch are the brachiocephalic, left common carotid, and left subclavian.

The *brachiocephalic artery* is about 4 or 5 cm. long and divides into the right common carotid artery and the right subclavian artery. Each vertebral artery arises from the origin of the corresponding subclavian artery.

The circulatory system **245**

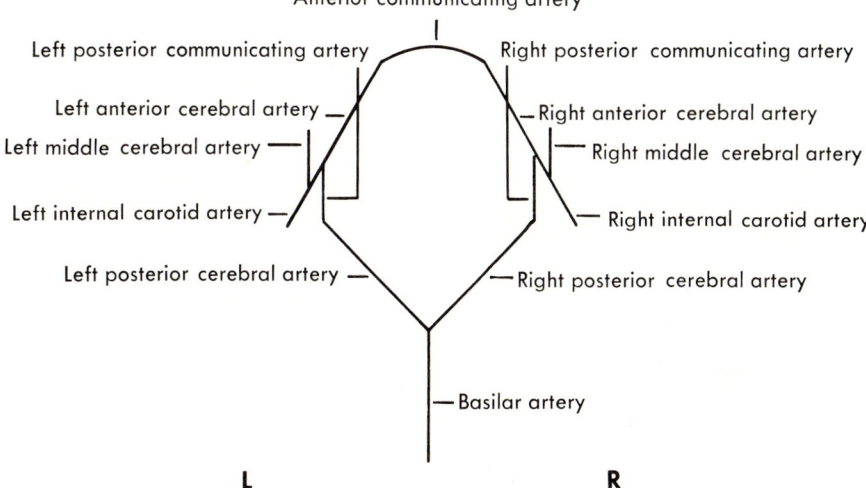

Fig. 6-5. Arterial circle, viewed from above (schematic).

The *vertebral arteries* ascend the neck through the transverse foramina of the cervical vertebrae to join above the first cervical vertebra in the foramen magnum and form the basilar artery.

The *common carotid arteries* ascend the neck, one deep on each side, and terminate in their respective *carotid sinuses* even with the upper border of the thyroid cartilage and immediately below the angle of the mandible.

The *internal and external carotid arteries* arise on each side from the corresponding carotid sinuses. The internal carotid arteries ascend, one on each side, to the base of the brain to aid in the formation of the *arterial circle* (circle of Willis) (see Fig. 6-5).

ARTERIAL CIRCLE. The arterial circle lies directly inferior to the brain and is formed by the following arteries and arterial branches. The *basilar artery* extends forward from its origin beneath the base of the brain and terminates by dividing into the two *posterior cerebral arteries*. Each posterior cerebral artery receives the corresponding *posterior communicating artery*, which anastomoses anteriorly with the corresponding *internal carotid artery*. The *anterior cerebral arteries* arise from the internal carotid arteries near the junctions of each internal carotid artery with the corresponding *middle cerebral artery*. From their origins the *anterior cerebral arteries* extend forward and medialward to anastomose with the short *anterior communicating artery*. Many arterial branches arise from the arterial circle to supply the brain, especially the anterior two-thirds of the brain; thus, there is collateral circulation* provided to prevent permanent interruption of the blood supply to the brain.

Through many branches the external carotid arteries supply those parts of

*Collateral circulation is that circulation of blood established through anastomoses of vessels with those of adjacent parts when the principal artery or vein is obstructed.

the skull and brain not supplied from the internal carotid and vertebral arteries. The principal structures supplied from these branches are situated outside the cranial and orbital cavities. Among the several major branches of the external carotids are the *maxillary, mandibular,* and *lingual arteries.* These arteries are not visualized in cerebral angiography.

ARTERIES, VEINS, AND BLOOD SINUSES VISUALIZED IN CEREBRAL ANGIOGRAPHY. Arteries visualized (demonstrated) in cerebral angiograms include the anterior cerebral, anterior choroidal, anterior communicating, anterior inferior cerebellar, anterior spinal, anterolateral ganglionic, anteromedial ganglionic, angular, ascending frontoparietal, basilar, callosomarginal, frontopolar, internal auditory, internal carotid, middle cerebral, ophthalmic, posterior cerebral, posterior communicating, posterior inferior cerebellar, posterior parietal, posterior temporal, pericallosal, posterolateral ganglionic, posteromedial ganglionic, pontine branches of the basilar, superior cerebellar, and vertebral.

Veins visualized in cerebral angiograms include the descending temporal, occipital, ascending frontal, internal cerebral, ascending parietal, great cerebral, posterior anastomotic, basal, and great anastomotic.

Blood sinuses visualized in cerebral angiograms include the inferior sagittal, superior sagittal, straight, and transverse sinuses. Each transverse sinus drains into the corresponding sigmoid sinus (see p. 255).

SUBCLAVIAN ARTERIES AND BRANCHES. The subclavian arteries and branches supply the upper extremities and parts of the pectoral girdle. Since the circulatory pathways of one side closely resemble those of the other side, the arteries and veins of a single side will be discussed.

The *subclavian artery* as its name indicates, extends lateralward beneath the clavicle to the axilla where the subclavian artery is continuous with the axillary artery. The subclavian artery gives rise to four arteries: the vertebral, previously discussed; the thyrocervical, supplying the neck and thyroid gland; the internal mammary, supplying the breast; and the costocervical, supplying the upper ribs and tissues of the intercostal spaces.

The *axillary artery,* from its origin at the inferior border of the axilla, extends downward in the upper arm to the lower end of the biceps muscle. At this point the axillary artery is continuous with the brachial artery.

The *brachial artery* extends to the elbow and divides into the radial and ulnar arteries. The *radial artery* extends along the radius to the wrist, and the *ulnar artery* extends along the ulna to the wrist.

In the wrist the radial and ulnar arteries give rise to both deep and superficial branches, which supply the hand, thumb, and fingers. These branches anastomose very often and form the *palmar arches.*

Arteries arising from the thoracic aorta. The arteries arising from the thoracic aorta are divided into the visceral division of four branches and the parietal division of three branches. The *visceral* branches are the pericardial, to the pericardium; the bronchial, to the bronchi; the esophageal, to the esophagus; and the mediastinal, to the various viscera of the mediastinal region.

The *parietal* branches of the thoracic aorta are the intercostal, to the intercostal muscles; the subcostal, to the ribs; and the superior phrenic, to the superior surface of the diaphragm.

Arteries arising from the abdominal aorta. The arteries arising from the abdominal aorta are divided into the visceral division of seven branches and the parietal division of three branches. Some anatomists include a third division, the terminal aorta, which bifurcates into the two common iliac arteries.

The *visceral* branches of the abdominal aorta are the celiac, superior mesenteric, inferior mesenteric, middle suprarenal, renal, internal spermatic (in the male), and ovarian (in the female).

The celiac artery gives rise to the left gastric artery, which supplies the stomach; the hepatic artery, which supplies the liver; and the splenic (lienal) artery, which supplies the spleen. The right gastric artery arises from the hepatic artery (or, as in about 12 percent of bodies examined, from the superior mesenteric artery) supplies the stomach, and finally anastomoses with the left gastric artery.

The hepatic artery gives rise to the gastroduodenal and cystic arteries. The gastroduodenal artery gives rise to the right gastroepiploic and superior pancreaticoduodenal arteries.

The splenic artery gives rise to the pancreatic, short gastric, and left gastroepiploic arteries.

The superior mesenteric artery branches into the inferior pancreaticoduodenal, intestinal, ileocolic, right colic, and middle colic arteries.

The renal arteries arise as two large trunks from the sides of the abdominal aorta immediately inferior to the superior mesenteric artery and at the level of the intervertebral disc between the first and second lumbar vertebrae. Each renal artery extends across the crus of the diaphragm and forms an angle of nearly 90 degrees with the aorta. The right renal artery is somewhat longer than the left renal artery and passes from the aorta posterior to the inferior vena cava, the right renal vein, the pancreatic head, and the descending duodenum to break up into four or five branches immediately prior to entering the right kidney hilum. The left renal artery is situated slightly higher in the abdomen than the right renal artery. The left renal artery passes posterior to the left renal vein, the pancreatic body, and the splenic vein, and is crossed by the inferior mesenteric vein prior to the breakup of this renal artery into four or five branches (as with the right renal artery), which enter the left kidney hilum.

The internal spermatic arteries in the male (and ovarian arteries in the female) arise anteriorly on the aorta slightly inferior to the origins of the renal arteries. The internal spermatic arteries supply the testes and are much longer than the ovarian arteries, which supply the ovaries.

The inferior mesenteric artery branches into the left colic, sigmoid, and superior hemorrhoidal arteries.

The *parietal* branches of the abdominal aorta are the inferior phrenic, lumbar, and middle sacral arteries.

In sequential order of origin along the aorta from above downward, the arteries are as follows:
1. Inferior phrenic
2. Celiac trunk
3. Hepatic
4. Left gastric
5. Splenic
6. Superior mesenteric
7. Renal
8. Internal spermatic
9. Lumbar
10. Inferior mesenteric
11. Middle sacral

The two small inferior phrenic arteries do not always have the same origin in different persons. Each inferior phrenic artery may have individual origin anteriorly on the aorta, or both may arise from a common origin in the same locus or from a common origin from the celiac artery. Other variations have been found. The inferior phrenic arteries supply the diaphragm and give rise (one each) to the superior suprarenal arteries, which supply the suprarenal glands. The inferior phrenic arteries assist in supplying other viscera in the regions beneath the diaphragm.

There are usually four pairs of lumbar arteries, one pair arising posteriorly from the aorta opposite each of the first four lumbar vertebrae. For the most part these arteries supply the posterior and more external structures in this region.

The middle sacral artery arises posteriorly on the aorta slightly superior to its bifurcation. This artery supplies some of the posterior and external structures in the region of the sacrum.

Arteries of the pelvis and lower extremities. Each *common iliac artery* divides into the corresponding external iliac and hypogastric (internal iliac) arteries. The *hypogastric artery* has two trunks, the anterior and posterior. From the anterior trunk arise the superior vesical, middle vesical, inferior vesical, middle hemorrhoidal, obturator, internal pudendal, and inferior gluteal arteries and in the female, the uterine and vaginal arteries.

From the posterior trunk arise the iliolumbar, lateral sacral, and superior gluteal arteries.

The internal pudendal artery gives rise to the muscular, inferior hemorrhoidal, perineal, urethral bulb, and urethral arteries and to the deep artery of the penis and the dorsal artery of the penis.

Outside the pelvic cavity the inferior gluteal artery branches into the muscular, coccygeal, sciatic, anastomotic, articular, and cutaneous arteries.

The *external iliac artery* gives rise to the inferior epigastric and deep iliac circumflex arteries and continues into the femoral artery. The distal part of the femoral artery continues into the popliteal artery shortly above the knee. The popliteal artery divides into the anterior and posterior tibial arteries in the lower part of the knee joint space.

The femoral artery gives rise to the superficial epigastric, superficial iliac circumflex, superficial external pudendal, deep external pudendal, muscular, deep femoral, and the highest genicular arteries.

The deep femoral artery branches into the medial and lateral femoral circumflex arteries and the perforating and muscular arteries.

In addition to dividing into the two tibial arteries, the *popliteal artery* gives rise to the superior and sural muscular arteries, the cutaneous artery, and the medial superior, lateral superior, middle, and medial inferior genicular arteries.

The *anterior tibial artery* gives rise to the posterior tibial recurrent, anterior tibial recurrent, fibular, muscular, anterior medial malleolar, and anterior lateral malleolar arteries.

The anterior tibial artery continues into the *anterior dorsal artery of the foot,* which branches into the lateral tarsal and medial tarsal arteries, the arcuate artery, the first dorsal metatarsal artery, and the deep plantar artery.

The *posterior tibial artery* branches into the peroneal, nutrient, muscular, posterior medial malleolar, communicating calcaneal, and lateral calcaneal arteries.

The plantar arch forms from the anastomosis of the lateral plantar artery and the deep plantar branch of the dorsal artery of the foot. This arch branches into the perforating and plantar metatarsal arteries.

Arterioles

Throughout the body the smaller arteries continue to branch, finally, into the arterioles. *An arteriole is any minute arterial branch, especially one just proximal to a capillary. A precapillary is a vessel that terminates in a capillary.*

Average-sized arterioles measure approximately 0.2 mm. in outside diameter. They contain the *continuous tunica intima* surrounded by an outer wall containing varying quantities of nonstriated muscle fibers innervated from the autonomic nervous system. The nonstriated muscle fibers encircle the arteriole lumen and are thus capable of both decreasing and increasing the lumen diameter through contraction and relaxation. This fact is significantly important in the control of blood pressure.

Capillaries

Capillaries, the tubular structures between the very smallest arterioles and venules, serve to connect the arterial and venous parts of the circulatory system. These minute vessels convey blood to all the body cells and, in extending throughout the skin, permit the normal, healthy appearance of the skin.

A single capillary seldom is more than 0.5 mm. long; it has been estimated that if *all* the capillaries in the body were placed end to end the combined length would extend many thousands of miles. Capillaries are less than 0.1 the diameter of arterioles, and the capillary lumina are from 8 to 10 microns in diameter. The diameter of a capillary is usually large enough to permit the free passage of only a single blood cell at a time. Some capillaries, as in the brain, are smaller; others, as in skin and the bone marrow, are as much as 20 microns in diameter. The total combined area of the cross sections of all the capillaries is from 600 to 800 times greater than the cross-sectional area of the aorta. This total area comprises the *capillary bed* (lake).

The capillary walls are extremely thin and consist of a single layer of endothelial cells with their edges joined together by an interstitial cement substance. The total area of capillary walls has been estimated to exceed 6,000 square meters, an area of about 1.5 acres. In many structures, the single-cell wall of the capillary is common with the wall of adjacent tissue. The capillary walls are the semipermeable membrane through which water, crystalloids, and some plasma proteins pass; the capillary walls are impermeable to large molecules.

The diameter of the capillary lumen is relatively constant from arterial to venous end, and the general structure of the capillary wall is the same throughout. However, the arterial ends of the capillaries permit the passage of oxygen and nutritional substances, and the venous ends of the capillaries permit the passage of waste products and carbon dioxide. The capillaries respond to the same stimuli as the arterioles, thus, the diameters of each may increase or decrease according to necessity.

The interchange of substances through capillary walls is made possible because of certain physical laws controlling filtration and osmosis.* In addition, the rate of flow of blood in the capillaries is between 1/600 and 1/800 of that in the aorta. Blood pressure decreases about 15 to 20 mm. Hg (millimeters of mercury) as the blood flows from the arterial ends to the venous ends of the capillaries. Through the walls of these tiny vessels (capillaries) *all* the work of assimilation of oxygen, nutrition, and elimination of waste products between the circulatory system fluids and the body cells occurs. The arterial part of the circulatory system conveys the oxygen and nutrition to the capillaries, and the venous part of this system collects the waste products from the body cells and tissues via the capillaries. The waste products are then conveyed via the venous system to the several excretory structures.

Veins

A vein is one part of the system of tubular branching vessels that arises from the union of the capillaries in the various organs and tissues and that carries the

---

*Blood hydrostatic pressure forces fluid through the capillary walls into the interstitial fluid. Blood colloid osmotic pressure (see p. 297) draws interstitial fluid through the capillary walls into the blood. Conversely, interstitial fluid hydrostatic pressure forces interstitial fluid through the capillary walls into the blood, and interstitial fluid colloid osmotic pressure draws fluid through the capillary walls into the interstitial fluid. In this discussion, Starling's hypothesis applies. The hypothesis states that the direction and rate of transfer between plasma in the capillary and fluid in the tissue spaces depend on the hydrostatic pressure on each side of the capillary wall, on the osmotic pressure of protein in plasma and in tissue fluid, and on the properties of the capillary wall as a filtering membrane.

blood toward or back to the heart. With the exception of the pulmonary veins between the lungs and the left atrium, the veins carry blood that is quite low in oxygen content.

Structure

Like the arterial walls, the venous walls are composed of three layers, which, from inside to outside, are the tunica intima, tunica media, and tunica externa. The principal differences between venous and arterial walls are twofold: venous walls are thinner than arterial walls, and there are fewer muscle fibers in the tunica media of the venous walls. Veins will collapse when cut; arteries will not.

Another difference between veins and arteries is in the internal (lumen) structure. Many of the veins, especially the longer ones, have semilunar valves in their lumina to aid the return of blood toward the heart. Since there is no muscular layer in the tunica media, there is no blood pressure in the veins; therefore, backflow of the blood is prevented by the valves and by the contraction of muscles in the limbs. The muscle contractions, valve actions, and replacement of blood volume with additional volume from the capillaries all contribute to the movement of blood toward the heart.

The final difference between arterial and venous networks is in number rather than structure. In general, there are two veins for each artery. Also, one of the two veins usually has the name of the corresponding artery.

Names

Blood is forced from the right ventricle via the pulmonary artery* to the lungs and returns via the *four pulmonary veins* to the left atrium. Blood is forced from the ventricle via the coronary arteries to the myocardium and returns via the coronary veins to the right atrium and ventricle. Blood is forced from the left ventricle via the aorta to the remainder of the body and returns via the superior and inferior venae cavae to the right atrium (see Fig. 6-6).

Major veins of the lower extremity. The major veins arising from the lower extremity include the intercapitular veins, which arise from the plantar cutaneous venous arch. The intercapitular veins join with the dorsal digital veins to form the short common digital veins. The latter unite across the distal ends of the metatarsal bones to form the dorsal venous arch. Medial and lateral marginal veins at the sides of the foot join an irregular venous network proximal to the dorsal venous arch. Branches from the superficial part of the sole of the foot are involved in the formation of the medial and lateral marginal veins.

The *great saphenous vein* arises in the medial marginal vein of the dorsum of the foot, ascends ventrally to the medial malleolus, and extends upward in fairly close relation to the saphenous nerve, always medial in the leg. The great

---

*The pulmonary artery branches at the inferior curve of the aortic arch into the right and left pulmonary arteries.

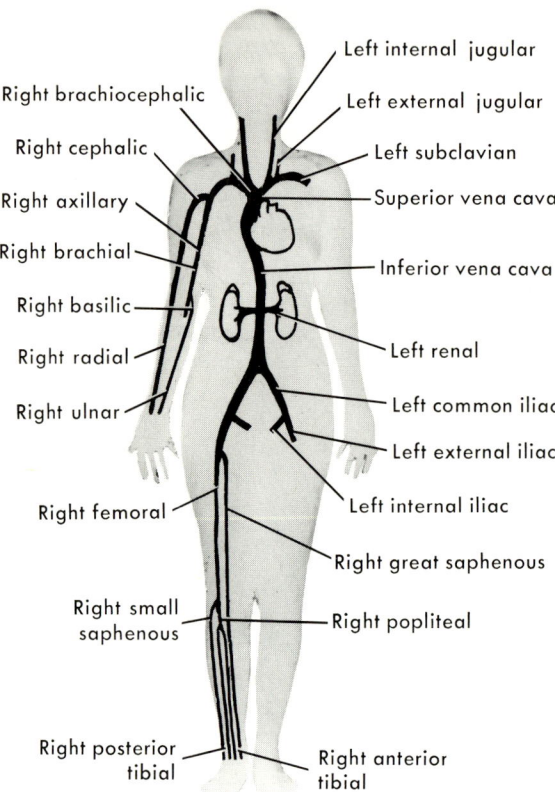

Fig. 6-6. Some major veins (schematic).

saphenous vein terminates in the femoral vein, approximately 3 cm. inferior to the inguinal ligament; thus, it is the longest vein in the body.

The *small saphenous vein* arises dorsally to the lateral malleolus and is a continuation of the lateral marginal vein. The small saphenous vein extends upward laterally to the calcaneal tendon and then crosses the tendon to extend to the midportion of the dorsal surface of the lower leg. At this point the small saphenous vein perforates the deep fascia in the popliteal fossa and joins with the popliteal vein.

Four metatarsal veins arise from the plantar digital veins and extend toward the heel to form the deep plantar venous arch. The medial and lateral plantar veins arise from this arch, extend toward the heel, communicate with the saphenous veins, and unite posterior to the medial malleolus to form the *posterior tibial vein*. This vein is adjacent to the posterior tibial artery and receives the *peroneal vein*.

The *anterior tibial vein* arises from the venae comitantes of the dorsal artery of the foot as an upward continuation. This vein passes between the tibia and fibula and unites with the posterior tibial vein to form the *popliteal vein*.

The popliteal vein ascends through the popliteal fossa to the aperture of the adductor magnus muscle and becomes the *femoral vein*.

Other veins of the lower extremity. In the upper two-thirds of the thigh the femoral vein spirals around the femoral artery. Inferior to the inguinal ligament and medial to the femoral artery, the femoral vein is in the same plane as the femoral artery. The femoral vein receives many tributary branches, particularly the deep femoral and the great saphenous veins. The femoral vein contains three semilunar valves.

Tributary veins corresponding to branches of the profunda artery supply the *deep femoral vein*. This vein communicates with the *inferior gluteal vein* and the popliteal vein and receives the *medial* and *lateral femoral circumflex veins*.

Major veins of the abdominal and pelvic cavities. The veins of the abdominal and pelvic cavities listed here commence with the external iliac vein as a continuation and upward extension of the femoral vein.

The *external iliac vein* arises posterior to the inguinal ligament and terminates in a union with the hypogastric vein opposite the sacroiliac articulation. The external iliac vein receives many tributaries, including the *inferior epigastric, deep iliac circumflex,* and *pubic veins*.

The *hypogastric (internal iliac) vein* arises close to the greater sciatic foramen and terminates in its junction with the external iliac vein. The hypogastric vein receives many tributaries, including the *gluteal, internal pudendal,* and *obturator veins,* which originate outside the pelvis; the *lateral sacral veins*, which are anterior to the sacrum; and the *middle hemorrhoidal, vesical, uterine,* and *vaginal veins,* which arise from venous plexuses in the pelvic viscera.

The union of the internal and external iliac veins forms the *common iliac vein*; this vein frequently contains one or more semilunar valves.

The right and left common iliac veins join each other on the level of the fifth lumbar vertebra to form the inferior vena cava. The right common iliac vein is much shorter than the left common iliac vein, and as a result, the right is more nearly vertical than the left. The junction of the common iliac veins is immediately anterior and slightly superior to the bifurcation of the aorta.

Inferior vena cava. From its origin at the level of L-5 the *inferior vena cava* ascends in the abdominal cavity anterior to the vertebral column along the right side of the aorta, receives several major tributaries, perforates the diaphragm in the vena caval foramen, and terminates in the inferoposterior part of the right atrium of the heart.

The inferior vena cava, in its ascent, receives the *lumbar, spermatic (or ovarian), renal, suprarenal, inferior phrenic,* and *hepatic veins*.

Other veins arising in the abdominal cavity. The most common origin of the *azygos vein* is the *right ascending lumbar vein*. The azygos vein arises opposite the first or second lumbar vertebra and ascends along the right side of the vertebral column to penetrate the diaphragm in the aortic hiatus. This vein ascends the right side of the thoracic vertebral column to a site opposite the fourth thoracic vertebrae, where the vein arches anteriorly and terminates in the *superior vena cava* prior to its penetration of the pericardium.

The *hemiazygos vein* usually arises from the *left ascending lumbar vein,* extends upward on the left side of the lumbar vertebrae, pierces the diaphragm through the left crus, and extends upward parallel with the left side of the thoracic vertebrae, often as far as the ninth thoracic vertebra. At this level the hemiazygos vein crosses behind the aorta and other viscera to terminate in the azygos vein. Tributaries of the hemiazygos vein include the last four or five intercostal veins, the left subcostal vein, and branches of the esophageal and mediastinal veins.

Other veins of the abdomen function to drain the digestive viscera and comprise the *portal system* of veins (see pp. 271 and 272).

Blood from the meninges, the spinal cord, the vertebral bodies, and adjacent muscles is drained by numerous veins into plexuses that extend along the vertebral column. The plexuses anastomose very often and terminate in the intervertebral veins.

Major veins of the thoracic cavity. The major veins of the thoracic cavity are the brachiocephalic, superior vena cava, internal mammary, superior phrenic, inferior thyroid, highest intercostal, accessory hemiazygos, and bronchial veins. The azygos and hemiazygos veins are discussed in the preceding paragraphs.

Each *brachiocephalic vein* arises from the junction of the corresponding *internal jugular* and *subclavian* veins. The right brachiocephalic vein commences at a site that is posterior to the sternal end of the clavicle and extends for about 2.5 cm. downward to join the left brachiocephalic vein. The right brachiocephalic vein lies anterior to the brachiocephalic artery.

The left brachiocephalic vein is somewhat longer than the right, about 6 cm. in length. The left brachiocephalic vein commences at a site that is posterior to the sternal end of the clavicle and extends obliquely downward to its junction with its opposite vein.

The two brachiocephalic veins join at a site that is immediately inferior to the first rib cartilage and close to the right sternal margin to form the superior vena cava. The right brachiocephalic vein receives blood from the right vertebral, right internal mammary, and right inferior thyroid veins; in some persons the right brachiocephalic vein receives blood from the first intercostal space. The left brachiocephalic vein receives blood from the left vertebral, left internal mammary, left inferior thyroid, and left highest intercostal veins; in some persons the left brachiocephalic vein receives blood from the thymic and pericardiac veins.

Superior vena cava. The *superior vena cava* is approximately 7 cm. long. It extends from its origin (see the preceding paragraph) to terminate in the right atrium of the heart, slightly superior to the orifice of the inferior vena cava. This large vein penetrates the pericardium at about the midpoint of the length of the vein.

Major veins of the upper extremity. The veins of the upper extremity are grouped into one of two divisions: deep and superficial. Veins of each division have frequent anastomoses with the veins of the other division. The deep veins are situated near the arteries; the superficial veins are situated just beneath the integument. The veins of both divisions contain valves, although there are more valves in the deep veins than in the superficial veins.

In general, the deep veins are found in pairs, one on each side of the corresponding artery; they form the venae comitantes* of that artery.

Beginning with the hand, the deep veins of the upper extremity include the deep volar venous arches, the common volar digital veins, the proper volar digital veins, the volar metacarpal veins, the dorsal metacarpal veins, the venae comitantes of the radial and ulnar arteries, the brachial veins, the axillary vein, and the subclavian vein.

The superficial veins of the upper extremity include the digital, metacarpal, cephalic, basilic, and median veins. The superficial arterial arches are associated with corresponding venae comitantes to form the superficial volar venous arches.

The four major veins of the forearm arise from the arches and venae comitantes of the hand. These four veins are the radial and cephalic, laterally, and the ulnar and basilic, medially. The four veins ascend the forearm, each receiving numerous tributaries and anastomosing frequently with other veins. The *radial* and *ulnar veins* unite in the antecubital space (front of elbow) to form the brachial veins. The major *brachial vein* receives the *basilic vein* approximately even with the lower border of the teres major muscle to form the *axillary vein*. The axillary vein receives the *cephalic vein* immediately inferior to the clavicle and forms the *subclavian vein*. In some instances the axillary vein communicates directly with the external jugular vein.

Major veins of the head and neck. The veins of the head and neck are grouped into one of three divisions: the exterior of the head and face; the neck; and the diploë, brain, and venous sinuses of the dura mater.

The veins grouped together as those of the exterior of the head and face include the frontal, supraorbital, angular, anterior facial, occipital, superficial temporal, internal maxillary, posterior facial, and posterior auricular veins. For the most part these veins drain into the corresponding external jugular vein; the external jugular vein terminates in the corresponding subclavian vein, immediately lateral to the terminus of the internal jugular vein.

The veins grouped together as those of the neck include the external jugular, posterior external jugular, vertebral, anterior jugular, and internal jugular veins. The vertebral veins descend posteriorly along the vertebral artery and join in a plexus drained by a single trunk, which emerges through the transverse foramen of the sixth cervical vertebra to join with the corresponding brachiocephalic vein near its origin.

Blood from the brain, superficial parts of the face, and the neck drain into the internal jugular vein (the great vein of the neck). This vein commences in that part of the transverse sinus situated in the groove on the mastoid portion of the temporal bone; this groove is the *sigmoid sinus*.

The veins grouped together as those of the diploë, brain, and venous sinuses of the dura mater include veins from three subdivisions. The diploic veins in the channels of the diploic layer of the cranial bones communicate with the

---

*The venae comitantes are the veins accompaning the corresponding artery.

meningeal veins and with the sinuses of the dura mater. The frontal, anterior temporal, posterior temporal, and occipital veins comprise this subdivision.

The veins of the brain (the second subdivision) include the cerebral and cerebellar groups. These veins possess extremely thin walls; they extend through the arachnoid membrane and meningeal layer of the dura mater to open in the cranial venous sinuses.

The venous sinuses of the dura mater (the third subdivision) include the superior sagittal, inferior sagittal, occipital, straight, and the two transverse sinuses. The superior and inferior sagittal sinuses are associated directly with the falx cerebri.* The junction line of the falx cerebri with the tentorium cerebelli† is the site of the straight sinus. This sinus extends posteriorly and inferiorly from the inferior sagittal sinus and usually terminates in the left transverse sinus near the internal occipital protuberance. The right transverse sinus originates near the internal occipital protuberance from the superior sagittal sinus and terminates in the right internal jugular vein. The left transverse sinus terminates in the left internal jugular vein. Each jugular vein passes through the floor of the skull in the corresponding jugular foramen.

Venules

A venule is a minute vein or venous radicle. Venules comprise the beginnings of the venous network in the cells and tissues and correspond with the minute arterioles, which are the termini of the arterial network. The capillaries lie between the arterioles and the venules.

Venules possess the same three layers in their walls, as do the larger veins. Valves are quite uncommon in the venules and in any vein with a diameter of less than 2 mm.

Lymphatic system

The circulatory system includes, as a distinct part, the system of capillaries and veins that conveys the lymph from the head and body to the neck, where the lymph fluid returns to the bloodstream.‡ The lymphatic system also includes the numerous lymph nodes and the spleen. Of all the tissues in the body, lymphoid tissue evidences the greatest sensitivity to the effects of (ionizing) radiation.

The magnitude or extensiveness of the lymph system can be appreciated when its generalized establishment is fully realized (see footnote on p. 41). This reference is to the reticuloendothelial system; the name is derived from *reticulum* meaning "resembling a net" and *endothelium* meaning "the layer of epithelial cells that lines the cavities of the heart and of the blood and lymph vessels" (see Fig. 6-7).

*The falx cerebri is the sickle-shaped fold of dura mater that extends downward in the longitudinal cerebral fissure and separates the two cerebral hemispheres.
†The tentorium cerebelli is the extension of dura mater that forms a partition between the cerebrum and the cerebellum and covers the upper surface of the cerebellum.
‡Lymph fluid flows only toward the neck, where it empties into the venous blood.

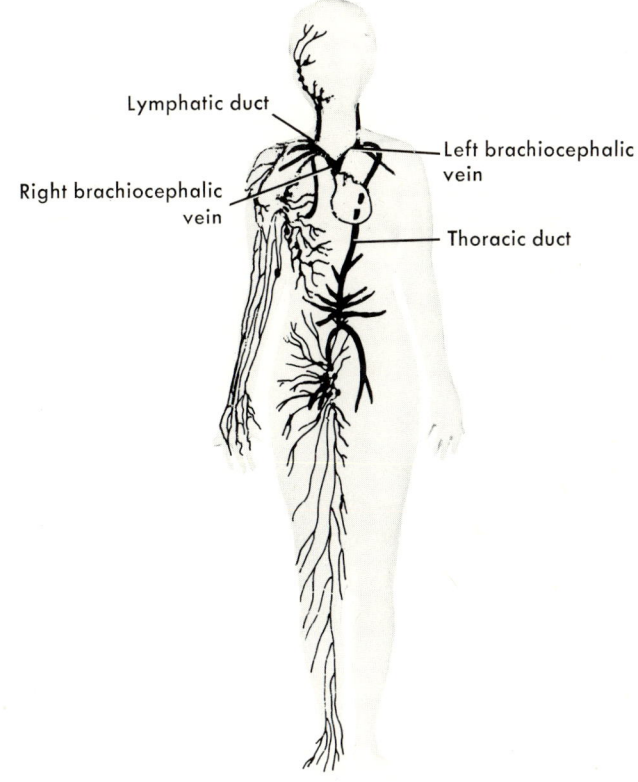

Fig. 6-7. Thoracic and lymphatic ducts.

Lymph nodes

A lymph node is one of the rounded masses of lymphoid tissue, surrounded by a capsule of connective tissue, which occurs in various parts of the body in the course of the lymphatic vessels. Lymph nodes are not true glands but consist of a reticulum of connective tissue fibers. In the meshes of the connective tissue fibers are contained numerous small round cells, the *lymphoid cells,* each having a large round deeply staining nucleus. When these cells are conducted away from the node by the lymph, the cells become *lymphocytes* in the circulating blood (see Fig. 6-8).

Lymph nodes are situated throughout the body both singly and in groupings of various numbers. Large concentrations of lymph nodes are found in the neck in both superficial and deep groupings extending from the base of the neck to the root of the tongue and the floor of the mouth. Other large concentrations occur beneath each clavicle, in the mediastinal region, in the axillae, around the breasts, throughout and intermingled with the digestive viscera, deep in the pelvis, and in the groin.

In general, the lymph nodes are rather small and bean shaped. Each node

258 *Textbook of anatomy and physiology in radiologic technology*

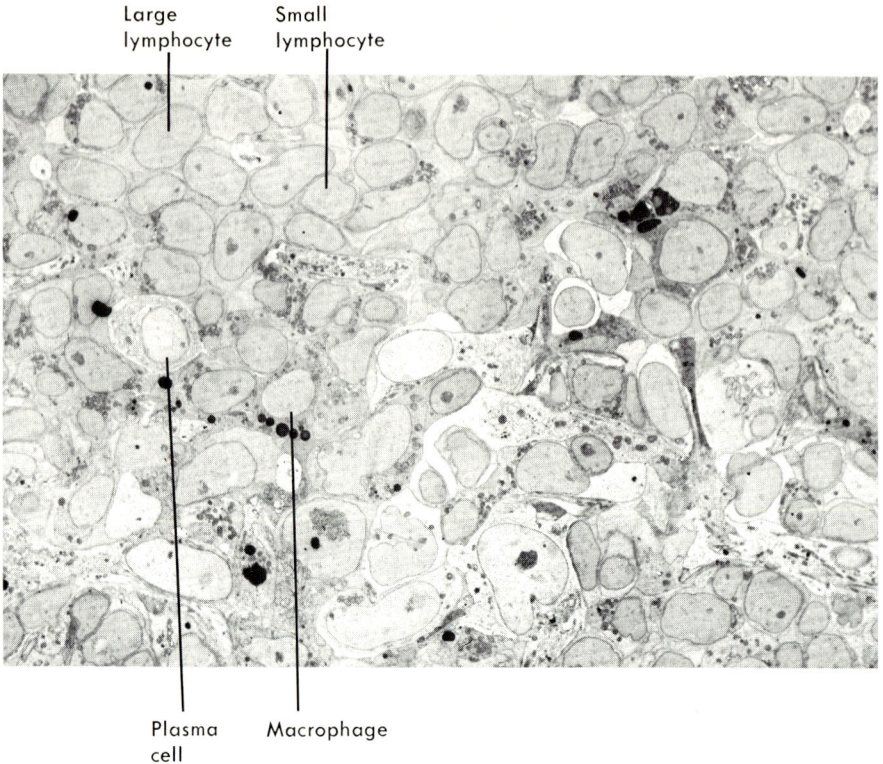

Fig. 6-8. Electron micrograph. Cortex of lymph node of mouse intestine ($\times$ 2,600). (From Rhodin: An atlas of ultrastructure, Philadelphia, 1963, W. B. Saunders Co.)

has a hilum through which the blood vessels and the efferent lymph vessels pass. The afferent lymph vessels enter a node usually in several sites on the periphery of the node. The nodes are situated in the course of lymphatic and lacteal* vessels to enable the passage of lymph and chyle† enroute to the circulating venous blood.

Entering the hilum along with the blood vessels are the nerves and nerve branches supplying each node. These nerves are believed to be from the larger vasomotor nerves in the particular region.

Lymphatics

A lymphatic is a vessel in which lymph is contained or conveyed, i.e., a lymphatic vessel. The lymphatics originate as interfibrillar or intercellular clefts or spaces in the various tissues and organs, the smaller ones having no distinct walls or having walls composed only of endothelial cells. The larger lymphatics are very similar to the blood veins in structure.

*A lacteal is any one of the intestinal lymphatics that take up chyle.
†Chyle is the milky fluid taken up by the lacteals from the food in the intestine after digestion. It consists of lymph and emulsified fat.

The larger lymphatics are arranged, for the most part, in two sets: a superficial set immediately beneath the skin and a deep set. As a rule the lymphatics have numerous small dilations, anastomoses, and valves in their courses. In one of the concentrations of lymph nodes deep within the abdomen is the *cisterna chyli.*

The cisterna chyli is a dilated lymph channel usually situated opposite the first and second lumbar vertebrae. The *thoracic duct,* one of the two principal lymphatic ducts, arises from the cisterna chyli.

The thoracic duct is by far the larger of the two principal lymphatic ducts. From its origin the thoracic duct ascends in the abdominal cavity posterior to and on the right side of the aorta and passes through the aortic hiatus of the diaphragm into the thoracic cavity. In the posterior mediastinal cavity the thoracic duct ascends between the aorta and the azygos vein to a level opposite the fifth thoracic vertebra, where the duct angles to the left and enters the superior mediastinal cavity. In this region the thoracic duct ascends posterior to the aortic arch, the thoracic part of the left subclavian artery, and other viscera in the region to enter the neck through the upper orifice of the thoracic cavity. In the neck the duct forms an arch, which passes about 4 cm. superior to the clavicle and anterior to several arteries, veins, and nerves to open into the bloodstream at the junction of the left internal jugular vein with the left subclavian vein (see Fig. 6-7).

In the average adult the total length of the thoracic duct is from 38 to 45 cm. The duct receives the lymph fluid and chyle from all parts of the body below the diaphragm except the convex surface of the liver, and from the left side of the thoracic cavity, the left upper extremity, the left side of the head and neck, the left lung, and left side of the heart.

The other major duct of this system is the much shorter *lymphatic duct.* This duct is about 1¼ cm. long and extends from the root of the neck near the medial border of the scalenus anterior muscle to the junction of the right internal jugular vein with the right subclavian vein. The lymphatic duct receives the lymph from the convex surface of the liver; the right upper extremity; the right side of the head, neck, and thoracic cavity; the right lung; and the right side of the heart.

Lacteals

In each villus of the small intestine is a small lymph channel, the *lacteal.* The lacteals continue outside the villi as the lymphatics of the intestinal wall and the mesentery. As a result of fat digestion the majority of absorption of hydrolyzed fats occurs through the lacteals (see p. 229). The emulsified fat particles carried by the lymph from the lacteals give the lymph an appearance similar in color and consistency to milk; this is the *chyle.* The fat is derived from the chyme of the small intestine and varies in quantity according to the amount of fat in the diet and the stage of digestion.

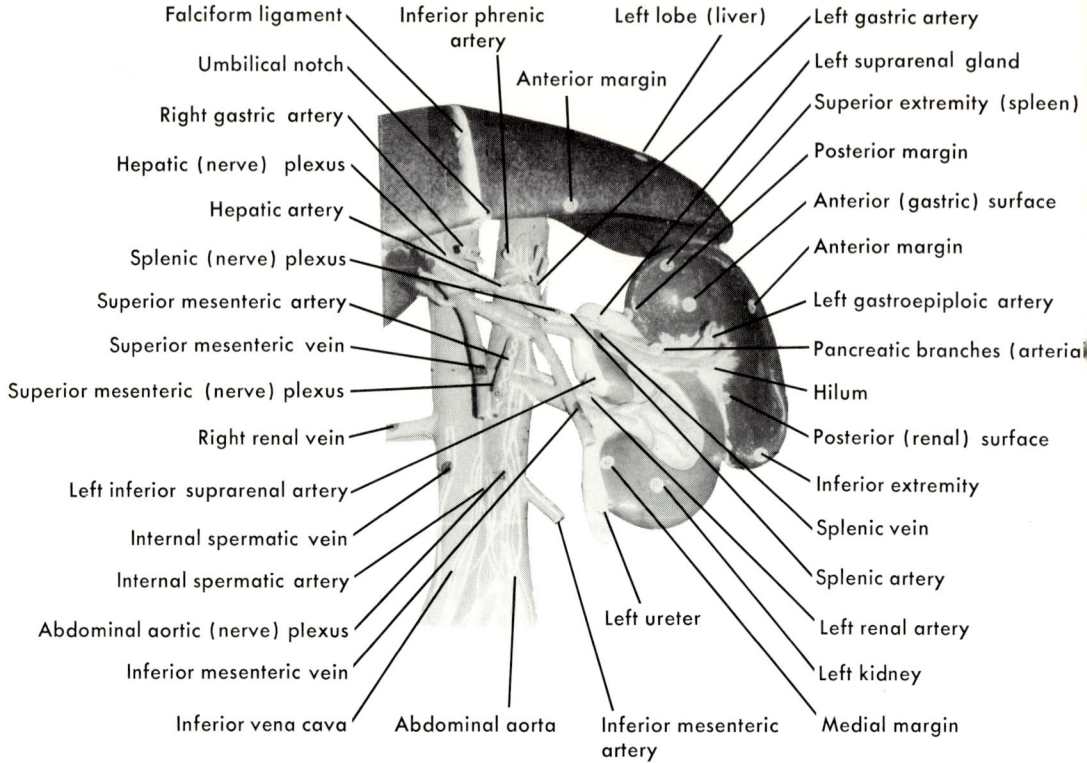

Fig. 6-9. Spleen and related structures, anterior aspect.

Spleen

The spleen in adults is quite variable in both size and weight. Usually the spleen is about 12 cm. long, 7 cm. deep, and 3.5 cm. thick. The weight of the spleen increases from young adulthood to middle age, after which the weight decreases with advancing age. The spleen varies in different persons from a minimum of 100 grams to as much as 250 grams.

The spleen lies, for the most part, in the left hypochondrium; some of it extends into the epigastrium. The spleen is oblong and flat, quite friable,* and highly vascular (see Fig. 6-9).

The spleen is a glandlike ductless organ, composed for the most part of loose adenoid tissue traversed by trabeculae derived from the fibrous outer capsule. Malpighian capsules are contained within the splenic pulp. Various functions have been ascribed to the spleen, and all of its functions are as yet unknown. During intrauterine life and for a short time following birth, the spleen is the site of erythrocyte formation. It is possible that this organ is involved in the destruction of old erythrocytes, and it is most certainly associated with the digestive system since there is a distinct increase in size following a meal.

*Friable means easily pulverized.

The spleen is the site of manufacture of both lymphocytes and monocytes. It contains numerous phagocytic cells of the reticuloendothelial system (see p. 41) to destroy bacteria and other foreign matter that might be harmful to the body.

Also, the spleen is one of the most important blood reservoirs of the body. As much as 350 ml. of blood can be contained within the splenic pulp. This volume can be reduced to as little as 200 ml. within an extremely short time to increase the volume of circulating blood; this is an important contribution toward homeostasis in shock prevention.

BLOOD, CELLS, FLUIDS, AND INTERRELATED FUNCTIONS

The general functions of the circulatory system have been discussed previously (see pp. 235 and 237). More specific functions of the cells and fluids of the circulatory system and the associated effects upon homeostasis are presented in the following paragraphs.

Blood

Blood is fluid connective tissue. The fluid part of blood is the plasma, accounting for approximately 55 percent of the total volume. Solids, chiefly cells, account for the remainder of the blood volume. In males the ratio range is from 40 to 54 percent cells to from 60 to 46 percent plasma; in females this ratio range is from 37 to 47 percent cells to from 63 to 53 percent plasma. In the average-sized adult the blood accounts for approximately one-twelfth of the total body weight, or about 5 or 6 liters. The blood and tissue fluid combined account for approximately 20 percent of the body weight.

Osmotic pressure

The osmotic pressure, the pressure produced by osmosis (see pp. 39 to 41), is about the same as that of a 0.9-percent NaCl (physiologic saline) solution. The various substances dissolved in the plasma are largely responsible for the osmotic pressure. Plasma proteins also influence osmotic pressure. The osmotic pressure is maintained nearly constant in healthy persons through selective renal elimination and/or retention of the dissolved substances.

Acid-base balance

The pH of blood varies only slightly, and any major degree of variation is usually associated with rather grave illness. Normally, the pH of the blood is slightly alkaline, and the greatest variation in healthy persons is between 7.3 and 7.5, although the usual range is considered to be from 7.35 to 7.45.

Human blood contains about twenty times as much $NaHCO_3$ (sodium bicarbonate) as H_2CO_3 (carbonic acid), a ratio that maintains the blood at about pH 7.4. $NaHCO_3$ is the principal alkali in the blood, and H_2CO_3 is the *only* free acid in the blood. The blood also contains Na_2HPO_4 (disodium phosphate) in a quantity about one-thirtieth of $NaHCO_3$ and a corresponding amount of potassium salts in the erythrocytes. Proteins are capable of uniting with either acids or bases and are very useful buffers in the blood. Hemoglobin is the most important protein buffer of the blood; hemoglobin averages about 16 grams/100 ml. of blood in males, and about 14 grams/100 ml. of blood in females. Hemoglobin is contained in the erythrocytes for the most part and much of it is in combination with potassium within the erythrocyte. The erythrocyte membrane is impermeable to the passage of hemoglobin and other proteins.

The several substances possessing buffering capabilities mentioned in these paragraphs (and on pp. 297 to 299) present a brief explanation of the mechanisms effecting pH control of the body fluids, principally of the blood. The combined effect of the several substances in the blood that contribute to the neutralization of acids is called the *alkali reserve* of the blood and is equal approximately to a 0.5-percent solution of $NaHCO_3$.

The preceding paragraphs have presented an overall picture of some of the properties of whole blood. The first detailed discussion of blood constituents begins with the blood cells. The two major types of cells found in the peripheral blood in addition to the platelets (thrombocytes) are the erythrocytes and the leukocytes.

Cells

A modern concept of the origin of all blood cells is that they originate from undifferentiated mesenchymal (stem) cells. The primitive members of each family of cells so strongly resemble each other that differentiation by appearance in this stage is exceedingly difficult. As a result, the cell names derive from the tissues in which the cells are found, the cells with which they associate, and the definitive cells eventually produced.

Erythrocytes

Erythrocytes (normocytes) are the red cells found in the peripheral blood as biconcave discs, averaging from 6 to 7 microns in diameter. They are named erythrocytes as a result of the natural buff color of the hemoglobin within the cells. In peripheral blood these cells normally possess no nucleus. The cell number averages from about 4.5 to 5.5 million in each cubic millimeter of blood in adult males and from about 4 to 5 million in each cubic millimeter of blood in adult females; the number of erythrocytes in each cubic millimeter of blood in newborn infants and in babies is remarkably greater.

The erythrocytes carry oxygen from the lungs to the body tissues and carbon dioxide from these tissues to the lung alveoli. The hemoglobin* (iron-containing protein) in the erythrocytes combines in the lung capillaries with oxygen to form *oxyhemoglobin*. Oxyhemoglobin (HbO_2) is transported via the arterial blood to cells that need oxygen; and the oxygen, which is in loose combination with the hemoglobin, is removed from the oxyhemoglobin molecule for cellular

---

*The chemical expression of one molecule of hemoglobin (HB) is $C_{3032}H_{4812}N_{780}Fe_4O_{872}S_{12}$.

metabolism. Carbon dioxide in the tissues combines with the erythrocytes, which have given up one molecule of oxygen, as carbohemoglobin. Carbohemoglobin* is then transported via the bloodstream to the lung alveoli for elimination of carbon dioxide.

Leukocytes

Leukocytes, the white cells found in the peripheral blood as either spheres or ellipsoids, are generally somewhat larger than the erythrocytes. These cells are named leukocytes as a result of the appearance of the buffy coat† in centrifuged (settled) oxalated blood and in the coagulated blood. The nuclei of adult leukocytes take the chromatin stain and appear bluish purple when the blood smear slide is stained with Wright's stain. The cytoplasm of the different kinds of leukocytes also exhibits distinctive staining properties because of the specific affinity of the leukocytes for certain dyes. The three major kinds of leukocytes are *polymorphonuclears,* with pinkish staining cytoplasm and bluish-black or reddish-pink staining granules; *lymphocytes,* with light-blue staining cytoplasm and azurophilic staining granules; and *monocytes,* with bluish-gray staining cytoplasm.

The polymorphonuclears are further subdivided according to the shape and number of segments in the nucleus and according to the staining properties of the granules: basic stained granules (dark blue)—basophils; acidic stained granules (pink to red)—eosinophils; and other cells in which the granules take neither the basic nor acidic stain—neutrophils. The total number of leukocytes averages between 5,000 and 10,000 per cubic millimeter of blood.

The polymorphonuclears and monocytes function as phagocytes; cells that engulf foreign particles in the bloodstream. The lymphocytes function to produce immune bodies that help ward off diseases caused by certain bacteria and rickettsiae.

Lymphocytes constitute one of the major classes of leukocytes and account for approximately 25 to 35 percent of the total leukocytes in normal peripheral blood. Lymphocytes usually are smaller than other leukocytes; however, lymphocytes are larger than erythrocytes.

Increases in the number of lymphocytes are usually associated with chronic conditions and are termed "lymphocytoses." Decreases in number of peripheral blood lymphocytes are associated with certain acute infections, such as pneumonia; these decreases are termed "lymphopenia."

The number of leukocytes generally increases in acute infections and may reach a concentration of 13,000 to 15,000 per cubic millimeter. For the most part, these increases will be in numbers of neutrophils, usually the younger ones. Other conditions may cause mild to extreme increases in the number of leuko-

---

*See pp. 319 to 322 for a more detailed discussion.
†The buffy coat is the buff-colored layer of white cells atop the much thicker layer of erythrocytes as seen in oxalated blood.

cytes. The opposite condition of leukemia is leukopenia, a decrease in the total number of leukocytes. Leukemia is a malignant condition; leukopenia is not. Some cases of acute leukemia may present as many as 200,000 leukocytes per cubic millimeter.

Platelets (thrombocytes)

The platelets probably form in the red bone marrow. They are small, flat structures that resemble cells; platelets cannot classify as cells because they have no nuclei and have not possessed nuclei at any time. They are about one-half the size of erythrocytes, number from about 250,000 to 400,000 per cubic millimeter of blood, and function to assist in the clotting process by collecting in the fibrin network.

Fluids
Plasma

Plasma is the fluid portion of the blood in which the blood cells and platelets circulate. Plasma is a pale, straw-colored, slightly viscid fluid with a pH, like that of whole blood, normally between 7.35 and 7.45.

Plasma is the supernatant part of the oxalated blood out of which the cells settle. Plasma is very much like interstitial fluid, which fills the spaces between tissues and bathes the cells not bathed directly by the blood. Plasma consists of about 92 percent water and contains the following substances:
1. Proteins—fibrinogen, serum globulin, and serum albumin
2. Inorganic substances—chlorides, phosphates, carbonates, sodium, potassium, magnesium, and calcium
3. Nutritive materials—glucose, fats, and amino acids
4. Other beneficial substances—antibodies, hormones, and enzymes
5. Waste materials—urea, uric acid, etc.

Serum

Blood serum is the clear, yellowish fluid that separates from the blood after coagulation of the fibrin. When whole blood is removed from the circulatory system and placed in a receptacle containing no anticoagulants, it loses the fluid condition after a few minutes and becomes a semisolid, red mass. This is the result of fibrin formation and the entrapment within the fibrin network of the erythrocytes, leukocytes, and platelets. After standing in a test tube for one-half an hour or more, the clot *retracts;* i.e., the fibrin network contracts, pulling the clot away from the blood serum.

Interrelated functions
Blood pressure

The heartbeat and the cardiac cycle have been discussed previously. As the left ventricle enters systole, the walls of the ventricle contract and force the blood out of this chamber into the aorta. Very shortly after initiation of systole in the left ventricle, the greatest (systolic) arterial blood pressure occurs; correspondingly, the least (diastolic) arterial blood pressure occurs after ventricular diastole. The *pulse pressure* is the numerical difference between the systolic and diastolic blood pressures.

Blood pressure can be considered equal to the force exerted laterally by the blood against a unit area of the vessel wall. Since the initial pressure forcing the blood through the arteries originates in the heart, it is logical that blood pressure measured nearer the heart is greater than when measured farther from the heart.

Five factors exert a considerable amount of control over the pressure of the blood against the arterial walls. These factors are the *cardiac output, peripheral vascular resistance,* the portion of the *blood volume* found in the arterial system, *blood viscosity,* and the *elasticity* of the arterial walls.

If the other four factors remain constant, changes in the cardiac output (stroke volume) will change both the systolic and diastolic pressures correspondingly.

Since any given volume of blood forced by a corresponding left ventricular systole into the arterial system must leave this system via arterioles, capillaries, and certain arteriovenous shunts prior to the succeeding left ventricular systole, any changes in the vessels through which the blood leaves the arterial system affects the blood pressure in the system. Thus, both increases and decreases in peripheral resistance exert corresponding changes in the arterial blood pressure.

As stated in a previous discussion of the spleen (see pp. 260 to 261), the volume of blood contained within the spleen can be reduced from about 350 ml. to about 200 ml. in a very short time to force this additional volume of blood into the arterial system as an antishock measure. Losses of blood in appreciable amounts can reduce blood volume. Such reductions will have a corresponding effect upon the arterial blood pressure. Correspondingly, increases of significant quantity in blood volume will increase arterial blood pressure.

From previous paragraphs it is evident that blood consists essentially of a part that is plasma and a part that is cells and that the cells flow and circulate in the plasma. The plasma has very low viscosity, and the whole blood has quite high viscosity. Logically, the increased viscosity of whole blood is the property of the erythrocytes and other solid particles. Therefore, increases of significant quantity in the number of erythrocytes will cause corresponding increases in the whole blood viscosity. It is logical that greater pressure is required to force blood of greater viscosity through the arteries than blood of decreased viscosity. These statements lead to the following statement: blood having greater viscosity than normal will precipitate greater arterial blood pressure.

The final factor exerting influence upon arterial blood pressure is that of arterial wall elasticity. Various conditions cause changes in one or more of the three layers of arterial walls. During ventricular systole, the arterial walls exhibit a certain amount of systolic contraction. Some of the potential energy of ventricular systole converts to kinetic energy in the arterial walls during diastole.

This affects both the elasticity and the distensibility of the larger arteries and indirectly effects corresponding changes in arterial blood pressure.

Blood clotting mechanism

Whole blood forms a clot when it is collected in a clean and dry glass tube and allowed to stand in the tube in the presence of air. Upon standing the clot will retract, and the blood serum remains on top of and surrounding the clot (see p. 264). The clot forms because of a specific sequence of events involving several elements found normally in the blood. These elements are prothrombin, calcium, fibrinogen, fibrin, thromboplastin, and thrombin. Fluid blood changes into clotted blood when the soluble fibrinogen converts into insoluble fibrin. Under certain conditions fluid fibrinogen as a sol can change into clotted fibrin as a gel.

The liver secretes the inactive prothrombin into the circulating blood. Circulating blood normally contains calcium, which is derived from digestive processes and is necessary for homeostasis. The tissues contain thromboplastin, which is released when the tissues are broken (suffer trauma). Thromboplastin next contacts the prothrombin in the presence of calcium to form the enzyme, thrombin.

After the thrombin forms, it acts to convert the fluid fibrinogen into the clotted protein, fibrin. In the gel state, fibrin exists as numerous, small threadlike fibers that collect cells and/or other materials to occlude natural openings and thus stop hemorrhages. The clot formation sequence is simplified in the following two steps:

1. Inactive prothrombin + thromboplastin + calcium $\longrightarrow$ thrombin
2. Active thrombin + fibrinogen $\longrightarrow$ fibrin*

Clots do not form within the blood vessels under normal conditions because thromboplastin is not present to convert prothrombin to thrombin. Since the possibility of damage to the blood vessel walls will always exist and since such damage would release small quantities of thromboplastin, some counteragent is necessary. This agent, antithrombin, is present in a small but sufficient quantity in the circulating blood, under normal conditions, to neutralize the thrombin. This latter mechanism deters fatal intravascular clotting (thrombosis). Other conditions may exert forces that permit formation of thromboses, and these may, in turn, precipitate other pathologic conditions.

Lymph fluid

Lymph is a weakly alkaline, nearly colorless, coagulable fluid found only in lymph vessels (lymphatics). Lymph contains colorless cells (leukocytes) but no erythrocytes. Lymph consists chiefly of blood plasma that has passed through the walls of the blood capillaries of the various tissues and organs and has been taken into the lymphatics. Lymph also contains protein substances in varying amounts.

---

*Although not indicated in these steps, vitamin K is essential in blood clot formation.

The protein concentration of lymph may vary from as little as 0.5 percent in an extremity during exercise to as much as 3.2 percent in the thoracic duct. Salts and diffusible substances found in the lymph include calcium, chlorides, and nonelectrolytes and appear to be similar to plasma in concentration.

Many of the inflammatory fluids, harmful to the body and containing foreign particles, pass via the lymphatics into the circulating blood. Mature polymorphonuclear cells and macrophages in the blood may ingest and partially destroy bacteria from the lymph. When passing through a lymph node, the lymph carries with it large numbers of cells from the substance of the node; these cells later become lymphocytes. Experimental evidence indicates that the growing lymphocytes in the nodes form varying amounts of gamma globulin and/or specific antibodies that flow with the lymph into the bloodstream.

The blood within the capillaries, although under no arterial pressure, exhibits a true pressure of filtration equal to approximately 32 mm. Hg at the arterial end of the capillary. At this same end, the protein osmotic pressure of the blood is equal to approximately 25 mm. Hg. The difference of these pressures is 7 mm. Hg, a pressure difference sufficient to provide for passage of fluids from the blood through the capillary membrane to form the interstitial fluid.

At the venous end of the capillary, the filtration pressure has fallen to approximately one-third the initial pressure and the protein osmotic pressure of the blood has risen slightly as a result of fluid volume loss. The decreased difference between these two pressures permits reabsorption by the bloodstream of fluid nearly equal to the losses at the arterial end of the capillary.

Interstitial fluid

The interstitial fluid exists between the cells and the tissues and derives from the blood by one or more of the physicochemical processes described in Chapter 2.

Since many of the body cells lie deep within other structures and since each cell must receive nourishment and eliminate waste to carry on its part in total homeostasis, the interstitial fluid becomes the medium through which the deep cells are nourished and through which the waste products are eliminated. The interstitial fluid appears to be essentially the same as plasma with the protein substances precipitated out. Interstitial fluid pH, as with blood pH, varies from 7.35 to 7.45. Following the nourishing processes in which the interstitial fluid is involved, it usually returns to the circulating blood.

Interstitial fluid contains in solution such substances as glucose (and some other sugars) and the phosphate, carbonate, and chloride salts of calcium, sodium, magnesium, and potassium. Some fats, hormones, and vitamins are included in the list of components of this fluid.

Intracellular fluid

As with the interstitial fluid, the intracellular fluid derives from the body by a single or combined physicochemical process or processes. As the sub-

268 *Textbook of anatomy and physiology in radiologic technology*

stances of the interstitial fluid pass from the blood into the spaces between the cells and tissues, they pass also from the interstitial fluid into the cytoplasm of the cells. Some of this fluid exchanges across the cell membrane with the nutritive and waste substances. The fluid contained within the cell is intracellular fluid and is much the same as interstitial fluid or plasma without the proteins.

FETAL CIRCULATION

With the development of the fertilized ovum in the uterine lumen, the mucosa thickens and enlarges and surrounds the growing embryo with a two-layered membrane, the *placenta*. The placental layers are the chorion (outer) and amnion (inner). The chorion is closely associated with the uterine mucosa, and the amnion encases the embryo. When the embryo develops beyond the end of the first trimester (the first 3 months of pregnancy), it is properly called the fetus.

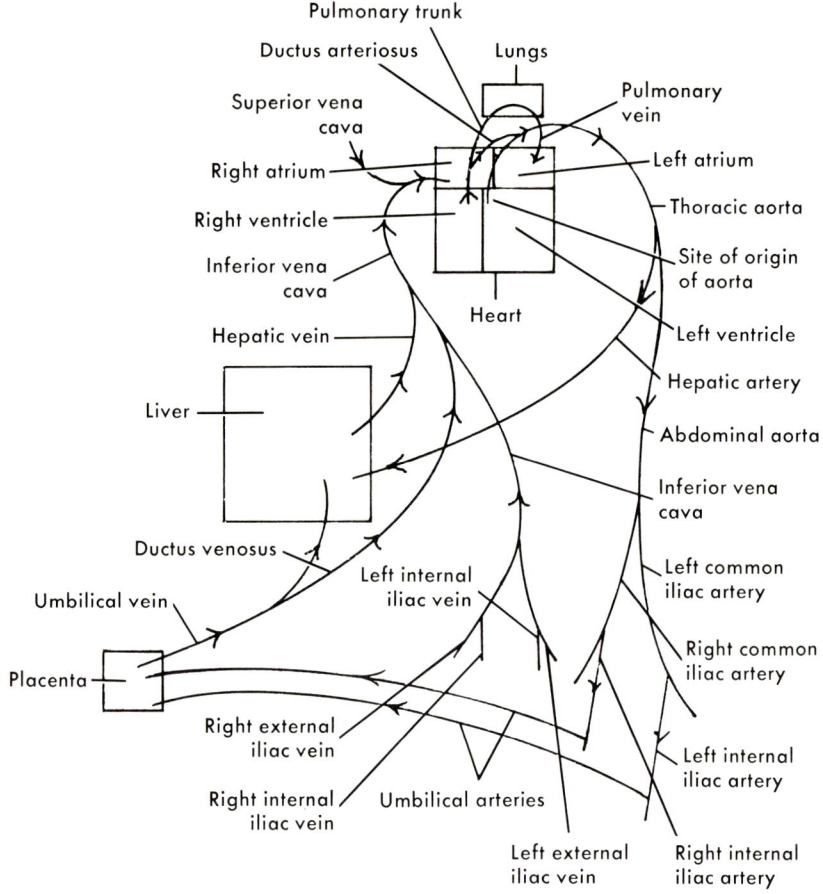

Fig. 6-10. Fetal circulation (schematic).

As the uterine mucosa thickens, blood sinuses develop in it to connect the maternal arteries and veins. The chorion extends numerous fingerlike projections, the chorionic villi, into these blood sinuses.

With the development of the embryonic circulatory system, two arteries (the umbilical arteries) extend from the fetal hypogastric arteries to the placenta. The umbilical vein commences from capillaries in the chorion and conveys blood from the placenta to the fetus. Specialized tissue envelops the two arteries and the one vein; the whole is called the umbilical cord. Most of the blood in the umbilical vein bypasses the fetal liver via the ductus venosus to flow eventually into the right atrium via the inferior vena cava (see Fig. 6-10).

Much of the blood from the fetal right atrium bypasses the pulmonary circulation via the foramen ovale, in the interatrial septum, to the left atrium. However, some blood passes into the right ventricle and into the pulmonary artery. Again, some of this blood bypasses the pulmonary circulation via the ductus arteriosus between the pulmonary artery and the aorta; the usual terminus is the descending aorta.

Blood from the left ventricle flows, via the aorta, throughout the fetal arterial system to the umbilical arteries and returns to the placenta. Blood that does not leave the fetal arterial system at a given instant or in a given cycle flows

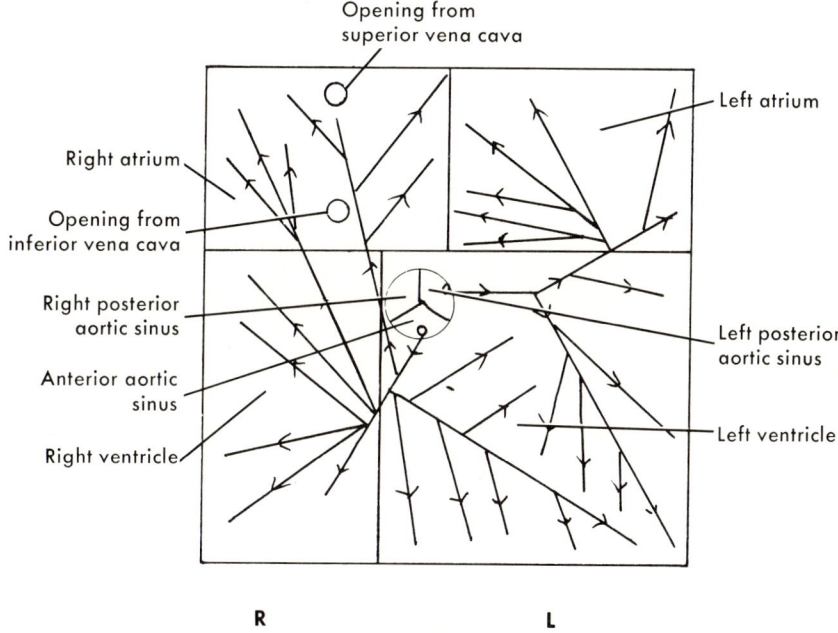

Fig. 6-11. Coronary circulation, arterial part (schematic). The left coronary artery arises from the left posterior aortic sinus, the right coronary artery arises from the anterior aortic sinus. Both arteries give rise to numerous lesser branches that supply the myocardium. Blood returns to the right atrium via numerous small veins that empty into the great coronary vein and coronary sinus, thence into the right atrium.

270 Textbook of anatomy and physiology in radiologic technology

throughout the arterial system in the lower extremities and returns to the fetal heart via the venous system to be recirculated.

CIRCULATORY SYSTEM DIVISIONS

Blood that leaves one side of the heart and passes through specific vessels to return to the opposite side of the heart flows through one of the *divisions* of the complete circulatory system. Three such divisions exist, and one of the divisions includes an important subdivision; the divisions are the *coronary circulation*, the *pulmonary circulation*, and the *systemic circulation* with its subdivision, the *portal circulation*. Blood in the portal circulation flows between two capillary networks.

Coronary circulation

The coronary circulation commences with the right and left coronary arteries, which arise from the aortic sinuses. Blood in these arteries nourishes the myocardium and returns to the right atrium via the right and left coronary veins and the coronary sinus (see p. 243 and Fig. 6-11).

Pulmonary circulation

The pulmonary circulation commences with the pulmonary artery, which arises in the upper right ventricle. Blood leaves the right ventricle via the pulmonary artery, which branches into the right and left pulmonary arteries and flows into the capillary network of each lung. Following the gaseous exchange that occurs between the blood in the capillaries and the air in the lung alveoli, the blood returns to the left atrium via four pulmonary veins, two from each lung (see Fig. 6-12).

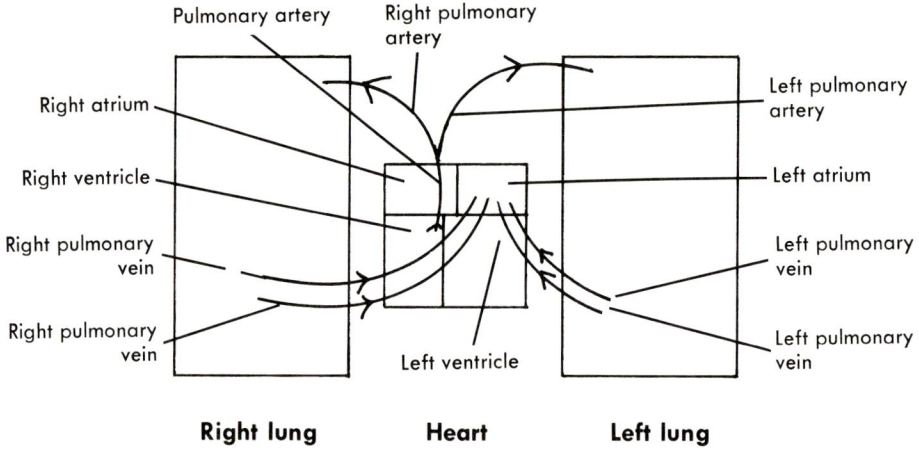

Fig. 6-12. Pulmonary circulation, anterior aspect (schematic). Blood flows from the lungs via the four pulmonary veins into the left atrium. Blood flows from the right ventricle via the pulmonary arteries into the lungs.

Systemic circulation

The systemic circulation commences with the ascending aorta and conveys blood via arteries and arterioles to the capillaries. This blood returns to the right atrium via the superior and inferior venae cavae. The aorta, throughout its length, gives rise to arteries supplying the various organs and other body structures. Among these arteries are the superior mesenteric and the celiac trunk and branches, which supply the digestive viscera either directly or indirectly through lesser arterial branches (see Fig. 6-13, A).

Portal circulation

The portal circulation commences with the splenic (lienal) and superior mesenteric veins. Blood from the digestive viscera, including the large intestine, flows through the aforementioned veins into the portal vein, which is formed by the junction of the splenic and superior mesenteric veins, and thence into the

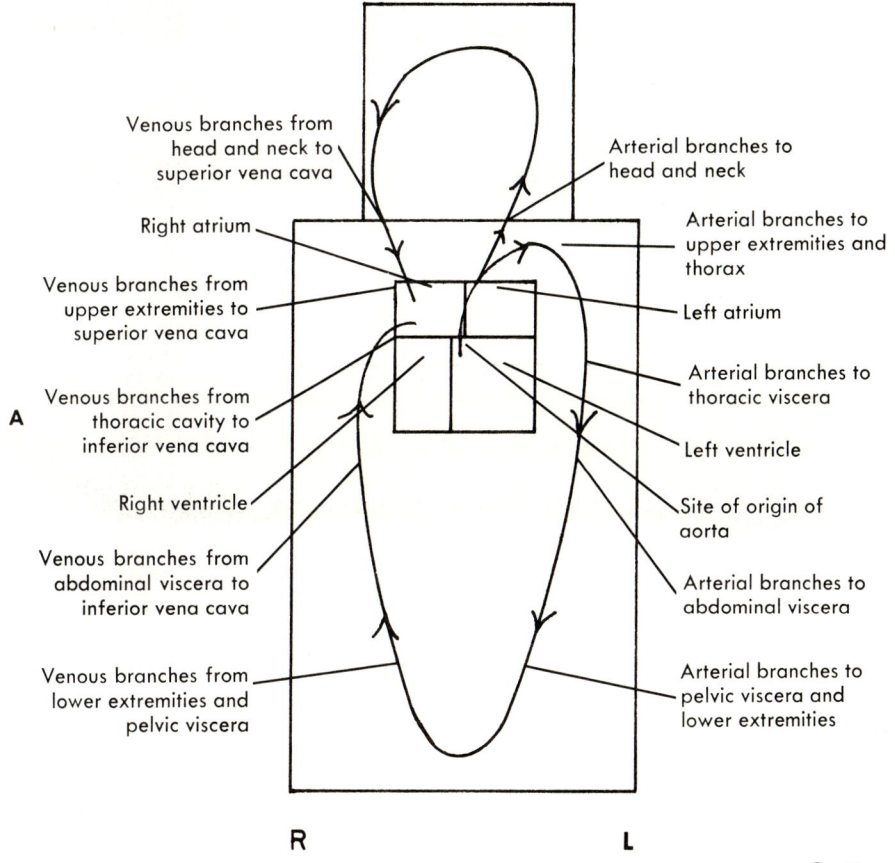

Continued.

Fig. 6-13. **A**, Systemic circulation, anterior aspect (schematic). **B**, Portal circulation (organs are arranged apart for clarity).

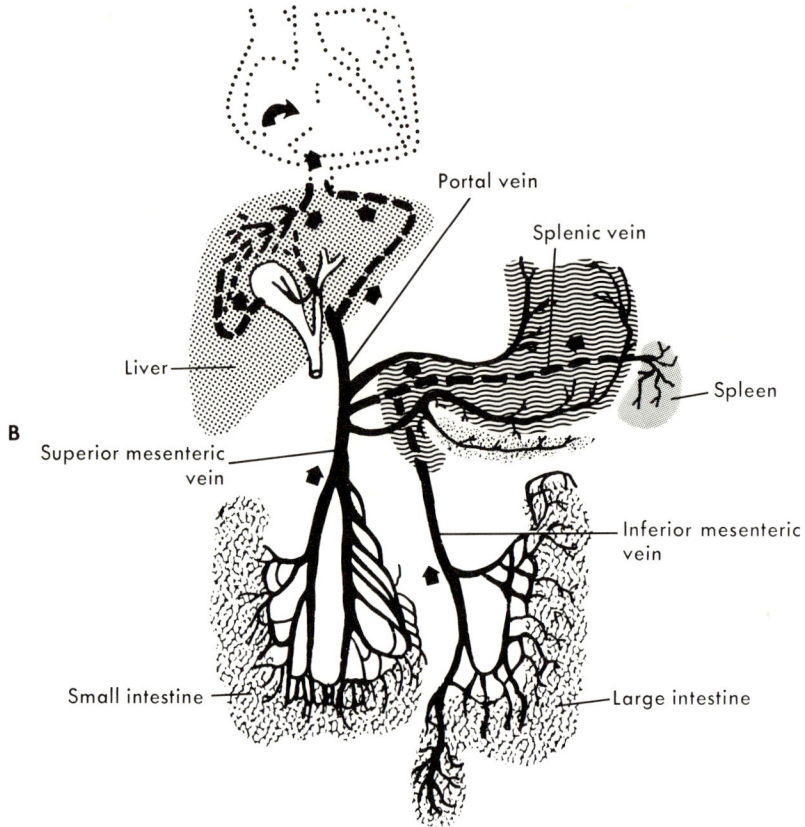

Fig. 6-13, cont'd. For legend see p. 271.

liver via the portal vein. The portal vein is approximately 8 cm. long; the junction of the splenic and superior mesenteric veins is on the level of L-2 between the inferior vena cava and the pancreatic neck. Blood from the large intestine and the pelvic parts of the small intestine flows via the inferior mesenteric vein into the splenic vein prior to its junction with the superior mesenteric vein as seen in Fig. 6-13, B (see also Fig. 5-15).

Following the various specialized chemical reactions between the portal circulation blood and the liver cells, this blood joins with the blood from the nourishing circulation throughout the liver to return to the inferior vena cava via the hepatic vein. *The portal circulation terminates with the portal vein entering the liver.*

RADIOGRAPHIC STUDIES

Radiographic studies in which the circulatory system is involved can be classed in two categories as follows:
1. The circulatory system, per se
2. The circulatory system as a transport mechanism to other structures

The circulatory system **273**

Circulatory system radiography

The following structures of the circulatory system can be examined directly by means of radiographic visualization: heart, aorta, arteries, veins, and lymphatics. Radiographic examination of the heart and great vessels after opacification with a suitable contrast medium is called *angiocardiography*.

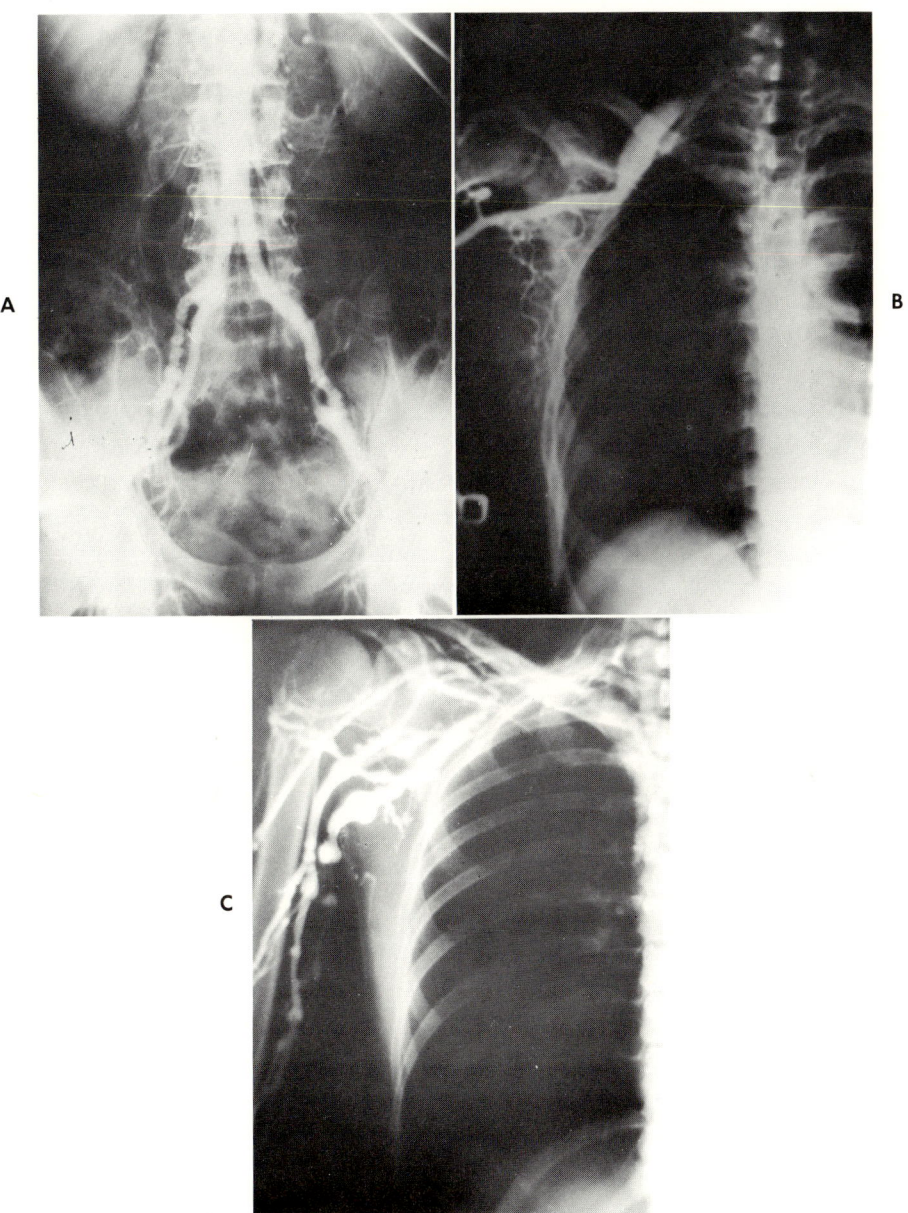

Fig. 6-14. A, Aortogram. **B,** Arteriogram. **C,** Venogram. (Courtesy Dr. G. J. Nicholson.)

The left side, the right side, or both sides of the heart may be examined radiographically. When the right side of the heart is to be examined, the contrast medium may be injected into the right brachial vein, after which the medium flows through the right subclavian vein into the right brachiocephalic vein and through it into the superior vena cava, thence into the right atrium. When the left side of the heart is to be examined, the contrast medium may be injected via the left brachial artery or one of the femoral arteries. In either event, a catheter is passed into the artery, through the artery into the aorta, and thence into either the left atrium or the left ventricle.

Patency of either the foramen ovale or the ductus arteriosus is demonstrable during angiocardiography and/or heart catheterization. Either patency permits continuous recirculation of nonoxygenated blood throughout the arterial system, resulting in a *blue baby*.

Radiographic examination of the aorta is called *aortography*. In aortography two needles may be inserted into the aorta very close to each other to permit the contents of each syringe to enter the same region of the aorta simultaneously. If only one quantity (twice that in the two syringes) were injected, the time element would prevent the excellent visualization necessary in this examination. The part of the aorta to be examined may be either immediately above the injection site or somewhat below this site. When the injection site is above or very slightly below the origin of the renal arteries, excellent visualization of the arterial network of each kidney is usually obtained (see Fig. 6-14, A).

Radiographic examination of other arteries is called *arteriography*. In arteriography, as in aortography, the injection site must be between the heart and the arterial segment to be examined (see Fig. 6-14, B).

Radiographic examination of the veins is called *venography*. Since the flow of blood in the veins is in a direction opposite to the direction of the flow in the arteries, the relation of the heart, segment of the vein to be examined, and the site of injection differs from this relationship in arteriography and aortography. In venography the segment of vein to be examined must lie between the site of injection and the heart (see Fig. 6-14, C).

Radiographic examination of the lymph vessels is called *lymphangiography*. This examination is performed similarly to venography in that the segment of a lymph vessel to be examined must lie between the site of injection and the superior vena cava; the second (radiographic) exposure usually is made one week following the first. (*Note:* In each of the four preceding examinations an ideal situation exists if a single *bolus* of contrast medium is obtainable in the segment being examined.)

The circulatory system as a transport mechanism

The category of the circulatory system as a transport mechanism in radiographic examination includes as structures of examination the brain, gallbladder, bile ducts, liver, and kidneys. Examination of the arteries of the brain is called *cerebral angiography*.

The circulatory system 275

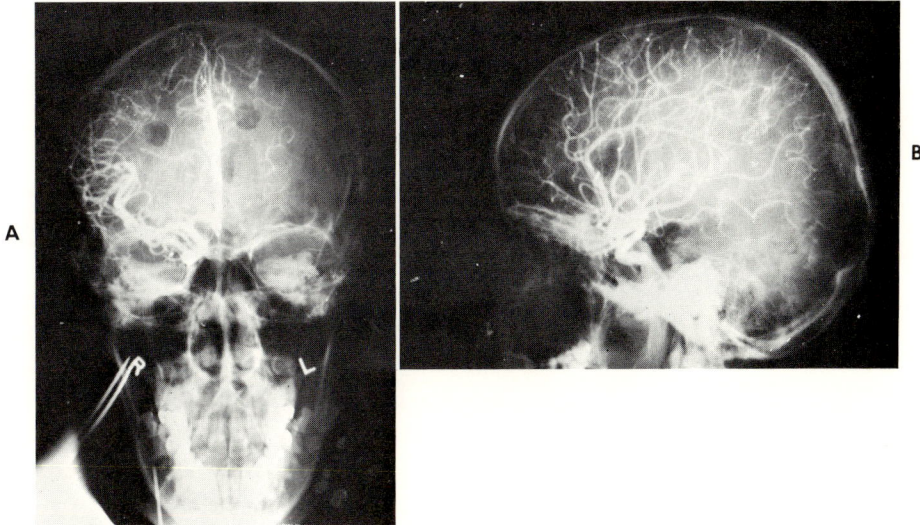

Fig. 6-15. Cerebral angiogram. **A,** Posterior view. **B,** Lateral view.

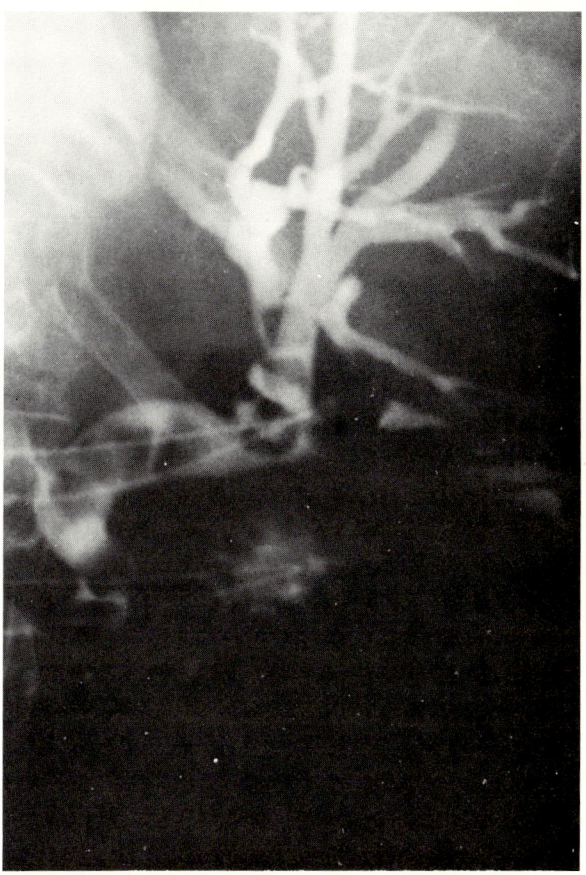

Fig. 6-16. T tube cholangiogram. (Courtesy Dr. J. M. Hilton.)

The usual site of injection for cerebral angiography is the internal carotid artery on the side to be examined (see Fig. 6-15).

The site of injection usually selected for cholangiography is one of the veins in the antecubital space (anterior surface of the elbow). The contrast medium flows with the blood through the brachial, axillary, subclavian, and brachiocephalic veins into the superior vena cava, thence into the right atrium of the heart. The contrast medium flows with the blood through the right heart to the lungs, returns to the left heart, and flows from the left heart via the aorta. Throughout the length of the aorta the contrast medium passes into several branches from the aorta and the systemic circulation to be recirculated from the left ventricle. However, a sufficient quantity of the contrast medium reaches the hepatic artery within a very few minutes after injection to permit visualization of the biliary ducts. The gallbladder may often be visualized in this examination (see Fig. 6-16).

Examination of the liver by radiographic visualization is not a common procedure; the site of injection and pathway of contrast medium is the same as in cholangiography.

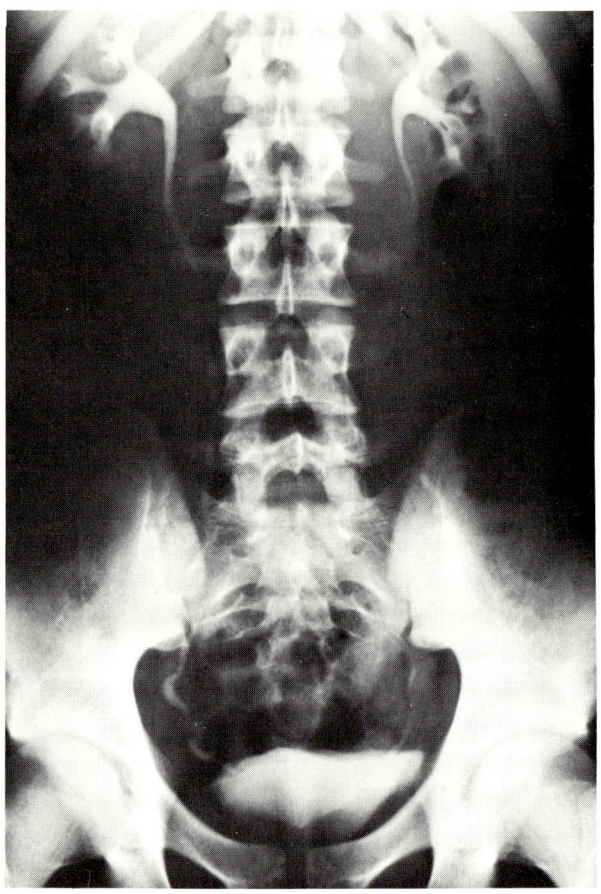

Fig. 6-17. Intravenous pyelogram (I.V.P.). (Courtesy Dr. J. M. Hilton.)

Examination of the kidneys is achieved by any one of three methods. One method, mentioned in conjunction with aortography, may be performed to study the renal arteries and branches. Another method is retrograde pyelography, which does not involve the circulatory system. The third method is called *intravenous pyelography (I.V.P.)*. In this examination the contrast medium is injected into the venous system as described in cholangiography. The concentration of contrast medium occurs in the kidneys after passing from the aorta into the large renal arteries. A time element of approximately 5 minutes after injection is required for sufficient concentration to permit radiographic visualization (see Fig. 6-17).

Heart margins

Teleroentgenography is a term used to describe radiography of the thoracic viscera at rather long focal-film distances to permit minimal magnification and subsequent possible distortion of each viscus. The usual position of the patient for this procedure is with the anterior surface of the body against the cassette holder, the anterior position, which enables an anterior view of the heart (see Fig. 6-18).

Increasing use of radiography in the diagnosis of heart disease requires that radiologic technologists have greater (than in the past) knowledge of the anatomy and associated pathology of the heart.

Since the heart is asymmetrical and since it is situated somewhat off true center in the body, demonstration of specific margins of the heart for radiologic visualization is made possible by rotating the patient into certain positions. These positions are usually either requested specifically by the radiologist (or other physician) or are a part of routine examination procedure.

Among the pathologic conditions having significance in radiographic positioning are the following associated items.

Left heart

1. Mitral stenosis results in left atrial and right ventricular enlargement.
2. Mitral insufficiency results in enlargement of the pulmonary artery and the left ventricle.
3. Aortic stenosis initially causes left ventricular enlargement only; as this condition continues, the preceding two conditions also develop.
4. Aortic incompetency repeats the preceding three conditions.

Right heart

1. Tricuspid valve disease is rare; however, if it is present, the end results are similar to those listed for the left heart and listed pathology of the mitral valve.
2. In pulmonary stenosis, the pulmonary semilunar valve thickens and causes dilation of the right atrium and ventricle.

Figs. 6-18 to 6-25 demonstrate the patient in position, the heart model in

Text continued on p. 286.

278 Textbook of anatomy and physiology in radiologic technology

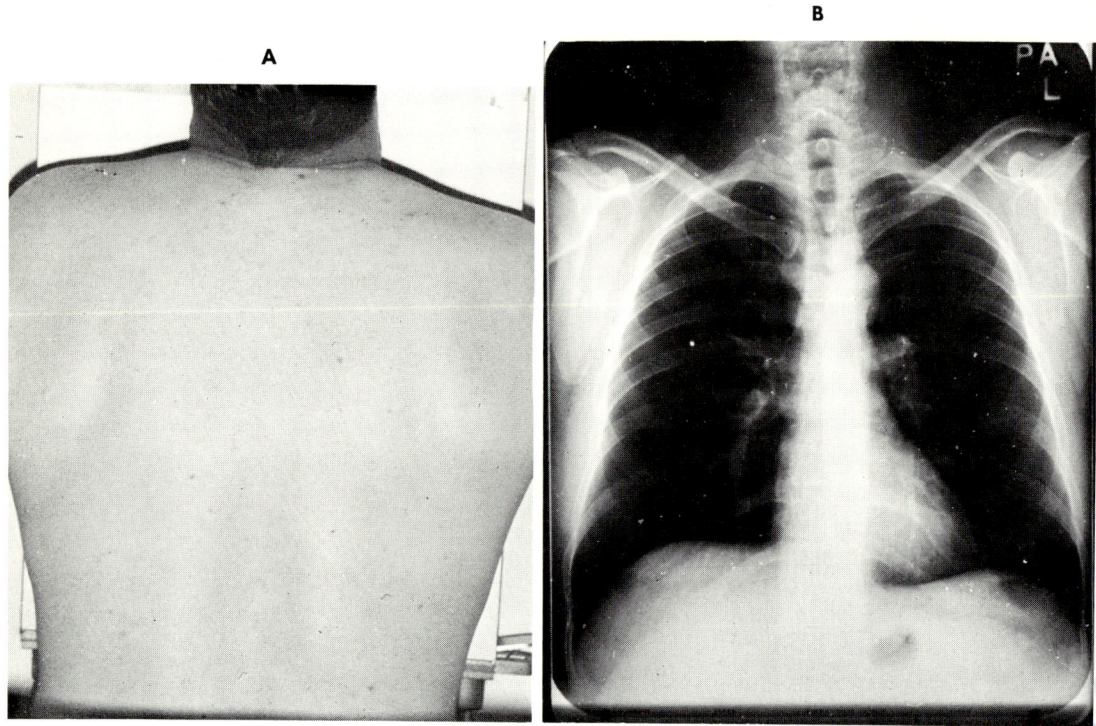

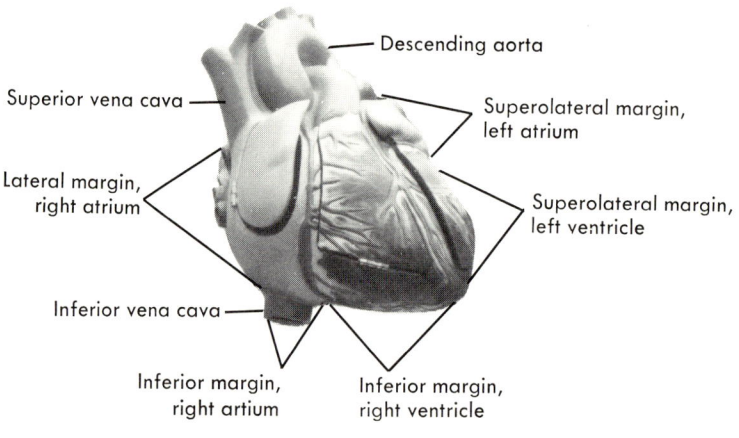

Fig. 6-18. A, Anterior position of patient. **B**, Anterior, heart margin radiograph. **C**, Heart model, patient in anterior position.

The circulatory system 279

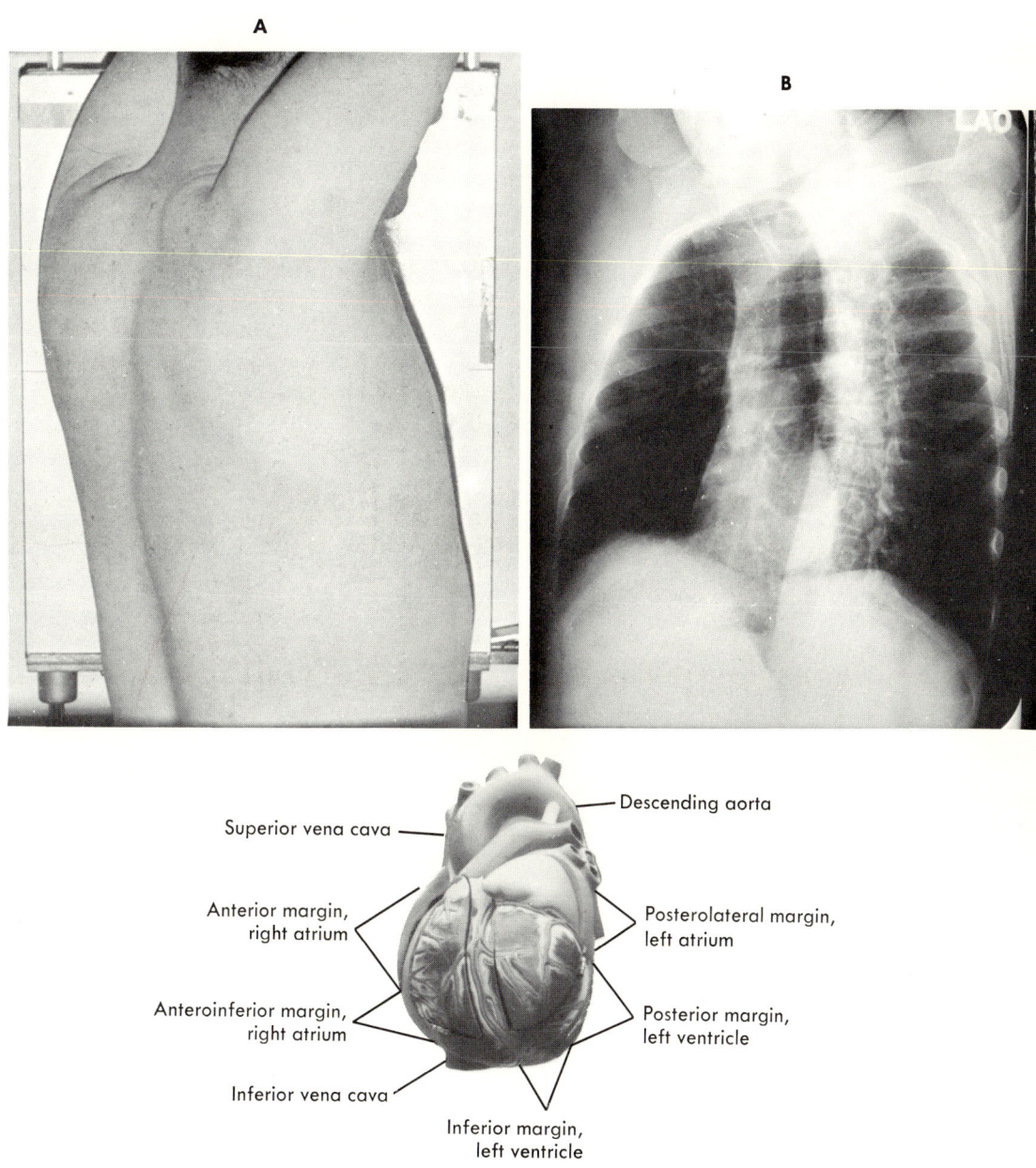

Fig. 6-19. **A**, Patient in L-A-O position. **B**, L-A-O position, heart margin radiograph. **C**, Heart model, patient in L-A-O position.

280 *Textbook of anatomy and physiology in radiologic technology*

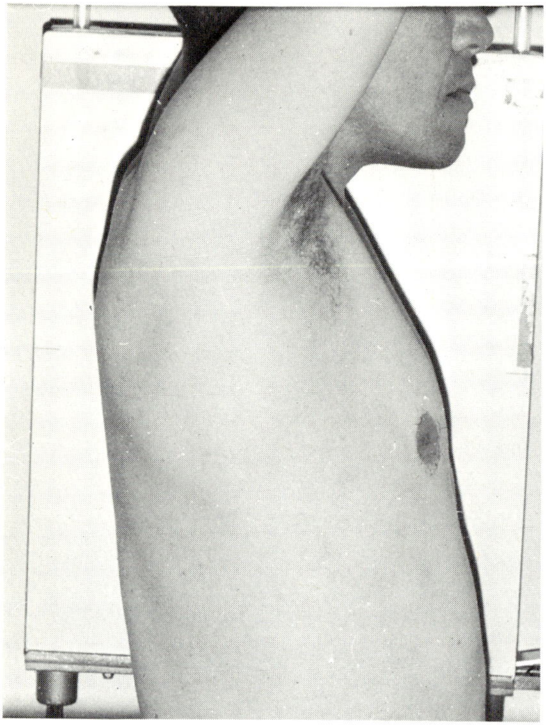

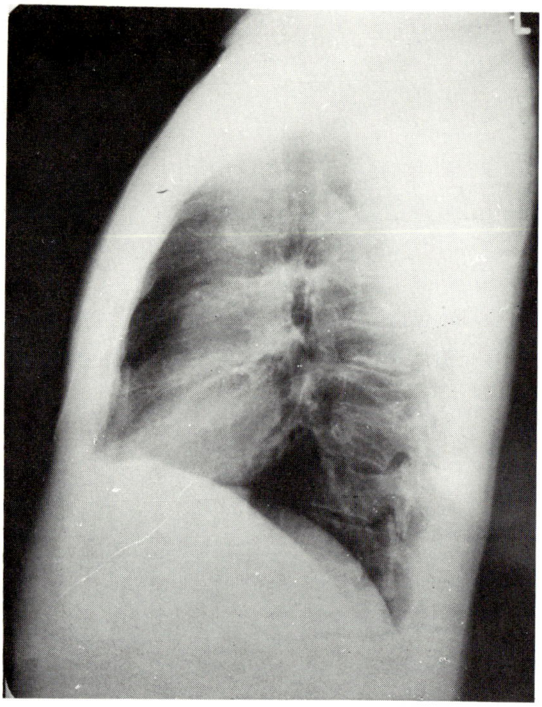

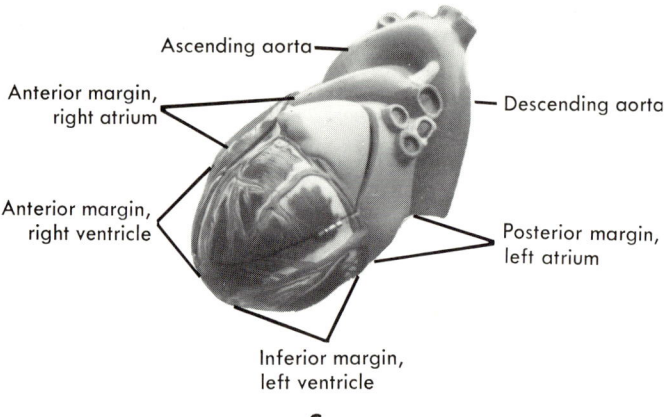

Fig. 6-20. **A**, Patient in left lateral position. **B**, Left lateral position, heart margin radiograph. **C**, Heart model, patient in left lateral position.

The circulatory system

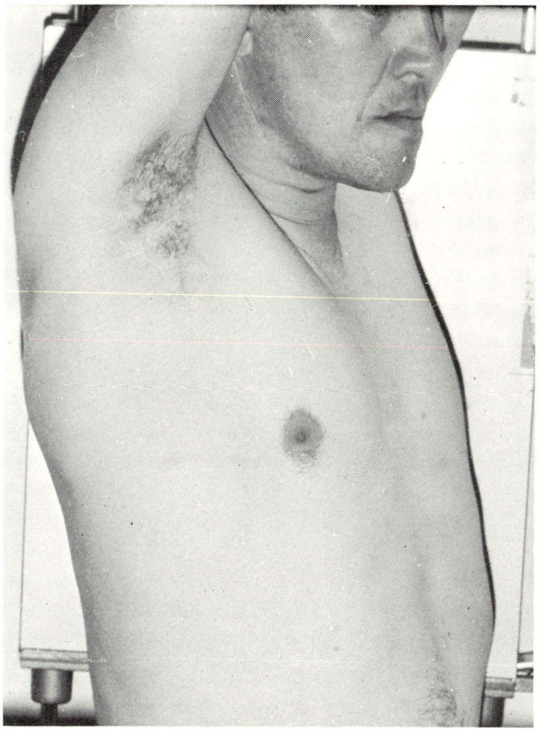

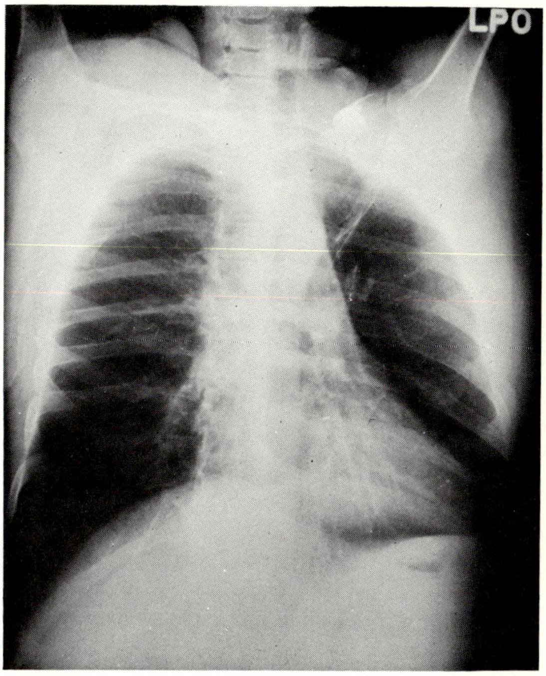

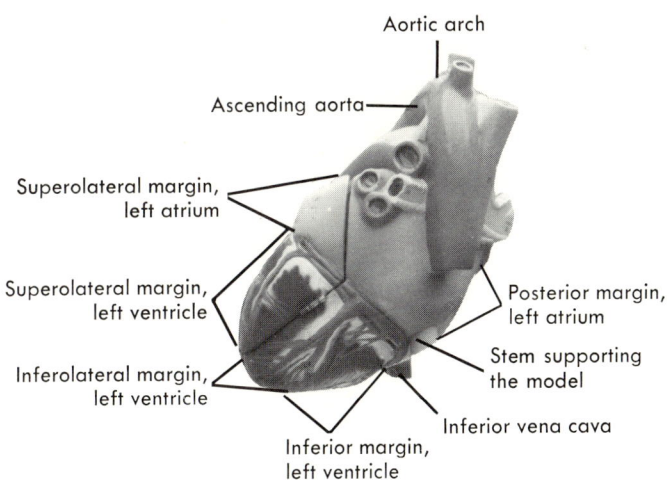

Fig. 6-21. A, Patient in L-P-O position. B, L-P-O position, heart margin radiograph. C, Heart model, patient in L-P-O position.

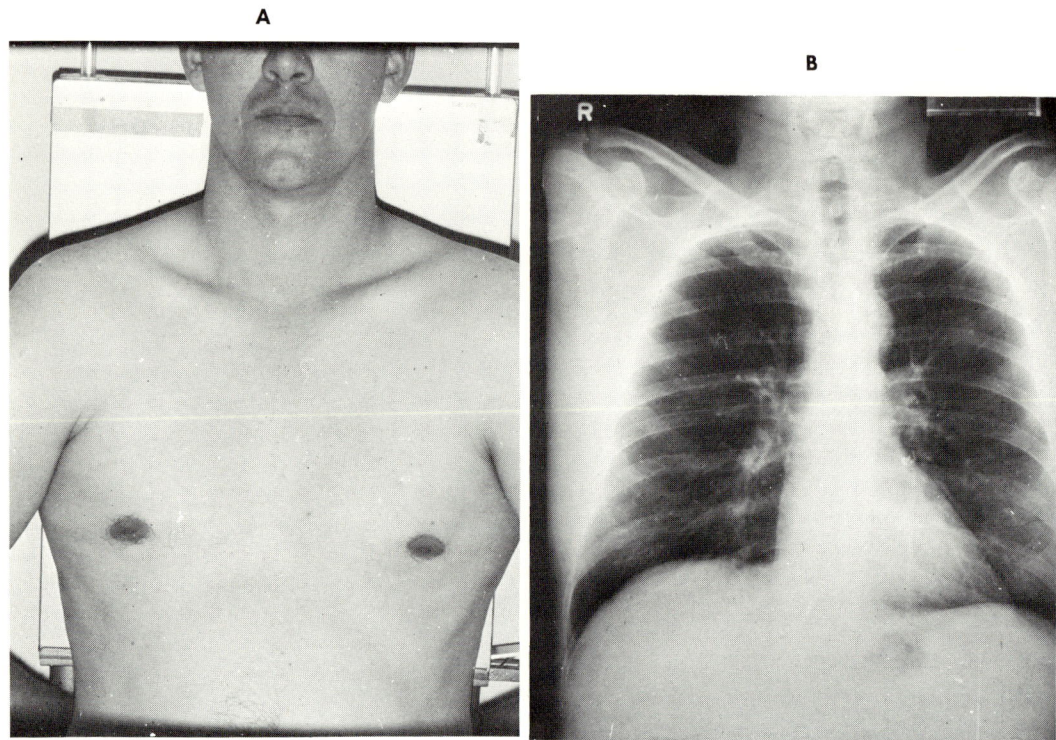

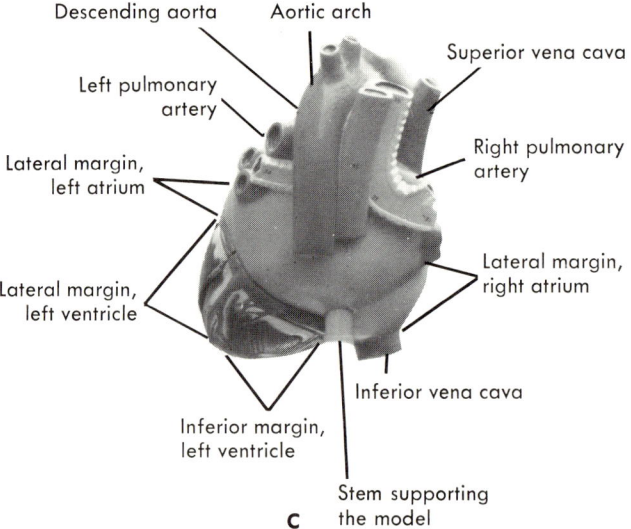

Fig. 6-22. A, Patient in posterior position. **B,** Posterior position, heart margin radiograph. **C,** Heart model, patient in posterior position.

The circulatory system   283

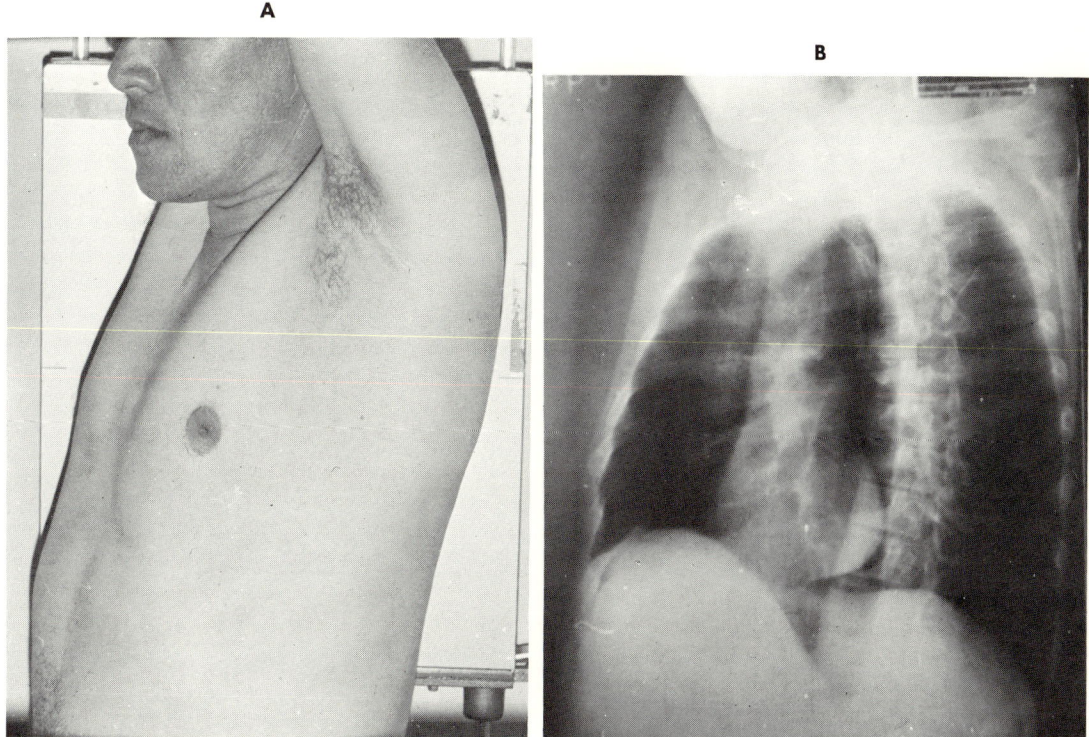

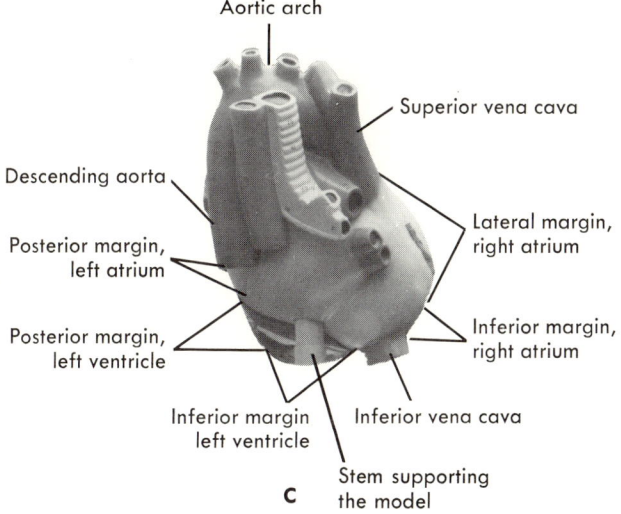

Fig. 6-23. A, Patient in R-P-O position. **B,** R-P-O position, heart margin radiograph. In situ, the descending aorta margin obscures the margins of the left atrium and the left ventricle. **C,** Heart model, patient in R-P-O position.

284   *Textbook of anatomy and physiology in radiologic technology*

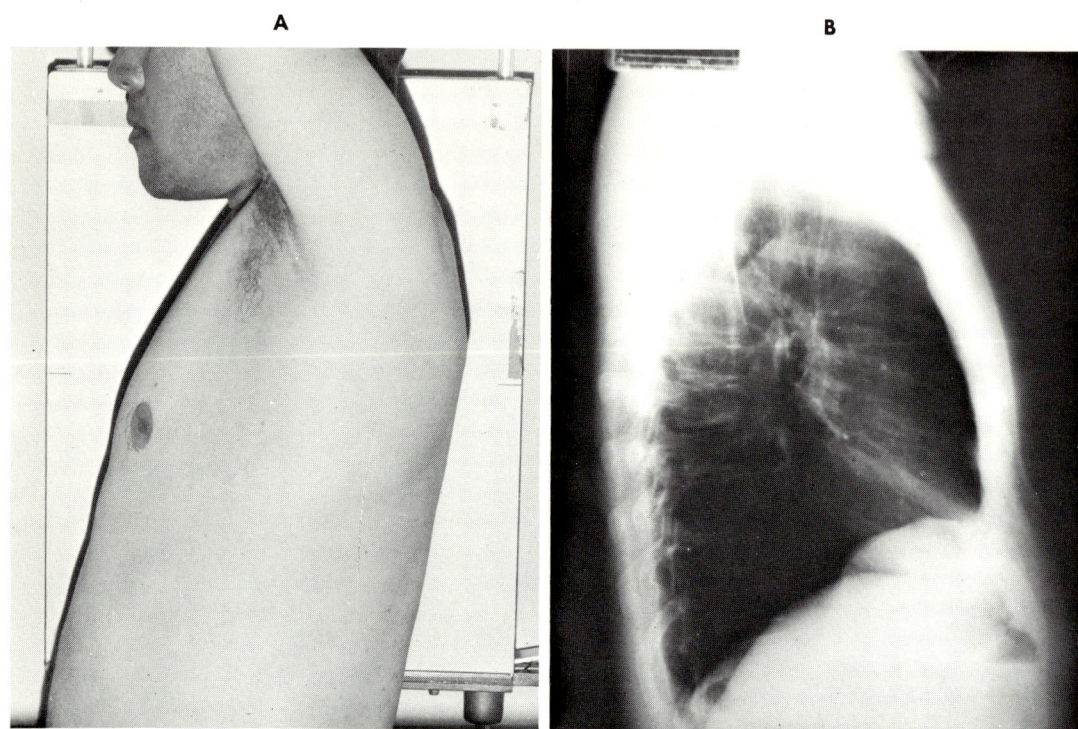

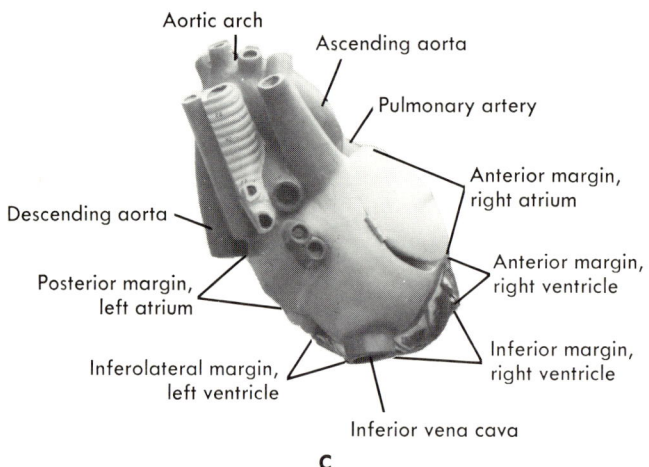

Fig. 6-24. A, Patient in right lateral position. **B,** Right lateral position, heart margin radiograph. In situ, the descending aorta obscures much of the posterior margin of the left atrium. **C,** Heart model, patient in right lateral position.

The circulatory system   285

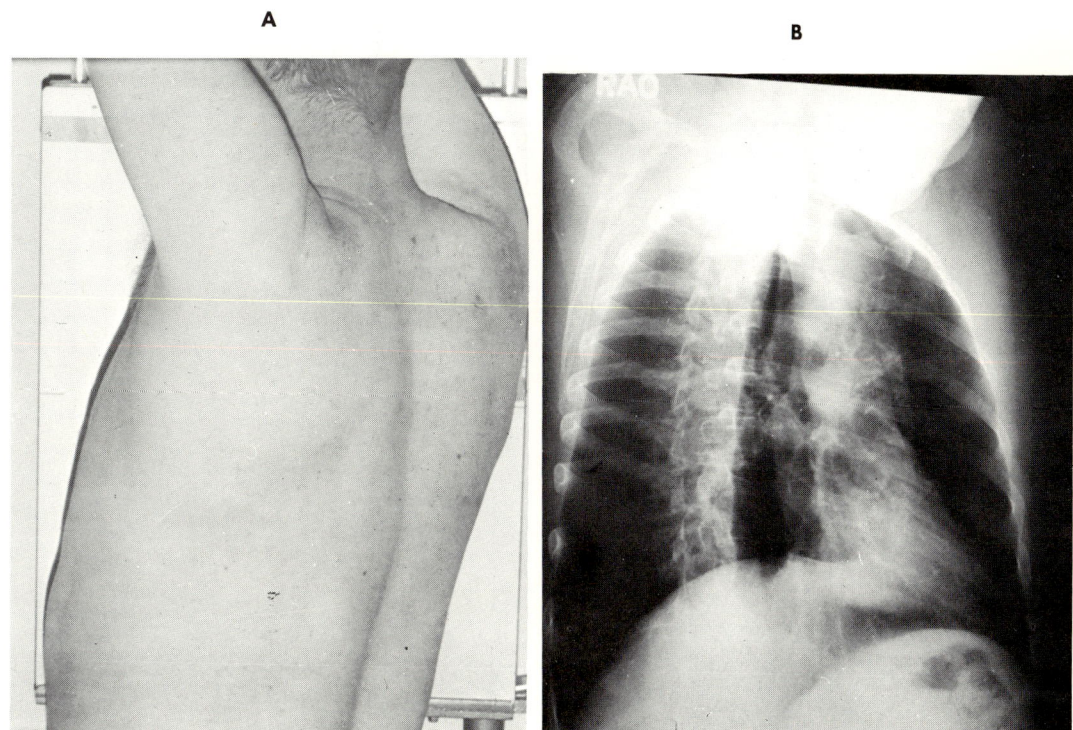

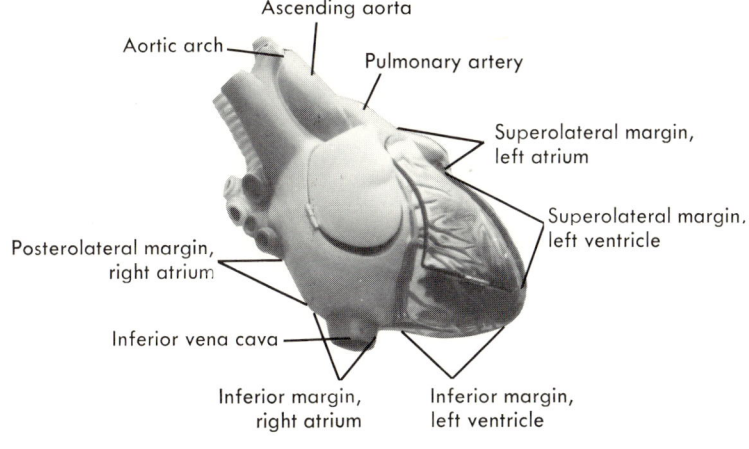

Fig. 6-25. A, Patient in R-A-O position. **B,** R-A-O position, heart margin radiograph. **C,** Heart model, patient in R-A-O position.

corresponding position in relation to the film, and the radiograph resulting from eight standard positions. These positions are not necessarily those of any routine and are included for the purpose of demonstrating normal anatomic structures.

REFERENCES

Burch, G. E., and DePasquale, N. P.: Primer of clinical measurement of blood pressure, St. Louis, 1962, The C. V. Mosby Co.

Gardner, E., Gray, D. J., and O'Rahilly, R.: Anatomy, Philadelphia, 1960, W. B. Saunders Co.

Goss, C. M.: Gray's anatomy, ed. 27, Philadelphia, 1959, Lea & Febiger.

Jacobi, C. A., and Paris, D. Q: Textbook of radiologic technology, ed 5, St. Louis, 1972, The C. V. Mosby Co.

Orten, J. M., and Neuhaus, O. W.: Biochemistry, ed. 9, St. Louis, 1975, The C. V. Mosby Co.

Schottelius, B. A., and Schottelius, D. D.: Textbook of physiology, ed. 17, St. Louis, 1973, The C. V. Mosby Co.

Windle, W. F.: Textbook of histology, ed. 3, New York, 1960, McGraw-Hill Book Co.

Wintrobe, M. W.: Clinical hematology, ed. 3, Philadelphia, 1961, Lea & Febiger.

CHAPTER 7

The urinary system and the skin

URINARY SYSTEM
Functions

Both the urinary system and the skin share the common function of assisting in the maintenance of the water balance of the body. Each functions to eliminate waste materials; however, the methods of elimination differ considerably. The urinary system will be discussed independently of the skin.

The urinary system filters the blood and extracts from it, for excretion from the body, the waste products that are carried in the circulatory system. In eliminating waste materials and water, the urinary system performs many important functions, each of which plays an important part in maintaining homeostasis. Of the five functions of the urinary system listed, the first three are elaborated upon under the headings of Filtration and Reabsorption. The five functions are as follows:

1. To assist in maintaining the water balance of the body
2. To assist in maintaining both the electrolyte balance and acid-base balance, thus assisting in maintaining the pH of the body fluids
3. To eliminate toxic wastes from the body
4. To detoxify certain substances (a relatively minor function)
5. To elaborate hormonal substances, such as renin (the only substance produced by the kidney that may be a hormone)

There is no general agreement as to any other functions of the kidneys.

The urinary system includes the kidneys, ureters, urinary bladder, and urethra. The organs and structures of the urinary system are the same in both sexes, except that the urethra is longer in the male. The right kidney is usually somewhat smaller and lower than the left kidney; the hilum of the right kidney is opposite the transverse process of the second lumbar vertebra (see Fig. 7-1).

The urine is extracted from the blood by an elaborate filtration process,

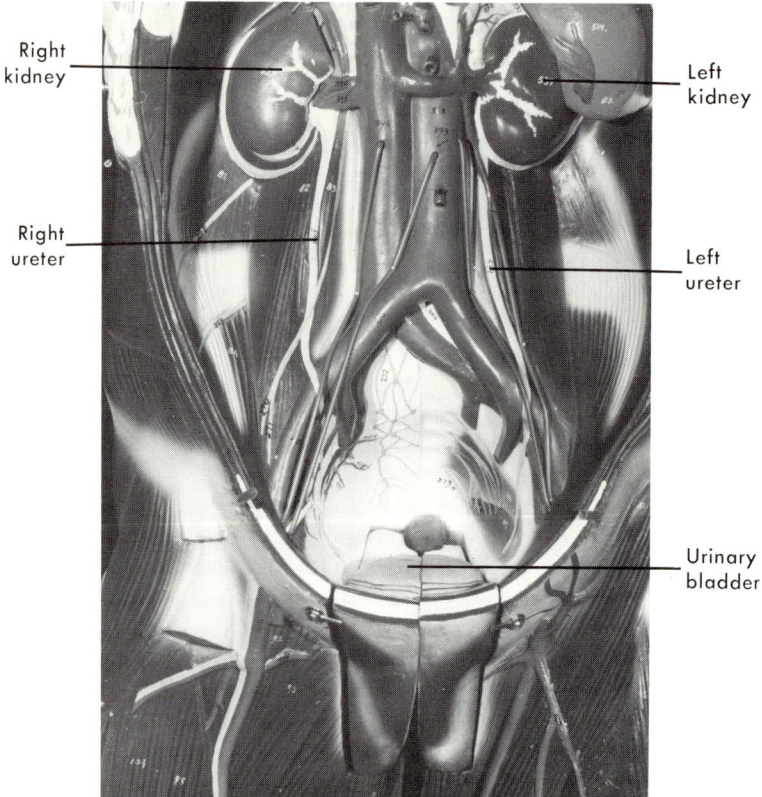

Fig. 7-1. Urinary viscera (female), relative positions.

eliminated from the kidneys into the ureters, contained in the urinary bladder, and excreted (voided) from the body via the urethra. True peristalsis, as in the digestive tube, forces the urine along the ureters from the kidneys into the urinary bladder. Elimination of the urine from the bladder is a voluntary act in healthy persons beyond very early childhood. The act of urine elimination is called *micturition*.

Anatomy

Urine is formed in the kidneys; the other structures of the urinary system serve either to convey or to store the urine.

Kidneys

The shape of each kidney closely resembles the shape of a bean, with the concavity of the kidney facing the midsagittal plane of the body. In the concavity of the kidney is the *hilum*, the slitlike opening through which the nerves and arteries pass into the kidney and through which the veins and lymphatics pass from the kidney. The renal pelvis extends through the hilar space and joins

The urinary system and the skin **289**

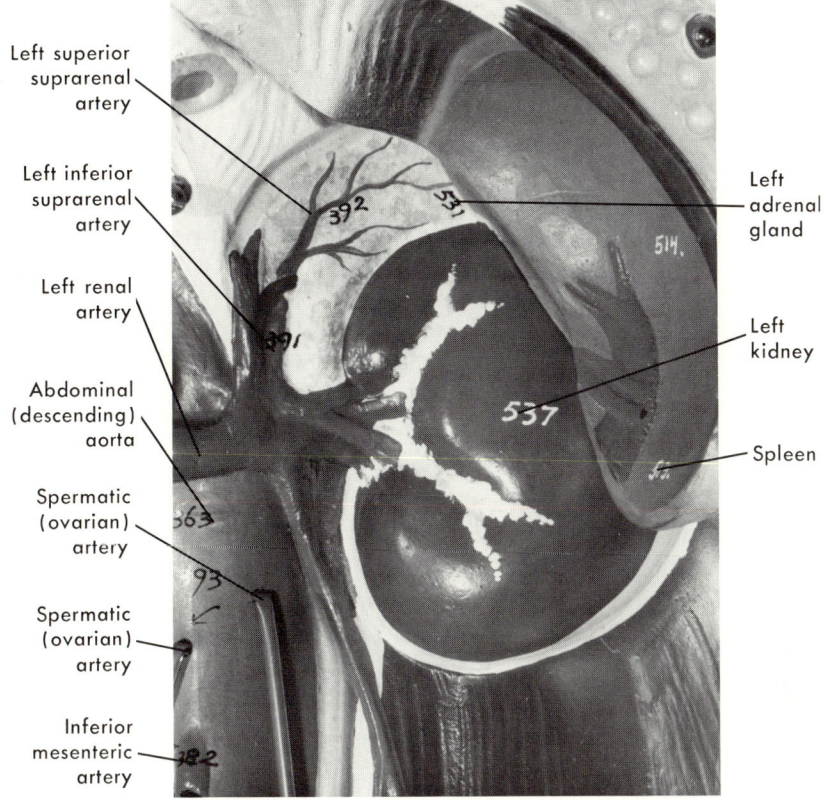

Fig. 7-2. Left kidney showing shape and associated viscera.

the ureter, which passes downward to join the urinary bladder. Each kidney is about 10.4 cm. (4.5 inches) in length, extending between the twelfth thoracic vertebra and the third lumbar vertebra, one on each side of the vertebral column (see Fig. 7-2).

The kidney is enveloped by a thin, tough fibrous capsule, *tunica fibrosa*, and is embedded in a layer of fatty tissue. The tissue surrounding the kidney is called the *perirenal tissue*. No rigid attachment fixes the kidney to the posterior abdominal wall, and since the kidney contacts the diaphragm superiorly, it moves with the diaphragm during respiration. A kidney may move up or down as much as 3 cm. during either phase of respiration and as much as 5 cm. between the recumbent and erect positions of the patient; these facts are important in certain aspects of renal radiography. The entire urinary system is retroperitoneal, i.e., behind (or below) the peritoneum; therefore, the organs of this system are situated outside the abdominal cavity.

All the organs of the body, including the kidneys, are divided into two portions: the inner *medulla* and the outer *cortex*. The medulla and cortex have both histologic and physiologic differences. The hilar space of the kidneys extends from the hilum into the medulla to form the *renal sinus*. The connective tissue capsule extends inwardly through the hilum and becomes continuous with the tunica adventitia of the renal pelvis and pyramids. The pyramids cause the floor of the renal sinus to be uneven.

The specialized glandular (epithelial) tissue of the kidney, the *parenchyma*, is quite dense and friable. The parenchyma comprises both the medullary and cortical portions of the kidney. The medulla consists chiefly of masses of renal tissue called the renal pyramids; each kidney contains from eight to eighteen pyramids. A pyramid is comprised of the straight parts of the many collecting tubules from the nephron units in a certain part of the cortex.

Usually one pyramid opens into a single *minor calyx*, although two and sometimes three pyramids may open into one minor calyx.

There are from four to thirteen minor calyces in each kidney. A minor calyx

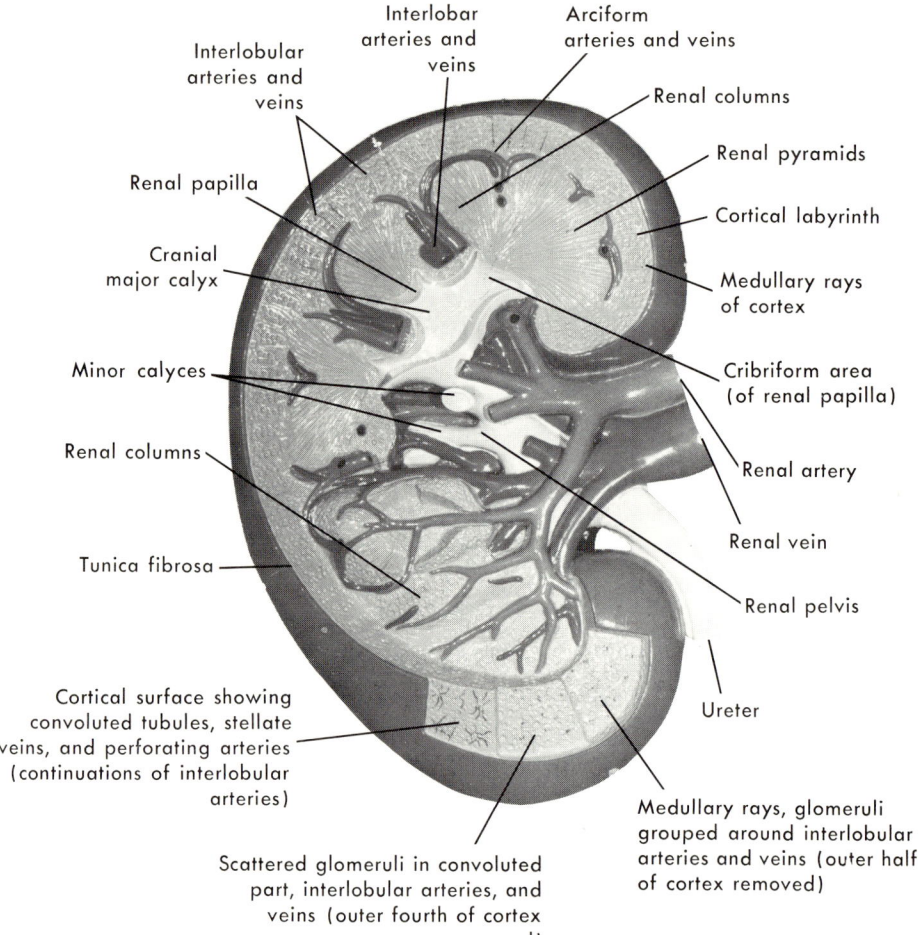

Fig. 7-3. Right kidney model, median section.

is a cup-shaped tube, embracing on one end its corresponding pyramid (or pyramids) and opening on the opposite end, along with two or three others, into a *major calyx*. There are usually from five to eight major calyces in each kidney. The major calyces join to form the renal pelvis,* a saclike arrangement of epithelial tissues that lines the renal sinus and extends through the hilum to be continuous with the corresponding ureter. The renal pelvis extends for a few millimeters beyond the hilum as a cone-shaped connection between the kidney and ureter (see Fig. 7-3).

The functional unit of the kidney, the *nephron*, commences in the cortex and connects with a *collecting tubule;* several nephrons connect with a single collecting tubule. The collecting tubules in a given region of the cortex extend into the medulla to form a renal pyramid. Urine from the nephron passes along a collecting tubule into the renal pyramid and out of the renal pyramid through the *renal papilla* at the summit of the pyramid into a minor calyx; the urine then passes into a major calyx and, thence, into the renal pelvis and ureter. Beginning with the major calyces, the succeeding structures are visualized radiographically in both intravenous pyelograms (I.V.P.) and retrograde pyelograms.

The cortical substance of the kidney extends inwardly between the renal pyramids to form the *renal columns*. The cortex contains the tortuous parts of the renal tubules, and the renal columns and pyramids contain the straight parts of these tubules. In a given pyramid, several straight collecting tubules are arranged around a *central duct* (duct of Bellini), which they join. From fifteen to twenty central ducts open into the minor calyx through the cribriform area of a renal papilla. The base of each pyramid lies essentially parallel with the outer surface of the kidney.

Nephron. As mentioned earlier in this chapter, the nephron is the functional unit of the kidney; i.e., urine is extracted from the blood in the nephrons. There are from 1,000,000 to 1,500,000 nephron units in each kidney. The nephron consists chiefly of tubules, of which there are about 75 miles in the two kidneys; approximately 20 percent of this tubular length is in the cortex, with the remaining 80 percent chiefly in the medulla.

A nephron is a complex tubular structure composed of columnar and, perhaps, other epithelial cells on a basement membrane of connective tissue. The renal tubule is about 14 mm. in length and 55 microns (μ) in diameter. The tubule commences in the cortical substance as a blind tube that invaginates at the cortical end to surround the glomerular tuft (capillaries between the afferent and efferent arterioles). The tubule, which makes a very tortuous course in the cortex, consists of a *proximal convolution* and a *distal convolution;* the convolutions are joined by two straight *limbs* in the renal column. The limbs are joined by a curved piece, also in the renal column. The two limbs and the curved piece are called the *loop of Henle*. Beyond the distal convolution the tubule joins

---

*The renal pelvis may be absent, and it may be replaced with an appropriate and additional number of major calyces.

with a collecting tubule that terminates in a pyramid. Each nephron unit measures approximately 0.2 mm. in diameter (see Fig. 7-4).

The saclike capsule of the renal tubule that surrounds the glomeruli is the renal capsule (capsule of Bowman). The totality of the capsule, glomeruli, and tubules forms a complete *renal (malpighian) corpuscle,* the nephron unit. The loop consists of the two straight parts, the *descending* and *ascending limbs,* connected by the curved piece. The walls of the limbs* consist of a single layer of broad epithelial cells, whereas the walls of the convoluted tubules consist of more than one layer of thicker epithelial cells having a more cuboidal shape.

Blood supply. The blood supply to each kidney is from the aorta via the corresponding *renal artery,* and the venous return is via the corresponding *renal vein* to the inferior vena cava. The renal arteries arise from the sides of the aorta immediately inferior to the superior mesenteric artery and about even with the space between the first and second lumbar vertebrae, and each renal artery normally commences with a funnel-shaped opening from the aorta. As each renal artery branches from the aorta, it forms an angle with the aorta of ap-

---

*Approximately 20 percent of the renal corpuscles are contained deeply within the cortex; their longer limbs project quite far into the medullary pyramids.

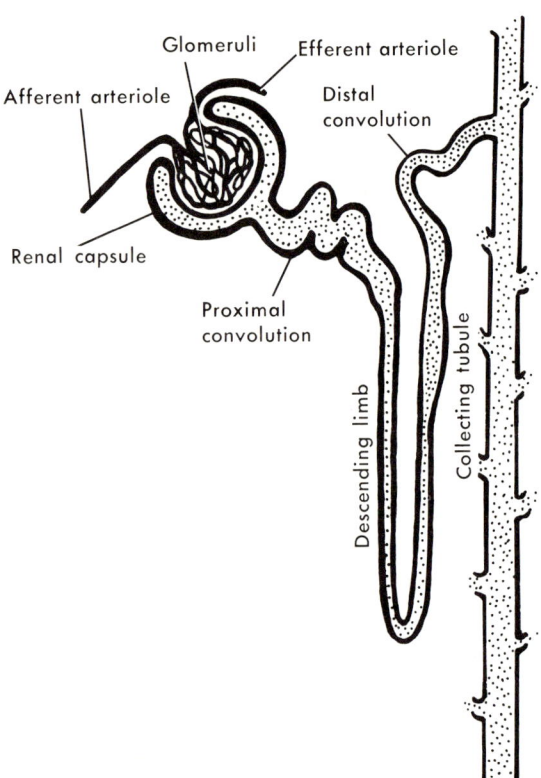

Fig. 7-4. Nephron unit (renal corpuscle).

proximately 90 degrees. The right renal artery is longer and lower than the left renal artery.

Each renal artery extends toward the hilum of the corresponding kidney and branches into four or five smaller arteries, which enter the hilum. The smaller arteries pass through the medulla and branch into the *interlobar arteries,* which pass between the pyramids. The interlobar arteries give rise to the incomplete arches of the *arciform (arcuate) arteries,* which cross the bases of the pyramids. *Straight arteries* extend from the arciform arteries into the medulla, and *interlobular arteries* radiate outwardly from the arciform arteries toward the surface of the kidney. From the interlobular arteries arise the *afferent arterioles,* one for each nephron. Upon entering the renal capsule, the afferent arteriole breaks up into an intricate network of capillaries, the *glomeruli,* which rejoin to form the *efferent arteriole* to convey the blood from the renal corpuscle to the network of capillaries that surrounds the renal tubules. Thus, the arterial blood entering each kidney passes through two capillary networks before returning as venous blood via the renal vein to the inferior vena cava. The second capillary network joins with a venous network. Veins of this network join to form the *interlobular vein.* The interlobular veins join to form the *arciform veins,* which receive the *straight veins.* The arciform veins join to form the *interlobar veins;* the interlobar veins join to form the *renal vein.*

Innervation. The kidneys are innervated from branches of the celiac and aortic plexuses and the lesser and lower splanchnic nerves; in all there are fifteen nerves supplying each kidney.

Other structures of the urinary system

The structures of the urinary system that are concerned with emptying and storing the urine are the ureters, urinary bladder, and urethra. The ureters convey the urine from the kidneys to the urinary bladder for storage. Elimination from the body at a later time is achieved via the urethra (see Fig. 7-5).

Ureters. The ureters are musculomembranous tubes extending for approximately 25 cm. between the renal pelves and the urinary bladder; the right ureter is approximately 1 cm. shorter than the left ureter. The ureters function to convey urine from the kidney to the bladder.

A mucous membrane of transitional epithelium lines the inner wall of the ureter; the membrane is continuous from within the renal sinus, through the ureter, and into the urinary bladder.

Each ureter descends at the corresponding side of the vertebral column to terminate in its junction with the bladder on the postero-inferior surface of the bladder and lateral about 3.5 or 4 cm. to the midsagittal plane of the body. The ureter joins the bladder in the lower pelvic cavity; the orifices in the posterior bladder wall are about 7.5 cm. apart. The lumen of each ureter may vary in an individual, depending upon his current physical condition, from as small as 1 mm. to as large as 1 cm. in diameter. Atresia (absence or closure) of the ureteral lumen prevents passage of urine through the ureter.

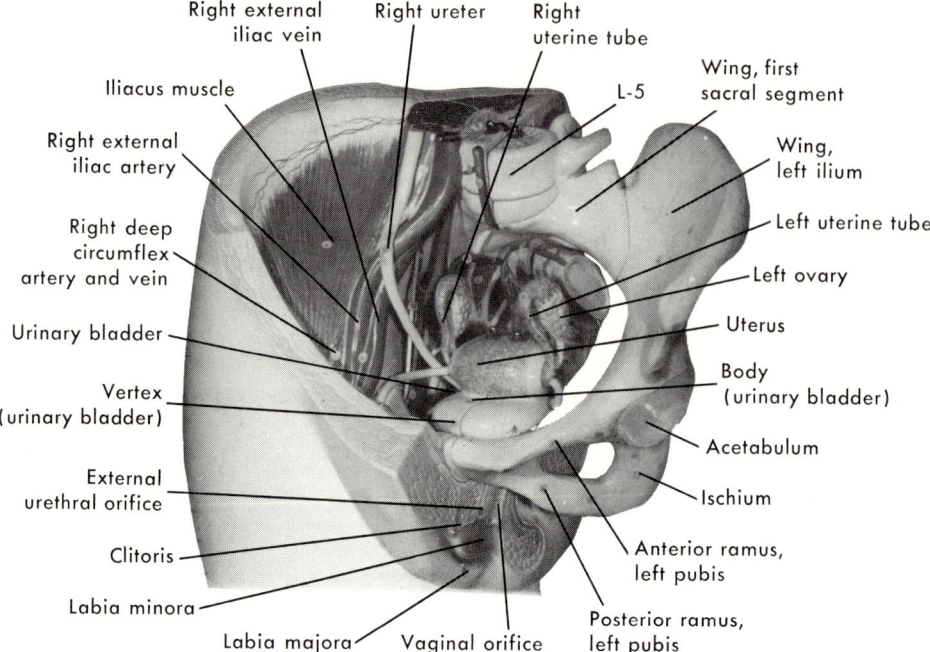

Fig. 7-5. Right ureter, relation to pelvic viscera.

Three coats of tissue comprise the ureteral wall. The outer coat is the *tunica adventitia,* which consists of connective tissue. The middle coat is the *tunica muscularis,* which consists of both circularly and longitudinally arranged nonstriated muscle fibers; connective tissue is interposed between the layers of muscle fibers. The inner coat is the *tunica mucosa,* which consists of several layers of epithelial cells and is continuous from within the kidney to within the urinary bladder. The tunica muscularis makes possible the peristaltic movement that forces the urine along the ureter.

The ureter receives arterial blood from branches of the renal artery and the spermatic artery and from branches that supply the urinary bladder. The venous return generally parallels the arterial supply. The ureters are innervated from the thoracolumbar and sacral autonomic outflows.

Urinary bladder. The urinary bladder is the musculomembranous sac in the pelvic cavity. The bladder receives urine via the two ureters and evacuates the urine via the single urethra. The part of the bladder that lies posterior to the orifices of the ureters and urethra is the *fundus;* the superior and anterior part of the bladder wall, which is covered with peritoneum, is the *vertex* (see Fig. 7-6).

The orifice of the urethra is anterior to the orifice of the ureters and is in the midline of the body in the lowest part of the floor of the bladder. The area marked by the three orifices is the trigone (trigonum). The ureteral orifices are

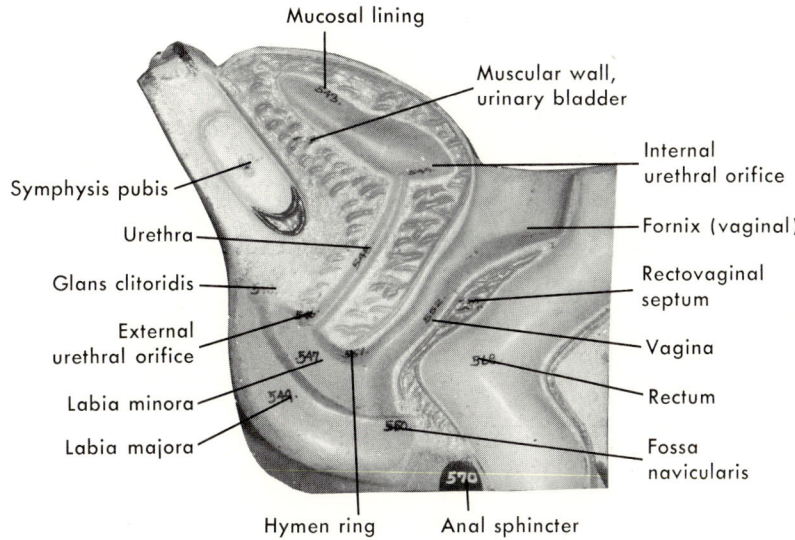

Fig. 7-6. Female urinary bladder and related structures.

oblique in the posterior wall; therefore, the orifices are closed during micturition, and refluxing* of the urine is made impossible.

Four coats of tissue comprise the bladder wall. The outer coat is the *tunica serosa,* beneath the peritoneum, and the *tunica adventitia,* on the remaining surface area; the latter consists of fibrous connective tissue. The first inner coat is the tunica muscularis, which is quite thick and consists of numerous interlaced bundles of nonstriated muscle fibers; this layer is continuous with the corresponding layer of the urethra. The second inner coat is the *tela submucosa,* which consists of fibrous connective tissue containing elastic fibers. The inner coat is the *tunica mucosa,* which consists of transitional epithelium.

The urinary bladder may contain as much as from 400 to 600 ml. of urine at a given time; however, sensory mechanisms cause the feeling of fullness and necessity of voiding when the urine volume reaches from 250 to 300 ml. The 24-hour volume of urine has been estimated to be from 1,000 to 1,800 ml.

The bladder receives arterial blood from branches of the internal iliac arteries. The venous return accompanies the arteries to the corresponding veins. The bladder has motor innervation from the thoracolumbar autonomic outflow and the sacral autonomic outflow. Sensory innervation of the bladder is achieved via sensory nerve fibers that accompany the motor nerve fibers.

Urethra. The urethra is the membranous tube that functions to convey the urine from the urinary bladder to outside the body. The male urethra is from 18.4 to 20.3 cm. (7.5 to 8 inches) in length, and the female urethra is from 3.8 to 5 cm. (1.5 to 2 inches) in length. The male urethra must pass immedi-

*Reflux of urine is considered to be pathologic at any age.

ately from the bladder through the prostate gland; thus the first 2 or 3 cm. of the male urethra are called the *prostatic urethra*. The next 0.6 cm. of the male urethra passes through the urogenital diaphragm and is called the *membranous urethra*. The remainder of the male urethra is enclosed by the corpus spongiosum of the penis and is called the *cavernous urethra*. The female urethra empties directly into the vestibule of the vulva above the vaginal opening, about 2.3 cm. dorsal and slightly superior to the glans clitoris.

In both sexes a sphincter muscle in the urogenital diaphragm acts as a voluntary sphincter; it is called the *external sphincter*.

Three coats of tissue comprise the urethral wall in both sexes. The outer coat is a thick layer called the *muscular* layer. The middle coat is thin and consists of spongy erectile tissue; it is called the *erectile* layer. The inner coat, which consists of squamous and transitional epithelium, is called the *mucous* layer.

The blood supply and innervation for the urethra derive from those of the bladder.

Urine formation

Urine forms as a result of two separate processes, each of which is achieved within the nephron unit. The first of these processes is filtration, which occurs in the renal corpuscle. The second of these processes is reabsorption, which occurs in the tubules. (Reabsorption begins with the proximal convolution and is completed prior to the collecting tubules.)

Filtration

Filtration (a part of urine formation) can be considered as a combined process of pressure and filtration through an exceedingly thin membrane, the renal capsule.

Arterial blood enters the glomeruli of a nephron unit via the afferent arteriole. The glomeruli are encased in the renal capsule in such a manner that any fluid permeating the walls of the glomeruli will collect in the capsule. This fluid, which differs very little, if at all, from plasma, flows through the tubular structure, where reabsorption occurs.

The hydrostatic pressure (see pp. 38 and 39) is quite high in the glomeruli for two reasons. The arterial blood pressure is quite high in the aorta, and the renal arteries branch directly off the abdominal aorta; therefore, the blood pressure in the large renal arteries and their branches is quite high. Additionally, the diameter of the efferent arteriole lumen is considerably less than the diameter of the afferent arteriole lumen. These two conditions cause the hydrostatic pressure of the glomerular blood to be quite high. This pressure is maintained at its high level by some obscure mechanism, even during times when the systemic blood pressure falls. The fact that the afferent and efferent arterioles constrict independently of each other is important in maintaining the glomerular blood pressure. However, when the systemic blood pressure falls appreciably (to 50 mm. Hg or less), all renal secretion may cease for a time or permanently; this is commonly named *renal shutdown.*

Experimental evidence demonstrates that the hydrostatic pressure in the renal capsule is approximately one-half the pressure in the afferent arteriole of the nephron. The thickness of the capsular membrane is approximately 0.1 micron (μ). Because of the great difference in pressure exerted on the opposite sides of this extremely thin membrane, it is believed that the fluid portion of the blood is forced through the membrane into the uriniferous tubule.

Expressed in milligrams of mercury (mm. Hg), the pressure of the blood in the glomeruli has been measured to be from 60 to 70 mm. Hg. The hydrostatic pressure of the filtrate in the renal capsule is 20 mm. Hg, and the colloid osmotic pressure* is 30 mm. Hg. The effective filtration pressure is the result of the blood hydrostatic pressure minus the sum of the filtrate hydrostatic pressure plus the colloid osmotic pressure, as follows:

$$70 - (20 + 30) = 20 \text{ mm. Hg of effective filtration pressure}$$

Finally, the blood volume delivered in unit time to the kidneys exerts great influence upon the filtration process. It has been determined that approximately from 1,000 to 1,400 ml. of blood are delivered to each kidney each minute. The rate of glomerular filtration is from 125 to 130 ml. of filtrate each minute. The filtrate from the capsule is very much like blood plasma, except that the filtrate normally contains no colloids.†

Reabsorption

The filtrate from the renal capsule and the glomerular filtrate are one and the same. The pH of this filtrate is approximately 7.4, or about the same as the pH of plasma. In passing through the tubules, base is reabsorbed from the alkaline filtrate into the interstitial substance and then into the blood according to pH requirements; the result is that the filtrate pH decreases as base is removed, and the urine formed is usually acid.‡ Ammonium is produced in the kidneys to replace sodium in combination with acid; either ammonium or sodium combines with acid to neutralize it.§ When ammonium replaces sodium in combination with acid, valuable sodium is conserved for reabsorption into the blood. Ammonium is produced from glutamine by the renal cells. These combined actions help to maintain the alkalinity of the body tissues and the pH of body fluids generally.

On preceding pages it is seen that the kidney receives by far the largest

---

*Colloid osmotic pressure is the pressure of osmosis resulting from concentrations of protein in the blood and interstitial fluid.
†Under certain conditions some proteins, perhaps less than 0.2 percent of albumin in the blood, may be found in the urine of healthy persons.
‡Although the urine is usually acid, it may be normal and have a pH value of between 8.2 and 4.5.
§Urine pH is not found to be less than 4.5; therefore, there must be a provision to prevent increased acidity.

supply of blood of any organ in the body and that the total quantity of the plasmalike filtrate obtained from the kidneys is approximately 180 liters in a 24-hour period as follows:

$$125 \text{ ml.} \times 60 \text{ (min.)} \times 24 \text{ (hrs.)} = 180 \text{ liters}$$

From this very large volume of fluid, the kidneys reabsorb about 99 percent and excrete about 1 percent of the volume as urine in a 24-hour period.

Because of the difference in content of the solute substances, such as glucose, mineral salts, etc., the gradient between osmotic pressures on the two sides of the tubular wall cause 80 percent of the water to pass through the tubular wall into the interstitial substance and then through the venous wall into the blood as the filtrate passes through the proximal convolution. *All glucose and most of the amino acids are reabsorbed in the proximal convolution along with many electrolytes.* The remainder of the filtrate passes on through the descending and ascending limbs of the renal loop and the distal convolution and, finally, passes into the collecting tubules. *Reabsorption in the proximal convolution is not subject to control, whereas reabsorption in the distal convolution is subject to control.* As a result, internal mechanisms (ADH and ATP* with perhaps others) effect the normal urine elimination volume by regulating the quantity of water reabsorbed in the distal convolution. The delicate pH balance is finally effected here also, although electrolytes are reabsorbed throughout the length of the tubule to the collecting tubule.†

As the filtrate moves from the proximal convolution into the descending limb of the loop, the filtrate retains its identity with plasma, but its volume is reduced to 20 percent of the original. The cells of the tubular walls are entirely passive to the passage of water; i.e., water diffuses through the walls according to the laws of osmosis. The cells of the walls of the distal convolution play an active part in the retention or selection of the electrolytes and certain other solute substances. The cells of the descending limb walls are active in some parts of urine formation; this may be true regarding the cells of the walls of the ascending limb of the renal loop. For the most part, we may say that the cells of the proximal convolution are passive in urine formation and that the cells of the distal convolution are active in urine formation. However, glucose is actively reabsorbed along with amino acids in the proximal convolution. In general, those substances most essential to maintaining homeostasis are actively selected and/or refused by the cells of the tubular walls. The substances of vital importance include sodium ions (Na^+), chloride ions (Cl^-), carbonate ions (HCO_3^-), glucose ($C_6H_{12}O_6$), and water (H_2O).

In addition to the foregoing discussion of filtration and reabsorption functions of the kidneys, it is to be noted that a third function is now considered to exist.

---

*ADH is the abbreviation for antidiuretic hormone; ATP is the abbreviation for adenosine triphosphate.
†Evidence indicates that electrolytes are probably reabsorbed through the collecting tubule walls.

Recent evidence indicates that the cells in the walls of the distal tubules actively secrete certain substances (potassium and acid) into the urine. In man glomerular filtration and tubular reabsorption dominate over tubular secretion.

Urine

The volume, contents, color, specific gravity, and pH of urine are subject to considerable variation according to the intake of food and water and depending upon the age and activity of the individual.

We have seen that the 24-hour volume of urine may vary from 1,000 to 1,800 ml.; however, the total volume (under normal conditions) will be 95 percent water and 5 percent solids. The solid contents consist of both organic and inorganic substances.

The specific gravity of urine may range between 1.003 and 1.025; however, the midrange between 1.011 and 1.018 is more common. First specimens upon arising are more concentrated and will consequently have higher specific gravity values.

The pH of urine is usually acid; however, diet or medication may cause the urine pH to become alkaline. The usual pH range is from 5.0 to 6.7; see footnote on p. 297.

In addition to the normal functions of metabolism, the kidneys are capable of removing large quantities of foreign substances of a poisonous nature that may find their way into the bloodstream. These substances include mercury, barbituric acid derivatives, alcohol, and most medicinal products administered in treatment. Also, the kidneys aid in the detoxification and subsequent excretion of certain substances, such as the conversion of benzoic acid into hippuric acid. Conditions of pathology in the kidneys present a urine that is completely different from the normal urine.

Pathology may occur in the kidney in either of two loci. Structural breakdown in the glomerulus and capsule would permit proteins to pass from the blood into the filtrate; thus, essential proteins and other colloidal substances would be lost. The other site of pathology in the kidney is the tubules. Physiologic failure of the cells of the tubules may result in inability of these cells to reabsorb or to reject essential substances in the filtrate. Either condition is readily determined in certain clinical laboratory procedures.

Hypertension (high blood pressure) is a pathologic condition developing from one or more of many causes. At least one kind of hypertension results from occlusion of the renal arteries. In some cases the splenic artery may be severed and anastomosed to the kidney to reduce hypertension.

In nephritis a large proportion of the glomeruli are the site of either acute or chronic inflammatory processes. As a result, many of the glomerular capillary loops are either temporarily or permanently destroyed, which reduces the total filtering surface of the kidney. When capillary loop destruction is extensive, serious impairment of renal function results. The tubules are supplied with blood that has previously traversed the glomerular capillaries. It follows that glomerular damage precipitates interference with essential nourishment of the tubules. Undernourishment of the tubules impairs the power of selective reabsorption and subsequent concentration of the filtrate by the capillary network surrounding the uriniferous tubules. In such cases the urine is more dilute that normal.

RADIOGRAPHIC INTEREST

Many pathologic conditions in the urinary system are demonstrable radiographically. Routine radiographic examinations include flat exposures of the abdomen, with the patient in supine position for visualization of the kidneys, ureters, and urinary bladder; this examination is called a K.U.B. and is performed without the use of contrast media. A K.U.B. often visualizes concretions of various elements in the kidneys, ureters, and/or the bladder, although approximately 20 percent of ureteral calculi are radiolucent. More extensive studies of the urinary system require contrast media to demonstrate certain structures; these studies are I.V.P.* (intravenous pyelograms) and retrograde pyelograms.

The performance of an I.V.P. requires that a physician inject the contrast medium into one of the veins of the body; a vein in the antecubital space is usually selected. The contrast medium carries a soluble form of iodine in an isotonic solution. The solution must pass from the upper extremity veins (or others if another injection site is selected) via the superior vena cava into the heart. From the right heart the contrast medium passes into the lungs and returns to the left heart. The medium passes from the left heart via the aorta to the renal arteries and their several branches into each nephron. Within a period of approximately 5 minutes after the injection, sufficient quantities of the contrast medium will concentrate in each kidney to render the major calyces and the renal pelves opaque to roentgen rays (x rays). If the kidneys function normally, the contrast medium will be expelled from each kidney through the corresponding ureter and will commence to appear in the urinary bladder within 15 to 20 minutes after injection. A useful routine for radiographic examination of this system is presented as follows:

Exposure #1–routine K.U.B. prior to injection

Exposure #2–5 minutes following injection

Exposure #3–15 minutes following injection

Exposure #4–30 minutes following injection

Compression applied just superior to the urinary bladder on the abdominal wall will hold back the contrast medium in the ureters and make possible better visualization of the renal structures and upper ureters. The compression should be removed after Exposure #3. The series of exposures is made with the patient in supine position; the patient should not move between exposures. Gas in the intestines must be eliminated prior to examination, and the patient should be dehydrated (have no fluids for the preceding 12 hours). The bladder must be empty prior to the examination (see Fig. 6-17).

*I.V.P. examinations are qualitative rather than quantitative.

The urinary system and the skin 301

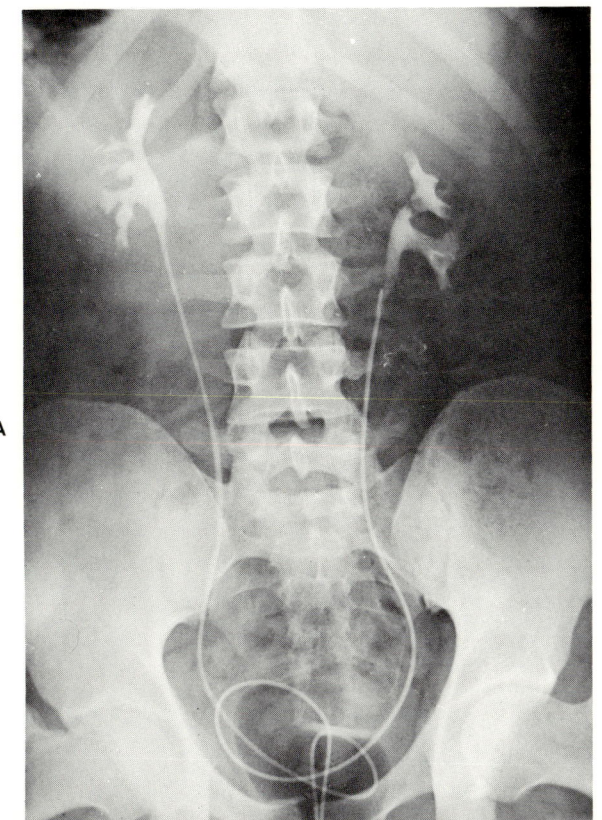
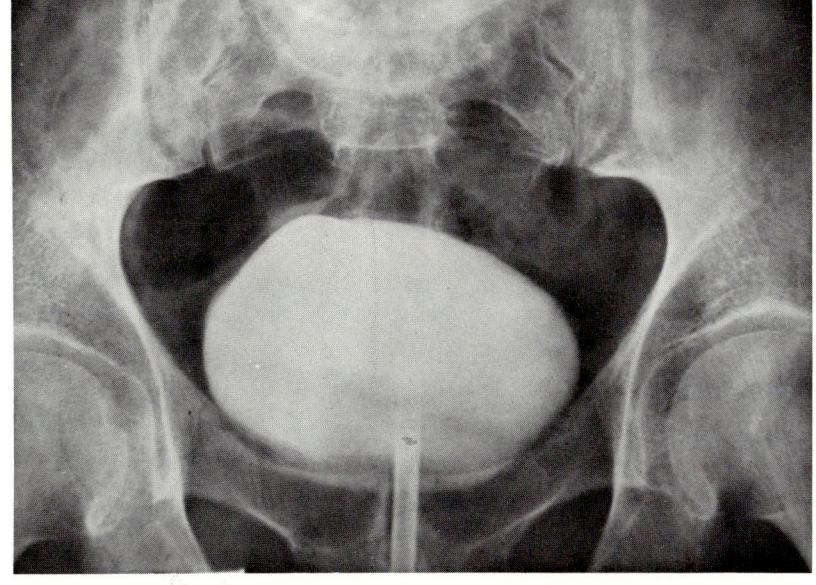

Fig. 7-7. A, Retrograde pyelogram. **B**, Cystogram, posterior view.

Retrograde pyelography (see Fig. 7-7, A) may be performed for the entire urinary system or for the bladder or the urethra. Such examinations of the urinary bladder are called *cystography* (see Fig. 7-7, B); for the urethra they are called *urethrography*. Female urethras are rarely examined by means of radiography. In all of these examinations the physician who injects the contrast medium will usually describe, prior to the examination, the number and type of exposures required for the examination series.

It is neither infrequent nor indicative of pathology to visualize multiple ureters and kidneys on one side and even on both sides. In such cases an increased number of calyces is usually visualized. Multiple ureters and extra calyces occur, also, without the presence of extra kidneys.

A lower-than-normal position of one or both kidneys is called *nephroptosis*. In such cases the corresponding ureter usually loops over itself and may even kink and close the lumen, at least partially. In such instances there could be no visualization of the lower ureter in an I.V.P. *Atresia* (absence or closure of a normal body orifice or passage) of the ureters is readily visualized in intravenous pyelography.

SKIN
Functions

The skin presents a very large surface area to the atmosphere. The skin, as a large and important organ, is the principal site of body heat regulation, presenting from 2,500 to 3,000 square inches of surface to the atmosphere. (These are figures for adults.)

The unbroken skin protects those structures beneath it from bacterial invasion, heat rays, dehydration, damage from minor trauma, and many other conditions.

The skin, by means of its glands, functions as an important excretory organ of water and certain mineral salts.

The extremely large number of sensory nerve endings located in the skin and mucous membranes make the skin a very important part of the sensation-receiving organ for transmission of these impulses to the central nervous system.

In excreting water and electrolytes, the skin functions importantly in assisting the body to maintain both the water balance and electrolyte balance so necessary to homeostasis.

In excreting water and oil, the skin functions most importantly in maintaining the normal temperature of the body. The skin excretions include, in addition to water and mineral salts, a considerable quantity of nitrogenous wastes.

Three kinds of glands are found in the skin, each performing its peculiar function assisting in total homeostasis, as follows:
1. The *sebaceous glands* secrete oil for the hair. There are two glands for each hair, one on each side of the hair follicle. The secretion of these glands is called *sebum;* it maintains the proper degree of softness of both hair and skin and prevents overevaporation and overabsorption of water. Sebum also assists in preventing excessive heat losses (see Fig. 7-8).

The urinary system and the skin **303**

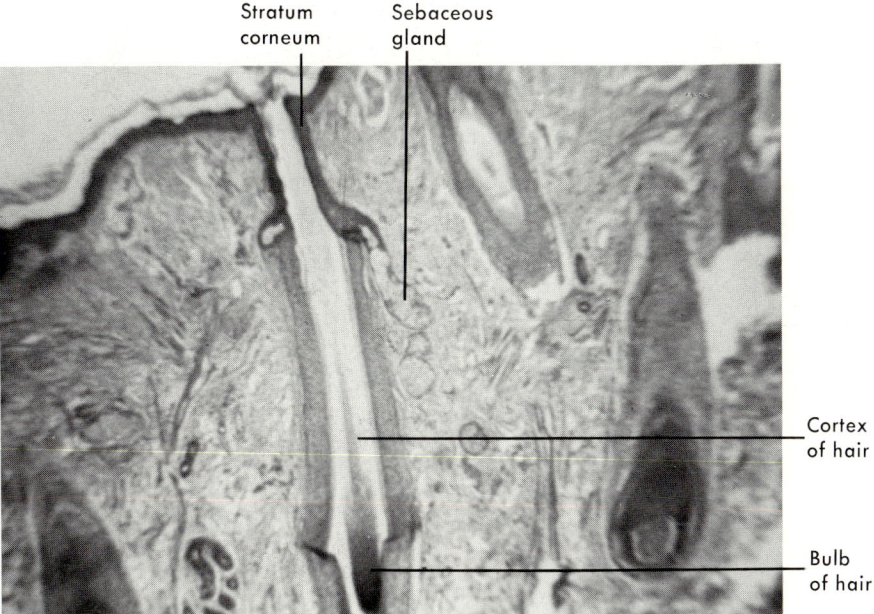

Fig. 7-8. Cross section of integument showing hair follicle.

2. *Sweat glands* are found in most parts of the skin, with as many as 3,000 glands in 1 square inch in the skin of the palms of the hands. The sweat glands make a valuable contribution to homeostasis in excretion of water and electrolytes. When body heat increases above normal (not fever), the sweat glands secrete more sweat to evaporate and cool the body. These glands also are excretory organs for nitrogenous wastes.
3. *Ceruminous glands* are modified sweat glands found in the walls of the external ears. These glands secrete the oily substance called *cerumin*, which helps to maintain the health of the auditory canal and the tympanic membrane.

Constriction of the arteries supplying the venous plexuses of the body can reduce the rate of blood flow in the skin to as little as from 20 to 30 ml. each minute, whereas complete dilation of these arteries can increase this rate to from 2 to 3 liters of blood each minute.

Anatomy*

The embryonic ectoderm gives rise to the epidermis, hair, nails, sweat glands, sebaceous glands, and mammary glands. Around certain openings of the body, the anus, vulva, nose, and mouth, the ectoderm reflects inwardly and is modified to form the mucous membrane. Around the mouth, further specialization of the reflected ectoderm gives rise to the formation of tooth enamel.

*See pp. 23 and 24.

The mesoderm, which lies beneath the ectoderm, gives rise to the formation of connective tissue, which becomes the dermis. The dermis contains the rich vascular and nervous tissues in the supportive network of connective tissue.

From inside to outside, the epidermis contains the following layers:
1. Stratum germinativum, the basal cell layer (The cells are columnar, and new cells form at the rate of loss in the corneum. Melanin [keratin] is found in the deepest part of this layer. Keratin is a protein secreted by the cells of ectodermal origin and by cells of entodermal origin, as in the esophagus.)
2. Stratum spinosum, a layer with several sublayers of irregularly shaped cells (It is often called the prickle cell layer.)
3. Stratum granulosum (The cells die in this layer.)
4. Stratum lucidum, found only on the palms of the hand and soles of the feet (It contains a transparent compound called *eleiden;* keratin forms from the eleiden.)
5. Stratum corneum (In this layer the dead cells are converted into a water-repellent substance called keratin. Keratin is in a state of continuous desquamation.)

REFERENCES

Anthony, C. P., and Kolthoff, N. J.: Textbook of anatomy and physiology, ed. 9, St. Louis, 1975, The C. V. Mosby Co.

Best, C. H., and Taylor, N. B.: The human body, ed. 4, New York, 1963, Holt, Rinehart and Winston.

Goss, C. M.: Gray's anatomy, ed. 27, Philadelphia, 1959, Lea & Febiger.

Guyton, A. C.: Function of the human body, ed. 2, Philadelphia, 1964, W. B. Saunders Co.

Jacobi, C. A., and Paris, D. Q: Textbook of radiologic technology, ed. 5, St. Louis, 1972, The C. V. Mosby Co.

Renal function, Transactions of Second Conference, October 19 and 20, 1950, Joshua Macy Foundation, Publisher.

Schottelius, B. A., and Schottelius, D. D.: Textbook of physiology, ed. 17, St. Louis, 1973, The C. V. Mosby Co.

CHAPTER 8

The respiratory system

FUNCTIONS

Respiration is the exchange of gases between an organism and its environment. Respiration includes three closely associated processes as follows:
1. Pulmonary ventilation, the process of moving atmospheric air into and out of the lungs
2. Gaseous exchange, the reciprocal processes called external and internal respiration (In these processes there is reciprocal exchange of gases between the air in the lungs and the blood and between the blood and the tissue cells.)
3. Cellular respiration, the process in which oxygen is used in metabolism to release energy for bodily activities (Actually, this process varies considerably depending upon air temperature and humidity.)

The totality of respiratory processes enables the respiratory system to achieve its dual functions of supplying the body with oxygen and eliminating waste from the body. The eliminated waste includes carbon dioxide and water; approximately 350 ml. of water are exhaled each 24 hours. The exhaled water volume is nearly constant under normal conditions.

ANATOMY

The respiratory system includes structures that are external to the lungs and structures that are contained within the lungs. The external and internal structures are named in reference to their positions in relation to the lungs rather than to their positions in relation to the thoracic cavity.

External structures

External strutcures of the respiratory system include the nose, nasal cavity, pharynx, larynx, trachea, and bronchi.

306 *Textbook of anatomy and physiology in radiologic technology*

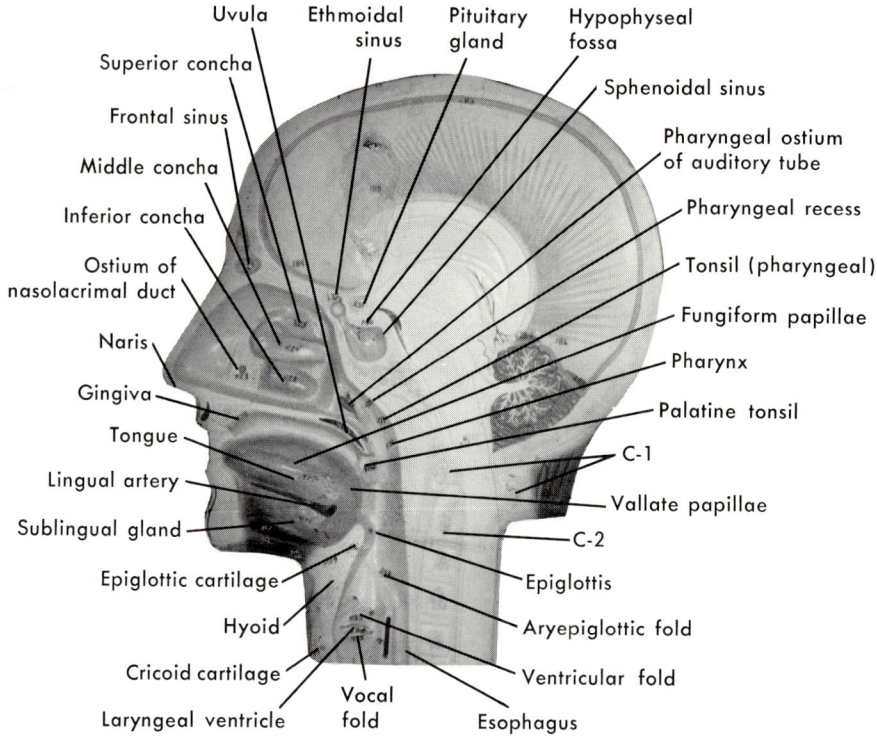

Fig. 8-1. Nasal cavity and pharynx, median aspect.

Nose

The nose is the olfactory organ of man. It is the prominent part of the face that contains the *nostrils* (anterior nares) and covers the anterior part of the nasal fossae. The nose also functions as the respiratory passage from the outside into the nasal cavity (fossae).

The root of the nose is its upper angle; the base of the nose contains the anterior nares; the alae of the nose are the lateral walls; the bony bridge of the nose extends between the root and the apex, which is the tip of the nose. The nose is continuous posteriorly with the nasal cavity (see Fig. 8-1).

Nasal cavity

The nasal cavity is separated into two parts by the nasal septum, which extends in the midsagittal plane from anterior to posterior. The nasal cavity is situated between the oral cavity and the cranial floor. From anterior to posterior, the following bones and/or bone parts form the *roof* of the nasal cavity: nasal bones, frontal spine, cribriform plate of the ethmoid bone, body of the sphenoid bone, sphenoidal conchae, alae of the vomer, and the palatine sphenoidal processes.

From anterior to posterior, the following bones and/or bone parts form the *floor* of the nasal cavity: palatine processes of the maxillae and the horizontal parts of the palatine bones. The floor of the nasal cavity is also the *hard palate*.

The nasal septum forms the *medial wall* of each half of the nasal cavity. From anterior to posterior, the following bones and/or bone parts form the nasal septum: nasal crests and frontal spine, perpendicular plate of the ethmoid bone, sphenoidal rostrum, vomer, and maxillary crests.

From anterior to posterior, the following bones and/or bone parts form the lateral wall of each side of the nasal cavity: frontal process of the maxilla, lacrimal bone, ethmoid bone, body of the maxilla, inferior nasal concha, vertical plate of the palatine bone, and medial pterygoid plate of the sphenoid bone.

The nasal cavity opens posteriorly into the nasopharynx through two bony openings, the *choanae* (posterior nares). Stratified squamous epithelium forms the lining immediately within the nose. However, pseudostratified ciliated columnar epithelium upon connective tissue (tunica propria) forms the richly vascular lining of the nasal cavity. The cilia beat outwardly, toward the anterior nares. The tunica propria contains numerous branched tubuloalveolar glands. The olfactory receptors lie in the mucosa of the medial surface of the superior conchae and the adjacent surfaces of the nasal septum (see p. 347).

The lateral wall of the nasal cavity presents three scroll-like processes, the superior, middle, and inferior conchae. The upper two conchae are largely cartilaginous; the inferior concha (turbinate) is largely osseous and is one of the facial bones. The superior and middle conchae are parts of the ethmoid bone.

The nasal cavity communicates directly with the *paranasal sinuses* (maxillary [nasal antra], frontal, sphenoidal, and ethmoidal) and indirectly with the mastoid air cells.

Pharynx

The pharynx (throat), common to both the respiratory and digestive systems, is described on p. 195.

Larynx

The larynx is the voice organ in man and other mammals. The larynx is the modified upper part of the trachea, which opens superiorly into the pharynx, and is continuous inferiorly with the trachea. The larynx consists of three single cartilages and three pairs of cartilages (a total of nine).

The arrangement of the nine cartilages forms a hollow box. The *thyroid cartilage*, the largest single cartilage, forms the anterior part of the larynx. This cartilage presents the deep thyroid notch in its superior margin between the two laminae. The rounded superior margins of the laminae extend posterolaterally from the notch; a slight elevation extends inferiorly in the midline below the notch. The two laminae of the thyroid cartilage form an angle (the prominence called the Adam's apple) that is approximately 90 degrees in the male and as much as 120 degrees in the female. Extending superiorly from the posterior surfaces of the laminae are the superior cornua. The inferior cornua (horns) extend inferiorly from the posterior surfaces of the laminae (see Fig. 8-2).

The *cricoid cartilage,* the second single cartilage, is ring shaped, with the larger part (the signet) posterior. The larger part extends between the inferior cornua of the thyroid cartilage. The cricoid cartilage rests upon and attaches to the first tracheal ring.

The third single cartilage is the leaflike *epiglottis.* The epiglottis is situated posterior to the root of the tongue, superior to the thyroid cartilage, and posterior to the body of the hyoid bone. The epiglottis is anterosuperior to the superior laryngeal opening; it is so situated as to deflect the swallowed food bolus to either side within the cavity (food trough). The stem of the epiglottis extends downward to attach to the thyroid cartilage.

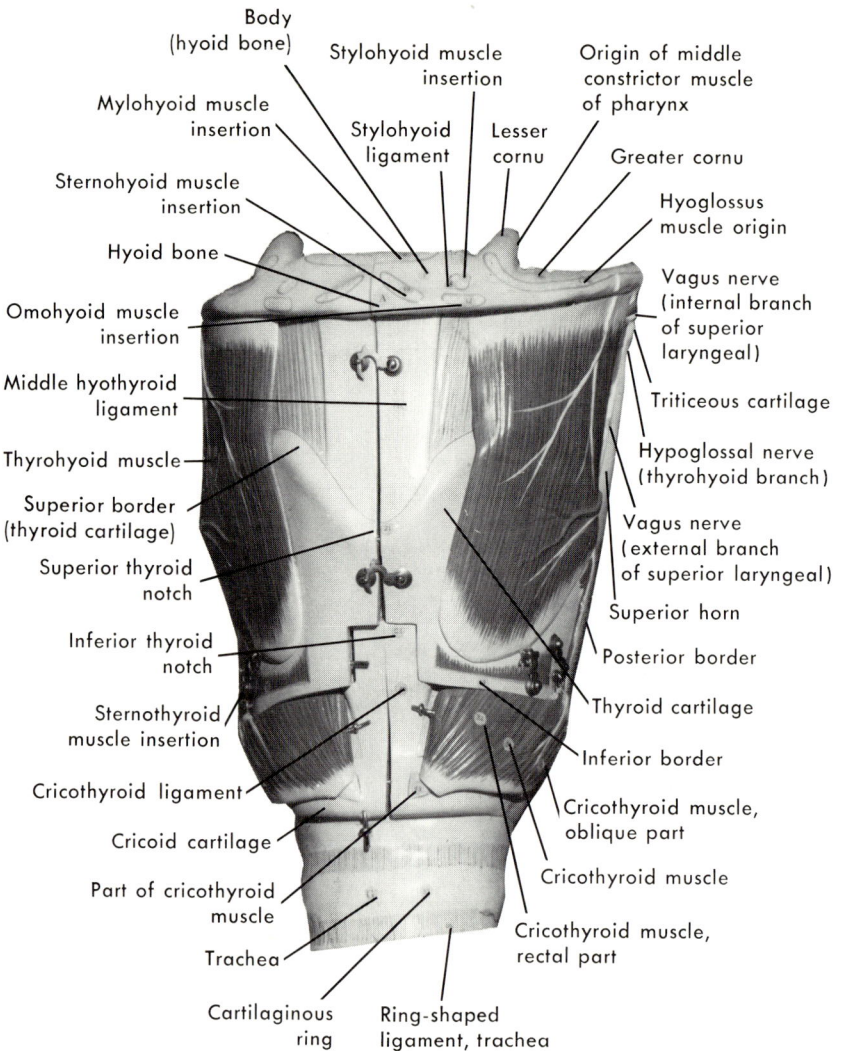

Fig. 8-2. Larynx model. **A,** Anterolateral aspect. **B,** Superoinferior aspect. **C,** Median section.

The respiratory system 309

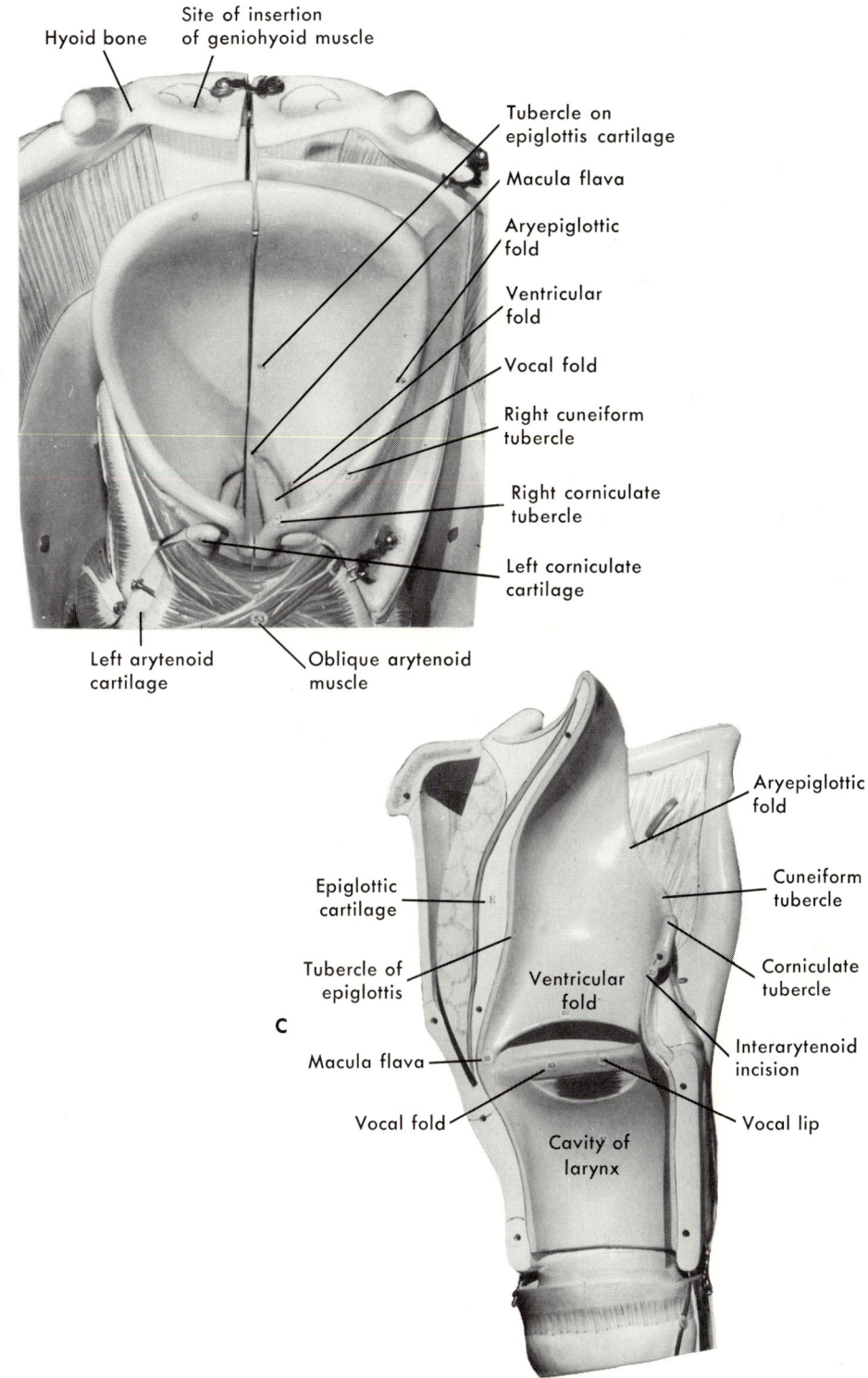

Fig. 8-2, cont'd. For legend see opposite page.

310 *Textbook of anatomy and physiology in radiologic technology*

Two small pyramidal cartilages, the *arytenoid cartilages,* rest on top of the signet part of the cricoid cartilage.

One of the paired, tiny *corniculate cartilages* is situated on each arytenoid cartilage.

The paired *cuneiform cartilages* resemble very small rods and are situated, when present, in the folds of mucous membrane between the epiglottis and arytenoid cartilages.

Two pairs of mucous membrane folds situated in the lateral walls of the larynx subdivide its cavity into three parts. The *ventricular folds* are superior, and above them is the vestibule of the larynx. The *true vocal cords (folds)* lie below the ventricular folds. The *aryepiglottic membrane* folds extend between the epiglottis anteriorly and the arytenoid cartilages posteriorly to separate the vestibule from the *piriform recess* (food trough) on each side. The *ventricle* (middle part of the laryngeal cavity) is the space between the ventricular and vocal folds. The inferior part of the laryngeal cavity is inferior to the vocal folds, and it continues into the trachea. The *glottis* is the space between the true vocal cords.

The several laryngeal muscles are either intrinsic or extrinsic; the latter group

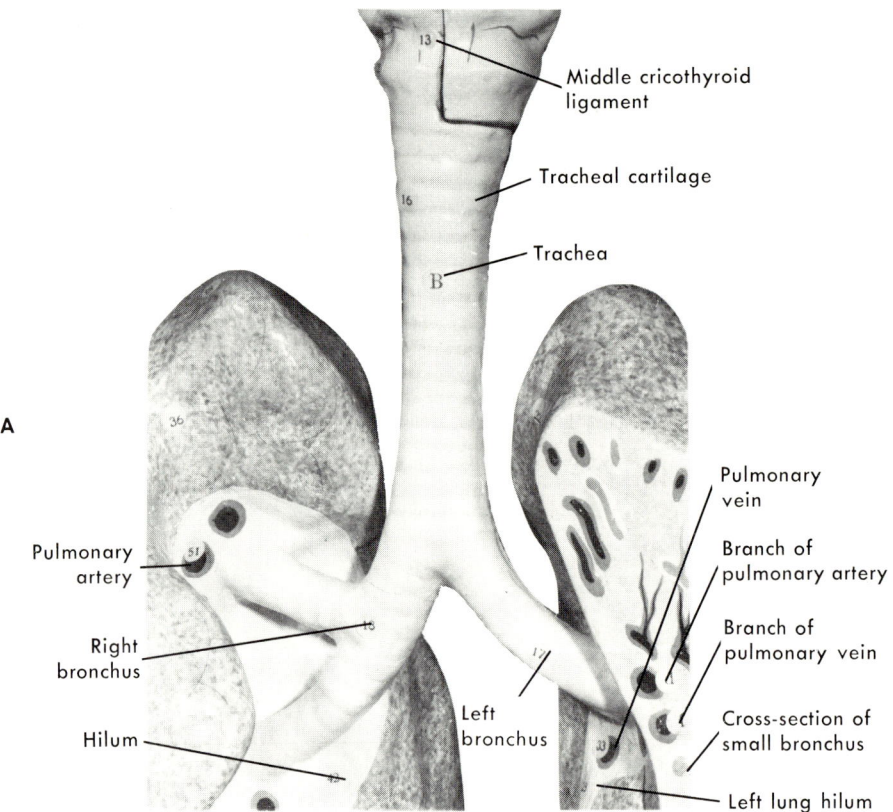

Fig. 8-3. A, Trachea, bronchi, and lungs. **B,** Bronchial tree shadow, anterior view of chest. Note the outline (in white ink) of tracheal and bronchial shadows.

includes all the muscles attached to the hyoid bone, thus giving this bone functional assistance to the larynx.

Within the larynx a mucous membrane, covered almost completely with ciliated columnar epithelium, forms the lining; however, stratified squamous epithelium covers the vocal cords and epiglottis. The lining of the larynx is continuous with the lining of the trachea.

Trachea

The trachea is a hollow tube extending from the laryngopharynx at the level of the inferior surface of the sixth cervical vertebra downward to the bifurcation into the two bronchi. In children the bifurcation occurs at the approximate level of the inferior border of the fourth thoracic vertebra; in adults the bifurcation is somewhat lower, at the approximate level of the superior border of the sixth thoracic vertebra. The trachea consists of several incomplete cartilaginous rings superimposed upon each other and connected by fibromuscular tissue; fibromuscular tissue also comprises the posterior (incomplete) parts of the rings. The arrangement permits swallowing of articles actually larger in size than the

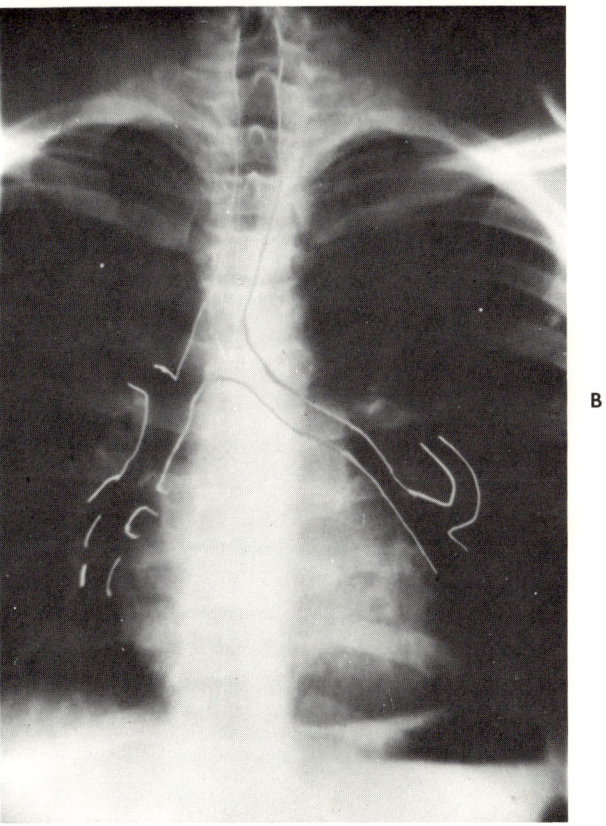

Fig. 8-3, cont'd. For legend see opposite page.

esophageal lumen; thus the food particles can press the esophageal wall forward into the posterior wall of the trachea. Since the tracheal walls are, for the most part, cartilaginous, the walls cannot collapse and occlude the air passage into the lungs (see Fig. 8-3, A).

The thyroid gland lobes are bilateral to the trachea in the neck, and the thyroid isthmus, between the lobes, lies anterior to the trachea. The trachea descends in the thoracic cavity in the midsagittal plane posteriorly in the mediastinum behind the heart and great vessels of the heart. The esophagus is sandwiched between the trachea anteriorly and the thoracic spine posteriorly.

It is of the utmost importance that the tracheal air shadow (see Fig. 8-3, B) be visualized in the midsagittal plane in normal chest radiographs. It is also important to know that the trachea and right bronchus form an almost straight line by their junction and that the angle formed by the juction of the trachea with the left bronchus is, comparatively, quite sharp. In the newborn infant the angle of the right bronchus with the trachea varies from 10 to 35 degrees, and the angle of the left bronchus with the trachea varies from 30 to 65 degrees. In the adult male the right angle is approximately 20 degrees, and the left angle is approximately 40 degrees. In the adult female the right angle is approximately 19 degrees, and the left angle is approximately 51 degrees.

Bronchi

The two bronchi are similar to the trachea in construction; the mucous lining, with ciliated columnar epithelium, is continuous through the bronchi and bronchioles. Each bronchus enters the corresponding lung through the lung hilum, where lesser bronchioles branch off from the main bronchus, which terminates, per se, very shortly. The left bronchus is smaller than the right bronchus, and all of the branches of the left bronchus arise inferior to the left pulmonary artery. The single branch of the right bronchus that arises superior to the right pulmonary artery is named the *eparterial bronchus*.

Internal structures

The right bronchus extends approximately 2.5 cm. from its commencement into the lung, where the lesser bronchi branch off. The left bronchus, although smaller in diameter than the right bronchus, is approximately twice as long as the right bronchus. Extending throughout the lung tissue, the lesser bronchi branch into the smaller bronchioles, eventually into the terminal bronchioles. The terminal bronchioles give rise to the respiratory bronchioles. Alveolar ducts arise from the respiratory bronchioles, and, in turn, each alveolar duct gives rise to several alveolar sacs. The individual alveoli extend outward from an alveolar sac, the whole structure having considerable resemblance to a cluster of grapes arising from the stem. Fig. 8-4 is an electron micrograph of a lung alveolus and its capillary.

The walls of the trachea, bronchi, and bronchioles are muscular and support the mucous membrane lining. In order to effect the reciprocal exchange of gases,

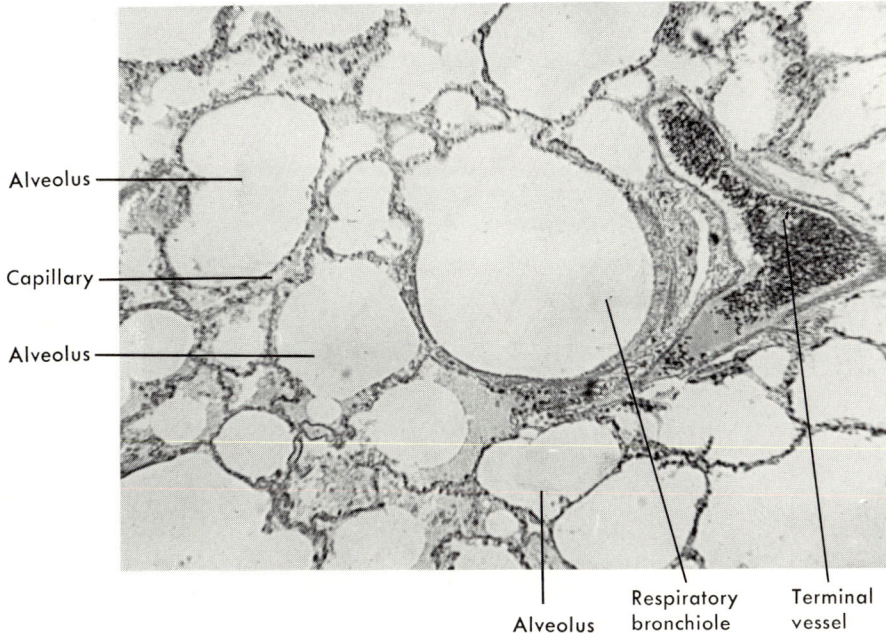

Fig. 8-4. Alveolar septum with capillary.

the structures beyond the respiratory bronchioles have walls consisting of a single layer of flat epithelium.

Each (of the numerous) small alveolus has its own rich blood supply in the form of a pulmonary artery capillary continuing into the very small pulmonary vein venule. The functional unit of a lung consists of the respiratory bronchiole, associated alveolar ducts, alveolar sacs, and alveoli and includes the capillaries and nerves supplying these structures. All of these structures together are called a *pulmonary lobule.*

Each lung consists of spongy tissue interspaced with the various bronchioles and branches. Prior to the initial inspiration (after birth), the lung tissue is solid and will sink in water. After the initial inspiration, there will always be some air trapped within the lung substance, and the lungs will float if placed in water. The lungs are somewhat pyramidal in shape. The base rests upon the diaphragm; the apex is rounded and extends into the neck base above the first rib. It is always important to demonstrate the lung apices and the costophrenic angles* in chest radiographs. Routine positioning technique requires that the shoulders be "rolled" forward to move the scapulae laterally away from the lung fields. The *lordotic* position is used to visualize the lung apices. The concave area in the median surface of each lung is the hilum, the slit through which the blood vessels, nerves, and bronchi pass into and out of the lungs. These structures form the root of the lung (see Fig. 8-5).

---

*A costophrenic angle is formed by the junction of the diaphragm with the ribs of one side of the body.

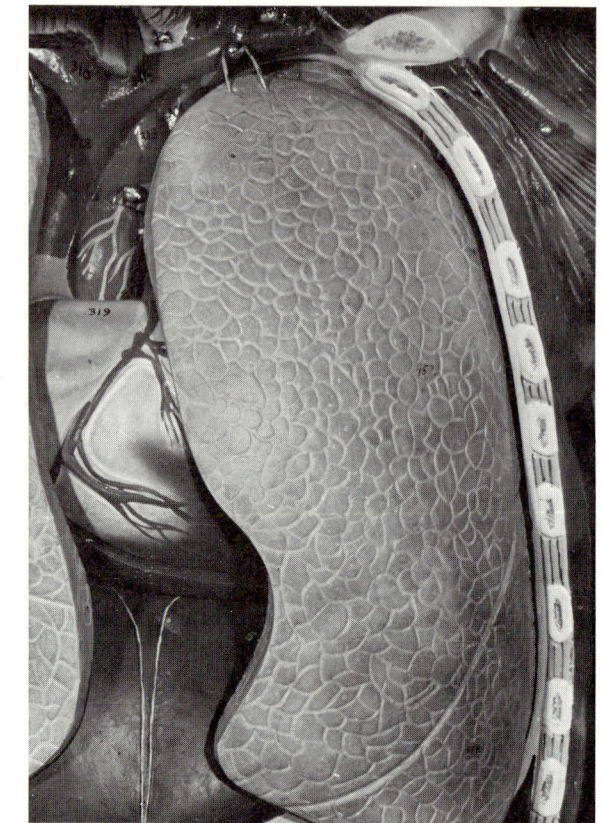

Fig. 8-5. Left lung in mannikin. **A,** Ventral surface. **B,** Medial surface.

The right lung is somewhat larger than the left lung; it consists of three lobes—the upper, middle, and lower. The left lung consists of an upper lobe and a lower lobe. The lobes are further subdivided into bronchopulmonary segments. The number of these segments corresponds with the number of bronchial branches. The right lung usually contains ten bronchopulmonary segments, and the left lung usually contains eight bronchopulmonary segments. The medial surface of the left lung is depressed inferiorly to contain the apex of the heart; this space is the *cardiac notch*.

THORACIC CAVITY

The thoracic cavity is one of the two great ventral cavities. The ribs and intercostal muscles comprise the lateral walls of the thoracic cavity. The ribs attach anteriorly to the sternum to form the anterior wall and posterioly to the vertebral column to form the posterior wall. The diaphragm forms the floor of the thoracic cavity. The walls extend upwardly and inwardly to form a narrowed superior part from which the neck extends. A fan-shaped layer of fascia envelops the trachea, esophagus, and blood vessels of the neck and attaches

The respiratory system 315

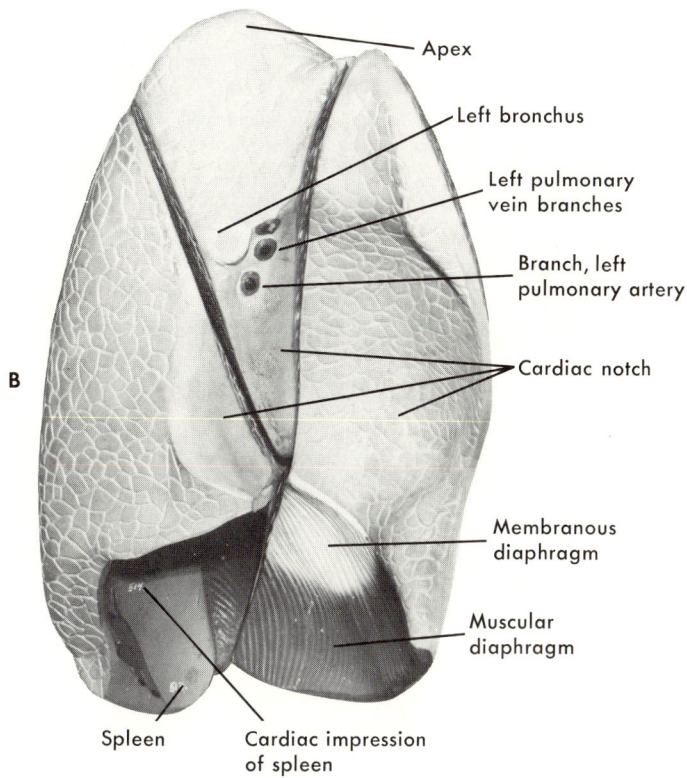

Fig. 8-5, cont'd. For legend see opposite page.

bilaterally to the inner edge of the first ribs and medially to the cervical vertebrae. Their fascia forms the seal between the neck and each upper half of the thoracic cavity (see Fig. 8-6).

In addition to the lungs, the heart and the various mediastinal structures comprise the thoracic cavity viscera. The twelve pairs of ribs, the thoracic vertebrae, and the sternum form a bony (and cartilaginous) protective covering for these structures. The intercostal muscles between each two ribs, in addition to functioning in respiration, also serve as a part of the protective shield.

The intercostal muscles arise from the lower borders of the first eleven ribs and insert into the upper borders of the next successive ribs (second through twelfth) beneath. These muscles are found in layers, principally the external and internal layers. Several other muscles are involved in chest wall formation and especially in respiration; the principal muscle of this category is the diaphragm.

The diaphragm, the chief muscle of respiration, presents a total surface area of approximately 270 square centimeters. Early research in this subject indicates that the diaphragm controls approximately 60 percent of the total volume of air inhaled in deep breathing. During inspiration the diaphragm descends approxi-

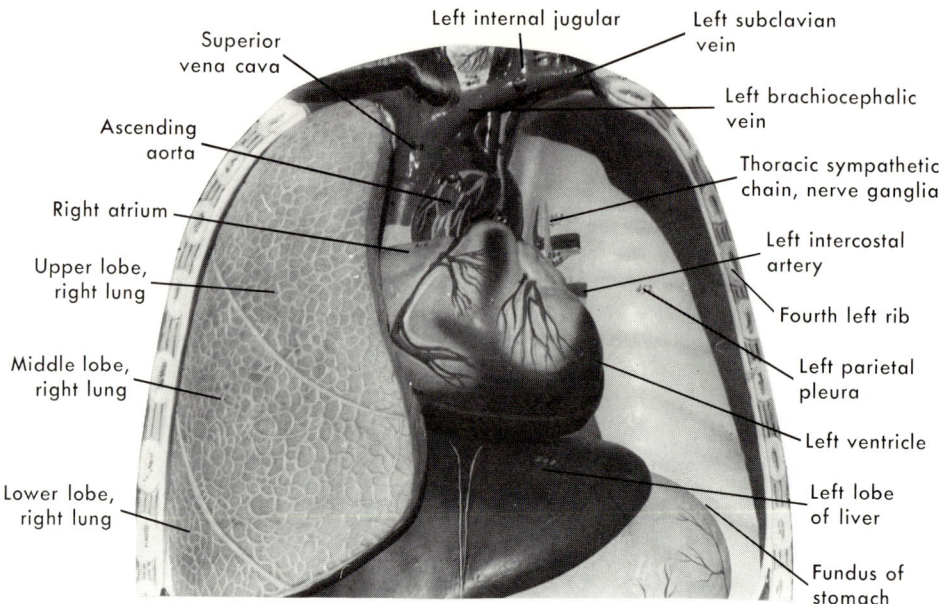

Fig. 8-6. Thoracic cavity with the left lung removed.

mately 3 cm. below the costal cartilages of the fifth pair of ribs. The diaphragm separates the thoracic cavity from the abdominal cavity. In chest radiography it is necessary to have the patient inhale deeply to depress the abdominal viscera and diaphragm in order to demonstrate at least ten ribs above the diaphragm on each side of the thoracic cage.

Pleura

Within the thoracic cage and above the diaphragm is a serous membrane, the *pleura,* which forms a closed sac of two layers, the pleural lining. The outer layer of the pleural lining is continuous and lies adjacent to the rib muscles and other structures comprising the rib cage; this layer is the *parietal pleura.* The inner layer is named the *visceral pleura.* The visceral pleura completely envelops each lung. The two layers are closely related; the space between the layers is a potential space, the *pleural cavity.* Other reflections of the parietal pleura cover the mediastinal structures, which separate the right and left pleural sacs (around each lung) except for a short distance anteriorly on a level with the body and xiphoid process of the sternum. The heart and pericardium are entirely contained within the thoracic cavity and parietal pleura; however, the pericardium and pleura are entirely separate membranes.

Mediastinum

The *mediastinum* is the space in the chest between the pleural sacs of the lungs. It contains the heart and all the viscera of the chest except the lungs. The four parts of the mediastinal cavity are the *superior mediastinum,* which lies above the pericardium; the *middle mediastinum,* which contains the heart and pericardial sac; the *anterior mediastinum,* which lies in front of the pericardium; and the *posterior mediastinum,* which lies behind the pericardium.

Specifically the mediastinal cavity contains the following structures: thoracic aorta, ascending aorta, aortic arch, brachiocephalic artery, beginnings of the left common carotid and left subclavian arteries, pulmonary artery and veins, parts of the superior and inferior venae cavae, vestigial thymus gland, thoracic duct, lymphatic duct, trachea, esophagus, parts of the primary bronchi, parts of the brachiocephalic veins, numerous lymph nodes and vessels, the heart and pericardium, and the vagi, left recurrent, laryngeal, cardiac, and phrenic nerves. The aforementioned nerves serve as the principal innervation of the mediastinal cavity and its contents.

The principal arteries supplying the thoracic cavity and its contained viscera include, in the visceral division, the pericardial artery to the pericardium, the bronchial artery to the bronchi, the esophageal artery to the esophagus, and the mediastinal artery to the several structures in the mediastinal cavity. The principal arteries supplying the thoracic cavity include, in the parietal division, the intercostal artery to the intercostal muscles, the subcostal artery to the ribs, and the superior phrenic artery to the superior surface of the diaphragm.

RESPIRATION

A general explanation of the processes involved in respiration is contained previously in this chapter. A more comprehensive treatment of the reciprocal processes of the gaseous exchange follows.

Respiration is both external and internal. *External respiration* involves the exchange of gases between the air in the lung alveoli and the blood in the lung capillaries. Oxygen in the alveolar air passes into the blood in exchange for carbon dioxide from the blood. *Internal respiration* involves the exact reverse of this process and occurs in the body tissues. Carbon dioxide in the tissues and cells passes into the blood in exchange for oxygen from the blood.

Mechanics of respiration

A comparison of the thoracic cage and lungs with a glass bell jar having a rubber diaphragm base and two rubber balloons within the jar illustrates the mechanics involved in external respiration. To assemble the essential mechanical elements for this demonstration, attach the balloons to a glass or plastic Y tube, which in turn is inserted through a rubber stopper in the mouth and neck of the bell jar, and attach a string to the lower center surface of the rubber diaphragm (see Fig. 8-7).

A downward pull on the string enlarges the capacity of the bell jar. Air will move into each balloon through the Y tube in the rubber stopper to satisfy the *negative pressure* (pressure less than standard atmospheric pressure) developed inside the bell jar but *outside* the balloons. An upward push on the rubber

318 *Textbook of anatomy and physiology in radiologic technology*

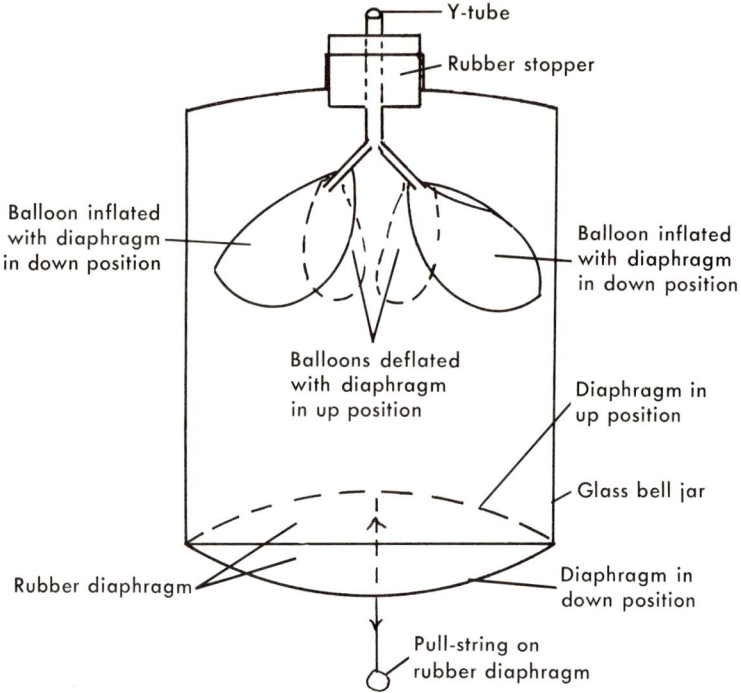

Fig. 8-7. Mechanics of respiration, bell jar demonstration.

diaphragm decreases the capacity of the bell jar. Air moves out of each balloon until an equilibrium in pressure exists between the inside pressure of each balloon and the inside pressure of the bell jar. This compares with the physical forces involved in both inspiration and expiration.

As the diaphragm contracts it moves downward* into the abdominal cavity and enlarges the capacity of the thoracic cavity. Simultaneously, the intercostal muscles between each two ribs raise the anterior (ventral) parts of the ribs, an action which causes anterior movement of the ribs. The combined results of the two aforementioned acts cause a considerable increase in the capacity of the thoracic cavity, which normally has a slight negative pressure. Atmospheric air pressure, under standard conditions, is equal to 760 mm. Hg. Intrathoracic pressure at the commencement of any quiet inspiration is equal to approximately 757.5 mm. Hg. Compared with atmospheric air pressure, intrathoracic pressure decreases further to equal between 752 and 753 mm. Hg as quiet inspiration continues; i.e., the negative pressure increases. As a result, air rushes into each lung to expand the lungs in order to overcome the excessive negative pressure in the thoracic cavity. This is inspiration.

Expiration is the reverse of inspiration. In expiration the diaphragm relaxes

---

*A downward movement of the diaphragm of 1 cm. causes the capacity of the thoracic cavity to increase approximately 270 ml.; thus, a like volume of air is caused to enter the lungs.

and resumes its normal domed shape, the dome extending upward into the thoracic cavity. Simultaneously, the intercostal muscles relax and permit the rib cage to lower; i.e., the ventral parts of the ribs move downward. The combined results of these two acts decrease the capacity of the thoracic cavity and raise (decrease) the negative pressure to near normal within the cavity; the entire process forces air from the lungs.

Physiology of respiration

The physiologic phenomena involved in the transport of gases across membranes and in the transport of gases by the blood are quite complicated and in-depth treatment is intentionally omitted since they have no direct radiographic application. The transport of gases across membranes is called the *gaseous exchange*. The gaseous exchange results from differences in gas pressures on opposite sides of the membrane, from the changeable conditions in which the gases are transported in the blood, and from the presence of an enzyme, *carbonic anhydrase*, in the blood.

In addition to the foregoing, the importance to the gaseous exchange of the anatomic relationship of the lung alveoli to the capillaries and to the total respiratory surface must be considered. A capillary wall is often common with an alveolar wall and is approximately 0.004 mm. in thickness. Also, the very large number of alveoli produces the large respiratory surface needed. Considered estimates place the number of alveoli at approximately 400,000,000 and the total surface involved in the gaseous exchange at approximately 125 square meters, which is about 100 times the entire surface area of the average human body. Under normal conditions of rest the heart pumps from 4 to 6 liters of blood to the lung alveoli via the pulmonary arteries each minute. Since the blood flows through the capillaries in a very thin film, approximately one erythrocyte in thickness, the thin film of blood presents the erythrocytes directly to the membrane between each alveolus and its capillary for the gaseous exchange. The blood-film surface compares in size with the total surface involved in the gaseous exchange. The actual movement of gases across the semipermeable membranes is in accordance with some of the gas laws and/or parts thereof.

The *general gas law* states that the volume of a given mass of a gas is inversely proportional to the pressure and directly proportional to the absolute temperature. The concentration of a gas depends on the number of its molecules in the volume (or per unit volume), and the pressure of a gas is either high or low depending on whether the concentration is high or low (whether the number of molecules of that gas is great or small).

Blood flowing into the lungs via the pulmonary arteries (see p. 270) is high in carbon dioxide and relatively low in oxygen, whereas the alveolar air is high in oxygen and low in carbon dioxide. Carbon dioxide passes from the arterial ends of the capillaries into the air of the lung alveoli, and oxygen passes from the air in the lung alveoli into the blood in the venous ends of the capillaries. Oxygen and carbon dioxide each pass through semipermeable membranes by means of diffusion, which always occurs from an area of high pressure or concentration to an area of lower pressure or concentration.

Both oxyhemoglobin and hemoglobin are acid, but oxyhemoglobin is the stronger of the two. As a result, when more oxyhemoglobin is present in the blood, the reaction direction is to the left; i.e., carbon dioxide is released. This phase occurs in the lungs as oxygen is taken on and carbon dioxide is given up. In the tissues (in venous blood), the hemoglobin is reduced (is less acid) and more carbon dioxide is combined. As a result, venous blood can (and does) transport more carbon dioxide as carbaminohemoglobin from the tissues to the lungs. Upon flowing into the lung capillaries, the reaction reverses again and the carbaminohemoglobin is forced to give up carbon dioxide across the membrane and into the lung alveoli.

Of great importance in the total process of respiration is the chloride shift; i.e., the moving of the chloride ion into and out of the erythrocyte to assist in maintaining pH and electroneutrality.

Carbonic anhydrase, an enzyme, is mentioned on p. 319. This enzyme acts as a catalyst and causes (or permits) readily reversible reactions involving carbon dioxide, water, and carbonic acid (H_2CO_3).

In the discussion of acid-base balance on pp. 261 and 262, it is mentioned that the erythrocyte membrane is highly permeable to water, to carbon dioxide, and to the following ions: carbonate (HCO_3^-), chloride (Cl^-), hydroxyl (OH^-), sodium (Na^+), potassium (K^+), and hydrogen (H^+); the erythrocyte membrane is impermeable to hemoglobin and to plasma proteins. Also, most sodium ions are contained in the plasma and most potassium ions are contained in the erythrocytes. Finally, it is mentioned on p. 262 that most of the hemoglobin is in combination with potassium in the erythrocytes as potassium hemoglobin.

The exchange of chloride ions in relation to the rapid exchange of oxygen and carbon dioxide occurs in the lungs and in the tissues. The chloride ion may be in combination in the plasma as sodium chloride (NaCl), and it may be in combination in the erythrocyte as potassium chloride (KCl). Some of the potassium ions in the erythrocytes are in combination with the carbonate ions as potassium carbonate ($K^+HCO_3^-$). Some of the reduced hemoglobin (HHb*) and some potassium ions in the erythrocytes are uncombined.

Lungs

In the lungs, oxygen enters the erythrocytes from the alveoli because of the higher oxygen pressure in the lung alveoli. As a result, reduced hemoglobin in the erythrocytes takes on oxygen and becomes oxyhemoglobin. Reduced hemoglobin takes on oxygen to form oxyhemoglobin. Since oxyhemoglobin is a stronger acid than reduced hemoglobin, the maintenance of equilibrium in the

---

*Reduced hemoglobin (Hb) may be written as HHb; also, oxyhemoglobin (HbO_2) may be written as $HHbO_2$.

acid-base balance of the blood requires that potassium carbonate and oxyhemoglobin form carbonic acid; this releases a potassium ion to combine with oxyhemoglobin to form potassium oxyhemoglobin (K^+HbO_2).

These reactions cause a decrease in carbonate ion concentration in the erythrocytes, which, in turn, causes carbonate ions to diffuse from the plasma, where carbonate ion concentration is higher, into the erythrocytes.

The preceding paragraphs deal with the chemical aspects of maintaining balance (equilibrium). It is equally necessary that electrical neutrality be maintained; i.e., the number of positive charges equals the number of negative charges ($+ = -$). To achieve electroneutrality, the following action is necessary. As each negative (HCO_3^-) ion diffuses from the plasma into the erythrocytes an equal number of negative ions diffuse from the erythrocytes into the plasma. We have seen previously that the chloride ion diffuses readily through the erythrocyte membrane and that chloride is a negative ion. It follows that chloride ions diffuse from the erythrocytes into the plasma and combine with sodium ions to form sodium chloride (Na^+Cl^-). The total basic ion (K^+ and Na^+) ratio of the erythrocytes continues, for the most part, to be static.

As carbonic acid forms, it decomposes almost instantly into carbon dioxide and water.

Carbon dioxide pressure in the lung alveoli is much lower than the carbon dioxide pressure in the blood from the pulmonary arteries; as a result, a favorable condition exists for the escape of carbon dioxide from the plasma and the erythrocytes into the alveoli. The escape of carbon dioxide from erythrocytes also causes a reduction in the number of osmotically active particles present in the erythrocytes; there are inevitable losses of water from the erythrocytes.

Tissues

Cells and tissues require oxygen to metabolize nutriments. The oxygen is derived from the arterial blood, and the oxidation of nutriments produces rather large quantities of carbon dioxide. As a result, the oxygen pressure in tissues is low. Because of low oxygen pressure in the tissues, the oxyhemoglobin in the erythrocytes gives oxygen to tissue fluids and becomes reduced hemoglobin. Thus, in the tissues, oxyhemoglobin gives up oxygen to the tissues and forms reduced hemoglobin. Since reduced hemoglobin is a weaker acid than oxyhemoglobin, the maintenance of equilibrium in the acid-base balance of the blood requires that carbonic acid be converted into a carbonate ion; thus, one hydrogen ion diffuses from carbonic acid, which leaves a carbonate ion. This pulls one potassium ion from the reduced potassium hemoglobin to combine with a carbonate ion and form potassium bicarbonate.

The preceding reactions cause an increase in carbonate ion concentration in the erythrocytes, which, in turn, causes carbonate ions to diffuse into the plasma from the erythrocytes.

Again there is a shift of chloride ions to maintain electroneutrality, but this shift is in the opposite direction; i.e., chloride ions diffuse from the plasma into the erythrocytes in exchange for the carbonate ions that diffused from the erythrocytes into the plasma. Thus, the basic ion ratio of the erythrocytes continues, for the most part, to be static.

Carbon dioxide forms continuously in the tissues as a result of oxidation and moves from the tissues, first into the plasma and then into the erythrocytes. Again the presence of carbonic anhydrase makes possible the formation of some carbonic acid,* and the procedures described above reoccur.

The increases in number of carbon dioxide molecules within the erythrocytes cause an increase in the number of osmotically active particles within the erythrocytes; hence, some water enters the erythrocytes.

In the erythrocytes, hemoglobin has a high affinity for oxygen; about nineteen or twenty parts of oxygen are absorbed for each 100 parts of blood. Oxygen is transported by hemoglobin as oxyhemoglobin. The plasma has a very slight absorbing power for oxygen when compared with hemoglobin, and the quantity of oxygen that the plasma can transport is far less than the requirements of the body. However, oxygen enters the plasma quite readily from the lung alveoli, and oxygen is then absorbed from the plasma by the erythrocytes according to the oxygen pressure. Normally, the erythrocytes absorb about 96 percent of the oxygen possible from the plasma. It is in this form that oxygen is transported from the lungs to the body tissues and cells.

Pulmonary ventilation

At least five different air volumes are considered to make up the air inhaled (pulmonary ventilation) into the lungs:
1. The *vital capacity,* the maximum volume of air that can be exchanged in a single respiration cycle
2. The *residual air,* that volume of air retained in the lungs at the termination of forced expiration
3. The *supplemental air,* that expiratory reserve volume of air still in the lungs that can be expelled, remaining following normal quiet breathing
4. The *tidal air,* that volume of air inhaled in a normal quiet inspiration
5. The *complemental air,* that additional volume of air possible to inhale following a quiet inspiration

The composition of air in pulmonary ventilation varies according to several conditions, each of which affects (within certain limits) the gaseous exchange.

The degree of exchange of alveolar air with blood depends upon the tidal air volume and the respiratory rate.† The degree of exchange also depends upon the volume of air in the dead space, the capacity of the trachea, bronchi, and bronchioles, and the volume of air remaining in the lungs at the end of respiration.

The dead space volume is approximately 140 ml. The normally inhaled volume is 500 ml. Subtracting 140 from 500 leaves 360 ml., the volume of air nor-

---

*A state of equilibrium is more favorable to more carbon dioxide than carbonic acid.
†The respiratory rate is the number of complete respirations each minute; normally this is twenty.

mally mixed with the 3,000 ml. in the lung alveoli, or approximately 12 percent of the air volume in the lungs is renewed in a quiet respiratory cycle. The warming effect on the air in the dead space and the fact that approximately 12 percent of the total air volume is renewed in a cycle prevents inspiration of excessive quantities of air that is cold enough to cause shock or other pathologic conditions.

Types of respiration

Some of the following types of respiration are normal and some are pathologic.

Eupnea, normal, quiet respiration
Hyperpnea, an increase in depth of inspiration
Dyspnea, difficult or labored breathing
Polypnea, a rapid (panting) rate of respiration
Tachypnea, is an exceedingly high or fast rate of respiration
Apnea, is the transient cessation of the breathing impulse that follows forced breathing
Cheyne-Stokes respiration, characterized by rhythmic variations in intensity

Respiratory center

Some of the preceding types of respiration result from indirect stimuli in the respiratory center, while others have more direct relation to this important part of the cerebellum. This is discussed in Chapter 9; however, a brief explanation follows.

Situated within the walls of the carotid sinuses are certain specialized nerve receptors (chemoreceptors), which are highly sensitive to an excessive quantity of acids.* The receptors send stimuli to the respiratory center to increase respiration, thus reducing the presence of excesses in carbon dioxide and other acids and/or acid formers. The principal control of respiration resides within the respiratory center in the medulla oblongata and responds to small changes in the carbon dioxide volume in the blood—thus helping the body maintain an almost constant level of carbon dioxide in the blood at all times (pp. 262, 263).

RADIOGRAPHY

The general purpose of radiography of the thoracic cage and its contents is to make the lung detail contrast against other structures or to make pathologic conditions contrast against normal structures of this region. This is accomplished through the use either of air as the contrast medium or of some opaque compound as the contrast medium.

Air contrast

Normal chest radiographic procedures use air as the contrast medium. Exposure factors are selected to penetrate the total thickness, yet avoid overex-

---

*Both reduced hemoglobin (HHb) and oxyhemoglobin ($HHbO_2$) are acids.

posing (burning out) the lung detail contrasted against air. The general technique employs a minimum of 6 feet as the focal-film distance so that the thoracic viscera are visualized in the film emulsion in their approximate normal sizes. This procedure is called *teleroentgenography* (see Figs. 6-18, C, and 6-20, C).

When making exposures of the chests of patients who have undergone pneumothorax,* compensation must be made for the increase in density of the side of the pneumothorax. This can be achieved by placing additional aluminum filters over the normal side of the chest or over the corresponding side of the cassette during exposure, or in the tube port.

Opaque media contrast

Various opaque media such as propyliodine (Dionosil), iodized oil (Lipiodol), and chloriodized oil (Iodochlorol) are employed in the procedure called *bronchography*. Bronchography is the radiographic examination and study of the bronchial tree after injection of suitable radiopaque material. The medium employed is instilled into the trachea either by dripping or spraying (see Fig. 8-8).

Some of the common pathologic conditions of the lungs lending to radio-

*Pneumothorax is an accumulation of air or gas in the pleural cavity, outside the lung.

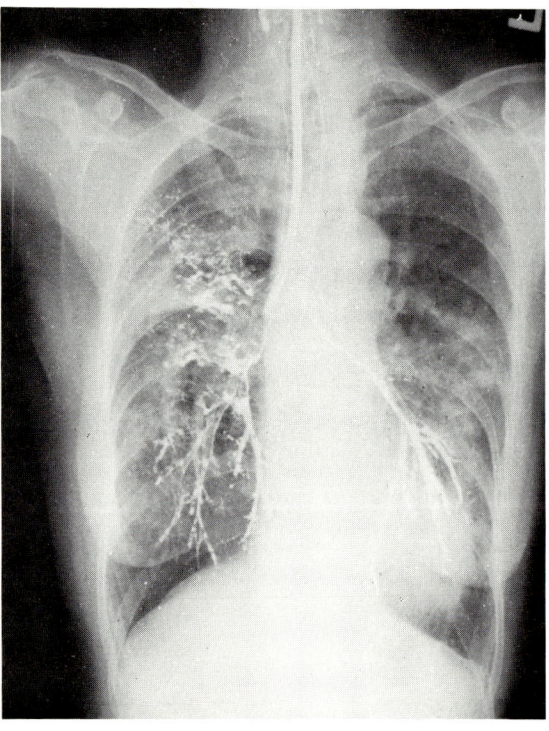

Fig. 8-8. Bronchogram.

graphic visualization include atelectasis and bronchiectasis. Atelectasis is the imperfect expansion or collapse of the air vesicles of the lungs. Bronchiectasis is a dilation of the walls of the bronchi. A pathologic condition called *emphysema* is becoming more common. Emphysema is a swelling or inflation due to the presence of air, applied especially to a morbid condition of the lungs.

Other pathologic conditions (aside from tuberculosis and cancer) that are diagnosable by means of radiography include pneumoconiosis and silicosis. *Pneumoconiosis* is a chronic fibrous reaction in the lungs to the inhalation of dust. *Silicosis* is pneumoniconiosis resulting from the inhalation of the dust of stone, sand, or flint containing silicon dioxide.

REFERENCES

Anthony, C. P., and Kolthoff, N. V.: Textbook of anatomy and physiology, ed. 9, St. Louis, 1975, The C. V. Mosby Co.

Best, C. H., and Taylor, N. B.: The human body, ed. 4, New York, 1963, Holt, Rinehart and Winston.

Goss, C. M.: Gray's anatomy, ed. 27, Philadelphia, 1959, Lea & Febiger.

Jacobi, C. A., and Paris, D. Q: Textbook of radiologic technology, ed. 5, St. Louis, 1972, The C. V. Mosby Co.

Orten, J. M., and Neuhaus, O. W.: Biochemistry, ed. 9, St. Louis, 1975, The C. V. Mosby Co.

Schottelius, B. A., and Schottelius, D. D.: Textbook of physiology, ed. 17, St. Louis, 1973, The C. V. Mosby Co.

CHAPTER 9

The nervous system

Nerve tissues (cells, fibers, etc.) are generally radiolucent; radiography of nerve tissues requires the use of various contrast media. Techniques for optimum demonstration of the nerve tissues to permit radiologic interpretation increase in number and diversity as research develops new contrast media. As a result, neuroradiography becomes increasingly important in the scope of radiologic technologists' responsibilities. Knowledge of the general anatomy and appreciation of the physiology of the nervous system prepares allied health technologists for efficient assistance to radiologists, neurosurgeons, and orthopedic surgeons.

The nervous system includes the following structures:
1. Central nervous system (C.N.S.)—brain and spinal cord
2. Peripheral nervous system (P.N.S.)—the twelve pairs of cranial nerves, the thirty-one pairs of spinal nerves, and the autonomic nervous system

FUNCTION

The central and peripheral parts of the nervous system function to transmit messages throughout all the body cells, tissues, glands, and organs to aid in maintaining homeostasis.

Although the total function of the nervous system is principally that of maintaining homeostasis, the individual activities of the various parts of this system in corresponding viscera and other parts of the body are so specialized that general description of function is omitted. Specific function is treated in the section titled Physiology.

GENERAL ANATOMY
Central nervous system

In the embryo the central nervous system arises from the ectoderm, a layer that exhibits a relatively high rate of metabolism in its midline (see p. 52). As a result, the cells of the midline of the ectoderm differentiate quite rapidly in comparison with other cells of this layer. The increased quantity of cell differentiation causes a definitely thickened plate of cells to extend along the midline from head to tail in the new embryo. The growth rate of the outer cells of this plate is greater than that of the inner cells, resulting in an indented groove, the *neural groove*, in the center. As the cells increase in number the groove continues to deepen and forms the *neural fold*, a tubular structure. Growth and maturation continue throughout adolescence; the end result is that the central nervous system continues as a tube for the individual's lifetime. With the exception of some supporting cells and the vascular supply, the entire nervous system develops from the neural tube and the associated crest cells.

Brain

The brain is that part of the neural tube that develops in the cranial case. During the very early stages of development, three enlargements of the neural tube present themselves as the *forebrain, midbrain,* and *hindbrain*. The adult brain develops from these original subdivisions and consists of the cerebrum, midbrain, cerebellum, and brain stem. Situated in the cerebrum and brain stem are four cavities (ventricles) containing the cerebrospinal fluid.

The four ventricles of the brain result from the original cavity of the neural tube. The cerebrum and diencephalon develop from the forebrain. The midbrain of the adult develops from the midbrain of the neural tube. The pons, medulla oblongata, and cerebellum all develop from the hindbrain.

Collectively, the midbrain, pons, medulla oblongata, and often the diencephalon are called the *brain stem*. As a group these structures are arranged as a stem (or stalk) between the spinal cord posteriorly and the cerebral hemispheres anteriorly.

Spinal cord

The spinal cord is the more common name for the medulla spinalis. Through the foramen magnum, the brain is continuous with the spinal cord, which extends from the base of the skull to the approximate level of the first and second lumbar vertebrae. At this point the spinal cord terminates per se and continues as the filum terminale interna to the terminus of the dura mater approximately on a level even with the first sacral vertebra. Beyond the dura mater the spinal cord continues to the level of the first coccygeal segment as the filum terminale externa.

Nerve cells

Supporting evidence of the high degree of nerve cell specialization and knowledge of nerve cell characteristics has accumulated from exhaustive studies. In addition to the phenomena of irritability and conductivity, nerve cells exhibit those characteristics of structure common to most cells. Nerve cells possess long protoplasmic processes, which permit connections of a functional nature between the cells.

328 Textbook of anatomy and physiology in radiologic technology

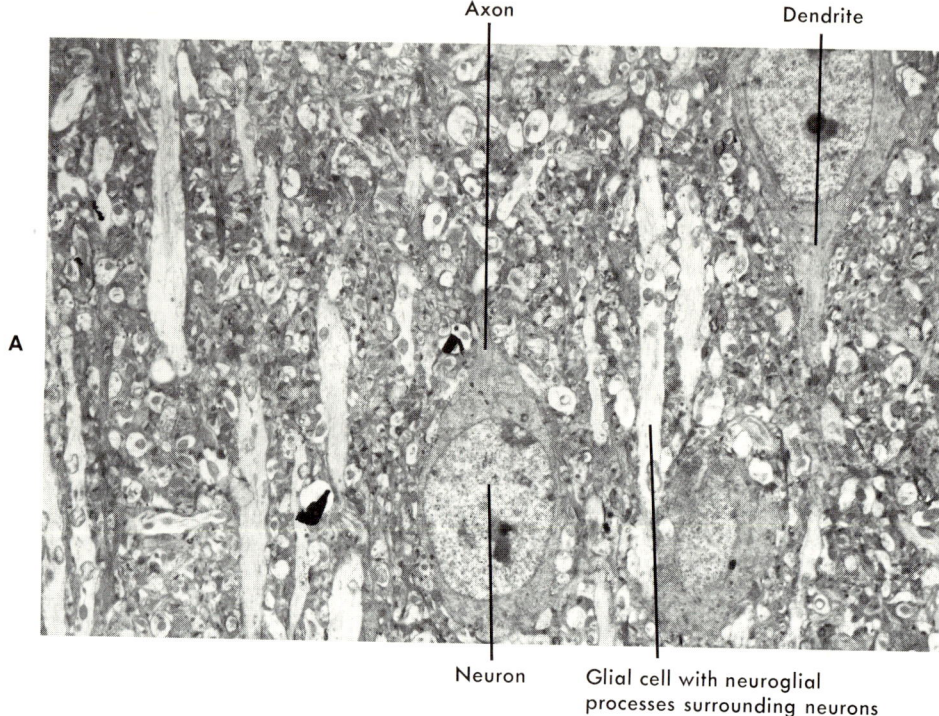

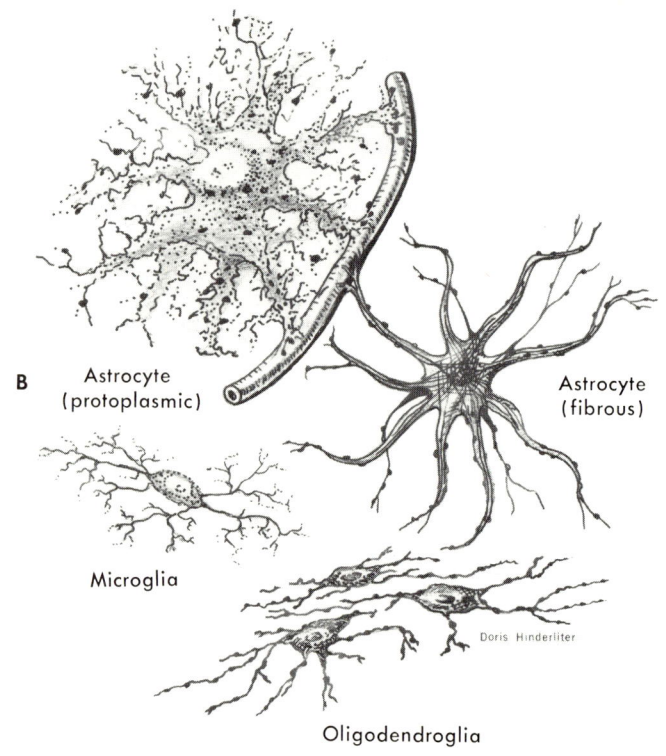

Fig. 9-1. For legend see opposite page.

The cells and associated functional connections throughout the processes comprise units called neurons. *A neuron is a nerve cell with all its processes.* Throughout the central nervous system the neurons are supported by the *neuroglia,* nonconducting cells. Neuroglia is the supporting structure of nervous tissue and consists of a fine web of tissue made up of modified ectodermic elements, in which are enclosed peculiar branched cells—the neuroglia cells. Many of the neuroglia attach to blood vessels by means of specialized extensions and, as such, may act as the *blood-brain barrier* (see p. 344). Additional supportive tissue cells called *microglia* function as phagocytes to remove the particles of normal cellular activity and, especially, to act upon nervous system infectious agents (see Fig. 9-1).

Nerve cell processes are of two kinds: *dendrites* (dendrons) and *axons.* The dendrites conduct messages (impulses) toward the cell body. The axons conduct messages away from the cell body. Experimental evidence indicates that electrical polarity at the points of functional connection between specific neurons determines the direction of impulse conduction. Along a given nerve the impulse always travels *from dendrite to cell body to axon (DCBA) to the next cell dendrite,* etc.

Neurons may be classed according to function and include three types—the *afferent, efferent,* and *internuncial neurons.* The afferent neurons conduct messages of a sensory nature toward the central nervous system. Efferent neurons conduct messages of a motor or secretory nature away from the central nervous system. Efferent motor impulses travel to muscles and stimulate muscle activity. Efferent secretory impulses travel to glands and stimulate glandular secretions. Within the central nervous system are the specialized cells called the internuncial neurons, which conduct impulses from the afferent neurons to the efferent neurons.

Neurons may be classed according to structure and include three types—the *multipolar, bipolar,* and *unipolar* neurons. Multipolar neurons possess multiple dendrites and a single axon. Bipolar neurons possess no dendrites and two axons. Perhaps only in the embryo are the true unipolar neurons, which have a single axon and no dendrite. Many pseudounipolar neurons are found to comprise the sensory neurons; these were bipolar in early stages of development, after which

---

Fig. 9-1. A, Electron micrograph of a neuron of mouse cerebral cortex tissue ($\times$ 2,000). **B,** Non-conductive nerve cells. Protoplasmic astrocyte: adult (star-shaped) cell of the neuroglia. Note the large nucleus, abundant granular cytoplasm, and the numerous, thick plasmatic expansions. Fibrous astrocyte: adult cell of the neuroglia. Note the long, relatively smooth expansions with little branching. The fibrillar structures are embedded in the cytoplasm and the expansions. Microglia: small interstitial cells that are migratory. Oligodendroglia: vinelike prolongations from an incomplete investment for myelin sheaths in the white matter along with the microglia form the perineuronal satellites in the gray matter. (**A** from Rhodin: An atlas of ultrastructure, Philadelphia, 1963, W. B. Saunders Co.)

the two processes fused for a short distance permitting a single process to extend from the cell body. This process branches after a very short length into a distal branch (dendrite) and a peripheral branch (axon).

The diameters of nerve cell bodies vary in different cells from 4 to 5 microns to as much as 50 and, sometimes, 100 microns. The processes of the nerve cells in different neurons may vary from a few microns to several feet in length. Variations in diameter and length are seen in the different parts of the nervous system.

In addition to the central nervous system there are two other systems of nerves: the *peripheral nervous system* and the *autonomic nervous system*.

Peripheral nervous system

Peripheral nerves are those nerves that extend from the body toward the central nervous system and from the central nervous system toward the remainder of the body. This system includes the cranial and spinal nerves, the associated peripheral combinations, and the peripheral portions of the autonomic nervous system.

The twelve pairs of cranial nerves arise from the inferior surface of the brain. Radiographic requests sometimes specify demonstrations of a particular foramen through which a certain cranial nerve passes (see p. 144 and Table 4-8). The cranial nerves and their functions are presented in Table 9-1.

The thirty-one pairs of spinal nerves arise from the spinal cord and pass from the spinal cord through the intervertebral foramina. The spinal nerve origins are described later in the chapter.

Autonomic nervous system

The nerve supply from the central nervous system to a skeletal muscle is a direct supply; i.e., the nerve fiber connects the spinal cord, without interruption, directly to a particular muscle. The nerve supply to the visceral muscles differs in its pathway from that of the skeletal muscles.

The nerve supply to the visceral muscles from the central nervous system is via an indirect pathway: i.e., a relay of two neurons. The preganglionic fiber (of one neuron) extends from its cell body within the spinal cord to terminate in a ganglion*; this fiber passes from the spinal cord in the ventral root. In the ganglion the preganglionic fiber makes synaptic connection with a second neuron whose *postganglionic* fiber extends to a visceral muscle, such as the walls of a blood vessel, or to the heart; the postganglionic fiber may terminate in a gland to stimulate secretory processes. These several indirect nerve relays and their associated ganglia comprise the autonomic nervous system.

The autonomic nervous system serves to innervate nonstriated and cardiac muscle tissues and the glands. This system includes two parts: the *craniosacral*

---

*A ganglion is a knot or a knotlike mass; the word is used in anatomic nomenclature as a general term to designate a group of nerve cell bodies located outside of the central nervous system.

outflow of ganglia, nerve plexuses, and trunks and the *thoracolumbar outflow*. The former is frequently called the *parasympathetic* nervous system; the latter is called (correspondingly) the *sympathetic* nervous system.

The craniosacral outflow connects the following structures: (1) midbrain with the eye, (2) medulla oblongata with the salivary glands, vasomotor and cranial mucous membrane, heart, bronchi, stomach, liver, pancreas, small intestine, kidneys, large intestine, urinary bladder, and genital organs.

The thoracolumbar outflow is far more complex and connects many more structures than the craniosacral outflow. The thoracolumbar outflow connects the following structures:

1. Through the first cervical ganglion are connections to the eyes, salivary glands, vasomotor and cranial mucous membranes, and the heart.
2. Through the second and third cervical ganglia are effected the connections to the heart.
3. Through the last four cervical ganglia are effected the connections to the bronchi.
4. Through the first eight thoracic ganglia are effected the connections to the pancreas and kidneys.
5. Through the superior mesenteric ganglia the sixth, seventh, and ninth thoracic ganglia also effect connections to the same viscera as in 4 above, and to the colon and rectum.
6. Through the inferior mesenteric ganglion the eighth, ninth, tenth, and eleventh thoracic ganglia effect connections to the colon, rectum, urinary bladder, and genital organs.
7. Through the last five cervical ganglia, the twelve thoracic ganglia, and the first four lumbar ganglia are effected connections to the cutaneous vessels and sweat glands.

Protective coverings

The central nervous system is well protected by the hard, bony cranial case (for the brain) and the vertebral column (for the spinal cord). The *meninges*, connective tissue structures that contain the cerebrospinal fluid and surround the central nervous system, are located within the osseous covering. The individual linings are continuous through the foramen magnum and surround the brain and the spinal cord.

Osseous protection

The osseous protection provided for the central nervous system is presented in two parts: protection for the brain and protection for the spinal cord.

Skull. The part of the skull that is involved in providing protection for the brain is called the cranial case. It consists of the eight cranial section bones: the frontal, both parietal, both temporal, the occipital, the ethmoid, and the sphenoid bones (see Fig. 9-2).

The cranial capacity may vary in different persons and with the various skull

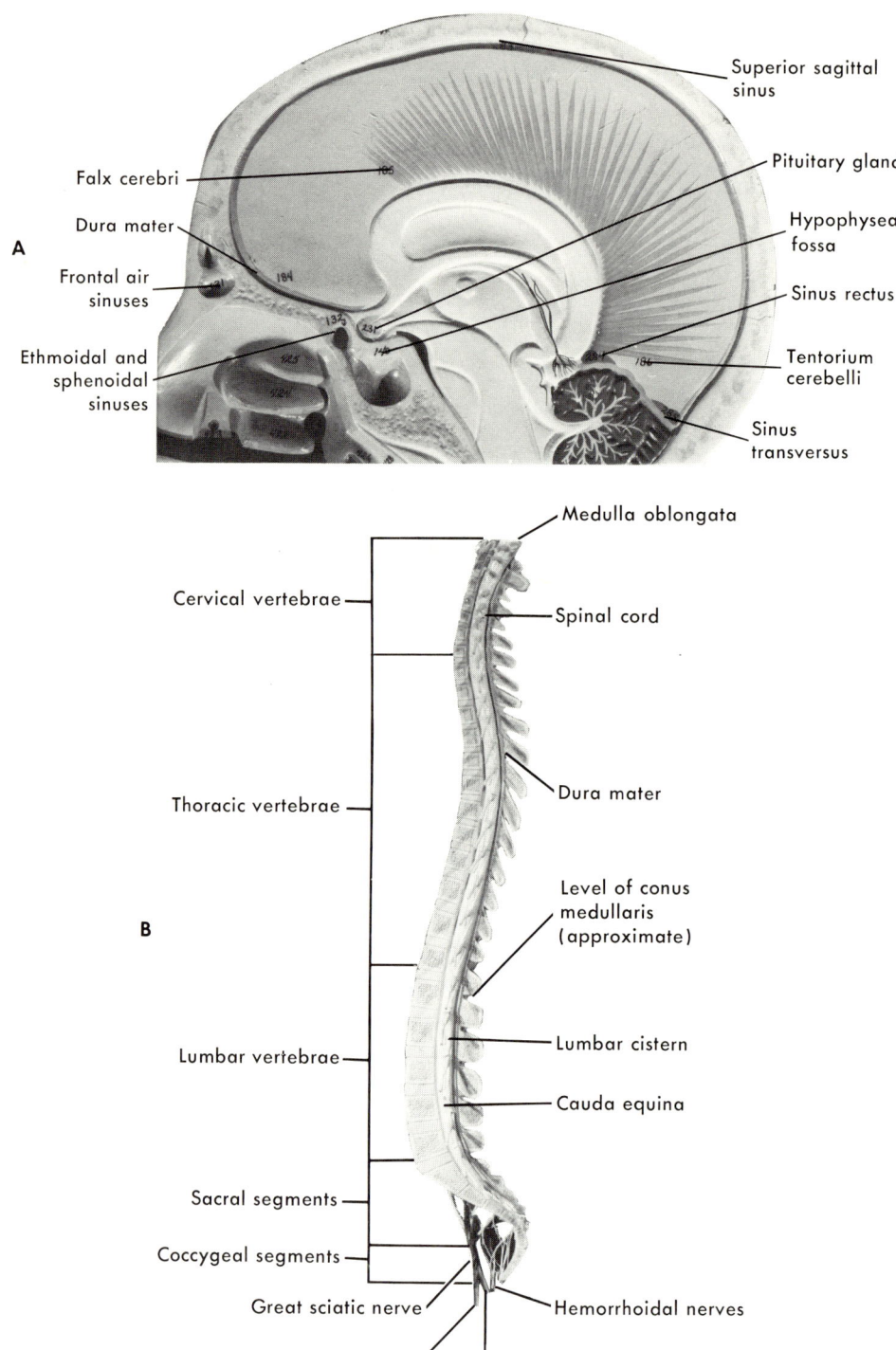

Fig. 9-2. A, Brain and skull, midsagittal plane showing meninges and blood sinuses. B, Vertebral column, midsagittal plane showing meninges and some nerves.

types. In some individuals the cranial capacity may be as little as 1,200 ml.; in other persons the cranial capacity may be as great as 1,500 ml. In either instance the cranial capacity is much greater in human beings than in other animals, permitting the greatly enlarged cerebral hemispheres that are common to man and that enable him to reason.

Six of the cranial bones enter into the formation of the cranial floor: the sphenoid, ethmoid, occipital, and frontal bones and both temporal bones. Numerous foramina and other openings in the cranial floor permit passage of the twelve pairs of cranial nerves and the arterial, venous, and lymphatic supplies to and from the brain. These foramina are visualized and identified in Figs. 4-57, *A*, and 4-58 and in Table 4-6. The largest of the foramina is the foramen magnum, through which the brain and spinal cord are continuous.

Vertebral column. The vertebral column provides a partial bony covering for the spinal cord. The vertebral foramen in any vertebra is formed by the centrum anteriorly, the pedicles bilaterally, and the laminae both bilaterally and posteriorly; the laminae fuse posteriorly in the midline to close the neural arch.

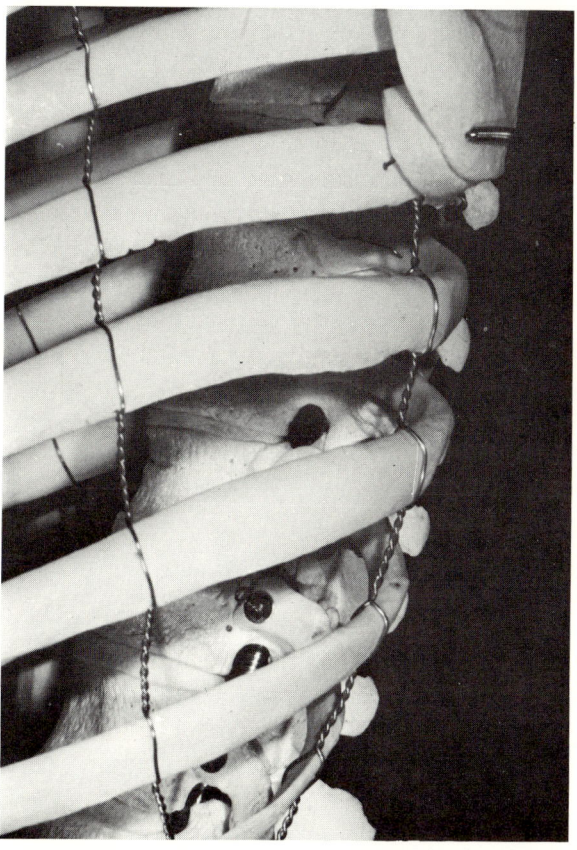

Fig. 9-3. Thoracic vertebrae, articulated skeleton showing intervertebral foramina.

When the several vertebrae are articulated, the corresponding vertebral foramina form the spinal canal to contain the spinal cord (see Fig. 9-4).

Directly posterior to the centrum and contained in a part of the pedicle on each side are two notches; the upper is the *superior vertebral notch,* and the lower is the *inferior vertebral notch.* When two vertebrae are articulated, the inferior vertebral notch of the body above and the superior vertebral notch of the body below form the *intervertebral foramina* (one on each side). The spinal nerves pass from the spinal cord through these foramina (see Fig. 9-3).

Meninges. Adjacent to the inner surface of the bones that form the cranial case is a tough, fibrous membrane, the *dura mater.* The membrane is continuous with the periosteum on the outer surface of these cranial bones through the sagittal suture. A large fold of the dura mater extends inward between the cerebral hemispheres to form the *falx cerebri,* the sickle-shaped fold of dura mater that extends downward in the longitudinal cerebral fissure and separates the two cerebral hemispheres. In the process of making the aforementioned fold, the dura mater forms two sinuses. One sinus is situated at the most distal (ventral) part of the fold and is contained within it; this is the *inferior sagittal sinus.* The other sinus is situated directly inferior to the vault of the skull and is contained within the dura mater; this is the *superior sagittal sinus* (see Fig. 9-2, A).

The meninges consist of three membranes with potential space between the outer two and a distinct space between the inner two. From outside to inside, the meninges are named the *dura mater, arachnoid membrane,* and *pia mater.* The dura mater is very tough and is difficult to penetrate even with a sharp needle. Although the cranial dura mater is next to the bone, the spinal dura mater is separated from the wall of the spinal canal by the *epidural space;* this space contains small veins and some loose areolar tissue (see Fig. 9-2, B).

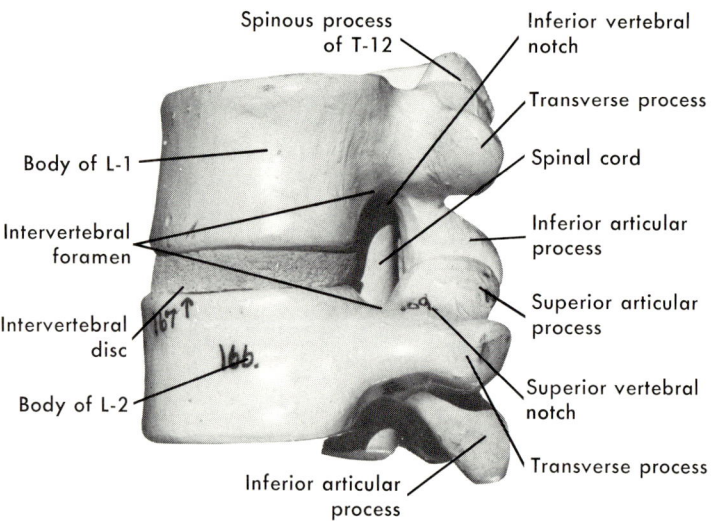

Fig. 9-4. First and second lumbar vertebrae, anterolateral (left) aspect.

The middle meningeal membrane is the arachnoid membrane, which is quite cellular and is less tough than the dura mater. The small potential *subdural space* contains minute quantities of cerebrospinal fluid and lies between the arachnoid membrane and the dura mater.

The innermost meningeal membrane is the pia mater. This membrane is quite loosely arranged in a network to permit passage of blood vessels to and from the central nervous system. The larger space between the arachnoid membrane and the pia mater is the *subarachnoid space,* which contains the cerebrospinal fluid.

In addition to the sinuses discussed in the preceding paragraphs, there are numerous other sinuses in the brain coverings. For the most part these sinuses have names that reflect their locations, shapes, directions, etc. These sinuses contain blood and usually act as collecting reservoirs for the blood flowing from the brain to the heart. One of these intramembranous sinuses is the large *sigmoid sinus,* which is immediately superior to the tympanic part of the temporal bone and gives rise to the *internal jugular vein.*

When the meninges of the brain are removed, the typical configurations of the brain tissue become evident. These configurations are the *gyri,* which are tortuous convolutions of brain tissue, and the *sulci,* which are intervals (depressions) between the gyri as seen in Fig. 9-5.

SPECIFIC ANATOMY

The preceding section of this chapter was almost entirely concerned with embryonic development and gross anatomy of the nervous system. This section deals with the more intricate structure and associated functions of the nervous system.

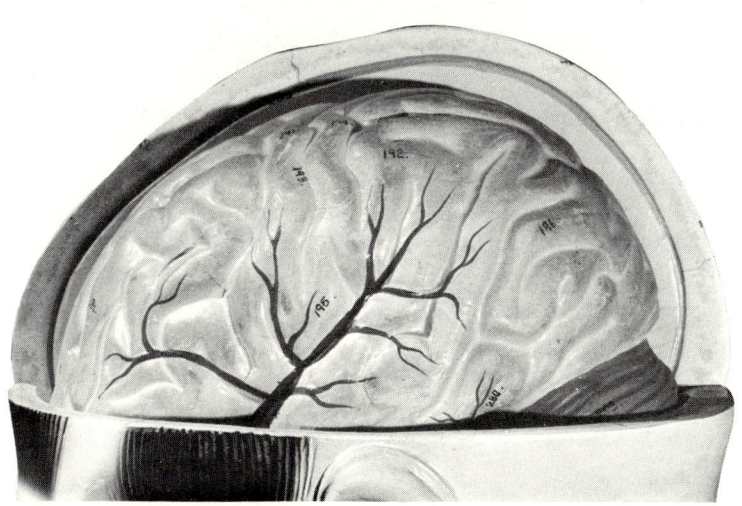

Fig. 9-5. Brain, left lateral surface showing gyri and sulci.

336 *Textbook of anatomy and physiology in radiologic technology*

Brain

The brain and spinal cord begin in the embryo as a tube. The tube persists and develops three enlargements. The first of these enlargements is the *forebrain,* which eventually forms the *cerebrum* and the *interbrain* (diencephalon). The cerebrum consists of the two cerebral hemispheres.

The second of these enlargements is the *midbrain,* which persists in the adult as the midbrain.

The third enlargement develops into the hindbrain and forms the *pons, medulla oblongata,* and *cerebellum.*

The *brain stem* contains the midbrain, pons, and medulla oblongata, and it is often said to include the diencephalon.

The brain tissue consists of gray matter and white matter. The brain cortex is composed largely of the bodies of nerve cells and, therefore, is gray in color.

The brain medulla is composed largely of processes and/or fibers of nerve cells and, therefore, is white (see Fig. 9-6).

Cerebrum

The two hemispheres of the cerebrum connect in many places by means of several bands of nerve tissue, the commissures. The outstanding commissure is the corpus callosum. The corpus callosum (great commissure of the brain) is an arched mass of white matter situated at the bottom of the longitudinal fissure.

Four lobes comprise each cerebral hemisphere: the *frontal, parietal, temporal,* and *occipital lobes.* Although there are certain, specific functions performed by

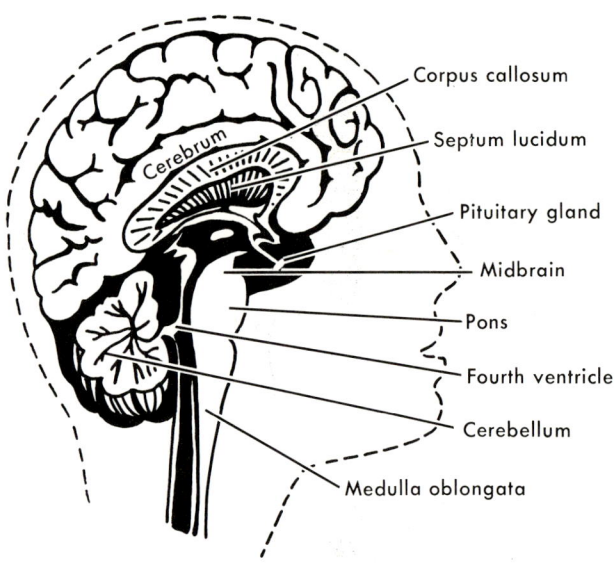

Fig. 9-6. Median section (aspect) of brain.

each lobe, for the most part the functions are rather well disseminated throughout more than one lobe. The occipital lobe is said to be concerned with vision; the frontal lobe is possibly concerned with voluntary actions and thinking; the parietal lobe is concerned with the senses of touch, heat, and cold; and the temporal lobe is concerned with hearing. The sense of smell is thought to be realized in the hippocampus below the temporal lobes.

Five pairs of cranial nerves arise from the two cerebral hemispheres. These are the olfactory (first), optic (second), oculomotor (third), trochlear (fourth), and abducens (sixth) cranial nerves.

The optic nerves pass from the brain through the optic foramina, which are demonstrated radiographically in the anterior view of the facial bones (Rhese view). Each optic foramen is formed by the small and great wings of the sphenoid bone on that particular side. The optic nerves cross to the opposite side of the midsagittal plane beneath the olivary eminence of the sphenoid bone in the chiasmatic groove. The crossing over is called the *optic chiasma.*

Diencephalon

The diencephalon lies between the cerebrum and the midbrain. The diencephalon consists of two parts: thalamus and hypothalamus. The thalamus is the larger of the two parts and is concerned with sensory functions. The hypothalamus is below the thalamus and is concerned with visceral functions.

Midbrain

The two cerebral peduncles are fibrous bands that attach the cerebellum to the brain stem. The cerebral peduncles form the midbrain. A peduncle is a supporting part (stem).

On the posterior surface of the midbrain are two *superior colliculi,* which are concerned with visual function. A colliculus is a small elevation. The two superior and two inferior colliculi comprise the corpora quadrigemina. The inferior colliculi also are situated on the posterior surface of the midbrain and are concerned with auditory function.

Situated at the posterior limit of the diencephalon is the pineal body, a small oval-shaped tissue of unknown function. Calcification of the pineal body often occurs with advancing age, making the pineal body radiopaque.

The paired oculomotor nerves attach to the front of the midbrain. The paired trochlear nerves arise in the midbrain and leave from its posterior surface.

Pons

The midbrain is continuous inferiorly with the pons, a large round structure that functions to connect the cerebellar hemispheres and contains numerous nerve tracts and specialized collections of nerve cells.

The trigeminal (fifth), facial (seventh), and acoustic (eighth) pairs of cranial nerves attach to their respective sides of the pons. The abducens nerves attach to the front of the pons.

Medulla oblongata

The pons is continuous inferiorly with the medulla oblongata, which is continuous inferiorly with the spinal cord at the level of the foramen magnum.

The anterior surface of the medulla oblongata contains the anterior *median fissure*. An elevation, the *pyramid*, lies on each side of the fissure. The glossopharyngeal (ninth), vagus (tenth), and spinal accessory (eleventh) pairs of cranial nerves attach to their respective sides of the medulla oblongata between the *olive* and the inferior cerebellar peduncle. The olive is a rounded elevation situated laterally to the upper part of each pyramid. The olive consists of an irregular mass of gray substance. The hypoglossal (twelfth) pair of cranial nerves emerges from the medulla oblongata between the pyramid and the olive from the corresponding anterior lateral sulcus.

The medulla oblongata contains specialized collections of nerve cells concerned with the vital functions of respiration, circulation, and the special senses.

The two vertebral arteries ascend the neck and pass through the foramen magnum anterior to the medulla oblongata. At this point these arteries unite to form the basilar artery. The basilar artery ascends anteroventral to the pons and midbrain to enter the arterial circle at the base of the brain.

Cerebellum

The cerebellum is a very important organ, concerned with the various aspects of muscular activity control. It consists of two hemispheres connected by the *vermis*. Paired peduncles connect the cerebellum to the brain stem. The space between the splenium of the corpus callosum (see p. 336) and the superior surface of the cerebellum contains the *great cerebral vein* and is called the *cisterna*

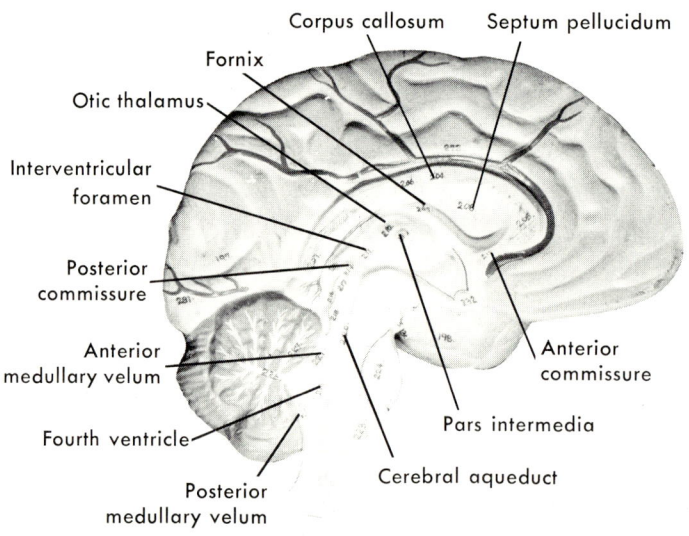

Fig. 9-7. Four ventricles of the brain and related structures.

magna. The cisterna magna extends between the layers of tissue, forming the choroid plexus of the third ventricle.

Ventricles

The four ventricles of the brain contain much of the cerebrospinal fluid. There are two *lateral ventricles,* one in each cerebral hemisphere. The *corpus callosum* unites the roofs of the two lateral ventricles, and the *septum pellucidum* forms the medial walls of these ventricles. Each lateral ventricle joins the *third ventricle* posteriorly and inferiorly through one of two interventricular foramina. Each foramen is situated anteriorly in the third ventricle.

The third ventricle is situated in a narrow space within the diencephalon. Each *otic thalamus* is situated on either side of the third ventricle. The *pars intermedia* is a mass of grey matter that joins the lateral walls of this ventricle. The third ventricle is continuous with the *cerebral aqueduct* (aqueduct of Sylvius), a narrow channel in the midbrain; the cerebral aqueduct is continuous inferiorly with the fourth ventricle. The fourth ventricle lies between the cerebellum posteriorly and the pons and medulla oblongata anteriorly.

The fourth ventricle continues inferiorly into a narrow channel, the *central canal,* which is in the lower part of the medulla oblongata and extends through the length of the spinal cord (see Figs. 9-6 and 9-7).

The fourth ventricle is formed by the pons and the medulla oblongata. A thin membrane forms the roof of the fourth ventricle and contains an opening for communication between the fourth ventricle and the subarachnoid space. Openings are found also on each side of the ventricle. Small, complex tufts of minute blood vessels, the *choroid plexuses,* are situated in each of the four ventricles. These plexuses are concerned with the formation of the cerebrospinal fluid. The *anterior medullary velum* and the *superior cerebellar peduncles* form the roof of the upper part of the fourth ventricle. The *posterior medullary velum* forms the roof of the lower part of the fourth ventricle (the velum is a thin sheet of ependymal epithelium and pia mater).

Spinal cord

The protective coverings of the spinal cord compare with the protective coverings of the brain and are discussed on pp. 331 and 334.

The spinal cord is continuous superiorly with the medulla oblongata and commences at the foramen magnum of the occipital bone. The cord of nerve fibers extends through the spinal canal of the vertebral bodies to the level of the inferior surface of the first lumbar vertebra and terminates per se in the *conus medullaris,* the conical terminus of the spinal cord. The *filum terminale interna* is a filamentous continuation of the spinal cord to the terminus of the dura mater at about the level of the first sacral vertebra. Extending beyond the dura mater and continuous with the filum terminale interna is the *filum terminale externa,* which usually terminates in the first coccygeal segment (see Fig. 9-8).

The large space surrounding the filum terminale interna and below the conus

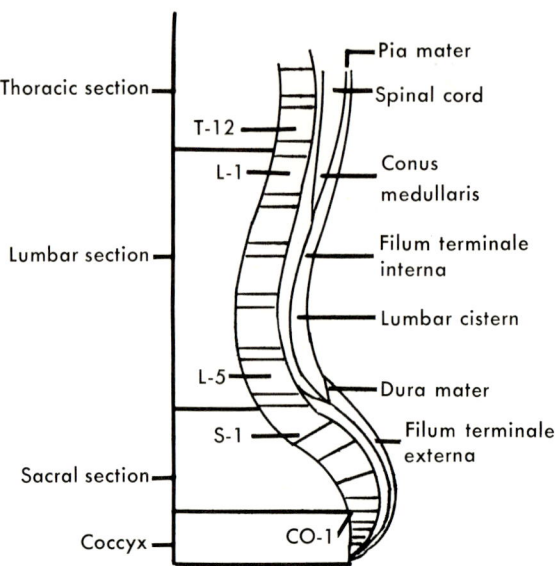

Fig. 9-8. Lower vertebral column, median section showing the relation of the medulla spinalis and the meninges to the vertebral bodies (schematic). The spinous processes are not diagrammed to avoid confusion.

medullaris is called the *lumbar cistern;* it contains much of the cerebrospinal fluid and is the usual site of spinal puncture. The spinal needle is passed between the spinous processes of two of the last three lumbar bodies and into the lumbar cistern. Spinal punctures are performed for many purposes, among which are the following:
1. To obtain cerebrospinal fluid for examination
2. To remove a specific quantity of cerebrospinal fluid and to replace it with an equal quantity of contrast medium for myelography

The spinal cord, throughout its length, gives rise to thirty-one pairs of spinal nerves that pass outward through the corresponding intervertebral foramina. The nerves arise from the outer parts of the spinal cord.

As opposed to the structure of the brain, the outer parts of the spinal cord are white, and the inner parts of the spinal cord are composed of gray matter. A cross section of the spinal cord reveals an H-shaped part of gray matter in the center of the structure. The gray matter is arranged in two anterior horns and two posterior horns connected by a bar of gray matter in the center. The thoracic part of the spinal cord contains a lateral horn. The central canal is situated in the connecting bar of gray matter; the canal (or parts thereof) may be obliterated in adults.

The white matter of the spinal cord includes many nerve tracts that connect dorsal and ventral root fibers with parts of the central nervous system. The cord gray matter consists of cells that are concerned with such functions as sensory impulses, muscle activity, and visceral and blood vessel activity.

Table 9-1. Cranial nerves

Nerve	Type	Function
Olfactory (first)	S	Smell
Optic (second)	S	Sight
Oculomotor (third)	M, S, P	Moves eyeball, proprioception from extrinsic eye muscles and contraction of nonstriated muscle in iris and ciliary body
Trochlear (fourth)	M, S	Moves eyeball, proprioception from extrinsic eye muscle
Trigeminal (fifth)	M, S	Mastication, sensations from face and head
Abducens (sixth)	M, S	Moves eyeball, proprioception from extrinsic eye muscle
Facial (seventh)	M, S, P	Facial expressions, salivation, lacrimation, and taste
Auditory (eighth)	S	Hearing and equilibrium
Glossopharyngeal (ninth)	M, S, P	Moves pharynx, salivation, taste, visceral reflexes, and sensations from small part of external ear
Vagus (tenth)	M, S, P	Moves pharynx and larynx; parasympathetic to thoracic and abdominal viscera, taste, and visceral reflexes
Spinal accessory (eleventh)	M, P	Moves larynx and shoulder; parasympathetic to all viscera
Hypoglossal (twelfth)	M	Tongue movements

Peripheral nerves

The peripheral nerves comprise the peripheral nervous system and include all the cranial and spinal nerves.

Cranial nerves

There are twelve pairs of cranial nerves. In Table 9-1 each nerve is listed, with an indication of the kind of nerve and its principal function or functions. In the table, M = motor, S = sensory, and P = parasympathetic.

The optic nerve passes through the optic foramen. The oculomotor nerve, trochlear nerve, ophthalmic division of the trigeminal nerve, and abducent nerve all pass through the superior orbital fissure. The facial and auditory nerves pass through the internal auditory meatus. The glossopharyngeal and vagus nerves pass through the jugular foramen. The hypoglossal nerve passes through the hypoglossal canal. The spinal accessory nerve passes through the foramen magnum (see Fig. 9-9).

Spinal nerves

The thirty-one pairs of spinal nerves consist of eight pairs of cervical nerves, twelve pairs of thoracic nerves, five pairs of lumbar nerves, five pairs of sacral nerves, and one pair of coccygeal nerves.

Each member of the first pair of cervical nerves emerges from the spinal canal

342 *Textbook of anatomy and physiology in radiologic technology*

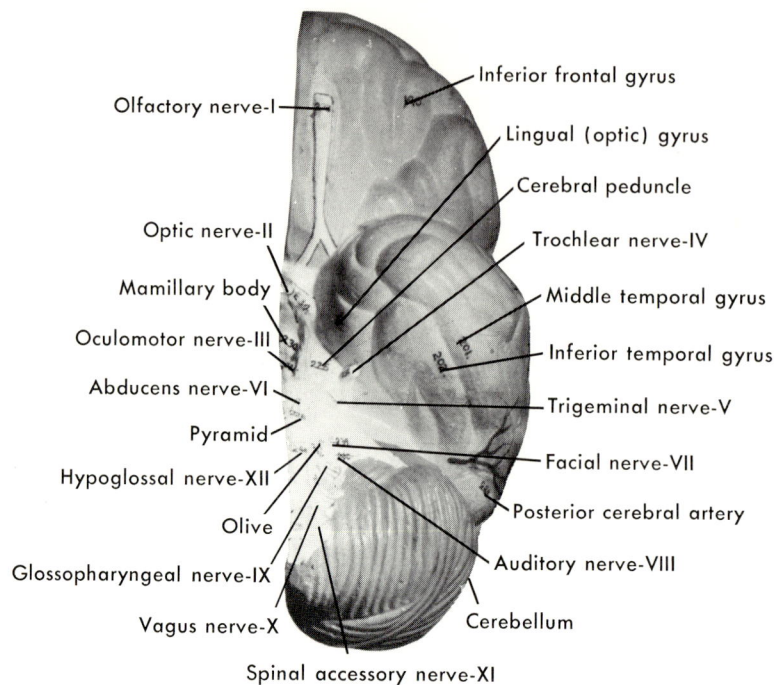

Fig. 9-9. Brain base, left half showing cranial nerves.

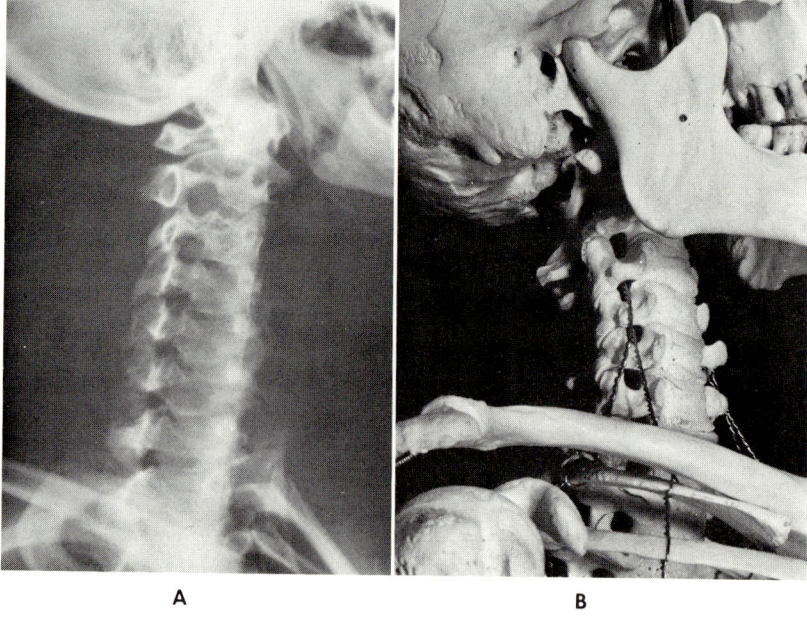

Fig. 9-10. A, Cervical spine, R-A-O view. **B,** Cervical vertebrae (articulated skeleton), right antero-oblique aspect showing intervertebral foramina.

between the occipital bone and the first cervical vertebra; these nerves are called the *suboccipital nerves*. The members of the second pair of cervical nerves emerge from the spinal canal between the first and second cervical vertebrae. The members of the eighth pair of cervical nerves emerge from the spinal canal between the seventh cervical vertebra and the first thoracic vertebra (see Fig. 9-10).

Throughout the length of the spinal cord two nerve roots attach each nerve to it; one root is anterior (ventral), and one is posterior (dorsal). A spinal ganglion is present in each posterior root.

PHYSIOLOGY
Cerebrospinal fluid

The cerebrospinal fluid circulates around and in the brain and spinal cord. The fluid derives from the blood and eventually flows back into the blood following an intricate process of filtration.

Formation

The four ventricles of the brain are described on p. 339. The ventricles are lined by a very special membrane, the *ependymal epithelium,* which is of extremely thin structure. Around the choroid plexuses, where the ependymal epithelium separates the blood vessels of the plexuses from the ventricular cavities, this membrane is called the *choroidal epithelium.*

The blood vessels of the choroidal epithelium of the lateral and third ventricles arise from the arterial circle, whereas the blood vessels of this epithelium in the fourth ventricle arise from branches of the vertebral arteries.

Arterial (capillary) blood flows through the several choroid plexuses as a result of normal arterial circulation. As the blood flows through the plexuses, the blood, in a thin film, is in contact with the choroidal epithelium. As a result of normal pressures forcing the blood against the membrane, an almost *protein-free* fluid (filtrate of blood) flows through the choroidal epithelium into each of the four ventricles.

The cerebrospinal fluid flows from the lateral ventricles via the interventricular foramina into the third ventricle. From the third ventricle the cerebrospinal fluid flows via the cerebral aqueduct into the fourth ventricle. From the fourth ventricle the fluid flows through the openings in the walls into the subarachnoid space.

Active circulation of the cerebrospinal fluid exists within the entire subarachnoid space around the brain and spinal cord. During the circulating process, the cerebrospinal fluid filters through small tufts (villi) of arachnoidal tissue projecting into the dural venous sinuses; in this manner the fluid is absorbed into the bloodstream. The cerebrospinal fluid is in all places between the dura mater and the brain and spinal cord.

The aforementioned circulation and production of cerebrospinal fluid must not be considered complete without introduction of other sources of production and other methods of reabsorption into the blood. Some evidence exists to substantiate the production of cerebrospinal fluid by certain nerve cells.

Quantities of fluid produced by nerve cells probably pass along the blood vessels by seeping through the loose tissues surrounding the blood vessels. Eventually this fluid reaches the subarachnoid space.

Correspondingly, certain quantities of cerebrospinal fluid* seep around the cranial and spinal nerves into the lymphatic tissues. Finally, these quantities of fluid are reabsorbed into the venous blood.

Circulating force and function

Numerous experiments can be performed to demonstrate the active pressure of the cerebrospinal fluid. Among the several factors effecting the active pressure is the pressure gradient that exists between the sites of formation and absorption of the fluid. The pressure at the site of formation is considerably higher than the pressure at the site of absorption; when the difference in pressure decreases to less than 10 mm. Hg, normal reabsorption ceases.

All the factors effecting the regulation of the circulating force and all of the functions of the cerebrospinal fluid are not presently known or understood. However, one function is the protection of the delicate tissues of the central nervous system against trauma to the head and neck. The fluid may also be responsible for transfer to the blood of metabolic substances resulting from physiologic activity of the central nervous system.

Additional protection to the central nervous system and the cerebrospinal fluid is furnished through an incompletely understood mechanism. Certain dyes injected into the blood fail to stain brain tissue, although these dyes stain most other tissues of the body. Evidence points to the possibility of blood-brain barriers and blood-cerebrospinal fluid barriers.

The estimates of the volume of cerebrospinal fluid in the body vary from as little as 60 ml. to as much as 150 ml. Since the fluid forms continuously, certain conditions may cause volumes greater than 150 ml. to be present.

Synaptic and neuromuscular impulse conduction

The information that all structures of the neuron function in the transmission of impulses in the overall effort to effect homeostasis is conveyed on p. 326. Several theories of nerve impulse conduction have been suggested and explored. The earlier electrical and chemical theories have been succeeded by perhaps the most widely accepted theory, the *neurohumoral theory of transmission* of nerve impulses.

This theory incorporates several chemical compounds into the sequence of events. For better understanding, definitions of these compounds and other pertinent information are presented in the following paragraphs.

---

*Although the quantities are believed to be considerable by some authorities, no definite quantity has been established.

Cholinesterase is an enzyme (esterase) present in all body tissues; it hydrolyzes acetylcholine into choline and acetic acid.

Choline is a vitamin, hydroxyethyl trimethyl ammonium hydroxide, $CH_2OH \cdot CH_2N(CH_3)_3 \cdot OH$, derivable from many animal and some vegetable tissues. It prevents deposition of fat in the liver.

Esterase is an enzyme that catalyzes the hydrolysis of an ester into its alcohol and acid.

An *ester* is any compound formed from an alcohol and an acid by the removal of water.

Acetylcholine is a reversible acetic acid ester of choline, $CH_3CO \cdot O \cdot CH_2 \cdot CH_2 \cdot N(CH_3)_3 \cdot OH$, normally present in many parts of the body. It is extremely important in the transmission of nerve impulses across a synapse.

Synapse is the name of the anatomic relation of one nerve cell to another; the region of contact between processes of two adjacent neurons or between the axon of one cell "clasping" the body of another cell.

Refractory period (or phase) is the brief period immediately following the response of a muscle, nerve, or other irritable element before it recovers its capacity to make a second response.

In conjunction with the chemical theory of nerve-impulse transmission, Dr. H. H. Dale made important contributions to the present neurohumoral theory. In essence, he theorized that in order for a chemical compound to function as the immediate and direct stimulator of striated muscle fibers, the compound would, of necessity, appear instantaneously with the arrival of the impulse at the motor nerve ending; also, the compound would have to disappear with equal suddenness during the very brief refractory period.

Neurohumoral theory of transmission

At a given synapse, minute neurovesicles in the axon end-foot (motor nerve ending) are stimulated to release acetylcholine upon the arrival of the nerve impulse. By some physiologic mechanism* acetylcholine alters cell membrane permeability, which produces a potential difference between the membranes of the synaptic cells. The existence of a potential difference permits the impulse to flow across the synapse. Almost instantaneously (within milliseconds) cholinesterase, which is always present in all body tissues, initiates inactivation of acetylcholine by hydrolysis, thus terminating the conditions that make impulse transmission possible.

Motor nerve axon terminals (motor end-plates, or boutons terminaux) attach to the muscle cell membrane to form the neuromuscular junction (see Fig. 9-11). The way in which transmission of a nerve impulse is achieved was discussed previously. Muscular contraction is practically instantaneous following conduction of the motor impulse across the synapse. Conduction of nerve impulse to

---

*There is a possibility that acetylcholine lowers the blood pressure.

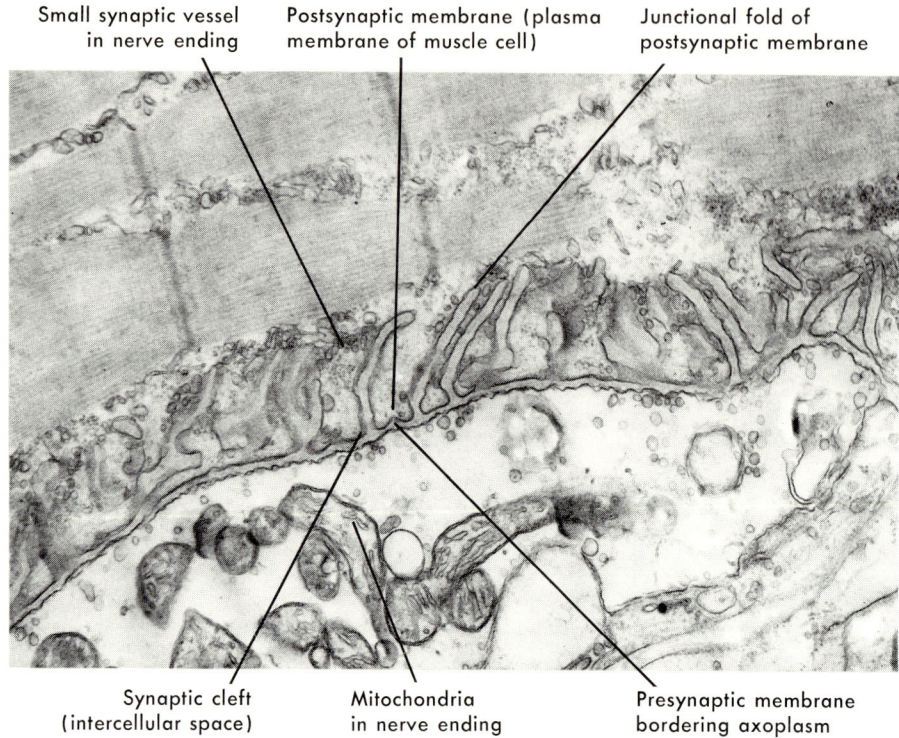

Fig. 9-11. Electron micrograph. Mouse thigh muscle, junction between motor nerve ending and striated muscle ($\times$ 33,000). (From Rhodin: An atlas of ultrastructure, Philadelphia, 1963, W. B. Saunders Co.)

muscle fiber has been proven experimentally. It is presumed that the same mechanics (as presented in the preceding paragraph) apply to the conduction of nerve impulse from nerve axon to nerve cell dendrite.

The fact that acetylcholine is released from minute neurovesicles located in the axon end-feet, and apparently only in them in the synapse, precludes the possibility of impulse transmission in the opposite direction. From the foregoing paragraphs it is seen that cholinesterase prevents the transmission of more than a single impulse during the time any acetylcholine is present.

SPECIAL SENSES

Previously (pp. 336 to 337) we learned that the various senses are localized in specific parts of the cerebrum, recognizing that there is some generalization. Prior to the study of the five special senses, the radiologic technology student should understand (at least in part) that the sense organs may be grouped according to source of stimulation and location of receptors. In this classification we refer to *exteroceptors* and *interoceptors*.

Changes (in temperature, sound, etc.) that occur outside the body stimulate exteroceptors. Changes (in balance, muscle tension, hunger, thirst, etc.) that occur within the body stimulate interoceptors. Interoceptors are further subdivided into three classifications: *proprioceptors, labyrinthine sense organs,* and *visceroceptors.*

Sense organs within muscles, tendons, and ligaments (joints) are proprioceptors; these enable the person to "sense" the stretch or tension existing within these structures. These feelings are also known as *kinesthetic sensations.*

Sense organs within the labyrinths (of the temporal bones) are labyrinthine sense organs; these enable the person to "sense" position and changes in position of the head.

Sense organs within the visceral walls (digestive, respiratory, urinary, etc.) are visceroceptors; these are stimulated by chemical and mechanical changes, and enable the person to "sense" hunger, thirst, pain, need for micturition and evacuation, etc. Impulses from these visceroceptors adapt and regulate the activities of the various viscera to body requirements.

Smell

The sense of smell is one of the chemical senses. Odors of a diverse nature and composed of certain chemicals act to stimulate the receptors in the olfactory bulbs. Odors enter the nose through the anterior nares to diffuse throughout the regions of the olfactory bulbs. The olfactory cells are found in the mucous membrane of a blind passage in the superior parts (above the superior conchae) of both halves of the nasal cavity directly inferior to the floor of the skull; these cells are situated in the septum and in those parts of the nasal conchae opposite the septum. Inspired air containing the chemical odors is drawn through the three inferior nasal passages (between each two nasal conchae) into the blind passage, where the odors stimulate the olfactory cells.

An olfactory cell has a long protoplasmic extension (thread) connecting outwardly to the mucous membrane. The opposite end of the olfactory cell presents a delicate nerve fiber. The nerve fibers combine to form approximately twenty bundles, which pass upward through the holes in the cribriform plate of the ethmoid bone to terminate in the olfactory lobe of the brain.

Sight

Sight is the process, function, or power of seeing; the end-organ of sight is the eye, by which the position, shape, and color of objects are received or perceived as stimuli through the medium of light proceeding from them. The physical laws of light rays affect the physiologic process of seeing (see Fig. 9-12).

As rays of light pass from one medium to another medium of different optical density, the rays are bent or refracted. When the rays pass from a rare medium to a more dense medium, the rays bend toward the perpendicular. Conversely, when light rays pass from a dense medium to one that is optically less dense, the rays bend away from the perpendicular. Since air is less dense than the aqueous humor, the light rays bend toward perpendicular.

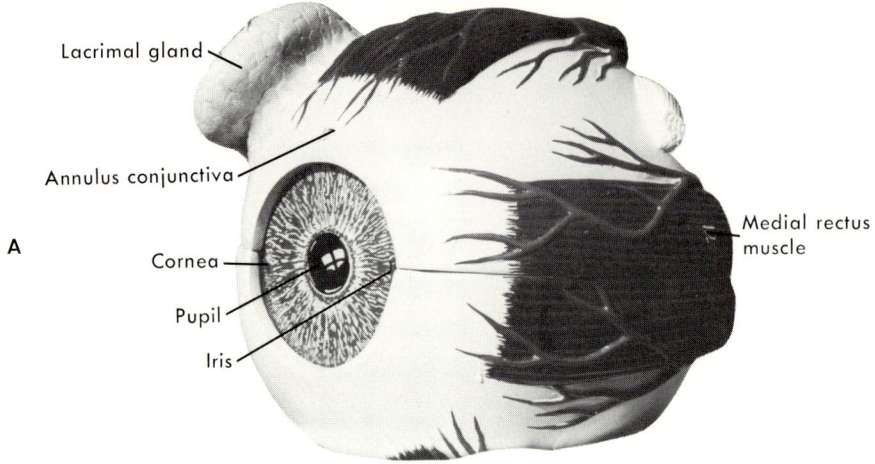

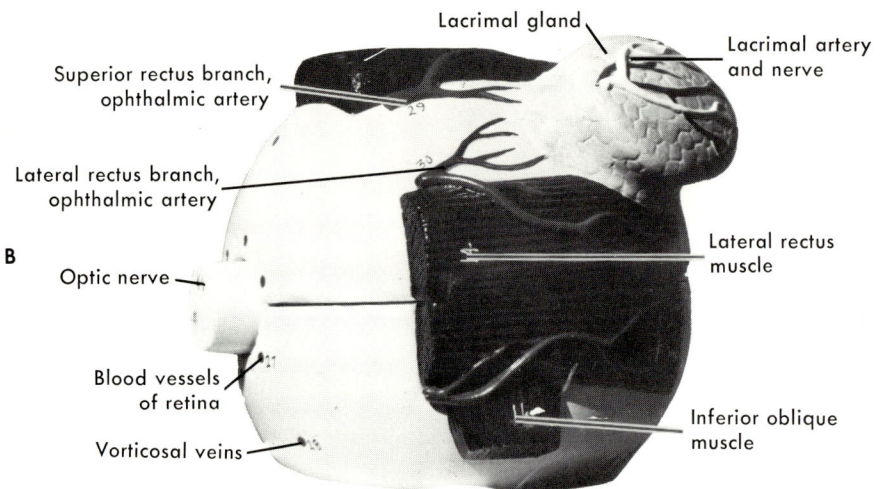

Fig. 9-12. Right eye. **A,** Anteromedial aspect. **B,** Posterolateral aspect.

The shape of the human eye closely resembles a sphere, somewhat smaller than the bony cavity that contains the eyeball. Connective tissue, fat, and the extrinsic muscles assist in keeping the eyeball in the cavity.

A tough, fibrous coat covers the eyeball; the anterior one-sixth of this coat is the *cornea*, which is transparent to permit entrance of light rays. The remaining five-sixths of this coat constitute the *sclera*, which is opaque and white to bluish in color. Directly posterior to the cornea lies the *iris*, a muscular diaphragm. Except for the actual iris opening *(pupil)*, the iris is opaque. The *crystalline lens* lies directly posterior to the iris. Between the muscular structure of the diaphragm and the cornea is the *anterior aqueous chamber*, which is filled with *aqueous humor*. *Vitreous humor* fills the *posterior aqueous chamber*, which is posterior to the lens. The remaining internal surface of the eyeball is covered with a highly vascular pigmented coat, called the *choroid*. The choroid is coated over almost its entire surface* with the specialized *retina*. The retina connects to the brain via the optic nerve. The artery of the retina lies in the center of the optic nerve. Three layers of neurons comprise the substance of the retina: the neuroepithelium, the bipolar cell layer, and the ganglionic cell layer. The choroid coat connects through the retina to the optic nerve via a chain of three neurons. A layer of flattened, pigmented cells separates the choroid from the neuroepithelial layer. The outer ends of the cells of the neuroepithelium form the rods and cones. The other ends of these cells connect to the bipolar cells. The bipolar cells and the ganglion cells connect by synapse.

Act of seeing

As one gazes upon an object, light rays from this object fall upon the *macula lutea* (yellow spot), a very special structure situated in the center of the retina directly posterior to the lens. The macula, about 0.6 mm. in diameter, contains in its center the *fovea centralis*. The fovea contains only cones. (In the most dense parts of the fovea, there are approximately 150,000 cones in each square millimeter of surface.) Various authors have estimated that there are approximately 400,000 fibers in each optic nerve and that each retina contains about 7,000,000 cones and about 125,000,000 rods. It appears that each cone cell of the fovea has a "private line" to the brain. One ganglion cell and its axon cell serve one cone; however, it is possible for there to be interaction between neighboring receptors. In the outer parts of the retina are found as many as ten cone cells to one ganglion cell and as many as 250 rod cells to one ganglion cell.

It is generally considered correct that cones are necessary for fixed vision (and color) and that rods are necessary for assistance (or additions) to the main object vision. The eyes are sensitive to an extremely wide range of wavelengths; however, maximum sensitivity to light rays lies between wavelengths of 5,000 and 6,000 Å. An angstrom unit (Å) is a measure of wavelength and is 1×10^{-8} cm.

Hearing

Hearing is the act or power of perceiving sound; the characteristic end-organ of sound is the ear, which is responsive to a characteristic stimulus—sound waves.

The external, middle, and inner ears comprise the complete ear. All the parts of the ear external to the *tympanic membrane* (eardrum) comprise the external ear. These parts are the *auricle (pinna), tragus, lobe,* and *external auditory meatus.* The latter is the canal leading to the tympanic membrane (see Fig. 9-13).

The middle ear, the tympanic cavity, is contained within the temporal bone. The tympanic cavity may be compared with a small box in that the cavity

---

*The retina covers all the anterior surface of the posterior choroid and extends anteriorly almost to the ciliary muscles.

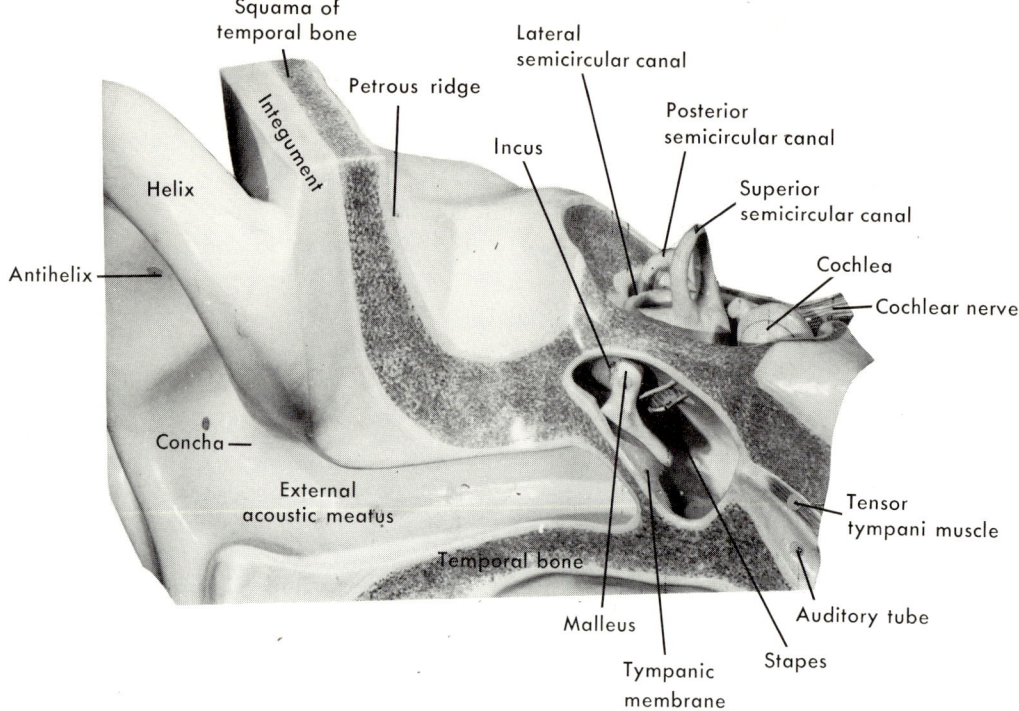

Fig. 9-13. Ear model showing the middle ear and related structures.

has six sides. The lateral wall of the tympanic cavity consists of the tympanic membrane, a thin, translucent disc of fibrous tissue that separates the external and middle ears. The medial wall of the tympanic cavity is entirely osseous except for two openings: the *round window* and the *oval window*. The two windows afford communication between the middle and inner ears. The roof of the tympanic cavity is a thin sheet of bone and separates this cavity from the middle cranial fossa. The floor of the tympanic cavity is thin and osseous and separates the cavity from the jugular fossa. The front wall of the tympanic cavity contains the opening of the *auditory (eustachian) tube*. The auditory tube connects the middle ear to the nasopharynx and functions to maintain equal air pressure on the two sides of the tympanic membrane. Infections travel in either direction between the nasopharynx and the middle ear. In the back wall of the tympanic cavity is a small opening between the cavity and the mastoid air cells of the temporal bone. It is possible for infections to travel in either direction through this opening between the cavity and the mastoid cells.

Also contained in the middle ear and extending between the lateral and medial walls as very important parts of the ear are the three auditory ossicles; the *malleus* (hammer), *incus* (anvil), and *stapes* (stirrup). The malleus attaches off-center to the tympanic membrane; the other end of the malleus articulates superiorly with one end of the incus. The other end of the incus

The nervous system 351

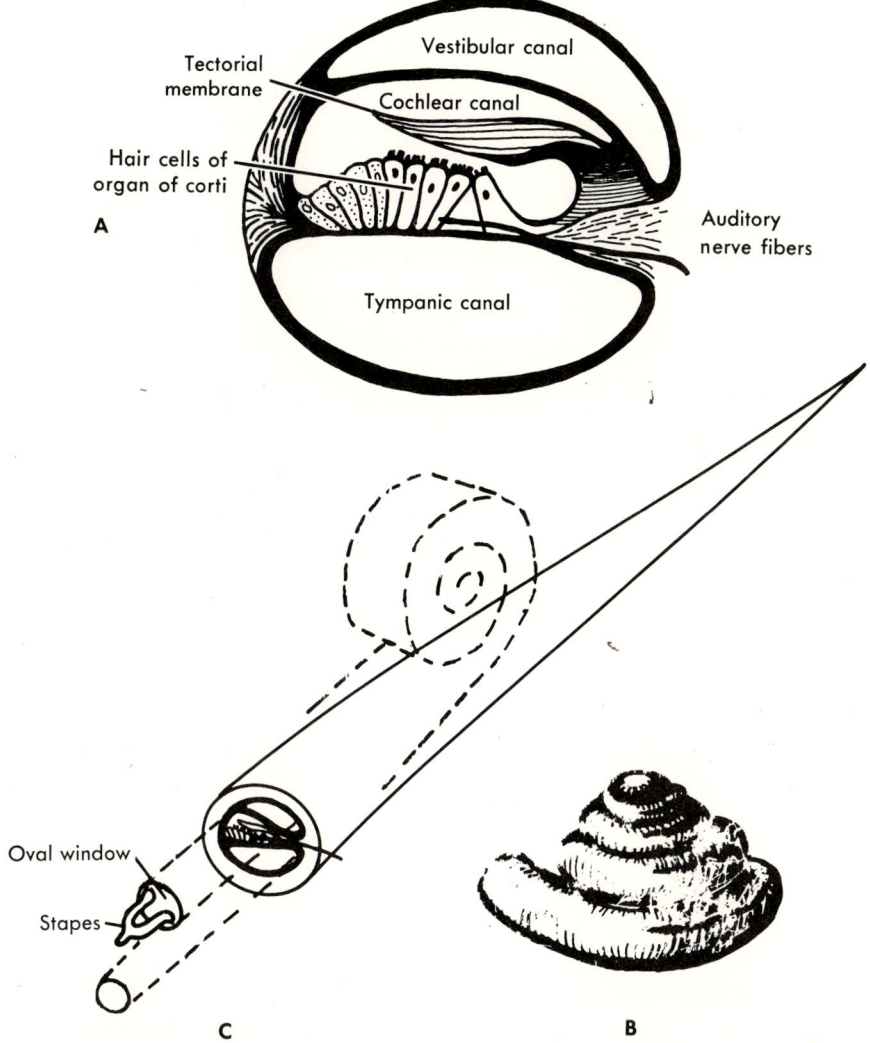

Fig. 9-14. A, Diagram of one turn of the cochlea showing the organ of Corti. **B,** The cochlea. **C,** Diagram of cochlea, partially straightened.

articulates inferiorly and medially with the free end of the stapes. The foot (opposite end) of the stapes fits into the oval window.

The inner ear is situated entirely within the petrous portion of the temporal bone. Complex bony passages containing smaller membranous passages comprise this structure. The bony passages comprise the *osseous labyrinth,* and the membranous passages comprise the *membranous labyrinth.* The inner ear has the dual functions of hearing and equilibrium.

The osseous labyrinth includes the three semicircular canals, the vestibule, and the cochlea. Communication is present between the vestibule and the semicircular canals and between the vestibule and the cochlea.

The semicircular canals are each in a plane at right angles to the planes of the other two canals. Also, each canal in one ear is nearly parallel with a canal in the opposite ear. The right superior semicircular canal is parallel with the left posterior semicircular canal; the left superior semicircular canal is parallel with the right posterior semicircular canal; the right lateral semicircular canal is parallel with the left lateral semicircular canal (see Fig. 4-57, C).

The cochlea makes two and one-half turns in a rather tight spiral around the central modiolus. A very thin layer of bone winds around the modiolus to separate the length of the cochlea into the *scala tympani* and the *scala vestibuli,* which communicate via a small opening in the apex of the modiolus (see Fig. 9-14).

Between the bony wall of the osseous labyrinth and its contained membranous labyrinth is a clear fluid. The two bodies of fluid do not intermix.

Within the bony vestibule are two saclike parts of the membranous labyrinth, the utricle and the saccule. Communication is present between the utricle and the saccule. The utricle also connects directly with the membranous counterpart of each semicircular canal. The scala media (cochlear duct) connects with the saccule via the very small canalis reuniens.

The ear has several innervations. The external ear has sensory innervations from the mandibular branch of the trigeminal, the vagus, and the great auricular nerves. The vagus nerve sends sensory branches to the pharyngeal mucosa. The chorda tympani nerve passes between the incus and malleus across the tympanic cavity. The tensor tympani muscle regulates the tension of the tympanic membrane and is innervated from the motor branch of the trigeminal nerve. The stapedius muscle (the smallest in the body) inserts on the stapes, is innervated from the facial nerve, and functions in the regulation of pressure of the stapes foot. The mucous membrane of the middle ear has sensory innervation from the glossopharyngeal nerve. The auditory nerve passes through the internal auditory meatus and consists of two parts: the cochlear part and the vestibular part.

Act of hearing

As sound waves strike the tympanic membrane, the membrane vibrates in the same frequency as the sound waves. Since the malleus attaches to the membrane, the sound waves pass down the length of the malleus to the incus and thence to the stapes. Because the malleus attaches off center in the tympanic membrane, a single vibration occurs in the membrane for each sound wave striking the membrane. This arrangement prevents continuation and reverberation of sound waves.

The sound waves pass from the foot of the stapes in the oval window to the perilymph* in the scala vestibuli of the cochlea. Vibrations in the perilymph are transmitted accurately across the membrane to the endolymph in the scala

---

*Perilymph is the fluid between the membranous and bony labyrinths of the ear.

media. The membrane separating the perilymph from the endolymph is the basilar membrane. The hair cells of the *organ of Corti* are the receptor cells; the hair cells rest upon the basilar membrane. Transmitted sound waves then drive the hair cells swiftly against the *tectorial membrane* floating above them. A series of taps is applied to the processes of the hair cells. The hair cells transmit the energy of the taps as mechanical stimuli to the terminals of the auditory nerve in the tectorial membrane. (Some authors theorize that the hairs attach to the tectorial membrane and that a succession of pulls upon these hairs provides the stimuli as the basilar membrane vibrates.)

Taste

Taste is the other (second) chemical sense. Taste is the sense by which certain attributes of bodies or substances are ascertained by contact with certain epithelial cells and organs occurring in the papillae on the surface of the tongue. The cells of the taste buds (papillae) are modified epithelial cells surrounded by nerve fibers from the seventh, ninth, and tenth cranial nerves. The fifth cranial nerve sends sensory fibers to ascertain touch and temperature for the tongue and mucosa of the mouth and nose.

Tastes are placed in the four categories of sweet, sour, salt, and bitter. Many combinations of these four are recognized, and some authors include a fifth taste—alkaline. Sweet and salt are best recognized at the tip of the tongue. Sour is recognized at each side of the tongue. Bitter is recognized at the back of the tongue. The facial nerve innervates the anterior two-thirds of the tongue. The glossopharyngeal nerve innervates the posterior third of the tongue.

Touch

Touch is the sense by which pressure or traction exerted on the skin or mucous membrane is perceived. Touch is called the tactile sense.

The sensations felt in the skin are, specifically, touch, heat, cold, and pain. Different numbers of specialized receptors are situated throughout the cutaneous regions to receive the special stimulus. This stimulus passes along a sensory nerve tract to the central nervous system, where the proper (usually) motor response originates. A motor nerve terminates in a gland, organ, muscle, or secreting surface. The usual effect of a sensory impulse is a complementary motor action. There is no sense of touch, heat, or cold in the viscera.

RADIOGRAPHIC STUDIES

A marked degree of cooperative effort exists among members of the clinical laboratory and radiologic department in that certain fluids are collected for laboratory analysis immediately prior to injections of contrast media into the patient for radiologic examination.

Cerebrospinal fluid is often sent to the laboratory for examination for bacteria, cells, and abnormal chemical compounds and for serology. Cerebrospinal fluid may be collected at the commencement of a myelography (see Fig. 4-

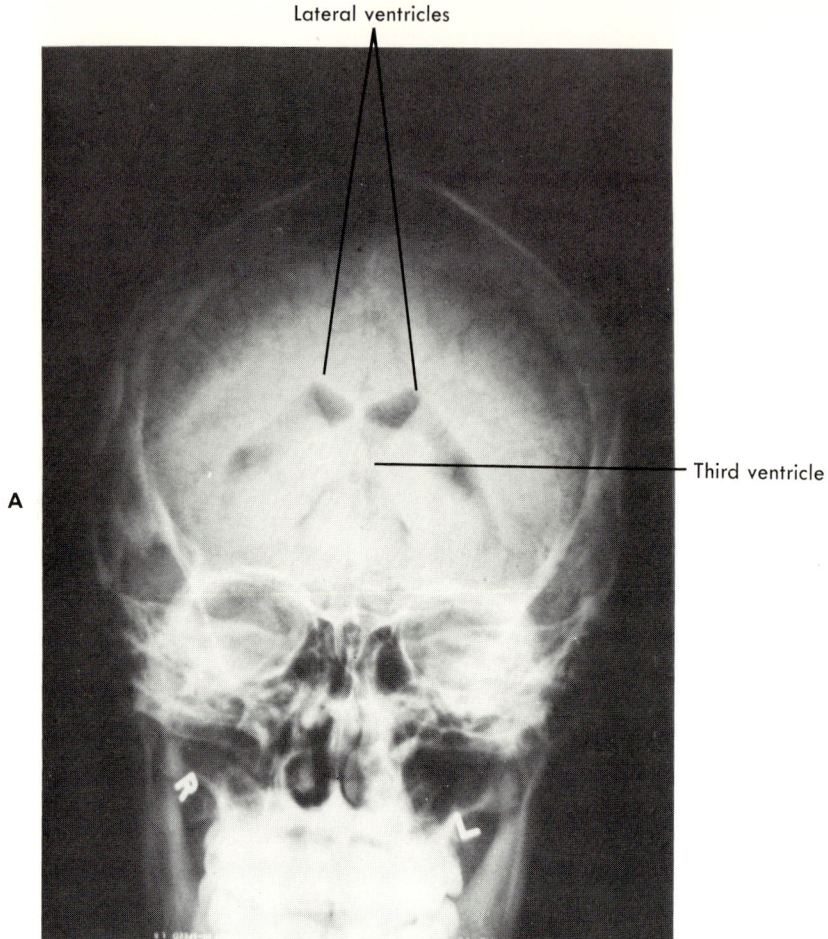
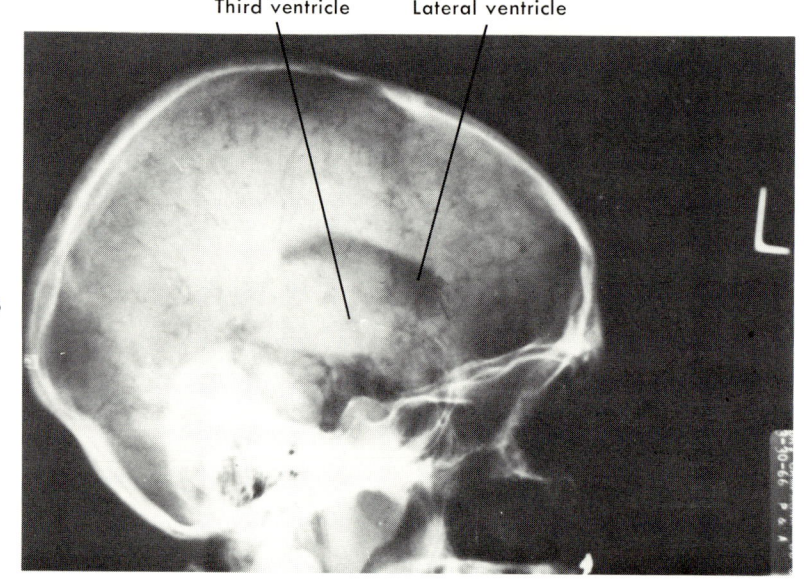

Fig. 9-15. Pneumoencephalogram. **A,** Anterior view. **B,** Lateral view.

14, A) using either oxygen, which is radiolucent, or some radiopaque compound as the contrast medium. A myelogram is a roentgenogram of the spinal cord and the subarachnoid space. The lumbar puncture for this procedure is made into the subarachnoid space, the puncture needle being inserted, usually, between the third and fourth lumbar vertebrae and into the lumbar cistern. When oxygen is used in this examination, the oxygen sometimes passes upward into the subarachnoid space of the cranial cavity. Such an examination is called *pneumoencephalography* (see Fig. 9-15).

Pneumoencephalography is also performed after injecting oxygen directly into the ventricles of the brain. This examination is also called *ventriculography*. The needles are passed into the ventricles through small *trephine* holes drilled through the skullcap (see Fig. 9-16).

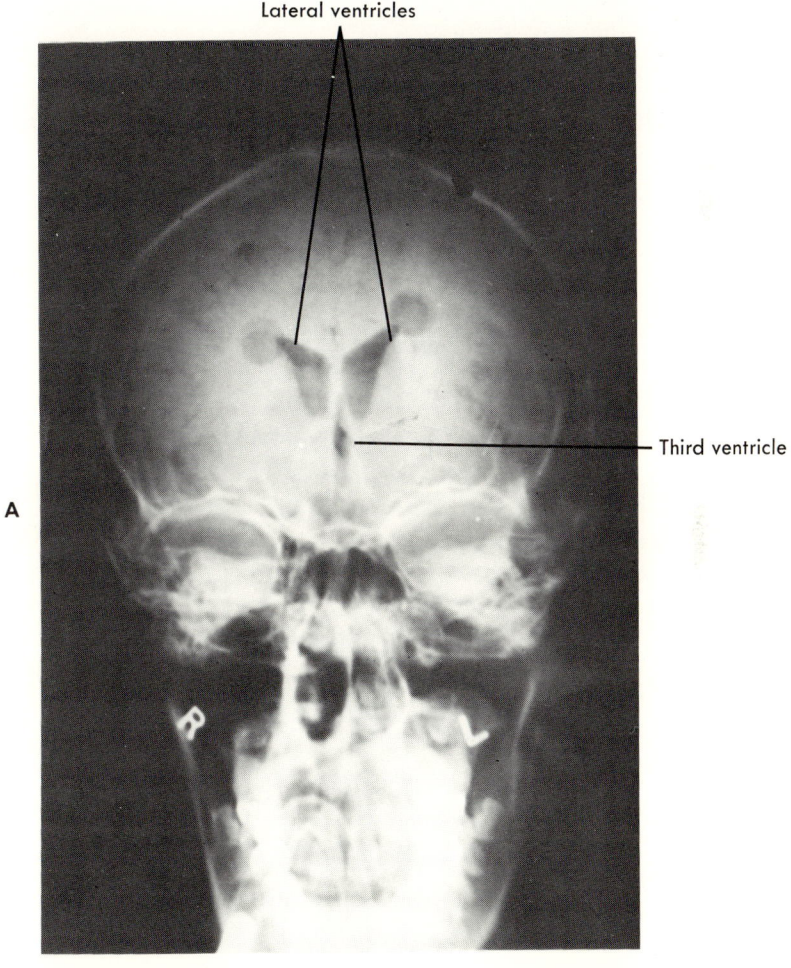

Continued.

Fig. 9-16. Ventriculogram. **A**, Posterior view. **B**, Lateral view.

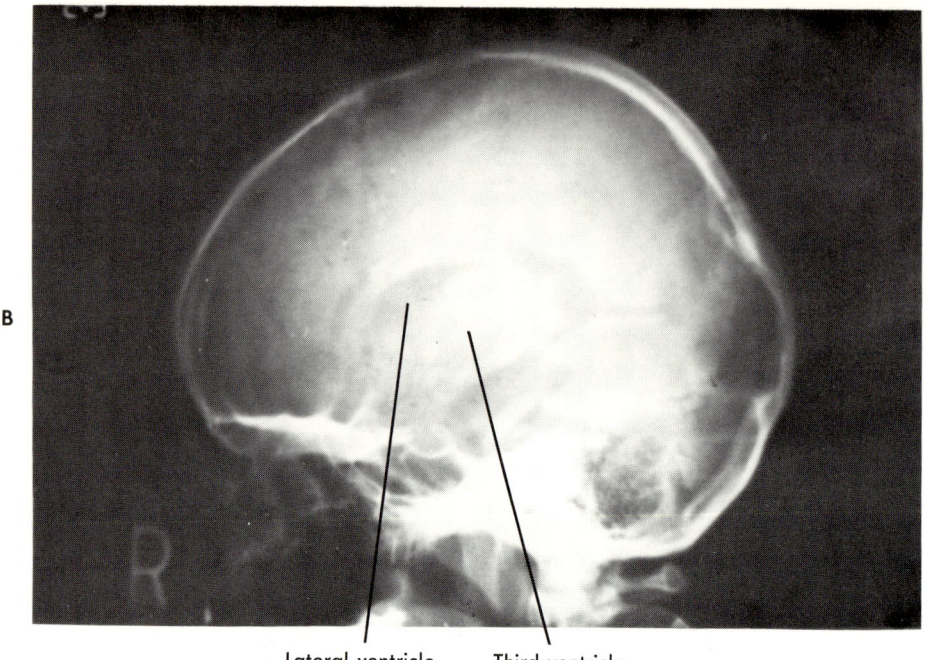

Lateral ventricle   Third ventricle

Fig. 9-16, cont'd. For legend see p. 355.

A fourth examination performed for radiographic study of the central nervous system is similar to myelography and is called discography (see Fig. 4-14, *B*). Discography is the radiographic examination of the intervertebral disc following direct injection of a radiopaque medium into the disc nucleus.

REFERENCES

Dyke, C. C., and Davidoff, L. M.: Roentgen treatment of diseases of the nervous system, Philadelphia, 1942, Lea & Febiger.

Francis, C. C, and Martin, A.: Introduction to human anatomy, ed. 7, St. Louis, 1975, The C. V. Mosby Co.

Gardner, E.: Fundamentals of neurology, ed. 3, Philadelphia, 1958, W. B. Saunders Co.

Goss, C. M.: Gray's anatomy, ed. 27, Philadelphia, 1959, Lea & Febiger.

Mountcastle, V. B.: Medical physiology, ed. 13, St. Louis, 1974, The C. V. Mosby Co.

Schottelius, B. A., and Schottelius, D. D.: Textbook of physiology, ed. 17, St. Louis, 1973, The C. V. Mosby Co.

CHAPTER 10

The muscular system

The more than 400 muscles of the body comprise approximately 35 to 45 percent of the total body weight. This figure varies between individuals according to nutritional status, age, sex, and other contributing factors. Striated muscles account for a large majority of all muscle tissue; in females, striated muscles account for approximately 35 percent of the total body weight; in males, striated muscles account for approximately 40 percent of the total body weight. In each sex the total percentage of body weight comprised by nonstriated and cardiac muscles is quite small. Other body tissues make up the principal difference in percentage of the total body weight (see Fig. 10-1).

All body tissues possess the common properties of irritability, growth, metabolism, and reproduction.* In addition, muscle tissue possesses the properties of tonicity, extensibility, and elasticity. *Tonus (tonicity) is the state of mild contraction exhibited by all healthy muscle tissue. The myocardium (heart muscle) exhibits extensibility during diastole;* when the chambers of the heart fill with blood, the walls expand to accommodate the increased volume. *Elasticity refers to the ability of muscle tissue to return to its normal length following cessation of the applied force.*

MUSCLE FUNCTIONS

In addition to the well-known function of movement, all body muscles possess two additional functions: posture maintenance and heat production; these are as important to homeostasis as is movement. Posture maintenance is a function limited to striated muscles. Heat production is a function of each of the three types of muscle tissue—striated, nonstriated, and cardiac. Muscle contractions release energy in proportion to the force applied.

Movement and heat production are explained together because of the over-

*The properties of conductivity and contractility are sometimes included with the four common properties of body tissues.

358 *Textbook of anatomy and physiology in radiologic technology*

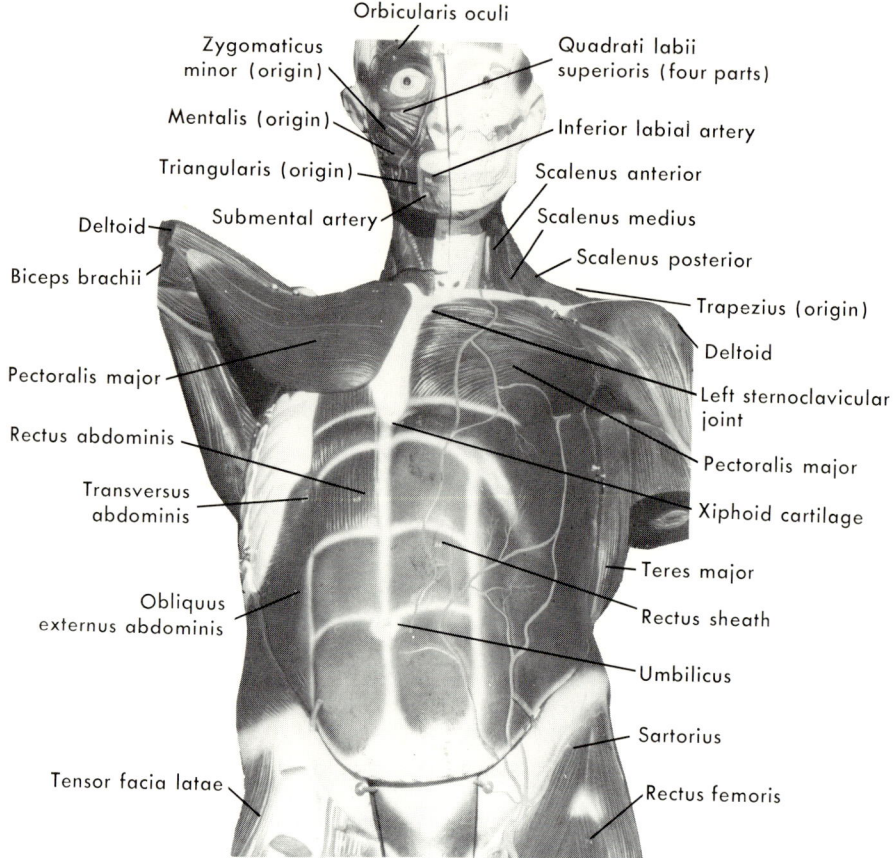

Fig. 10-1. Human mannikin torso, some anterior muscles.

lapping of various anatomic and physiologic processes. Posture maintenance is discussed singly, although certain aspects of anatomy and physiology also apply to movement and heat production.

Movement and heat production

In general, muscles move (contract and/or expand) as a result of nerve stimuli, which may be either voluntary or involuntary. Muscular movements invariably result from motor nerve stimulation. Each movement results in the consumption of some nutrition to afford the energy required for the movement. A brief explanation of the structures and processes involved in skeletal muscle movement and heat production is presented in the following paragraphs.

Striated (skeletal) muscle cells are usually much greater in length than in thickness. For this reason the cells are usually referred to as *muscle fibers*. However, it is uncommon for an individual muscle fiber to extend throughout the entire length of the muscle. In the larger muscles the fibers are bundled together, in conformity with their shapes, and encased in connective tissue

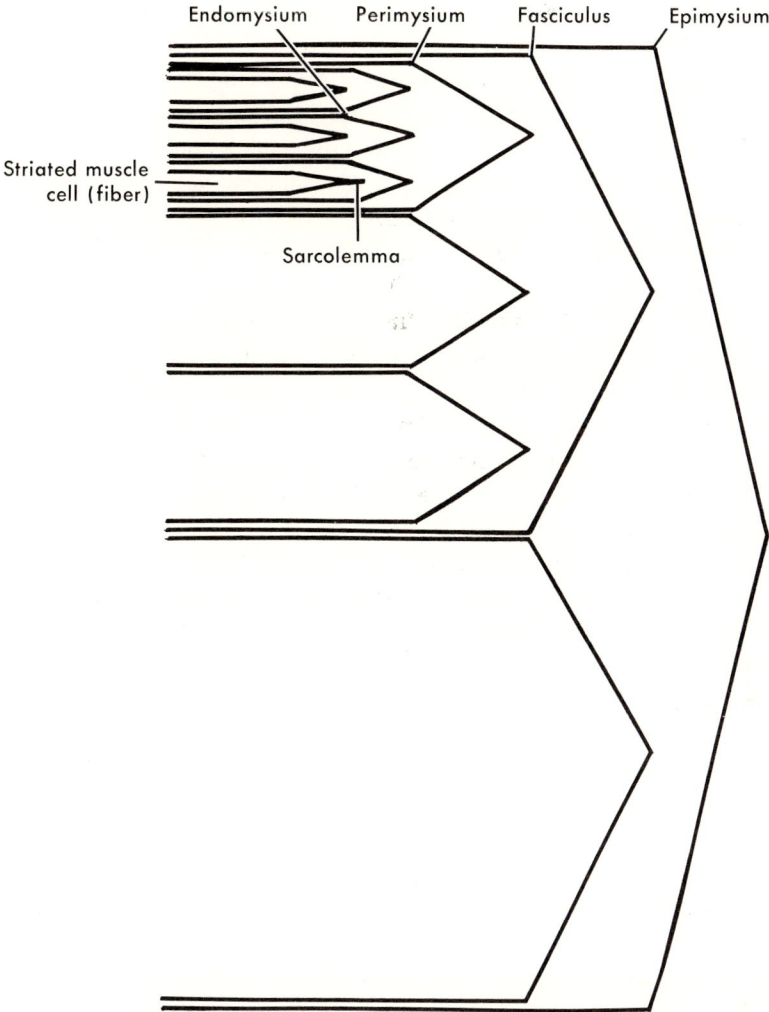

Fig. 10-2. Schematic drawing of the relation of connective tissue layers to striated muscle fibers.

sheaths; the whole bundle is a *fasciculus*. The connective tissue layer just outside the sarcolemma* is the *endomysium*. The *perimysium*, another connective tissue layer, encloses several muscle fibers. The *epimysium*, a third connective tissue layer, completely envelops an entire muscle. Extending inwardly from the epimysium are the *septa*, which are sheets of connective tissue separating the muscle fibers (see Fig. 10-2).

Each fiber of striated muscle resembles an elongated cylinder, and the fibers are situated within the muscle in parallel lines. The several nuclei of a fiber are usually situated on the outer surface of the individual fiber. Within and ex-

---

*The sarcolemma is the encasing membrane of striated muscle cells.

tending with the length of a muscle fiber are numerous fine, parallel filaments, the *myofibrils*. Experimental analysis of a myofibril reveals that it consists of a large number of very fine threads of a protein material named *actomysin;* actomysin consists of actin-tropomysin and myosin protein molecules. The threads of actomysin are the essential contractile components of muscle, the *micellae*. The number of micellae in a particular muscle determines the power of that muscle and the quantity of work the muscle can perform.

The single muscle fiber is the structural unit of muscle; the lengths of these fibers measure between 1.0 and 40 mm.,* and the diameters measure between 0.01 and 0.1 mm. One thousand or more myofibrils (each about 0.001 mm. in diameter) are arranged in loose bundles within a fiber and constitute much of the solid substance within the sarcoplasm (matrix) of the fiber. Microscopic study of a myofibril discloses transverse striations of the myofibril due to the presence of alternate light and dark bands. The myofibrils are so aligned that these bands coincide to present the fiber with a striated appearance—hence the name, striated muscle.

One pair of bands constitutes a *sarcomere*. The dark (A) band is anisotropic and consists of myosin molecules; the light (I) band is isotropic and consists of actin molecules. During contraction of a muscle the actin filaments slide inwardly among the myosin filaments, an action that results in the apparent shortening of the muscle.

From one to several small branching structures are situated on the surface of each muscle fiber; these are the *motor end-plates* to which the nerves attach. The motor end-plates are situated within the sacrolemma. There are many more muscle fibers in a given muscle than there are nerve fibers to supply the muscle. As a result, the nerves that supply a muscle branch within the muscle so that from 5 to 200 muscle fibers are innervated from a single nerve fiber. The processes involved in the transfer of nerve impulses to muscle are discussed in Chapter 9 under the neurohumoral theory of nerve impulse transmission.

In summary of the foregoing statements, muscle fiber is enclosed by the sarcolemma. Immediately beneath the sarcolemma is the true cell membrane of the muscle fiber, the plasma membrane. The plasma membrane transmits action potentials in the same manner in muscles as does the nerve membrane in nerves.

Experiment reveals that muscle tissues, as all excitable tissues, develop differences in electrical potential according to the degree of contraction of one muscle (or part) compared with that of an adjacent muscle (or part). The part of a muscle that contracts exhibits a negative electrical charge to the adjacent, noncontracted part. Use of the knowledge acquired concerning this phenomenon led to the development of the many instruments for measuring activity of various parts of the body by means of electrical currents: e.g., the electrocardiograph.

The activity of which a given muscle is capable has direct relation to the

---

*Some muscle fibers have been measured as long as 120 mm.

composition of the muscle. In general, fresh muscle tissue consists of about 20 percent protein, 75 percent water, and 5 percent minerals and various organic compounds such as steroids, lipids, glycogen, glucose, and nonprotein nitrogenous compounds. Potassium constitutes the principal mineral in muscle tissue; calcium, magnesium, sodium, and phosphorus are present in much lesser quantities. Creatine, phosphocreatine, adenosine triphosphate, and urea are the principal nonprotein nitrogenous substances of muscle tissue.

Muscle activity involves the phenomena of energy release, contracture, and lactic acid oxidation. The energy for muscular contractions derives from the breakdown of carbohydrates (glycolysis); carbohydrates are stored in muscle tissue as glycogen (see pp. 229 and 231). The final step in this process is the utilization of oxygen, resulting in the release of carbon dioxide and water.*

Muscle activity

Glycolysis without oxygen achieves an incomplete breakdown of the glucose molecule and thereby releases energy and forms two pyruvic acid molecules from the glucose molecule. (A cellular deficiency in oxygen content causes the production of two lactic acid molecules.)

With the release of energy by glycolysis from the glucose molecule, a small part (the small fraction, or one-nineteenth) of the energy transfers to two molecules of ATP (adenosine triphosphate), another compound within the muscle fiber. ATP is stored in muscle tissues in the form of phosphocreatine.

$$ATP + creatine + H_3PO_4 \rightleftarrows ADP + phosphocreatine$$
$$ADP\sim P + C + H_3PO_4 \rightleftarrows ADP + C\sim PO_4$$

(*Note:* The sign, $\sim$, denotes an energy bond.)

ATP is a high-energy molecule, which, when broken down, releases the energy to accomplish muscle contraction. The results of experimental stimulation of muscle cells indicate that the breakdown of ATP is triggered with the arrival of nerve impulses at the muscle cells. In addition to energy release, the breakdown of the ATP molecule results in the formation of ADP (adenosine diphosphate) and phosphoric acid (H_3PO_4).

Myosin, one of the major proteins in muscle tissues, acts as an enzyme to catalyze the breakdown of ATP into muscular action. The sliding-filament hypothesis of contraction, as proposed by Huxley, suggests that the energy from ATP breakdown is necessary for the fine actin filaments to slide inwardly between the coarser myosin filaments, thus contracting (shortening) the muscle cells. Actin is the other major protein found in muscle tissue. (See pp. 359 and 360 for a discussion of the actin and myosin filaments.)

Lack of muscle activity permits synthesis of excesses of ATP, hence its storage in the form of phosphocreatine. Excessive muscle activity causes ATP break-

---

*Glycolysis without oxygen will yield only one-nineteenth as much energy as glycolysis with the use of oxygen.

down faster than ATP can be synthesized, hence a buildup of lactic acid in the muscle tissues. Lactic acid buildup results because the acceleration of both respiration and circulation (although circulation is increased as much as 100 times normal by the dilation of blood vessels and by the increased flow from the heart) is insufficient to supply the muscle cells with sufficient oxygen to oxidize pyruvic acid.

Lactic acid forms when a deficiency in oxygen exists. The major quantity of lactic acid is converted back to pyruvic acid in the cell prior to further usage. Excessive accumulation of lactic acid in muscle tissue produces *fatigue*. (Fatigue often is used interchangeably with the phrase *oxygen debt*; a detailed discussion of oxygen debt is not pertinent in this text.)

During these several chemical reactions, one-fifth of the lactic acid formed is oxidized to carbon dioxide and water; the remaining four-fifths is resynthesized into glycogen. Each of these chemical reactions depends upon the presence of a specific enzyme; each reaction consumes energy. Each time energy is used, heat is released (law of conservation of energy and matter). *The muscles act as converters of chemical energy into mechanical energy to cause movement.*

Posture

Anatomic posture does not occur as a simple happening. It is the result of coordinated and combined functions of the several body systems. Gravity affects everything on the earth. Gravity exerts certain forces on each of the bones of each person. If the muscles did not hold the bones in specific relationship to each other, vertebrates would exhibit no distinct (typical) shape. The muscles exert a pull on the various bones in opposition to the pull of gravity. Muscle pull is made possible because of the tonicity of healthy muscles. The muscular system is not entirely responsible for posture; each part of the body must contribute toward homeostasis.

When muscles, bones, or nerves in a given region of the body are damaged or otherwise incapacitated, that part of the body may be rendered incapable of retaining its normal posture; after a time, this body part may atrophy or become otherwise useless. The health of the whole body exhibits a reciprocal relation to its posture; each is interdependent upon the other. Posture and the state of health of a given body part exhibit strong influence in diagnostic radiography; there is a definite relationship between the quantity and quality of muscle tissue and the amount of x-ray energy absorbed as roentgen rays (x rays) penetrate these same muscles.

The skeletal muscles are those involved principally with posture, although the visceral and cardiac muscles play very important parts.

MUSCLE ANATOMY
Skeletal muscles

During embryonic development and growth most of the skeletal (striated) muscle tissue arises from the mesoderm. For the most part, skeletal muscles attach to bones although some attach to cartilage, ligaments, skin, or other muscles. Skeletal muscles have unique features that set them apart as to function and histologic study when compared with the other types of muscle.

Striated muscles

One animal body substance is familiar to all; this substance is striated (skeletal) muscle tissue—lean meat. Striated muscle tissue consists of approximately 20 percent solid proteins and 80 percent water, containing numerous minerals in solution.* The color of healthy skeletal muscle (pink to red) results from the presence of the pigment, *myoglobin*.

Shape

Muscles of the skeleton differentiate into at least five shapes: *fusiform, pennate, rhomboid, straplike,* and *triangular*. The arrangement of the fibers within a muscle may lend further to its differentiation. Fusiform muscles are shaped like a spindle. Pennate muscles resemble a feather, especially since their fibers extend obliquely from the lateral surface toward the central tendon. Subdivisions of pennate muscles are *unipennate and bipennate*. Rhomboid muscles have four sides, the two members of each pair parallel with each other. Straplike muscles are usually long and shaped like a strap: e.g., the hamstring muscles. Triangular muscles have three sides (see Fig. 10-3).

Several muscle names are derived in part or in whole from particular shapes: e.g., the deltoid (triangular), the quadratus (quadrilateral or rhomboid), etc.

Attachment

Skeletal muscles have two attachments†; one is the *origin*, the opposite is the *insertion*. The major or more stationary muscle attachment is the origin, and the minor or more mobile muscle attachment is the insertion. With certain muscles (depending upon the movement), a particular action may reverse the origin and insertion of a specific muscle. In most of the extremity muscles, the origin is the proximal attachment.

In general, muscle origins are continuations of the periosteum of the bone, and insertions are continuations of the muscles via tendons into the periosteum of other bones.

In many instances the bone of origin or the region of the muscle lends itself to the name of that muscle: e.g., the quadriceps femoris—a four-headed muscle with origin from the femur.

Movement

Muscles move in response to nerve stimuli. The movements may be either voluntary or involuntary; both voluntary and involuntary movements may be

---

*See pp. 360 and 361 for greater detail.
†Some muscles have more than one origin, such as the biceps and triceps; muscles may also have more than one insertion.

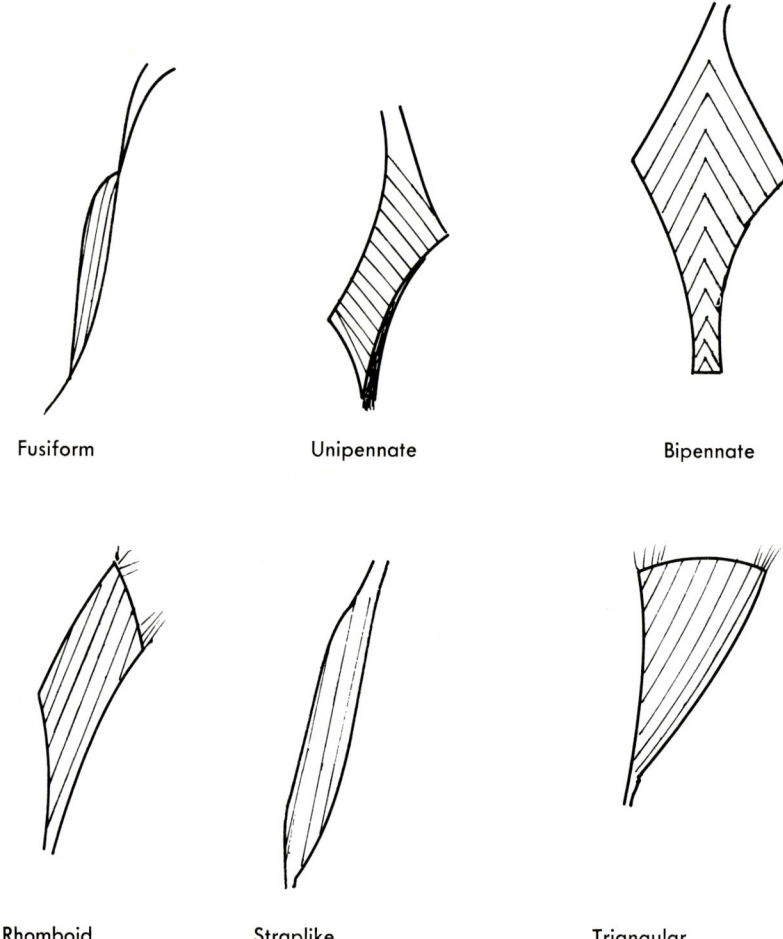

Fig. 10-3. Muscle shapes and fiber directions.

either contractions or extensions. Usually the muscle movement will cause corresponding change in position of one or more bones and will extend across one or more joints. However, muscles may move skin, or they may cause certain glands to secrete.

Motor nerves terminate in muscles and glands. In muscles the nerve endings cause the muscle to act in accord with the motor impulse: i.e., either contract or extend, abduct or adduct, etc. The movement of a muscle is often included in the name of that muscle: e.g., extensor, pronator, flexor, supinator.

The bones to which the moving force is applied are like the levers of a machine, and the articulations of these bones become the *fulcra*. In general, a single muscle is not solely responsible for a movement. Usually several muscles in one region contribute to a given motion. Such muscles are classified under group names: *prime movers, antagonists, synergists,* and *fixation muscles.*

Prime movers. In the prime mover classification are the muscles that cause the principal movement; e.g., the biceps muscle in the upper arm flexes the elbow when the biceps muscle contracts. The biceps muscle has its origin in two heads, as is indicated in its name. One origin (attachment) is the tip of the coracoid process and the other is an elevation superior to and on the upper rim of the glenoid cavity. The biceps muscle inserts into a tendon above the elbow joint; the tendon inserts posteriorly into the radial tuberosity. When the biceps muscle is flexed, the muscle of opposite function (triceps brachii) is relaxed. When the forearm is extended, the triceps muscle becomes the prime mover and the biceps muscle is relaxed. Other examples of similar conditions are quite common throughout the body.

Antagonists. In the antagonist classification are the muscles that act in opposition to the prime movers. Using the same examples as for the prime movers, the triceps muscle is the antagonist of the biceps muscle. When the triceps muscle is flexed it extends the lower arm so that the upper and lower arms form an angle of 180 degrees (a straight line). During this time the biceps muscle is relaxed and is the antagonist.

Synergists. In the synergist classification are the muscles that function to support the action of other muscles. Synergistic muscles act to immobilize a joint, crossed by the tendon of a prime mover, in the best position for the prime mover to act. Synergistic action is exemplified by the wrist muscles in many of the normal movements and actions of the hand, wrist, and forearm.

Fixation muscles. In the fixation muscle classification are the muscles that act to hold stationary the bone (or bones) that serve as the site of origin of the prime mover. Fixation is exemplified in the act of holding an object with the hands and forearms when the elbows are fixed (maintained) at an angle of less than 180 degrees. Radiologic technologists often find it necesasry to lift or to assist in lifting patients from the stretcher to the radiographic table. Muscles that fix the elbows function in this action.

Muscle names

A minimum of seven factors contribute to the formation of skeletal muscle names. Each of these factors is listed with one example as follows:
1. Attachment—sternocleidomastoid
2. Shape—deltoid
3. Action and length—flexor tibialis longus
4. Location—carpi radialis
5. Number of divisions or heads—quadriceps femoris
6. Muscle size—gluteus maximus
7. Direction of fibers—transversus

Certain muscles will be involved either directly or indirectly in both traumatic injury and disease. When the name of a muscle is used regarding radiographic procedures, the muscle name should establish in the minds of radiologic technologists its approximate location, origin, insertion, and possible involvement

Text continued on p. 374.

Table 10-1. Muscles that make significant radiographic shadows

Muscle	Group membership	Function	Visualization	Origin	Insertion	Innervation
Buccinator (parotid duct opens through it opposite the upper second molar)	Mouth	Compresses cheeks—sucking and blowing	Sialography	Alveolar process of maxilla and mandible; pterygomandibular raphe	Both lips, with orbicularis oris	Facial nerve
Masseter	Mastication	Raises mandible and closes mouth	Sialography	Aponeurosis from zygomatic process of maxilla, lower border of zygomatic arch and posterior one-third and lower border of zygomatic arch	Angle and lower one-half of lateral surface of ramus of mandible and upper half of ramus and lateral surface of coronoid process of mandible	Mandibular nerve
Trapezius	Lateral cervical	Braces and raises shoulders, rotates scapulae, and moves shoulder girdle	Anterior and posterior views of upper T-spine and entire C-spine; lateral view of C-spine	External occipital protuberance and medial one-third of superior nuchal line of occiput	Posterior border of lateral one-third of clavicle, medial margin of acromion process, scapular spine, and (via aponeurosis) tubercle at apex of spine and clavicle	Accessory nerve and branches of third and fourth cervical nerves
Sternocleidomastoideus	Lateral cervical	Singly, draws head down to corresponding shoulder; together, raises thorax in inspiration.	Posterior, anterior, and lateral views of C-spine	Two heads: sternum and clavicle	A tendon into mastoid process and an aponeurosis into superior nuchal line of occipital bone	Accessory nerve and branches of second and third cervical nerves

The muscular system

Sacrospinalis	Deep back muscles	Moves back and trunk in various directions and holds back erect	Anterior and posterior views of T-spine and L-spine	A broad, thick tendon from T-12, all L-spinous processes, and iliac crests	Branches insert variously into ribs and transverse and spinous processes of C-vertebrae and T-vertebrae	Posterior primary divisions of spinal nerves
Diaphragm*	Thorax muscles	Rises (relaxes) with expiration and lowers (contracts) with inspiration	Anterior, lateral, and oblique views of chest; anterior and posterior oblique views of ribs below diaphragm; K.U.B.; anterior view of abdomen; lateral and posterior views of T-spine	Xiphoid process, cartilages and ribs 7 to 12, lumbocostal arches, and lumbar bodies	An aponeurosis, the central tendon	Last six intercostal nerves and phrenic nerve
Psoas major	Posterior muscles of abdomen	Moves trunk and flexes hip	Anterior and posterior views of L-spine	Transverse processes of L-vertebrae, sides of T-12 through L-5, intervertebral disc, and tendinous arches	Lesser trochanter of femur	Second and third lumbar nerves
Psoas minor	Posterior muscles of abdomen	Tensor of iliac fascia; flexes hip	Anterior and posterior views of L-spine	Sides of T-12 and L-1	Pectineal line, iliopectineal eminence, and iliac fascia	First lumbar nerve

*See pp. 374 and 375.

Continued.

Table 10-1. Muscles that make significant radiographic shadows—cont'd

Muscle	Group membership	Function	Visualization	Origin	Insertion	Innervation
Iliacus	Posterior muscles of abdomen	Moves trunk and flexes hip	Anterior and posterior views of L-spine	Iliac fossa and crest, anterior sacroiliac and iliolumbar ligaments, and anterior superior and inferior iliac spines, and the intervening notches	Psoas major tendon and femur	Second and third lumbar nerves
Trapezius	Connects upper extremity to vertebral column	Braces and raises shoulders, rotates scapulae, and moves shoulder girdle	Anterior and posterior views of upper T-spine and entire C-spine; lateral view of C-spine	External occipital protuberance and medial one-third of superior nuchal line of occiput	Posterior border of lateral one-third of clavicle, medial margin of acromion process, scapular spine, and (via aponeurosis) tubercle at apex of spine and clavicle	Accessory nerve and branches of third and fourth cervical nerves
Latissimus dorsi	Connects upper extremity to vertebral column	Elevates shoulder and moves scapula	Possible in anterior or posterior view of abdomen	Spinous processes T-7 through T-12, lumbodorsal fascia, crest of ilium, and ribs (10, 11, and 12)	Lower bicipital groove	Eleventh cranial nerve and third and fourth cervical nerves
Rhomboideus major	Connects upper extremity to vertebral column	Retracts scapula	Possible in posterior view of scapula	Spinous processes T-2 through T-5 and spinal ligament	Scapula	Dorsoscapular nerve from C-5

The muscular system

Muscle		Action	X-ray view	Origin	Insertion	Nerve supply
Rhomboideus minor	Connects upper extremity to vertebral column	Retracts scapula	Possible in posterior view of scapula	Ligamentum nucha and spinous processes C-7 and T-1	Scapula	Dorsoscapular nerve from C-5
Deltoideus	Shoulder muscles	Raises arm to horizontal position	Posterior shoulder and lateral axial shoulder views	Clavicle, acromion process, and scapular spine	Tendon into deltoid tuberosity of humerus	C-5 and C-6 via axillary nerve
Teres major	Shoulder muscles	Draws humerus down and back and rotates humerus inwardly	Posterior view of shoulder	Scapula	Lesser tubercle of humerus	C-5 and C-6 through lowest subscapular nerve
Biceps brachii	Arm	Flexes elbow	All views of humerus	Two heads: *short*—via tendon from coracoid process; *long*—from supraglenoid tuberosity	Tendon into radial tuberosity	Musculocutaneous nerve
Brachialis	Arm	Flexes forearm (defense of elbow)	All views of elbow	Anteriorly from lower humerus	Via tendon into ulnar tuberosity and coronoid process (anteriorly)	Musculocutaneous nerve
Triceps brachii	Arm	Extends forearm	All views of humerus	Three heads: *long*—via tendon from infraglenoid tuberosity of scapula; *lateral*—posteriorly from body of humerus; *medial*—posteriorly and medially from body of humerus	Via tendon posteriorly into olecranon process	From C-7 and C-8 through radial nerve

Continued.

Table 10-1. Muscles that make significant radiographic shadows—cont'd

Muscle	Group membership	Function	Visualization	Origin	Insertion	Innervation
Pronator teres	Forearm	Rotates radius upon ulna	All views of forearm	Two heads: *ulnar*—medially from coracoid process; *humeral*—from immediately above medial epicondyle	Laterally via a tendon into midbody of radius	Median nerve
Flexor carpi radialis	Forearm	Flexes and extends wrist and assists in pronation and in flexion of elbow	All views of forearm	Medial epicondyle of humerus	Bases of second and third metacarpals	Median nerve
Brachioradialis	Forearm	Flexes elbow	Posterior and anterior views of forearm	Supracondylar ridge of humerus	Base of styloid process of radius	C-5 and C-6 through radial nerve
Extensor carpi radialis longus	Forearm	Abducts hand and extends wrist	Posterior and anterior views of forearm	Lateral condylar ridge of humerus	Base of second metacarpal	C-6 and C-7 through radial nerve
Extensor carpi radialis brevis	Forearm	Abducts hand	Posterior and anterior views of forearm	Lateral epicondyle of humerus	Base of third metacarpal	C-6 and C-7 through radial nerve
Extensor digitorum communis	Forearm	Extends phalanges, wrist, and elbow	Posterior and anterior views of forearm	Lateral epicondyle of humerus	Middle and terminal phalanges of second through fifth digits	C-7 through deep radial nerve
Extensor carpi ulnaris	Forearm	Radial deviation; extends wrist	Posterior and anterior views of forearm	Lateral epicondyle of humerus	Tubercle on ulnar side of base of fifth metacarpal	C-7 through deep radial nerve

Abductor pollicis brevis	Hand	Draws thumb forward in a plane of 90 degrees to plane of palm of hand	Anterior, posterior, and lateral views of hand or thumb	Scaphoid and trapezium	Base of first phalanx of thumb	Median nerve
Opponens pollicis	Hand	Draws first metacarpal over palm of hand	Anterior, posterior, and lateral views of hand or thumb	Trapezium	First metacarpal	Median nerve
Flexor pollicis brevis	Hand	Flexes and abducts first phalanx of thumb	Anterior, posterior, and lateral views of hand or thumb	Two portions: *lateral*—trapezium; *medial*—first metacarpal	First phalanx of thumb (sometimes a sesamoid bone is in the tendon)	Median nerve (lateral portion) and ulnar nerve (medial portion)
Abductor pollicis obliquus	Hand	Pulls thumb toward palm of hand	Anterior, posterior, and lateral views of hand or thumb	Capitate and second and third metacarpals	Proximal phalanx of thumb via a tendon containing a sesamoid bone	Ulnar nerve
Adductor pollicis transversus	Hand	Pulls thumb toward palm of hand	Anterior, posterior, and lateral views of hand or thumb	Third metacarpal	Base of proximal phalanx of thumb	Ulnar nerve
Psoas major	Iliac region	Moves trunk and flexes hip	Anterior and posterior views of L-spine	Transverse processes of L-vertebrae, sides of T-12 through L-5, intervertebral disc, and tendinous arches	Lesser trochanter of femur	Second and third lumbar nerves
Quadriceps femoris 1. Rectus femoris	Thigh Thigh	Assists psoas major and iliacus in supporting pelvis and trunk	Posterior and lateral views of thigh and femur	Anterior inferior iliac spine and groove above brim of acetabulum	Base of patella	Femoral nerve

Continued.

Table 10-1. Muscles that make significant radiographic shadows—cont'd

Muscle	Group membership	Function	Visualization	Origin	Insertion	Innervation
Quadriceps femoris—cont'd						
2. Vastus lateralis	Thigh	Extends leg and draws patella upward	Posterior and lateral views of thigh and femur	Intertrochanteric line and other parts of upper femur	Patella	Femoral nerve
3. Vastus medialis	Thigh	Draws patella medialward and upward	Posterior and lateral views of thigh and femur	Lower one-half of intertrochanteric line	Patella and quadriceps femoris tendon	Femoral nerve
4. Vastus intermedius	Thigh	Draws patella upward	Posterior and lateral views of thigh and femur	Body of femur	Quadriceps femoris tendon	Femoral nerve
Adductor longus	Thigh	Holds knee inward (as in grasping saddle in horseback riding) and adducts thigh	Posterior views of femur	Symphysis pubis	Linea aspera	Obturator nerve
Adductor magnus	Thigh	Holds knee inward (as in grasping saddle in horseback riding) and adducts thigh	Posterior views of femur	Inferior rami of pubis and ischium and the ischial tuberosity	Linea aspera and adductor tubercle on medial condyle of femur	Obturator nerve
Gluteus maximus	Thigh	Extends femur, supports pelvis, and raises trunk from stooping position	Lateral and posterior views of hip and posterior views of pelvis	Posterior gluteal line of ilium; sacrum and coccyx	Fascia lata to gluteal tuberosity	Inferior gluteal nerve

Muscle	Location	Action	Origin	Insertion	Nerve
Biceps femoris (a hamstring muscle)	Thigh	Flexes leg upon thigh	Posterior and lateral views of femur	Two heads: *long*—ischial tuberosity; *short*—lateral lip of linea aspera	Head of fibula and lateral condyle of tibia	Common peroneal nerve
Semitendinosus (a hamstring muscle; has an extremely long tendon of insertion)	Thigh	Flexes leg upon thigh	Posterior and lateral views of femur	Ischial tuberosity	Body of tibia	Tibial nerve
Semimembranous (a hamstring muscle)	Thigh	Flexes leg upon thigh	Posterior and lateral views of femur	Ischial tuberosity	Medial condyle of tibia	Tibial nerve
Gastrocnemius	Leg	Extends foot at ankle and flexes femur upon tibia	Posterior and lateral views of lower leg (tibia and fibula)	Two heads: *medial*—medial condyle of femur; *lateral*—lateral condyle of femur	Aponeurosis into a tendon which unites with the tendon of the soleus muscle to the tendo calcaneus	Tibial nerve
Soleus	Leg	When standing, steadies leg on foot and prevents the body from falling forward	Posterior and lateral views of lower leg (tibia and fibula)	Head and body of fibula and popliteal line of tibia	Joins with the tendon of gastrocnemius to form the tendo calcaneus	Tibial nerve
Plantaris	Thigh	Accessory to gastrocnemius	Posterior and lateral views of lower leg (tibia and fibula)	Linea aspera and oblique popliteal ligament	Tendo calcaneus, into the calcaneus	Tibial nerve
Peroneus longus	Thigh	Extends foot on leg and everts foot	Posterior and anterior views of tibia and fibula	Head of fibula and sometimes lateral condyle of tibia	Tendon, laterally into base of first metatarsal and into medial cuneiform	Superficial peroneal nerve

or visualization in specific radiographs. Table 10-1 lists many of the skeletal muscles that make significant radiographic shadows. The muscle list includes group membership, general and specific functions, and the sites of principal origin and insertion.

Some of the surface and more prominent muscles are visualized in Fig. 10-1.

Among the several muscles that are distinctly visible in radiographs, the diaphragm is perhaps the most outstandingly visible. This large muscle is visualized in all radiographs involving the lower thoracic and upper abdominal cavities. *The diaphragm is that partition composed of muscles and sinews separating the thoracic and abdominal cavities.*

The diaphragm is large and shaped much like a dome; the sides are muscular and the upper midportion is tendinous. The diaphragm is innervated from the phrenic and sympathetic nerves. The muscles of the diaphragm originate from the inner surfaces of the xiphoid cartilage, the lower six costal cartilages and associated ribs, and some of the lumbar vertebrae. These several muscles of the diaphragm insert medially into the central tendon. The diaphragm either assists or is the principal muscle in several body functions; among these are respiration, defecation, and parturition.

In the diaphragm are three major openings: the *aortic hiatus, esophageal hiatus,* and *vena caval foramen.* The aortic hiatus is anterior to and slightly to the left of the vertebral column. It transmits the aorta, azygos vein, and thoracic duct (the azygos vein occasionally passes through the right crus). The esophageal hiatus is slightly to the left of the midsagittal plane and near the exact center of the coronal plane. It transmits the esophagus, vagus nerves, and some small esophageal blood vessels. The vena caval foramen is anterior to the esopha-

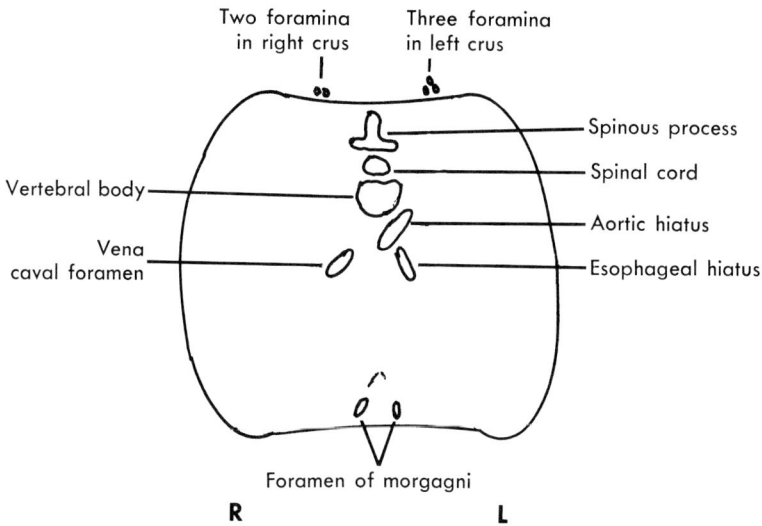

Fig. 10-4. Schematic drawing of the diaphragm, superior aspect.

geal hiatus and slightly to the right of the midsagittal plane (see Fig. 10-4). It transmits the inferior vena cava and some branches of the right phrenic nerve. The foramen of Morgagni is a small gap on either side of the midline, between the sternal and costal portion of the diaphragm, for the passage of the superior epigastric blood vessels and a few lymphatic vessels. Radiologists evidence interest in the region of the foramen of Morgagni in the upper gastrointestinal (UGI) examinations. The two foramina in the right crus transmit the greater and lesser right splanchnic nerves, and the three foramina in the left crus transmit the greater and lesser left splanchnic nerves and the hemiazygos vein.

The diaphragm is superior to both the stomach and liver; it is somewhat higher in the body (over the liver) on the right side than on the left side. Herniations through the diaphragm may occur as a result of trauma or disease, or they may be congenital. Hernias usually occur near natural openings; if there is herniation through a natural hiatus, the condition is called a *hiatus hernia.* Hernias occur in other parts of the diaphragm, usually on the patient's right side. Hernias may be determined radiographically during barium enema examination, fluoroscopy of the large intestine with the patient in the Trendelenburg position. The patient is asked to exhale deeply, then to inhale sharply.

In addition to the diaphragm and other muscles listed in Table 10-1, several other striated muscles are of interest to radiologic technologists, although they cast no significant, if any, radiographic shadows. Among these muscles are the sartorius (the longest muscle in the body), the stapedius (the shortest muscle in the body), the tensor tympani, the nine muscles of the tongue, and the four muscles of mastication.

The muscles of the tongue are classified as either extrinsic or intrinsic. The extrinsic muscles of the tongue have their origins outside the tongue on the hyoid and mandible* and their insertions within the tongue; these muscles are the genioglossus, hyoglossus, chondroglossus, palatoglossus, and styloglossus. The intrinsic muscles of the tongue have their origins and insertions within the tongue; these are the longitudinal superior, longitudinal inferior, transverse, and vertical muscles of the tongue. All the extrinsic and intrinsic tongue muscles receive their innervation from the hypoglossal nerve, except for the palatoglossus muscle, which receives its innervation from the accessory nerve.

The masseter, one of the four muscles of mastication, is listed in Table 10-1. A second muscle of mastication is the temporal muscle, which raises the mandible, retracts the lower jaw, and closes the mouth. The temporal muscle originates in the temporal fossa and associated fascia and inserts via a tendon into the medial surface, apex and anterior border of the coronoid process, and the anterior border of the ramus of the mandible. The temporal muscle is innervated from the mandibular nerve.

---

*The genioglossus originates via a short tendon from the superior spine, posteriorly on the symphysis mandibulus. The hyoglossus originates from the length of the greater cornu of the hyoid. The chondroglossus originates from the base and medial side of the lesser cornu of the hyoid.

A third muscle of mastication is the external pterygoid, which protrudes the lower jaw, enables a grinding action during mastication, and assists in opening the mouth. The external pterygoid muscle originates from the lateral surface of the great wing of the sphenoid bone and infratemporal crest and from the lateral surface of the lateral pterygoid plate. The external pterygoid muscle inserts into the anterior surface of the condylar neck and anterior margin of the articular disc of the temporomandibular joint. The external pterygoid muscle is innervated from the mandibular nerve.

The fourth muscle of mastication is the internal pterygoid, which raises the mandible, enables a grinding action in mastication, and closes the mouth. The internal pterygoid muscle originates from the medial surface of the lateral pterygoid plate, the grooved surface of the palatine pyramidal process, and the maxillary tuberosity. The internal pterygoid muscle inserts via a tendinous lamina into the lower and back part of the medial surface of the mandibular ramus and angle. The internal pterygoid muscle is innervated from the mandibular nerve.

Visceral muscles

In embryonic development all of the visceral (nonstriated) muscle tissue arises from the mesenchyme.

Shape

The fibers of visceral muscles are either fusiform or spindle shaped. The fibers are arranged in sheets (layers); the length of the visceral muscle cell usually is parallel with the long axis of the sheet or of the organ.

Attachment

Nonstriated muscle cells are found in the walls of most organs and in many of the blood vessels. In many instances, the nonstriated muscles comprise a major part of the organ wall and are maintained in place by means of connective tissues. Except in a few instances, there are no true origins or insertions of nonstriated muscle cells; they attach to adjacent muscle cells or to connective tissues.

Function

The visceral muscles function to supply the motive power (energy and motion) used in circulation, digestion, secretion, and excretion. The power used results from stimuli supplied from the autonomic nervous system.

The constant movement exhibited by the stomach and small and large intestines is peristalsis,* which is not limited to these digestive system structures.

Throughout the skin are small collections of nonstriated muscle fibers, the *arrector pili*. These small muscles cause goose flesh when a person is chilled;

---

*Peristalsis is the wormlike movement by which the alimentary canal or other tubular organs provided with both longitudinal and circular muscles fibers propel their contents. It consists of a wave of contraction passing along the tube.

they also cause the hair on the nape of a dog's neck to stand on end (bristle) when the dog is angry.

Cardiac muscles

In embryonic development all of the cardiac (branching) muscle tissue arises from the mesoderm.

Shape

The fibers of cardiac muscle cells are arranged in a syncytium. The fibers are long and multinucleate like the skeletal muscle cells; the cardiac muscle cells possess cross striations like those found in skeletal muscle cells.

Attachment

The cardiac muscle cells comprise the walls of the heart and, like the non-striated muscle cells, are attached only to each other or to connective tissues.

Function

Cardiac muscles function to maintain the flow of blood and, indirectly the flow of lymph throughout the entire body. Blood is forced from one chamber to the next and from the left and right ventricles by the contractions of the heart muscle when those parts of the heart are in systole. Blood returns to the heart and flows into the several chambers when the chambers are in diastole. The cells of the heart muscle (myocardium) are supplied by nerves from the autonomic nervous system.

Both visceral and cardiac muscles exhibit intrinsic contractility, which is regulated by impulses from the autonomic nervous system. In addition to autonomic innervation and control of both types of fibers, there is humoral control of these fibers.

REFERENCES

Francis, C. C, and Martin, A.: Introduction to human anatomy, ed. 7, St. Louis, 1975, The C. V. Mosby Co.

Goss, C. M.: Gray's anatomy, ed. 27, Philadelphia, 1959, Lea & Febiger.

Guyton, W. C.: Function of the human body, ed. 2, Philadelphia, 1964, W. B. Saunders Co.

Jacobi, C. A., and Paris, D. Q: Textbook of radiologic technology, ed. 5, St. Louis, 1972, The C. V. Mosby Co.

Schottelius, B. A., and Schottelius, D. D.: Textbook of physiology, ed. 17, St. Louis, 1973, The C. V. Mosby Co.

CHAPTER 11

The reproductive system

The organs and glands of the reproductive system are specialized according to sex. Most of the structures are comparable between the sexes and, in most instances, have the same blood and nerve supplies. Certain muscles are common to both sexes in the reproductive system; others are not. There are structures in each sex common to both the reproductive and urinary systems: the urethra in the male and the vulva in the female.

The reproductive system *gonads* (testes and ovaries) secrete hormones that influence normal body development. Some of the secondary sex characteristics develop as a result of the aforementioned hormones; other secondary characteristics develop from stimuli of other hormones.

The structures of the male reproductive system are the testicle (testis), epididymis, rete testis, scrotum, spermatic cord, vas deferens, seminal vesicle, ejaculatory duct, prostate gland, bulbourethral (Cowper's) gland, and penis (containing the urethra).

The structures of the female reproductive system are the ovary, uterine tube (fallopian tube or salpinx, including the fimbria and infundibulum), uterus (including the cervix), vagina, vestibular (Bartholin's) gland, and vulva.

The breast is common to both sexes and develops in the female at puberty. During pregnancy the female breast undergoes specific changes in preparation for parturition. After the cessation of lactation (milk secretion), the breast resumes its normal condition of a nonsecreting gland. The breast is influenced by both primary and secondary sex hormones.

Other than the erectile tissue in the breast nipple, the erectile tissue in both sexes is innervated from the sacral plexus via the pudendal nerve. The gonads receive arterial blood from either the spermatic or ovarian artery. The spermatic and ovarian arteries arise from the same locus on the abdominal aorta.

ANATOMY
Male

The *scrotum* is the pouch that contains the testes and their accessory organs. The scrotum suspends from the pubic and perineal regions directly behind the

The reproductive system

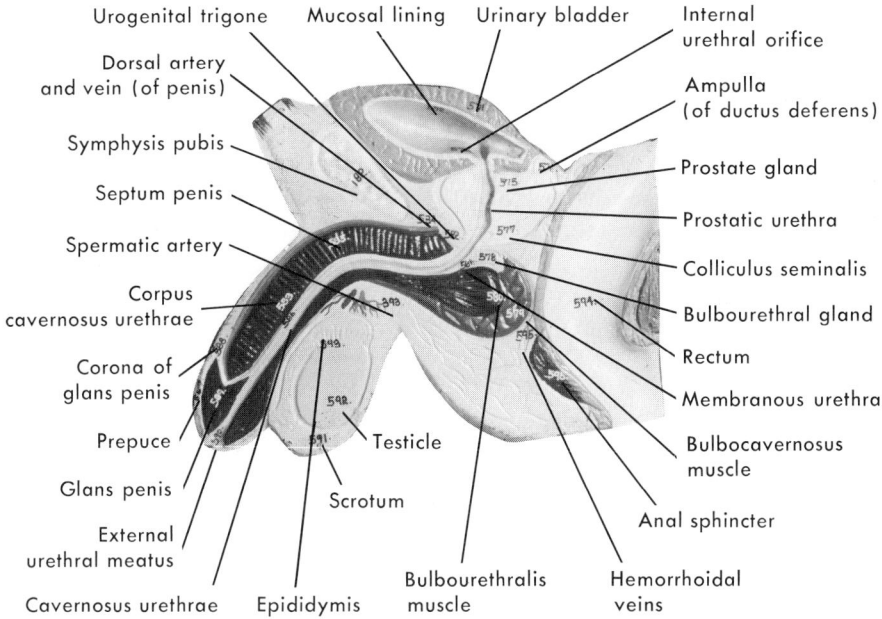

Fig. 11-1. Male genitalia and related structures, median aspect.

root of the penis. The scrotum consists largely of skin and subcutaneous tissue with nonstriated muscle fibers, the dartos muscle, in the subcutaneous tissue. The midline ridge on the scrotum is the *raphe*. The raphe begins at the inferior surface of the *glans penis* (head of the penis) and extends posteriorly across the scrotum and beyond to terminate at the anus. The pelvic floor and the associated structures occupying the pelvic outlet comprise the *perineum*, which is bounded anteriorly by the pubic symphysis, laterally by the ischial tuberosities, and posteriorly by the coccyx.* The rounded prominence at the symphysis pubis in the female is the *mons veneris*. The scrotal tissue in the male corresponds to the tissue of the labia majora (major or large lips) in the female (see Fig. 11-1).

The testes develop in intrauterine life on the genital ridge within the abdominal cavity and descend through the inguinal canal into the scrotum prior to birth. Failure of both testes to descend results in *cryptorchidism*. The spermatic arteries arise from the abdominal aorta to supply each testis. The vein from the right testis empties directly into the inferior vena cava. The vein from the left testis empties into the left renal vein. The lymphatic veins from both testes empty into lymph nodes surrounding the aorta. The spermatic cord includes the aforementioned vessels and the vas deferens.

As with the vascular supply, the nerve supply of the testes has a different origin than the nerve supply of the scrotum. The thoracolumbar autonomic out-

---

*An incision (to facilitate birth) extending dorsally from the vulva in the perineum is an *episiotomy*.

flow supplies nerve fibers to the blood vessels of the testes. Visceral sensory fibers also supply the blood vessels of the testes.

Each testis is an oval-shaped gland, about 5 cm. long. The duct of the testis arises posteriorly as the epididymis. Each testis is completely covered with the *tunica albuginea*,* which sends radiating septa into the substance of the gland to form the fibrous framework of this gland. Between each two septa is the glandular substance that contains the greatly convoluted seminiferous tubules.

The *epididymis* is the elongated, cordlike structure along the posterior border of the corresponding testis. It consists of a head, a body, and a tail. The ducts of the epididymis store the sperms (spermatozoa), the mature male germ cells secreted as the specific output of the testes. The *ductus (vas) deferens* is the efferent continuation of the duct beginning in the testis.

The ductus deferens is approximately 45 cm. long. From the lower end of the epididymis the duct extends upward in the spermatic cord, passing through the inguinal canal into the abdominal cavity. The duct passes over the brim of the pelvis and across the urinary bladder to descend a short way and dilates to form the ampulla and to receive the duct of the seminal vesicle. The ampulla continues as the *ejaculatory duct*. The *seminal vesicle* is the lobulated membranous pouch between the fundus of the urinary bladder and the rectum; the seminal vesicle secretes a fluid that is added to the total ejaculum via the ejaculatory duct. The two ejaculatory ducts enter the prostate gland and join the urethra in this gland.

The urethra originates in the neck of the urinary bladder and, in the male, passes through the prostate gland into the membranous structure at the root of the penis. The urethra then enters the *corpus spongiosum*† of the penis, extends the length of the penis, and opens externally through the meatus in the glans penis. The male urethra is approximately 21 cm. long and consists of three parts: the prostatic, membranous, and spongy urethra.

The *penis* consists of three longitudinal columns of erectile tissue bound together with fibrous connective tissue and contained within a thin layer of subcutaneous tissue and skin.‡ The ventral (median) column is the corpus spongiosum penis (corpora cavernosum urethrae). The dorsal (lateral) two columns are the *corpora cavernosa penis*.§ The latter two lie alongside each other and above the more centrally located corpus spongiosum penis. The corpus spongiosum penis expands in its terminal end to form the glans penis. The glans penis overlays the terminal ends of the corpora cavernosa penis.

The corpus spongiosum penis extends posteriorly into the perineum and ex-

---

*The tunica albuginea of the testis is the dense, white, inelastic tissue immediately covering the testis, beneath the visceral layer of the tunica vaginalis.

†The corpus spongiosum penis is the mass of erectile tissue in the ventral part of the penis containing the urethra.

‡The skin that extends over the glans penis in the noncircumcised male is the prepuce, or foreskin.

§The corpora cavernosa penis are the spongy bodies, the two erectile columns of the dorsum of the penis that fill with blood to produce turgidity.

pands to form the bulb of the penis. The bulbospongiosus muscle covers the bulb, the total structure functioning to aid both in micturition (passage of urine) and in copulation (sexual intercourse, coitus). One of the two small bulbourethral glands is situated on each side of the membranous urethra immediately behind the bulb. Each bulbourethral gland empties via a single, small duct into the spongy urethra.

The prostatic urethra lies within the prostate gland—hence, its name. The prostate gland surrounds the neck of the urinary bladder and consists of three lobes—one median lobe and two lateral lobes. Each lobe contains many small lobules separated from each other by septa. The glandular tissue of the lobules secretes a thin fluid that empties into the prostatic urethra during coitus via several small ducts. The prostatic secretion is strongly alkaline and functions to neutralize the acidic fluid from the testes. Alkalinity causes the sperm to become highly active, whereas sperms are largely inactive in the acidic environment of the epididymis. Prostatic acid phosphatase, an important secretion of the prostate gland, is considered normal if the quantity does not exceed 20 percent of the total acid phosphatase.

The secretions of the bulbourethral glands and other small mucous glands in the skin along the course of the urethra function as lubricant fluids during coitus. The secretion of the seminal vesicles functions in maintaining the health of the sperms.

Female

The female gonads, *ovaries*, lie within the abdominal cavity at all times. As in the male gonads, the ovaries arise from the genital ridge in the embryo. Differential growth situates the ovary directly above the pelvic inlet. After birth, the ovary gradually descends to its position on the lateral pelvic wall. It remains there until pregnancy, when the enlarging uterus may force the ovary into a new position. The lower part of the ovarian ligament extends through the inguinal canal between the uterus and the labium majus (see Fig. 11-2).

The tissue in the female that corresponds to the male scrotum is the paired labia majora. The ovarian arteries arise from the abdominal aorta to supply each ovary. Several veins (as a plexus) lie alongside each ovarian artery. The veins on the right side empty directly into the inferior vena cava. The veins on the left side empty into the left renal vein. The lymphatic veins from the ovaries pass upward with the ovarian vessels to empty into lumbar and aortic lymph nodes.

As in the male reproductive system, the thoracolumbar autonomic outflow supplies sensory nerve fibers and filaments to the blood vessels of the ovaries in the female reproductive system.

Each ovary is an almond-shaped gland, about 4 cm. long. The ovary is quite solid and, unlike the testis, has no directly connected duct. The ovary is attached to the broad ligament of the uterus by the mesovarium, a fold of the peritoneum, and is situated laterally in the pelvic cavity. The ovarian liga-

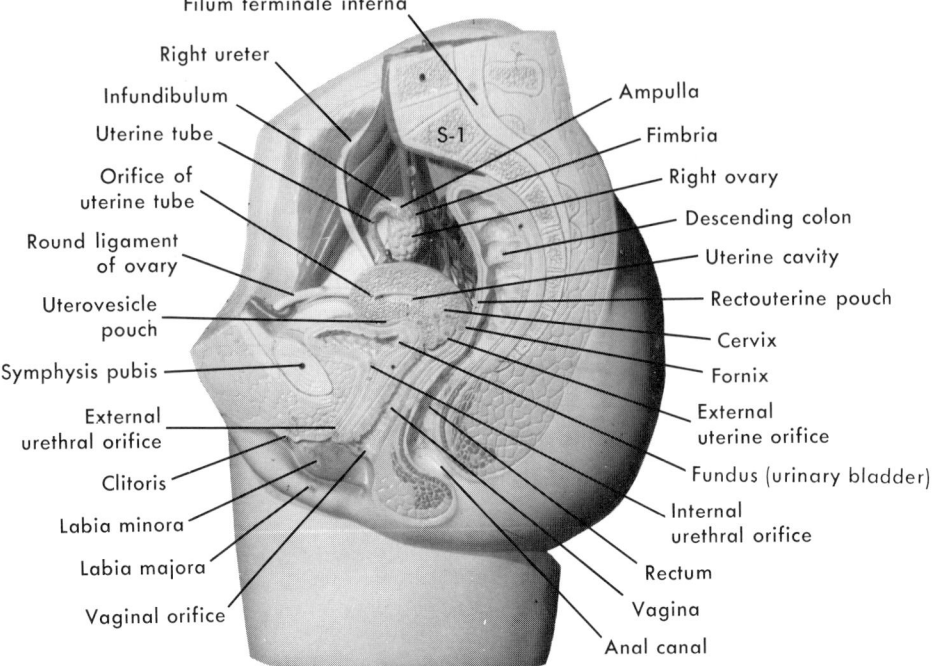

Fig. 11-2. Female genitalia and related structures, median aspect.

ment lies between the layers of the broad ligament and extends between the ovary and the upper part of the lateral border of the uterus. The infundibulopelvic ligament extends between the lateral pelvic wall and the ovary and fimbriated end of the uterine tube.

The uterine tube extends between the ovary and one cornu (horn) of the uterus. The tube connects directly to the uterus but has no direct connection to the ovary. The ovarian end of the tube is funnel shaped (the infundibulum), and the border of the infundibulum is fimbriated and called the *fimbria*. The inner wall of the uterine tube is composed of ciliated columnar epithelium, and nonstriated muscle fibers are found in these walls. Thus, the ova are propelled along the tubular lumen toward the uterus by peristaltic movement and by the movement caused by the cilia.

The uterus, a hollow, thick-walled, muscular organ about the size of a doubled fist, is situated in the anterior part of the pelvic cavity between the bladder anteriorly and the rectum posteriorly. The uterus is larger in its fundic end than in either the body or the cervical end. The uterine lumen is nearly triangular in shape, with its base situated toward the fundus, which is anterosuperior in the pelvis. The two uterine tubes enter the uterus bilaterally in the base through the cornua, and the fundus of the uterus lies above the tubular openings. The body of the uterus comprises about two-thirds of the total structure and terminates distally in the cervix (uterine neck). The cervix extends into

The reproductive system **383**

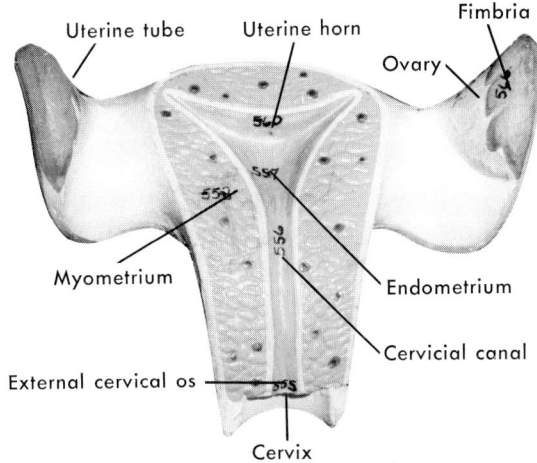

Fig. 11-3. Uterus, cross section showing walls, lumen, and ovaries.

the vagina. The apex of the uterus opens through the cervix into the vagina (see Fig. 11-3).

The uterus receives its blood supply via the uterine artery, a branch of the hypogastric artery. The uterus is innervated from the hypogastric plexus (thoracolumbar autonomic outflow) and from some of the sacral nerves (sacral autonomic outflow).

Six ligaments attach the uterus to the pelvic walls, rectum, and urinary bladder. The ligaments are the anterior and posterior ligaments, two broad ligaments, and two round ligaments. The fold of the peritoneum extending between the uterus and the bladder comprises the *anterior ligament*. The *posterior ligament* extends between the vagina and the rectum. Two wide folds of peritoneum extending between each side of the uterus and corresponding lateral pelvic wall comprise the *broad ligaments,* which, with the uterus, divide the pelvic cavity into the anterior and posterior compartments. Each broad ligament contains the corresponding uterine tube. Two narrow and flat bands of connective tissue extend between the layers of the broad ligaments from the upper uterus to the labia majora and comprise the *round ligaments.* These ligaments contain nonstriated muscle fibers. The *uterosacral ligaments* are not included in the six previously listed ligaments; the uterosacral ligaments are folds of peritoneum extending between the uterus and the sacrum.

Three layers of tissue comprise the uterine wall. The outer layer consists of longitudinal, circular, and oblique nonstriated muscle fibers. The inner layer, the *endometrium,* consists of a mucous coat of columnar and ciliated columnar epithelium and tubular glands. At the cervix the endometrium is continuous with the vaginal stratified squamous epithelium.

Surrounding the cervical neck in the vagina is a circular recess called the *fornix;* this recess is often subdivided into anterior, posterior, and lateral fornices.

The vaginal canal extends downward and forward from the uterus to the vulva. The sheath (vagina) is about 8 cm. long, the posterior wall being longer than the anterior wall. The anterior wall is 7.5 cm. long, and the posterior wall is 9 cm. long. The vaginal wall consists of a mucous membrane (stratified squamous epithelium) lining inside a circular layer of nonstriated muscle and an outer layer of longitudinal nonstriated muscle. The vagina opens externally between the labia minora.

Several structures combine to form the *vulva* (external genitalia). These are the labia majora, labia minora, mons pubis, clitoris, and vaginal vestibule.

The *labia majora* correspond to the male scrotum in structure but not in function. The labia majora are two folds of skin that extend longitudinally, backward and downward from the mons pubis,* to join about 3 cm. anterior to the anus.

The *labia minora* are also folds of the skin, but they are smaller than the labia majora. The labia minora are situated beneath the cover of the labia majora and extend posteriorly from the clitoris. Between the labia minora is a cleft called the *vestibule* (vaginal vestibule). The urethra opens superiorly in the vestibule above the vaginal opening.

The *mons pubis* consists principally of fatty tissue and is situated as a rounded mound anterior to the pubic symphysis.

The *clitoris* consists of erectile tissue and is situated between the anterior ends of the labia minora in front of and above the urethral orifice. This structure corresponds to the male glans penis, and the erectile tissue forms two corpora cavernosa and a small glans. The clitoris attaches on each side via a crus to the corresponding ischial and pubic rami. The bulbocavernosus muscle covers the crus. Two elongated masses of erectile tissue, one on either side of the vaginal orifice, comprise the vestibular bulb, which is the homologue of the bulb of the penis in the male. The vestibular (Bartholin's) glands are situated in the corresponding vestibular bulbs. The duct of each vestibular gland empties into the corresponding groove between the labium minus and the attached hymen† margin. The secretions of these glands function as lubricant fluids in the lower vaginal tract during coitus.

STRUCTURE AND FUNCTIONS OF THE GONADS
Male

During intrauterine life the peritoneum evaginates to form the vaginal process. This process passes into the inguinal canal and scrotum of the male and into the inguinal canal and labium majus of the female. Each testis descends through the inguinal canal behind the vaginal process and into the scrotum. As the testis descends, three layers of tissue form investments surrounding the

---

*The terms "mons pubis" and "mons veneris" are synonymous and are used interchangeably.
†The hymen is the membranous fold that partially or wholly occludes the external orifice of the vagina in most young women.

testis and spermatic cord: the *external spermatic fascia* from the aponeurosis of the external oblique muscle, the *cremasteric fascia* from the internal oblique muscle, and the *internal spermatic fascia*. The internal spermatic fascia is continuous with the transversalis fascia. In addition to these investments, the vaginal process separates from the peritoneum to form a closed sac, partially investing the testis.

The *tunica albuginea*, a dense, white, fibrous tissue, forms the outer coat of each testis. Extending from the inner surface of the tunica albuginea are numerous fibrous septa that pass deep into the glandular substance to form wedge-shaped lobules. The several septa continue beyond the lobules and join the *mediastinum testis* (the partial septum of the testis). This structure contains the arteries, veins, lymph veins, and nerves that supply the testis.

Each lobule of the testis contains from one to three tortuous convoluted tubules (seminiferous tubules). The tubules from several adjacent lobules join to form straight ducts. The straight ducts join to form the *rete testis*,* which leads into several small, straight ducts. The latter ducts enlarge and become quite tortuous and finally unite to form the epididymis.

Several layers of specialized cells form the lining of the seminiferous tubules. The outer layer of cells lies upon the basement membrane. Some of these cells are cuboidal epithelium; others are columnar epithelium. The columnar cells support the cuboidal cells and extend inwardly toward the lumen of the tubule; the columnar cells are the *cells of Sertoli*. The cuboidal cells are the *spermatogenic cells*, which, through the process of meiosis, eventually give rise to mature spermatozoa.

The inner layers of cells form the actual wall of the tubule and consist entirely of spermatogenic cells in advancing stages of meiosis; the more nearly mature cells are found closer to the lumen. As a spermatogenic cell matures into a spermatocyte, it detaches from the wall of the lumen. The spermatocyte, by means of its tail, begins to migrate along the tubules leading to the vas deferens.

Connective tissue fills the spaces between the seminiferous tubules and contains the *interstitial cells* (cells of Leydig). These cells contain yellow granules in their cytoplasm and are believed to secrete *testosterone*, the male hormone.

Fertilization of the ovum is the function of the mature spermatozoa. However, large numbers of healthy sperms are required for male fertility. The normal (average) quantity of ejaculum is approximately 3 ml. and each milliliter of ejaculum will contain approximately 120,000,000 (360,000,000 total) sperms for the man to be (considered) fertile. When the sperm total falls below 150,000,000 sterility is probable as a (theorized) result of insufficient quantity of *hyaluronidase* to disperse the granulosal cells covering the ovum. When the enzyme, hyaluronidase is present in sufficient quantity, the sperms are per-

---

*The rete testis is the network consisting of the vasa recta in the mediastinum testis.

mitted (enabled) to "attack" the ovum. For fertility a preponderance of normal and viable sperms is also necessary.

Female

Like the testis, the ovary consists of specialized epithelial tissue cells supported by connective tissue. Unlike the testis, this (germinal) epithelial tissue lies on the surface of the ovary rather than being arranged in tubules within the gland (as in the testis). Early in the developmental stages of the ovary some cells of this germinal epithelium migrate into the mass of the ovary. The body (stroma) of the ovary consists largely of connective tissue that contains the blood vessels and nerves.

With the development of both the ovarian stroma and the surrounding germinal epithelium, the epithelium extends inwardly into the stroma at irregular intervals; this results in a mass of epithelial cells becoming separated from the main layer.

A single cell of this mass gives rise to an oocyte (immature ovum), and the remaining cells arrange in a layer to surround the oocyte and form a follicle.* This is the *primordial*† *follicle*. The primordial follicles number in the thousands in the ovaries of female children and young women; the total number of immature ova in the two ovaries is approximately 400,000 and most of these follicles and ova are absorbed by the ovarian stroma. Those remaining for eventual ripening are limited to a relatively small number; estimates are that a human female may produce about 400 mature ova during her reproductive years.

During the reproductive years (from puberty‡ to menopause§) the FSH (follicle-stimulating hormone of the anterior lobe of the pituitary gland) causes the primordial follicles to ripen. Following puberty the follicles migrate more deeply within the ovarian stroma, and the cells surrounding the oocyte continue to increase in number. As the cells increase in number a follicular fluid *(liquor folliculi)* forms and separates those cells immediately around the oocyte from those cells that are more remote from the oocyte. This process results in the *graafian follicle,* a maturing ovarian follicle comprising a theca filled with follicular fluid; the inner part of one side of the graafian follicle bears the cumulus, which contains the ovum (germ cell).

The theca consists of two layers: the *theca externa* (outer layer), which is quite fibrous, and the *theca interna* (inner layer), which is more vascular. In addition to formation of the *cumulus oophorus* (egg-bearing mass) the liquor folliculi collection results in pressure against the remaining cells and forces them against the follicular wall, forming the *membrana granulosa* (the cell layer lining the inner surface of the graafian follicle). A single layer of cells, the *corona radiata,* remains in contact with the ovum.

---

*A follicle is a very small excretory or secretory sac or gland.
†Primordial means original or primitive; of the simplest and most undeveloped character.
‡Puberty is the age at which the reproductive organs become functionally operative and the secondary sex characteristics develop.
§Menopause is the period of life when menstruation normally ceases.

Table 11-1. Menstrual cycle

Stage	Length in days	When starts	Description of events
Resting	4 to 5	Immediately following cessation of menstrual flow (color)	The flow begins with a discharge and breakdown of uterine tissue (endometrium) linings and a subsequently very thin mucosa. At commencement, the ovary contains a corpus luteum, from the previous cycle, whose secreting strength is declining. A new graafian follicle is maturing, and filling with follicular fluid and estrone.*
Interval	Approximately 10	Immediately following resting stage	Adaptive changes occur in uterine mucosa in preparation to receive the next fertilized ovum. The endometrium undergoes rapid thickening with markedly increased blood supply. The uterine glands enlarge in size and increase in number. During the first 5 days, the graafian follicle ruptures (ovulation occurs about midway in cycle). The production of estrone reaches its maximum. Following ovulation, the follicle cells undergo rapid multiplication to form the new corpus luteum. New lutein cells commence secreting progesterone, which assists estrone to act on the uterus. The ovum next enters uterine tube where fertilization occurs.
Premenstrual	Approximately 10	Immediately following interval stage	Endometrium continues to thicken, and blood supply continues to increase to successfully contain and nourish fertilized ovum. Fertilization causes corpus luteum to continue maximum secretion toward final preparation of endometrium. If ovum is not fertilized, functions of both ovary and corpus luteum decline rapidly, and endometrium commences regression (commencement of destructive stage).
Destructive	Approximately 4	Immediately following premenstrual if fertilization failed to occur	*Dismantling process* of endometrium terminates, all detritus† discharges from uterus; subsequent cycle commences.

*Estrone is one of the estrogenic steroids. It has been isolated from the urine of pregnant animals and differs chemically to a minor extent from the hormone estradiol, another estrogenic steroid.
†Detritus is the cumulative remains of any broken-down tissue.

During the ripening process of the graafian follicle, it migrates toward the perimeter of the ovary. When the follicle is mature, it ruptures and expels the *nonmatured* ovum into the fimbriated opening of the uterine tube. The ovum continues to mature during its short migration down the uterine tube. The process of expelling the ovum, called *ovulation,* occurs midway between two menstrual periods.

Following ovulation, the cells lining the ruptured graafian follicle multiply and form the *corpus luteum.* The cells forming the new structure contain *lutein,* a yellowish fatty substance—hence the name corpus luteum, or yellow body. The blood clot that forms as a result of the follicular rupture is soon replaced with the corpus luteum, which now begins to secrete *progesterone,* the hormone of the corpus luteum. The female sex hormone that corresponds to the male testosterone is *estradiol.* This hormone, secreted in the ovarian stroma, is responsible for the secondary sex characteristics of females in addition to its role in both *proestrus* and *estrus.**

Menstrual cycle

The menstrual cycle varies in different mammals as to length of time involved. In women the total time involved may vary somewhat, but it is usually 28 days. The total period consists of four stages: resting, interval, premenstrual, and destructive (see Table 11-1).

Fertilization of the ovum

During coitus the spermatozoa are deposited (as a result of ejaculation) posteriorly in the vagina. The sperms,† by means of their tails, then move through the cervix and uterine lumen into one or the other of the uterine tubes.

Following ovulation, the ovum moves down the uterine tube while continuing and completing maturation. When the ovum has reached maturity, it has also migrated down about one-half the length of the uterine tube; for fertilizaation to occur, a normal and healthy sperm must have migrated up this tube to its midpoint at about the same time. For additional information regarding fertilization the student is referred to pp. 50 and 51.

FEMALE MAMMARY GLAND (BREAST)

From birth until puberty there is no visible difference (or histologic difference) between the mammary glands of males and females. Two stages of development of the breast in women exist. The first stage of development occurs with puberty, and the second occurs with pregnancy. In the female the hormone estradiol, the follicular hormone, stimulates the growth of the extensive duct system beginning with puberty. Other than growth of the breast corresponding to the increasingly extensive duct system, there is no further development until

---

*Estrus is a recurrent, restricted period of sexual receptivity in female mammals, marked by intense sexual urge.
†Sperm is a correct shortened term for spermatozoon; the pleural, then, is sperms.

The reproductive system

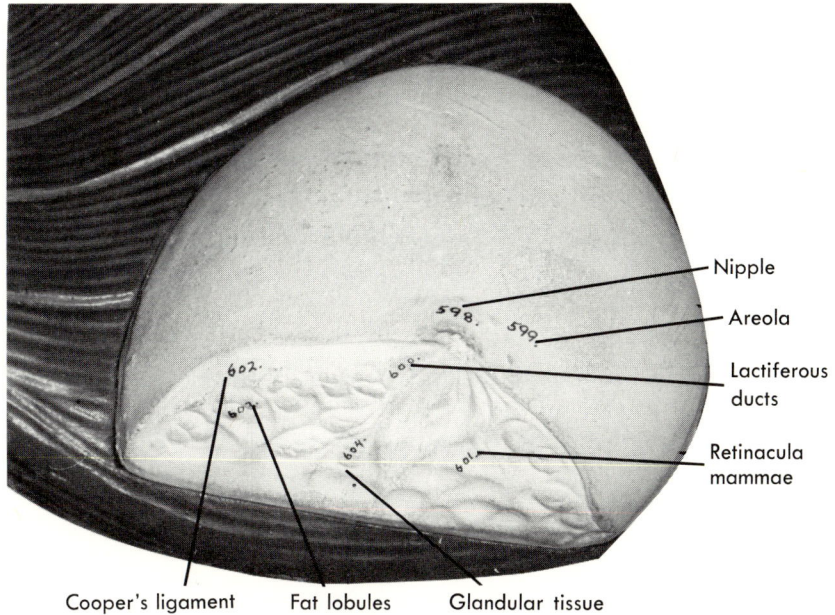

Fig. 11-4. Right breast oblique section.

pregnancy. The growth at puberty is one of the principal secondary sex characteristics of women.

The breast is situated within the subcutaneous tissue on the anterior chest wall and is "suspended" from the corresponding pectoralis major muscle, with its center on the approximate level of the fourth rib interspace. Each breast resembles a hemisphere, with a projection of breast tissue extending from the superolateral margin toward the corresponding axilla. The left breast is usually the larger of the two (see Fig. 11-4).

The breast is composed of epithelial tissue organized to form follicles. A duct from each follicle joins with the system of ducts, which collect into larger ducts eventually to empty through an opening in the nipple. There is one milk (lactiferous) duct for each lobe of the breast; each duct terminates in its opening in the nipple. Connective tissue in the form of septa separate the breast into about twenty lobes. Each lobe is further subdivided into several lobules that contain the follicles (alveoli). Fatty tissue is interspersed between each two lobes to give size to the gland.

In the center of the breast surface is a rounded area of highly pigmented skin called the *areola*. Erectile tissue comprises the nipple, which protrudes from the middle of the areola and contains the several minute openings from the milk ducts.

With pregnancy* progesterone stimulates the follicles to very active growth,

---

*See Chapter 12 for a discussion of lactogenic hormone.

with corresponding enlargement of the whole gland. During the early stages of the first pregnancy, the areola color changes from the prepregnant brownish-pink shade to a dark brown shade; this color persists thereafter. The full development of the breasts normally enables the secretion of milk, usually in the latter stages of pregnancy; however, no milk appears externally until after *parturition*.* The first secretion produced externally for 2 or 3 days after parturition is a viscid and yellowish fluid called *colostrum*. It is sometimes possible to force small quantities of colostrum from the nipples in the third or fourth month of pregnancy. Milk secretion follows when colostrum secretion terminates. Colostrum differs from milk principally in that the colostrum coagulates when heated and contains considerably more protein than milk contains. Human milk contains many of the essential minerals and the foodstuffs in the proper proportions to support good growth of the newborn infant for about the first 6 months of life. The percentages of the contained elements are as follows:

Substance	Approximate percentage
Lactose (milk sugar)	6.7
Protein	1.5
Fat	4.0
Calcium, phosphorus, potassium, sodium, and magnesium	Adequate amounts

Human milk is notably deficient in iron. Cow's milk is also deficient in iron. As a result, babies may become anemic if maintained exclusively on a milk diet much past 6 months of age.

RADIOGRAPHY

Radiography of the male urethra is called *urethrography*. The same is true with the female urethra, although it is seldom radiographed. The urethra is made radiopaque by the injection of contrast medium of relatively high viscosity through the meatus of the glans penis into the lumen of the urethra. The penis is usually strapped against the left thigh. The examination is useful in the diagnosis of certain pathologic conditions of the prostate gland.

Radiography of the uterine lumen and of the uterine tubes is called *hysterosalpingography*; the term is often shortened to *salpingography*. The radiopaque, viscid fluid (contrast medium) is forced through the cervix into the uterine lumen and into the uterine tubes. The examination is usually employed to visualize the tubes to determine if they are patent (open) and to determine if the woman can become pregnant (see Fig. 11-5).

Fetography is the term used for radiography of the unborn fetus. This examination is decreasingly requested and is performed with the recognition of the potential danger to the fetus and to the ovaries of the mother. The request is

---

*Parturition is the act or process of giving birth to the young of the species.

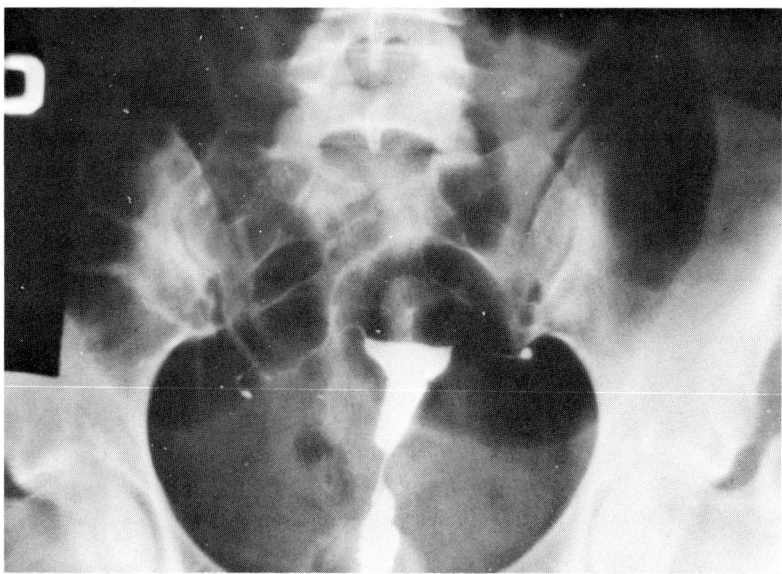

Fig. 11-5. Hysterosalpingogram. Note the radiopaque medium in both uterine tubes (shadows).

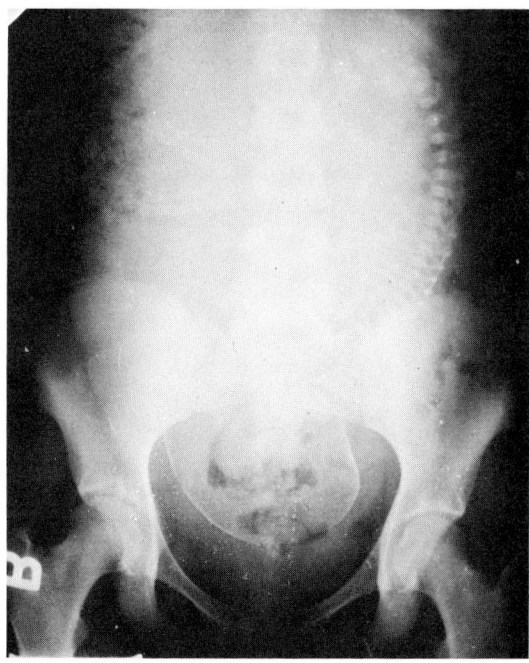

Fig. 11-6. Fetogram.

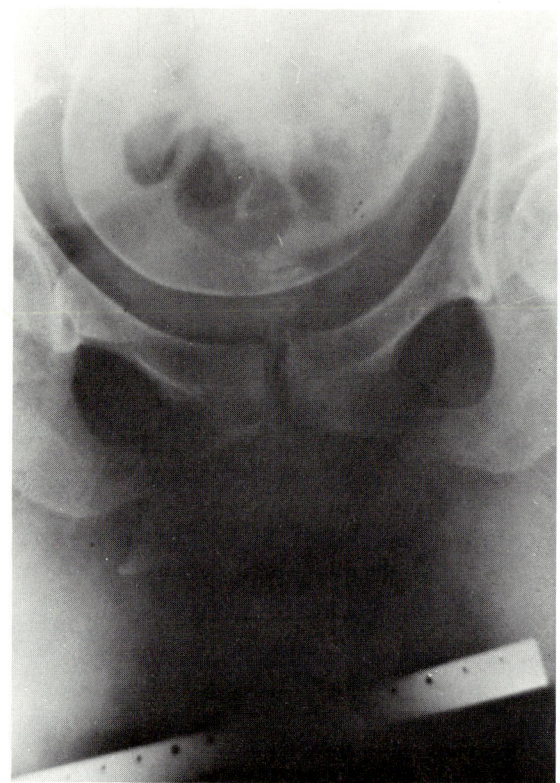

Fig. 11-7. Pelvimetry posterior view.

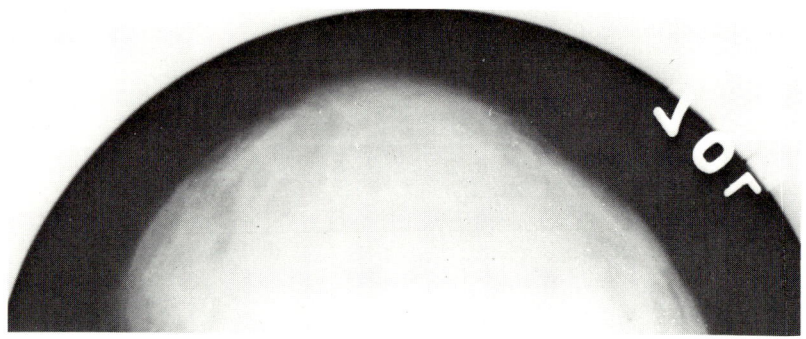

Fig. 11-8. Mammogram craniocaudad view.

usually for one exposure (a flat plate) of the abdomen, including the pubic arch, and is used to determine the position of the fetus. (see Fig. 11-6).

Pelvimetry is a radiographic examination of the abdomen of a pregnant woman made in both posterior and lateral positions using a ruled, metal scale placed near the patient's body and even with the midthickness to determine fetal head size in relation to the size of the pelvic outlet (see Fig. 11-7).

The breast is often radiographed to determine the presence or absence of malignant growths. Radiography of the breast may be performed both with and without contrast media. *Mammography* is the roentgenographic diagnosis of breast tumors. The contrast media employed, if any, may be either radiopaque (some fluid containing iodine) or radiolucent (some substance such as air or oxygen). With either of the types of contrast media, the medium employed is injected through the nipple into each of the lactiferous duct openings. Craniocaudad and lateral views are usually made, although anterior views are not unusual. Crainocaudad views without contrast media are quite common (see Fig. 11-8).

REFERENCES

Best, C. H., and Taylor, N. B.: The human body, ed. 4, New York, 1963, Holt, Rinehart and Winston.

Francis, C. C, and Martin, A.: Introduction to human anatomy, ed. 7, St. Louis, 1975, The C. V. Mosby Co.

Gardner, E., Gray, D. J., and O'Rahilly, R.: Anatomy, Philadelphia, 1960, W. B. Saunders Co.

Lowsley, O. S., Hinman, F., Smith, D. R., and Gutierrez, R.: The sexual glands of the male, New York, 1945, Oxford University Press.

Wharton, L. R.: Gynecology, Philadelphia, 1943, W. B. Saunders Co.

CHAPTER 12

The endocrine system

The endocrine system of glands and tissues functions to regulate and/or control all of the body functions and, therefore, to effect homeostasis. The endocrine system includes all the glands, parts of glands, and tissues that secrete chemical substances *(hormones)* directly into the circulatory system. Glands not included in the endocrine system secrete their substances onto a mucous surface and are called *exocrine* glands. Other glands have both endocrine and exocrine functions and are called *compound* glands. Another anatomic difference between endocrine and exocrine glands relates to the method of emptying; the endocrine glands have no duct through which to empty their secretions, whereas the exocrine glands have their own ducts (see Fig. 2-6).

The principal physiologic similarity among the endocrine glands is their production of hormones. *A hormone is a discrete chemical substance secreted into the body fluids by an endocrine gland; it exhibits a specific effect on the activities of other organs.*

Specific endocrine glands

The specific endocrine glands that belong solely to the endocrine system include the pituitary gland, thyroid gland, parathyroid glands, and adrenal glands (see Fig. 12-1).

Other hormone sources

Compound glands having endocrine (ductless) secretions are the gonads, pancreas, and, probably, the liver. Hormones are secreted also from the gastric and small intestine mucosa. There is inconclusive evidence of hormone secretion by the pineal and thymus glands and by the spleen.

Function

The endocrine system of glands functions to secrete hormones and thereby exerts control over and integration of the numerous and diverse activities of the

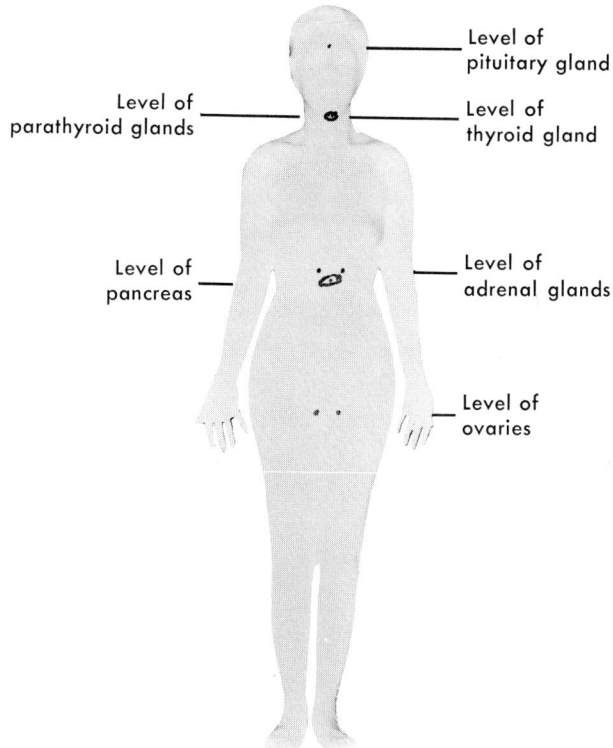

Fig. 12-1. Endocrine and other hormone secreting glands approximate loci.

body. Each gland possesses a rich blood supply into which the hormones are secreted. Since hormones enter the circulatory system directly, they are able to act on all tissues (quickly) as homeostasis necessitates.

ENDOCRINE GLANDS
Pituitary gland

If any one gland of the endocrine system can be singled out as exerting greater influence than the others, it is the pituitary gland. This gland, because of its numerous secretions, is called the master gland. Other names for the pituitary gland are the *hypophysis, hypophysis cerebri,* and *hypophyseal gland.*

The hypophysis lies in the sella turcica (Turk's saddle) of the sphenoid bone. The nervous part (pars nervosa) of the hypophysis develops from an evagination of the diencephalic part of the neural tube. The distal part (pars distalis) of the hypophysis forms from a diverticulum of oral ectoderm. These two parts arise separately in the embryo and fuse prior to birth. In man these parts remain fused (see Fig. 12-2).

The pia mater extends from the brain covering to encapsulate the hypophysis, but it does not extend any prominent trabeculae into the stroma of this gland. The stroma is found principally in the anterior lobe, which derives from the

396 *Textbook of anatomy and physiology in radiologic technology*

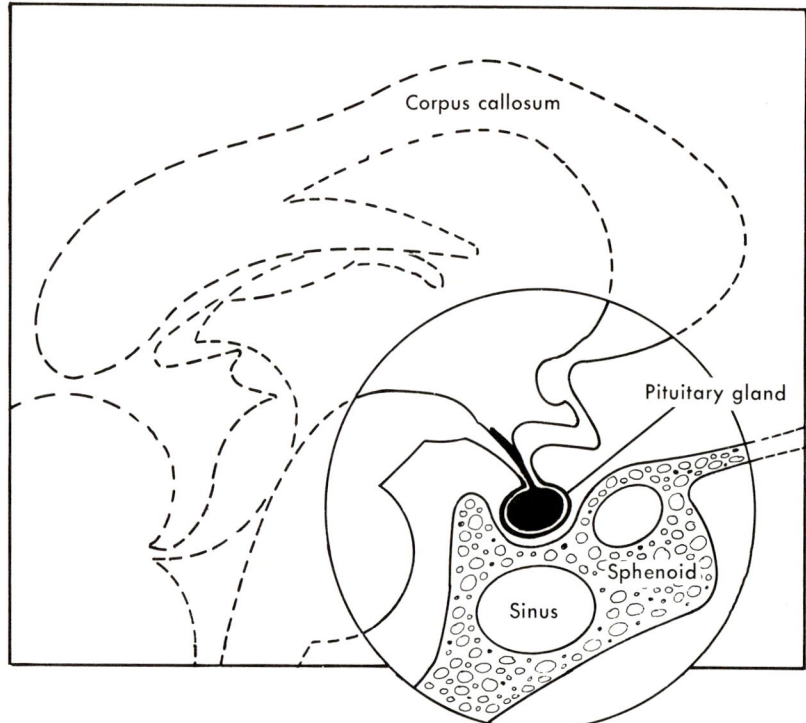

Fig. 12-2. Pituitary gland in sella turcica.

oral ectoderm.* Reticular tissue forms most of the stroma. A few collagenous fibers from the nervous part invade the stroma along the blood vessels, which are profuse in this gland. Sinusoids are quite extensive throughout the stroma. Arteries at the base of the brain (see pp. 245 and 246) give rise to arterioles that supply the two major subdivisions of this gland.

The hypophyseal gland consists of the anterior and posterior lobes, the intermediate part, and the stalk. The stalk attaches this gland to the base of the brain. The intermediate part is often called the intermediate lobe. The intermediate lobe connects the anterior and posterior lobes and has the same origin as the anterior lobe. These three parts of the hypophysis possess separate endocrine functions.

The **anterior lobe** of the hypophysis is more often called the master endocrine gland than is the entire hypophysis. This lobe secretes a large number of hormones, some of which exert direct regulatory control of metabolism in peripheral tissues while others exert differential stimuli upon various other endocrine glands and tissues.

---

*See pp. 52 to 53 for additional material on the origin and development of different body tissues.

STH (somatotropic hormone) is one of the secretions of the anterior lobe. This hormone is the growth-promoting hormone. Although STH exerts the principal control over normal growth, other hormones, hereditary traits, and nutrition also exert considerable effect upon both normal and abnormal growth. Growth in the length of the bones is influenced by STH and is described on p. 186.

Excessive quantities of STH injected into immature animals stimulate the long bones to continue growing in length; the eventual result is *gigantism*. Deficiency in quantity of STH permits early ossification of the metaphyses. As a result, the growth of the individual is stunted; this condition is called *dwarfism*.

Some stimulative effects of STH include soft tissue growth, protein storage, decreased urinary nitrogen excretion, and decreased amino acid content in blood. Other effects achieved as a result of abnormal STH quantity and/or stimuli include glyconeogenesis, hyperglycemia, and depressed respiratory quotient and correspondingly increased oxygen consumption.

Excessive STH resulting from hyperactivity of the anterior lobe of the hypophysis may result in the condition called *acromegaly* in which mature bones have responded to STH stimuli by increasing in size and, sometimes, by changing shape.

Gonadotropic hormones are secreted by the anterior lobe of the hypophysis. These hormones stimulate the gonads of either sex to final maturity.

Lactogenic hormone is also secreted by the anterior lobe of the hypophysis. This hormone stimulates the mammary glands to secrete milk (to lactate) during the latter part of pregnancy and throughout the nursing period of the infant.

TSH (thyrotropic hormone), another hormone secreted by the anterior lobe of the hypophysis, controls the rate of iodine uptake by the thyroid gland. The hormone also controls the synthesis of the thyroid hormone, thyroxine, from diiodotyrosine.

Adrenotropic hormone is also called ACTH (adrenocorticotropic hormone). This hormone also is secreted by the anterior lobe of the hypophysis and controls the activity of the adrenal gland cortex.

The **posterior lobe** of the hypophysis secretes three active principles: vasopressin, oxytocin, and ADH (antidiuretic hormone).

Vasopressin stimulates the arterioles and most of the nonstriated muscle fibers. The stimulating effect exhibited in the walls of the arterioles of the coronary and pulmonary vessels increases blood pressure (see pp. 264 to 266).

Oxytocin stimulates the muscle walls of the uterus. A mixture of both oxytocin and vasopressin may be injected in the muscles of the mother during parturition to assist the natural uterine wall contractions. Oxytocin has been demonstrated as important in the ejection of milk from the human mammary glands.

ADH (antidiuretic hormone) functions to regulate the quantity of urine eliminated to an amount proportional to the quantity of fluid taken into the body. An excessive quantity of ADH causes retention of urine. A deficient

398   Textbook of anatomy and physiology in radiologic technology

quantity of ADH results in the condition called *diabetes insipidus,* an overexcretion of urine containing no sugar and accompanied by great thirst.

The **intermediate lobe** secretion has no known function or effect upon human beings, but causes skin color to change in cold-blooded animals.

Thyroid gland

The thyroid gland, the largest endocrine gland, consists of two lobes situated bilaterally below the larynx and on the ventral surface of the upper part of the trachea. The two lobes are connected by the isthmus, a piece of thyroid gland tissue situated between the two lobes. A large number of small, closed follicles comprise the substance of the thyroid gland. A single layer of columnar or cuboidal epithelium forms the walls of the follicles, which are bound together by areolar tissue. A rich network of capillaries is interspersed throughout the areolar tissue to surround each follicle. A colloidal secretion of the epithelial cells occupies the follicular lumina. This secretion, named *colloid,* is rich in iodine and contains *thyroxine,* the hormone of this gland (see Fig. 12-3, A).

Thyroglobulin is the protein in the thyroid gland. Both thyroglobulin and thyroxine contain large amounts of iodine. The thyroid gland contains almost

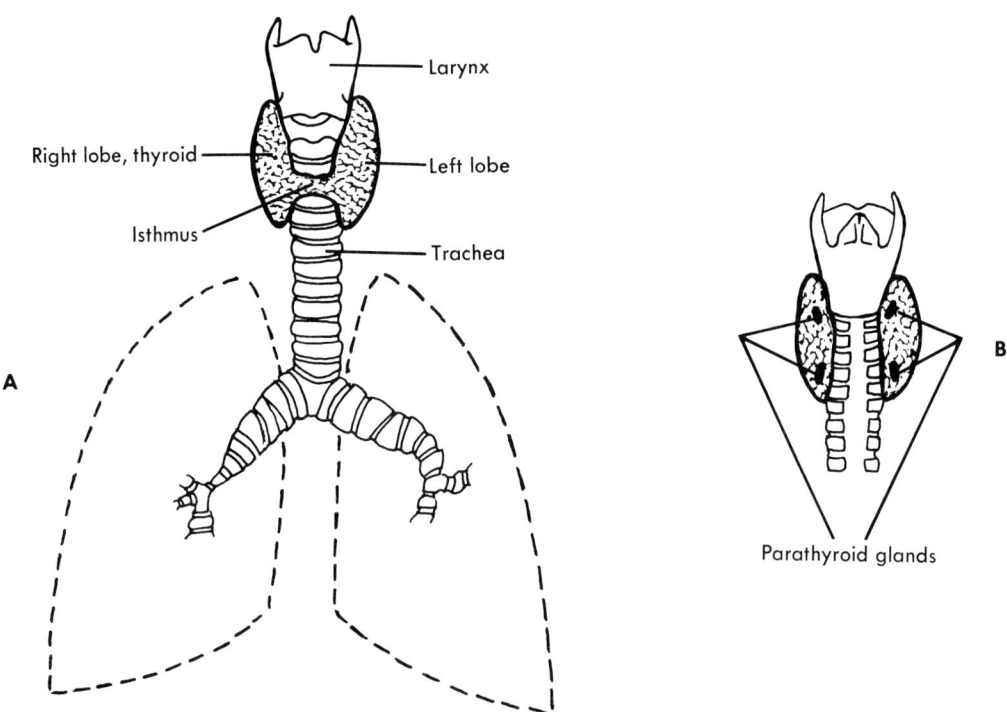

Fig. 12-3. Thyroid gland and related structures. **A,** Anterior aspect. **B,** Posterior aspect showing position of parathyroids.

50 percent of the iodine found in the body, although each living cell of the body contains a small quantity of iodine. The thyroid gland collects iodine from the blood either to store the iodine or to combine it, according to need, with tyrosine and amino acid. In either event, following several chemical steps, thyroxine forms from the iodine. Thyroxine is then available for secretion into the blood or for storage as thyroglobulin in the colloid of the thyroid gland.

Most of the many functions of thyroxin may be classed as those related to growth, development, and differentiation of tissues or as those related to metabolism. Both hyperthyroidism and hypothyroidism are related to growth, development, and differentiation of tissues.

True hyperthyroidism is an abnormal condition in which an excess of thyroxine is secreted. This condition may or may not be associated with enlargement of the thyroid gland (goiter).

Hyperthyroidism causes an increase in basal metabolic rate (B.M.R.) of 50 percent or more above normal.* Increased B.M.R. stimulates increased appetite and increased cardiac output (which is proportional to the B.M.R.). Blood pressure usually increases, and the nervous system becomes hyperirritable. Weight loss usually occurs, and in some instances the eyeballs protrude (exophthalmos).

Thyroid secretion deficiency (hypothyroidism) in infants prevents proper growth and development of the nervous system and bones. The degree of improper development depends upon the quantity of thyroxine secreted. *Cretinism* is the name of this condition in the very young. The bones fail to grow to the proper length, there is an excess of connective tissue (particularly in the face, hands, and feet), the reproductive organs fail to develop and mature, and mental development is severely retarded. If the condition is diagnosed early, remarkable improvement can be expected from the administration of thyroid extracts. Treatment will be necessary as long as the thyroid gland fails to secrete normally.

Reduction in B.M.R. from 30 percent or more in adults, *myxedema,* is associated with hypothyroidism. Persons with this condition experience corresponding body changes, such as decreased heat production, reduced rate of pulse, deficient muscle tonus, and increased susceptibility to fatigue. They may also become lethargic; and tissue fluids accumulate. The condition responds very well to administration of thyroxine.

Thyroxine increases protein synthesis in young animals and, therefore, influences growth in the young. A dissimilar effect is noted in adults in that thyroxine tends to increase the destruction of proteins. There is also an increase in oxidation of lipoids and carbohydrates, accompanied by increased energy release.

Thyroxine secretion increases as a result of stimulus from TSH (thyrotropic hormone) from the pituitary gland. Iodine uptake is accelerated following TSH

---

*The normal range of B.M.R. is from minus ten (-10) to plus fifteen ($+15$).

administration. This has been demonstrated effectively by use of radioiodine (^{131}I).

Parathyroid glands

Four quite small bodies of glandular (epithelial) tissue lie on the dorsal surface, superiorly and inferiorly, of the two lobes of the thyroid gland. These are the parathyroid glands, which are sometimes imbedded in the thyroid gland tissue. The parathyroid glands vary from 2 to 4 mm. in length in man. The hormone of these glands, called *parathormone*, functions to regulate the calcium-phosphorus ratio (Ca:P) in the circulating blood. The normal calcium level is from 9 to 11.5 mg. in each 100 ml. of blood serum. Complete removal of these glands causes a severe drop in the calcium level. A calcium level of 6 mg. or less per 100 ml. of blood serum usually results in a condition called *calcium tetany*. The symptoms resemble those of true tetanus and respond quickly to injections of an isotonic solution of calcium in combination with a suitable salt. The eventual outcome of great calcium deficiency is death. Decreases in calcium cause corresponding increases in phosphorus (see Fig. 12-3, *B*).

Deficiencies in calcium ions cause poor development of teeth and bones in growing animals. When calcium is deficient in the blood either from poor diet or from lack of vitamin D, the parathyroid glands are stimulated to secrete more hormone. This results in increased renal excretion of phosphates and an upset in normal Ca:P ratio. As a result, these calcium salts leave the bones to collect in the blood. It may be that osteoclasts (see pp. 178 and 179) become more active from parathormone stimulation, thus permitting release of calcium directly from the bones. The principle of *bone resorption* guarantees that the fetal bones (and the bones of breast-fed infants) will have a sufficient supply of calcium, even at the expense of the mother's bones.

Adrenal glands

The two adrenal glands are situated, one each, atop the corresponding kidney and are often called the suprarenal glands. The adrenal glands are independent of the kidneys, each having its own blood, lymph, and nerve supply. Each gland is separated from the kidney by fibrous connective tissue (see Fig. 12-4).

The adrenal glands, when compared with the thyroid gland, parathyroid glands, and pancreatic insular glands, are considerably more complex both in structure and in function. Each adrenal gland weighs about 5 or 6 grams and consists of two distinct parts: the *medulla* and the *cortex*. The medullary part of the adrenal gland and the sympathetic nervous system derive from the same embryonic tissue. The cortical part of the adrenal gland and the gonads (in both sexes) derive from the same embryonic tissue. Each part secretes a different substance, which helps maintain homeostasis.

The blood supply of the adrenal glands is exceedingly rich; more blood flows through these glands than through any other body tissue of comparable size. The adrenal gland is covered by a fibrous connective tissue, which passes into the gland at the hilum. The relatively large suprarenal vein leaves the

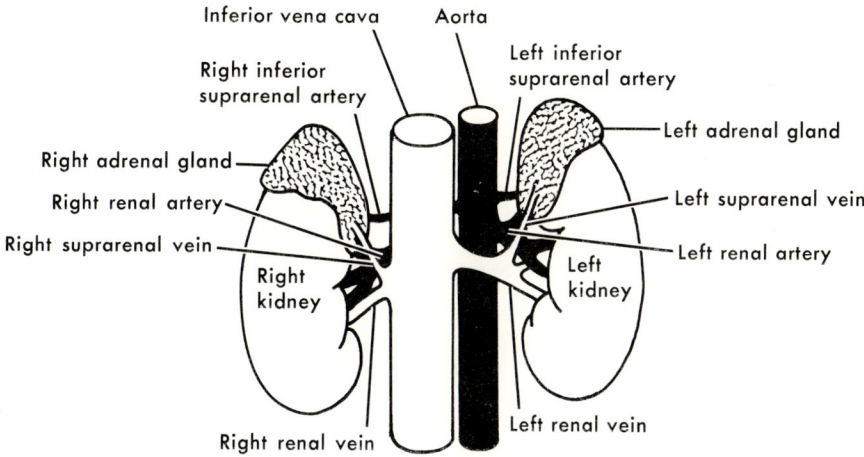

Fig. 12-4. Adrenal glands and related structures. Of the three arteries entering each adrenal gland, only the inferior arteries are shown.

gland through the hilum. Three arterial blood supplies send branches to enter the capsule at different points. The branches extend from these points to penetrate the stroma with delicate trabeculae from the capsule. Lymphatics lie beneath the capsule. The substance of the adrenal gland consists of epithelial tissue and an extensive network of reticular connective tissue, the very fine subdivisions of the trabeculae. There are numerous sinusoids and small venules in the medulla and large numbers of very small arterioles and capillaries in the cortex of this gland. The medulla is innervated by bundles of nerve fibers passing from the capsule through the cortex. Most of these fibers terminate in a diffuse pattern among the secreting cells of the medulla; however, some of the fibers supply the smooth muscles of the suprarenal blood vessels.

The adrenal medulla secretes at least two hormones: *epinephrine* and *norepinephrine*. Epinephrine is synthesized commercially. The proprietary name for these hormones is Adrenalin, which consists of a mixture of these two catecholamines.

Injections of epinephrine cause an increase in the rate and the force of the heartbeat. Epinephrine stimulates strong contractions in the arterioles of the skin and mucous membranes and dilation of the arterioles of the coronary system and, perhaps, the arterioles of the lungs and the skeletal muscles. As a result of the aforementioned reactions, the blood pressure rises throughout the body. Fortunately, the effects of excessive amounts of epinephrine are not of long duration; this is a result of the destructive action on this hormone by the enzyme, *monamine oxidase*.

Most of the visceral muscles are affected by epinephrine—some are stimulated and others are inhibited. The blood sugar level rises as a result of epinephrine injections. Epinephrine also increases the length of time that skeletal muscles may work before fatigue occurs. Epinephrine causes increased oxygen consumption; and heat production may increase sharply.

Norepinephrine differs from epinephrine in that norepinephrine is limited to only a constricting effect causing greater rises in blood pressure than epinephrine. Experimental evidence indicates that norepinephrine and sympathin (a chemical of the neurohumoral theory of nerve impulse transmission) are the same.

Although the medullary part of the adrenal gland has been proven to be unnecessary for life, this is not true with the cortical part of this gland. *Steroids* is the general name of the hormones secreted by the adrenal cortex. There remains much to be learned about the steroids. They are presently classed as *glucocorticoids* and *mineralocorticoids*. The corticoids affect homeostasis through the effect of mobilizing the many body resources.

A deficient quantity or a total lack of glucocorticoids results in lowered B.M.R. and in decreased glucose in the blood (and probably a decrease in glycogen in the liver and striated muscle tissues). These decreases effect muscular weakness and fatigue, lowered body temperature, lowered blood pressure, and numerous other progressive and terminal conditions. All the preceding conditions respond to administration of cortisone.

Addison's disease, a hypofunction of the adrenal glands, is usually fatal. It is characterized by a bronzelike pigmentation of the skin, severe prostration, progressive anemia, low blood pressure, diarrhea, and digestive disturbances. Associated with this disease is the increased excretion of sodium salts in the urine. This results in the loss of these salts from the blood, with subsequent lowering of osmotic blood pressure. Additional disturbance of the sodium-potassium ratio results from the decreased potassium elimination in the urine. These several conditions indicate that the cortical hormones exert the necessary influence upon maintaining both the water and mineral balances of the body.

The adrenal secretions exert a strong effect upon the gonadal hormone secretions. It is known that injections of ACTH (adrenocorticotropic hormone) destroys certain leukocytes, particularly the eosinophils and lymphocytes, and the lymphoid tissue. ACTH exerts a very stong influence upon the activity of the adrenal gland.

A quite remarkable function of the adrenal glands is their part in the "stresses and alarm reaction." Certain stimuli are believed to be transmitted to the adrenal medulla via the sympathetic nervous system. The adrenal medulla, in turn, secretes an excessive quantity of epinephrine, which stimulates the anterior lobe of the pituitary gland. This gland, in turn, secretes ACTH, which stimulates the adrenal cortex to secrete large quantities of corticoids (which stimulate the anterior lobe of the pituitary) and cortisone (which stimulates the tissues that first transmitted or received the alarm stimulus [target tissues]).

OTHER HORMONE SOURCES
Compound glands

The hormones secreted in both sexes by the gonads were discussed in Chapter 11. The male gonadal hormone is *testosterone,* and the corresponding female gonadal hormone is *estradiol.*

Throughout the pancreas the specialized islands of Langerhans secrete the hormone *insulin*. The actions of insulin were discussed in Chapter 5 in conjunction with the pancreas.

There is little evidence to support endocrine secretion in the liver; however, bile secretion is influenced by a hormone, secretin, that is secreted in the intestinal mucosa.

Hormones known to be secreted in the mucosa of the stomach and intestine are as follows:

1. *Gastrin*, secreted from the pyloric mucosa, is absorbed by the blood and transported to the fundic cells, where the hormone stimulates these cells to secrete hydrochloric acid.
2. *Enterogastrone*, secreted from the mucosa in the antrum of the stomach and, perhaps, in the intestine mucosa, acts as an inhibitor of hydrochloric acid secretion, particularly when excessive quantities of lipoids are ingested.
3. *Secretin*, secreted from the intestine mucosa, is liberated by action of the acid chyme; secretin is carried by the blood to the pancreas to stimulate the flow of pancreatic juice.
4. *Pancreozymin*, secreted from the mucosa of the upper small intestine, is acid stable and thermostable, but it is destroyed by alkali. It also stimulates the flow of pancreatic juices.
5. *Enterocrinin*, secreted from the intestine mucosa, stimulates the flow of intestinal juices.
6. *Cholecystokinin*, secreted from the intestine mucosa, stimulates the gallbladder walls to contract.

There is complete agreement that both secretin and cholecystokinin are truly hormones secreted in the gastrointestinal mucosa. There remains some doubt that the other four substances are truly hormones.

Glands of unknown function

The thymus gland is situated between the upper sternum and the pericardium. It increases somewhat in size until the person is about 14 to 16 years old and then gradually atrophies. Decreased sex gland activities prolong the atrophying process. The opposite is also true. It is known that the thymus gland manufactures lymphocytes in the very young person. Little else is known about the thymus gland.

The pineal gland (or body) is situated immediately superior and posterior to the pituitary gland in the center of the brain and on its dorsal aspect. The pineal gland usually commences to degenerate when the person is about 7 years of age. Normally, in the adult the pineal body consists of fibrous tissue. There is no known function of this gland.

RADIOGRAPHIC INTEREST

Since the endocrine glands and tissues are of essentially the same density as the surrounding tissues and since these glands have no lumina, there is very limited knowledge to be gained from radiography of these structures. Future research may disclose methods of opacifying these glands and tissues.

Calcification enabling visualization of the pineal gland is noted in some skull radiographs, and visualization of the adrenal arterioles has been noted in some aortograms. Increased activity and blood supply with enlargement of the thyroid gland sometimes is visualized in cervical spine radiographs, particularly the lateral views.

GENERAL INTEREST

Experimental evidence indicates that no bodily function (apparently) is lacking in hormonal influence, if not control. Some functions are influenced to a very high degree by hormones; these are the functions of *digestion, growth,* and *reproduction*. Additional experimental evidence suggests that hormonal control represents a very primitive mechanism since the secretions (hormones) are common in all species.

REFERENCES
Best, C. H., and Taylor, N. B.: The human body, ed. 4, New York, 1963, Holt, Rinehart and Winston.
Schottelius, B. A., and Schottelius, D. D.: Textbook of physiology, ed. 17, St. Louis, 1973, The C. V. Mosby Co.
Windle, W. F.: Textbook of histology, ed. 3, New York, 1960, McGraw-Hill Book Co.

Glossary

PREFIXES

a- Want or absence.
ab- From, off, or away from.
ad- To or toward, addition to, nearness, or intensification.
amphi- On both sides; around or about; double.
an- See *a-*; *an-* is used before stems beginning with a vowel.
angio- (also angi-) Relationship to a vessel, usually a blood vessel.
ante- "Before" in time or place.
anti- Against or over against.
arthro- Relationship to a joint or joints.
bi- Two or twice.
bio- Relationship to life.
broncho- Connection to or relationship with the bronchi.
cardio- Relationship to the heart.
cephalo- Relationship to the head.
chole- Relationship to the bile.
cysto- Relationship to a sac, cyst, or bladder.
derma- (also dermato-, dermat-, dermo-) Relationship to the skin.
di- Twice.
ec- Out of.
ecto- Situated on, without or on the outside.
en- In, into, forming verbs.
epi- On or upon.
gastro- Relationship to the stomach or to the abdomen.
hemo- (also hema-) Relationship to the blood.
hepat- (also hepato-) Relationship to the liver.
homo- The same.
hyper- Above, beyond, or excessive.
hypo- Beneath, under, or deficient.
ileo- Relationship to the ileum.
ilio- Relationship to the ilium or flank.
inter- Situated, formed, or occurring between elements indicated by the word stem to which it is affixed.
intra- Situated, formed, or occurring within the element indicated by the word stem to which it is affixed.
iso- Equal, alike, or the same.
lyso- Lysis or dissolution.
myelo- Relationship to marrow, often used in specific reference to the spinal cord.
myo- Relationship to muscle.
nephro- (also nephr-) Relationship to the kidney.
neuro- (also neur-) Relationship to a nerve or nerves, or to the nervous system.
odonto- Relationship to a tooth or to the teeth.
osteo- Relationship to a bone or to the bones.
oto- (also ot-) Relationship to the ear.
para- Beside, beyond, accessory to, apart from, against.
peri- Around.
pneumo- (also pneumono-) Relationship to the lungs.
post- After or behind.
pre- Before.
pro- Before or in front of.
proct- (also procto-) Relationship to the rectum.
pyo- Relationship to pus.
retro- Backward, or located behind.
sub- Under, near, almost, or moderately.
super- Above or implying excess.
supra- Above or over.
trans- Through, across, or beyond.

SUFFIXES

-ad Direction toward, as cephalad, caudad.
-ectomize Deprivation by excision.
-ectomy Deprivation by excision.

-graphy The act of (writing or) recording, as in radiography.

-ism State, condition, or fact of being, or the process or result of an action.

-itis Inflammation of the part indicated by the word stem to which it is attached.

-ose The substance is a carbohydrate.

-osis A process, often a disease or morbid process, and sometimes conveying the meaning of abnormal increase.

-stomy The surgical creation of an artificial opening into a hollow organ (colostomy), or a new opening between two such structures (gastroenterostomy).

-tomy The operation of cutting, or incision.

TERMS

A

abdomen (ab-do'men) That portion of the body that lies between the thorax and the pelvis. It consists of a cavity separated by the diaphragm from the thoracic cavity, and lined with a serous membrane, the peritoneum.

abduct (ab-dukt') To draw away from the median line or from a neighboring part or limb.

aberration (ab″er-a'shun) Deviation from the usual course or condition.

abscess (ab'ses) A localized collection of pus in a cavity formed by the disintegration of tissues.

acanthion (ah-kan'the-on) A point at the base of the anterior nasal spine.

acetabulum (as″e-tab'u-lum) The large cup-shaped cavity on the lateral surface of the os coxae (pelvic bone) in which the head of the femur articulates.

achlorhydria (ah″klor-hid're-ah) Absence of hydrochloric acid from the gastric secretions.

acid (as'id) 1. Sour; having properties opposed to those of the alkalis. 2. Any compound of an electronegative element with one or more hydrogen atoms that are replaceable by electropositive atoms; a compound that, in aqueous solution, undergoes dissociation with the formation of hydrogen ions.

acoustic (ah-koōs'tik) Pertaining to sound or to the sense of hearing.

acromegaly (ak″ro-meg'ah-le) A condition characterized by hyperplasia of the extremities of the skeleton—the nose, jaws, fingers, and toes; the converse of acromicria.

acromicria (ak″ro-mik're-ah) A condition characterized by hypoplasia of the extremities of the skeleton—the nose, jaws, fingers, and toes; the converse of acromegaly.

acromion (ah-kro'me-on) The lateral extension of the spine of the scapula, forming the highest point of the shoulder.

adduct (ah-dukt') To draw toward the median line of the body or toward a neighboring part.

adenoma (ad″ĕ-no'mah) An epithelial tumor, usually benign, with a glandlike structure.

adipose (ad'ĭ-pōs) 1. Of a fatty nature; fatty; fat. 2. The fat present in the cells of adipose tissue.

adrenal (ad-re'nal) Situated near the kidney.

afferent (af'er-ent) Centripetal; conveying toward a center.

alkali (al'kah-li) Any one of a class of compounds that form soluble soaps with fatty acids, turn red litmus blue, and form soluble carbonates.

allele (ah-lēl') 1. One of two or more contrasting genes, situated at the same locus in homologous chromosomes, which determine alternative characters in inheritance. 2. One of two or more contrasting characters transmitted by alternative genes.

allergy (al'er-je) A hypersensitive state acquired through exposure to a particular allergen, re-exposure bringing to light an altered capacity to react.

alveolar (al-ve'o-lar) Pertaining to an alveolus.

alveolus (al-ve'o-lus) A general term used in anatomical nomenclature to designate a small saclike dilatation.

ambulatory (am'bu-lah-to″re) Ambulant; walking or able to walk.

amnion (am'ne-on) The thin, transparent, silvery and tough membrane lining the chorion laeve and chorion frondosum (placenta), which produces, at the very earliest period of fetation, the amniotic fluid.

ampulla (am-pul'lah) A general term used in anatomical nomenclature to designate a flasklike dilatation of a tubular structure.

anabolism (ah-nab'o-lizm) Any constructive process by which simple substances are converted by living cells into more complex compounds, especially conversion of simple compounds into protoplasm; constructive metabolism.

anastomosis (ah-nas″to-mo'sis) 1. A communication between two vessels. 2. The surgical or pathological formation of a passage between any two normally distinct spaces or organs.

anatomy (ah-nat'o-me) The science of the structure of the animal body and the relation of its parts.

anemia (ah-ne'me-ah) A reduction below normal in the number of erythrocytes per cu. mm., the quantity of hemoglobin, or the volume of packed red cells per 100 ml. of blood that occurs when the equilibrium between blood loss and blood production is disturbed.

anesthesia (an"es-the'ze-ah) Loss of feeling or sensation.

anesthesiologist (an"es-the"ze-ol'o-jist) A specialist in anesthesiology.

aneurysm (an'u-rizm) A sac formed by the dilatation of the walls of an artery or of a vein and filled with blood.

angiolith (an'je-o-lith") A calculus or concretion in the wall of a blood vessel.

angioma (an"je-o'mah) A tumor whose cells tend to form blood vessels (hemangioma) or lymph vessels (lymphangioma); a tumor made up of blood vessels or lymph vessels.

ankylosis (ang"kĭ-lo'sis) Abnormal immobility and consolidation of a joint.

antecubital (an"te-ku'bi-tal) Situated in front of the cubitus or forearm.

anterior (an-te're-or) Situated in front of or in the forward part of, affecting the forward part of an organ, toward the head end of the body; in official anatomical nomenclature, used in reference to the ventral or belly surface of the body.

anteversion (an"te-ver'zhun) The forward tipping or tilting of an organ.

antiseptic (an"tĭ-sep'tik) 1. Preventing decay or putrefaction. 2. A substance that will inhibit the growth and development of microorganisms without necessarily destroying them.

antrum (an'trum) A cavity or chamber; used as a general term in anatomical nomenclature, especially to designate a cavity or chamber within a bone.

anus (a'nus) The distal or terminal orifice of the alimentary canal.

aorta (a-or'tah) The main trunk from which the systemic arterial system proceeds.

apex (a'peks) 1. A general term used in anatomical nomenclature to designate the top of a body organ or part, or the pointed extremity of a conical structure. 2. The point of greatest activity, or the point of greatest response to any type of stimulation.

aphagia (ah-fa'je-ah) Abstention from eating.

aphasia (ah-fa'ze-ah) Defect or loss of the power of expression by speech, writing, or signs, or of comprehending spoken or written language, due to injury or disease of the brain centers.

apnea (ap-ne'ah) The transient cessation of the breathing impulse that follows forced breathing.

aponeurosis (ap"o-nu-ro'sis) A white, flattened or ribbonlike tendinous expansion, serving mainly as an investment for muscle, or connecting a muscle with the parts that it moves.

apophysis (ah-pof'ĭ-sis) Any outgrowth or swelling, especially a bony outgrowth that has never been entirely separated from the bone of which it forms a part, such as a process, tubercle, or tuberosity.

arachnoid (ah-rak'noid) Resembling a spider's web.

areola (ah-re'o-lah) 1. Any minute space or interstice in a tissue. 2. A circular area of a different color, surrounding a central point, such as an area surrounding a pustule or vesicle, or the part of the iris surrounding the pupil of the eye.

areolar (ah-re'o-lar) Containing minute interspaces.

artery (ar'ter-e) A vessel through which the blood passes away from the heart to the various parts of the body, including the myocardium.

arthritis (ar-thri'tis) Inflammation of a joint.

arthroclasia (ar"thro-kla'ze-ah) The breaking down of an ankylosis in order to secure free movement in a joint.

arthrodesis (ar-throd'ĕ-sis) The surgical fixation of a joint by fusion of the joint surfaces; artificial ankylosis.

articulate (ar-tik'u-lāt) 1. Divided into or united by joints. 2. To divide into or to unite so as to form a joint. 3. In dentistry, to adjust or place the teeth in their proper relation to each other in making an artificial denture.

articulation (ar-tik-u-la'shun) The place of union or junction between two or more bones of the skeleton.

asterion (as-te're-on) The point on the surface of the skull where the lambdoidal, parietomastoidal, and occipitomastoidal sutures meet.

ataxia (ah-tak'se-ah) Failure of muscular coordination: irregularity of muscular action.

atelectasis (at"e-lek'tah-sis) 1. Incomplete expansion of the lungs at birth. 2. Collapse of the adult lung.

atlas (at'las) The first cervical vertebra.

atony (at'o-ne) Lack of normal tone or strength.

atresia (ah-tre'ze-ah) Absence or closure of a normal body orifice or passage.

atrium (a'tre-um) A chamber; used in anatomical nomenclature to designate such a chamber affording entrance to another structure or organ.

atrophy (at'ro-fe) A defect or failure of nutrition manifested as a wasting away or diminution in the size of a cell, tissue, organ, or part.

atypical (a-tip'e-kal) Irregular; not conformable to the type.

auditory (aw'dĭ-to-re) Pertaining to the sense of hearing.

auricle (aw're-kl) The portion of the external ear not contained within the head; the pinna, or the flap of the ear.

autopsy (aw'top-se) The postmortem examination of a body.

axilla (ak-sil'ah) The small hollow beneath the arm, where it joins the body at the shoulder.

axis (ak'sis) 1. A line about which a revolving body turns or about which a structure would turn if it did revolve. 2. The second cervical vertebra.

B

benign (be-nīn') Not malignant, not recurrent; favorable for recovery.

bilateral (bi-lat'er-al) Having two sides, or pertaining to both sides.

bile (bīl) A fluid secreted by the liver and poured into the intestine.

biliary (bil'e-ar-e) Pertaining to the bile, to the bile ducts, or to the gallbladder.

biopsy (bi'op-se) The removal and examination, usually microscopic, of tissue or other material from the living body for purposes of diagnosis.

blastocyst (blas'to-sist) The modified blastula of mammals.

blastoderm (blas'to-derm) Collectively, the mass of cells produced by the cleavage of a fertilized ovum, forming the hollow sphere of the blastula, or the cellular cap above a floor of segmented yolk in the discoblastula of telolecithal eggs.

blastoma (blas-to'mah) A true tumor; a tumor, not teratogenous, which exhibits an independent localized growth.

blastula (blas'tu-lah) The usually spherical structure produced by cleavage of a fertilized ovum, consisting of a single layer of cells (blastoderm) surrounding a fluid-filled cavity (blastocele).

bolus (bo'lus) A mass of food ready to be swallowed or a mass passing along the intestines.

brachial (bra'ke-al) Pertaining to the arm.

brachium (bra'ke-um) The arm; specifically the arm from shoulder to elbow.

bronchiectasis (brong"ke-ek'tah-sis) A chronic dilatation of the bronchi or bronchioles marked by a fetid breath and paroxysmal coughing, with the expectoration of mucopurulent matter.

bronchiole (brong'ke-ōl) One of the finer subdivisions of the branched bronchial tree.

bronchus (brong'kus) One of the larger air passages within the lungs.

bursa (bur'sah) A sac or saclike cavity filled with a viscid fluid and situated at places in the tissues at which friction would otherwise develop.

bursitis (bur-si'tis) Inflammation of a bursa.

C

C.A. Abbreviation for chronological age.

calcaneus (kal-ka'ne-us) The irregular quadrangular bone at the back of the tarsus.

calcareous (kal-ka're-us) Pertaining to or containing lime or calcium.

calcigerous (kal-sij'er-us) Producing or carrying calcium salts.

calculus (kal'ku-lus) An abnormal concretion occurring within the animal body, and usually composed of mineral salts.

callus (kal'us) 1. A callosity. 2. An unorganized meshwork of woven bone developed on the pattern of the original fibrin clot, which is formed following fracture of a bone and is normally ultimately replaced by hard adult bone.

calvarium (kal-va're-um) The calvaria; the calvaria is the domelike superior portion of the cranium, composed of the superior portions of the frontal, parietal, and occipital bones; also called the skullcap.

calyx (ka′liks) Calix; a cup-shaped organ or cavity.

canaliculus (kan″ah-lik′u-lus) An extremely narrow tubular passage or channel.

cancellous (kan′sĕ-lus) Of a reticular, spongy, or latticelike structure; used mainly of bony tissue.

cancer (kan′ser) A cellular tumor, the natural course of which is fatal, usually with the formation of secondary tumors.

canthus (kan′thus) The angle at either end of the slit between the eyelids; the canthi are distinguished as an outer or temporal and an inner or nasal.

capillary (kap′i-lār″e) 1. Pertaining to or resembling a hair. 2. Any one of the minute vessels that connect the arterioles and venules, forming a network in nearly all parts of the body.

carcinoma (kar″sĭ-no′mah) A malignant new growth made up of epithelial cells tending to infiltrate the surrounding tissues and give rise to metastases.

caries (ka′re-ez) The molecular decay or death of a bone, in which it becomes softened, discolored, and porous.

carpus (kar′pus) The joint between the arm and hand; the wrist, made up of eight bones.

cartilage (kar′tĭ-lij) A specialized fibrous connective tissue, forming most of the temporary skeleton of the embryo, providing a model in which most of the bones develop, and constituting an important part of the growth mechanism of the organism.

catabolism (kah-tab′o-lizm) Any destructive process by which complex substances are converted by living cells into more simple compounds; destructive metabolism.

catalyst (kat′ah-list) Any substance that brings about catalysis, which is the change in velocity of a reaction produced by the presence of a substance that does not form part of the final product.

caudad (kaw′dad) Directed toward a cauda or tail; opposite to cephalad.

caphalad (sef′ah-lad) Toward the head; opposite to caudad.

cephalic (sĕ-fal′ik) Pertaining to the head or the head end of the body.

cerebellum (ser″e-bel′um) That division of the central nervous system postero-inferior to the cerebrum and above the pons and fourth ventricle.

cerebrum (ser′e-brum) The main portion of the brain occupying the upper part of the cranium, the two cerebral hemispheres, united by the corpus callosum, forming the largest part of the central nervous system in man.

choana (ko′a-nah) 1. Any funnel-shaped cavity or infundibulum. 2. The paired openings between the nasal cavity and the nasopharynx.

cholecystitis (ko″le-sis-ti′tis) Inflammation of the gallbladder.

cholelithiasis (ko″le-lĭ-thi′ah-sis) The presence or formation of gallstones.

chorion (ko′re-on) The outermost envelope of the growing zygote or fertilized ovum that serves as a protective and nutritive covering.

chromosome (kro′mo-sōm) One of several small dark-staining and more or less rod-shaped bodies that appear in the nucleus of a cell at the time of cell division.

chyle (kīl) The milky fluid taken up by the lacteals from the food in the intestine after digestion. It consists of lymph and emulsified fat. It passes into the veins by the thoracic duct, becoming mixed with the blood.

chyme (kīm) The semifluid, homogeneous, creamy, or gruel-like material produced by gastric digestion of food.

cicatrix (sik-a′triks) The new tissue that is formed in the healing of a wound.

circumduction (ser″kum-duk′shun) The active or passive circular movement of a limb or of the eye.

circumflex (ser′kum-fleks) Curved like a bow.

coarctation (ko″ark-ta′shun) A straightening or pressing together; a condition of stricture or contraction.

coccyx (kok′siks) The small bone caudad to the sacrum in man, formed by union of four (sometimes five or three) rudimentary vertebrae, and forming the caudal extremity of the vertebral column.

collateral (kŏ-lat′er-al) 1. Secondary or accessory; not direct or immediate. 2. A small side branch, as of a blood vessel or nerve.

colloid (kol′oid) 1. Glutinous or resembling glue. 2. A state of matter in which the matter is dispersed in or distributed throughout some medium called the dispersion medium.

coma (ko′mah) A state of unconsciousness

from which the patient cannot be aroused, even by powerful stimulation.

compact (kom-pakt′) Dense; having a dense structure.

condyle (kon′dīl) A rounded projection on a bone.

congenital (kon-jen′ĭ-tal) Existing at, and usually before, birth; referring to conditions that are present at birth, regardless of their causation.

coracoid (kor′ah-koid) 1. Like a raven's beak. 2. The coracoid process.

cornu (kor′nu) A hornlike excrescence or projection.

coronal (ko-ro′nal) 1. Pertaining to the crown of the head, or to any corona. 2. Situated in the direction of the coronal suture; said of a transverse plane or section parallel with the long axis of the body.

coronoid (kor′o-noid) 1. Shaped like a crow's beak. 2. Crown shaped.

cortex (kor′teks) The outer layer of an organ or other body structure, as distinguished from the underlying substance.

cortical (kor′tĭ-kal) Pertaining to or of the nature of a cortex.

costal (kos′tal) Pertaining to a rib or ribs.

costophrenic (kos″to-fren′ik) Pertaining to the costal and diaphragmatic pleurae.

crepitus (krep′ĭ-tus), **bony** The crackling sound produced by the rubbing together of fragments of fractured bone.

cretinism (kre′tin-izm) A chronic condition due to congenital lack of thyroid secretion.

cribriform (krib′rĭ-form) Perforated with small apertures like a sieve.

crista (kris′ta) A projection or projecting structure, or ridge, especially one surmounting a bone or its border; also called crest and ridge.

crus (krus) A general term used to designate a leglike part.

cuboid (ku′boid) 1. Resembling a cube. 2. The cuboid bone.

cuneiform (ku-ne′ĭ-form) Shaped like a wedge.

cyanosis (si″ah-no′sis) A bluish discoloration, applied especially to such discoloration of the skin and of mucous membranes due to excessive concentration of reduced hemoglobin in the blood.

cyst (sist) Any sac, normal or abnormal, especially one that contains a liquid or semisolid material.

D

dactyl (dak′til) A digit; a finger or toe.

decubitus (de-ku′bĭ-tus) An act of lying down; also the position assumed in lying down.

deglutition (deg″loo-tish′un) The act of swallowing.

dermis (der′mis) Sometimes used with special reference to the corium.

dermoid (der′moid) 1. Resembling the skin. 2. A dermoid cyst.

desquamation (des″kwah-ma′shun) The shedding of epithelial elements, chiefly of the skin, in scales or sheets.

diaphragm (di′ah-fram) The musculomembranous partition separating the abdominal and thoracic cavities.

diaphysis (di-af′ĭ-sis) The portion of a long bone between the ends or extremities, which are usually articular and wider than the shaft.

diastole (di-as′to-le) The dilatation, or period of dilatation, of the heart, especially that of the ventricles. It coincides with the interval between the second and the first heart sounds.

digestion (di-jest′yun) The process or act of converting food into materials fit to be absorbed and assimilated.

digit (dij′it) A finger or toe.

dilatation (dil-ah-ta′shun) 1. The condition of being dilated or stretched beyond the normal dimensions. 2. Dilation.

distal (dis′tal) Remote; farther from any point of reference; opposed to proximal.

diverticulum (di″ver-tik′u-lum) A circumscribed pouch or sac of variable size created by herniation of the lining mucous membrane through a defect in the muscular coat of a tubular organ.

dorsal (dor′sal) 1. Pertaining to the back or to any dorsum. 2. Denoting a position more toward the back surface than some other object of reference; same as posterior in human anatomy.

dorsum (dor′sum) 1. The vertebral aspect of the body; the back. 2. The aspect of an anatomical part or structure corresponding in position to the back; posterior, in the human.

duodenum (du″o-de′num) The first or proximal portion of the small intestine, extending from the pylorus to the jejunum; so called because it is about twelve fingerbreadths in length.

dura mater (du′rah ma′ter) The outermost, toughest, and most fibrous of the three membranes (meninges) covering the brain and spinal cord.

dyspnea (disp′ne-ah) Difficult or labored breathing.

E

ecchymosis (ek″ĭ-mo′sis) An extravasation of blood under the skin.

edema (e-de′mah) The presence of abnormally large amounts of fluid in the intercellular tissue spaces of the body; usually applied to demonstrable accumulation of excessive fluid in the subcutaneous tissues.

efferent (ef′er-ent) Centrifugal; conveying away from a center.

effusion (ĕ-fu′zhun) 1. The escape of fluid into a part or tissue. 2. An effused material.

embolism (em′bo-lizm) The sudden blocking of an artery or vein by a clot or obstruction that has been brought to its place by the blood current.

embryo (ĕm′brĭ-o) The early or developing stage of any organism, especially the developing product of fertilization of an egg. In the human, the embryo is generally considered to be the developing organism from one week after conception to the end of the second month.

emesis (em′e-sis) Vomiting; an act of vomiting.

emetic (e-met′ik) 1. Bringing on or causing the act of vomiting. 2. An agent that causes vomiting.

emphysema (em″fĭ-se′mah) A swelling or inflation due to the presence of air, applied especially to a morbid condition of the lungs.

empyema (em″pi-e′mah) Accumulation of pus in a cavity of the body, especially in the chest.

encephalon (en-sef′ah-lon) The mass of nerve tissue contained within the cranium, including the cerebrum, cerebellum, pons, and medulla oblongata.

endocrine (en′do-krīn) 1. Secreting internally; applied to organs whose function is to secreete into the blood or lymph a substance that has a specific effect on another organ or part. 2. Pertaining to internal secretions.

endosteum (en-dos′te-um) The tissue lining the medullary cavity of a bone.

enteric (en-ter′ik) Pertaining to the intestines.

enzyme (en′zim) An organic compound, frequently a protein, capable of accelerating or producing by catalytic action some change in a substrate for which it is often specific.

epicondyle (ep″ĭ-kon′dĭl) An eminence upon a bone, above its condyle.

epidermis (ep″ĭ-der′mis) The outermost and nonvascular layer of the skin, made up of five layers from without inward: (1) stratum corneum, (2) stratum lucidum, (3) stratum granulosum, (4) stratum spinosum, and (5) stratum basale (germinativum).

epigastrium (ep″ĭ-gas′tre-um) The upper middle region of the abdomen, located within the sternal angle.

epiphyseal (ep″ĭ-fiz′e-al) Pertaining to or of the nature of an epiphysis.

epiphysis (e-pif′ĭ-sis) 1. The end of a long bone, usually wider than the shaft, and either entirely cartilaginous or separated from the shaft by a cartilaginous disc. 2. Part of a bone formed from a secondary center of ossification, commonly found at the ends of long bones, and at tubercles and processes; during the period of growth epiphyses are separated from the main portion of the bone by cartilage.

epipteric (ĕp″ĭp-ter″ĭk) **bone** Designating a small wormian bone sometimes present in the human skull between the parietal bone and the great wing of the sphenoid.

episiotomy (e-piz″e-ot′o-me) Surgical incision of the vulvar orifice for obstetrical purposes.

epithelium (ep″ĭ-the′le-um) The covering of internal and external surfaces of the body, including the lining of vessels and other small cavities. It consists of cells joined by small amounts of cementing substances. Epithelium is classified into types on the basis of the number of layers deep and the shape of the superficial cells.

erythema (er″ĭ-the′mah) A name applied to redness of the skin produced by congestion of the capillaries, which may result from a variety of causes, the etiology or a special type of lesion often being indicated by a modifying term.

erythrocyte (e-rith′ro-sīt) One of the elements found in peripheral blood. Normally, in the human, the mature form is a non-nucleated, yellowish, circular, biconcave disc, adapted, by virtue of its configuration and its hemoglobin content, to transport oxygen.

esophagus (e-sof'ah-gus) The musculomembranous passage extending from the pharynx to the stomach.

ethmoid (eth'moid) Cribriform; sievelike.

etiology (e″te-ol'o-je) The study or the theory of the causation of any disease; the sum of knowledge regarding causes.

eversion (e-ver'zhun) A turning outward or inside out.

exacerbation (eks-as″er-ba'shun) Increase in the severity of any symptoms or disease.

excise (ek-siz') To cut out or off.

excoriation (eks-ko″re-a'shun) Any superficial loss of substance, such as that produced on the skin by scratching.

excrete (eks-krēt') To throw off, as waste matter, by a normal discharge.

exhale (eks-hāl') 1. To expel from the lungs by breathing. 2. To give off a watery or other vapor.

exostosis (ek″sos-to'sis) A bony growth projecting outward from the surface of a bone.

expiration (eks″pĭ-ra'shun) 1. The act of breathing out or expelling air from the lungs. 2. Termination, or death.

extend (ĕks-tĕnd') To straighten out, as a limb or part.

external rotation (eks-ter'nal ro-ta'shun) The process of turning or rotating a part outwardly about its axis, as in rotating the humerus; opposed to internal rotation.

extirpation (ek″ster-pa'shun) Complete removal or eradication of a part.

extravasation (eks-trav″ah-sa'shun) 1. A discharge or escape, as of blood, from a vessel into the tissues. 2. The process of being extravasated. 3. Blood or other substance that has been extravasated.

exudate (eks'u-dāt) Material, such as fluid, cells, or cellular debris, that has escaped from blood vessels and been deposited in tissues or on tissue surfaces, usually as a result of inflammation. An exudate, in contrast to a transudate, is characterized by a high content of protein, cells, or solid materials derived from cells.

F

facet (fas'et) A small plane surface on a hard body, as on a bone.

fascia (fash'e-ah) A sheet or band of fibrous tissue such as lies deep to the skin or forms an investment for muscles and various organs of the body.

fasciculus (fah-sik'u-lus) A small bundle or cluster; used as a general term in anatomical nomenclature to designate a small bundle of nerve or muscle fibers.

femur (fe'mur) The bone that extends from the pelvis to the knee, being the longest and largest bone in the body.

fenestra (fĕ-nes'trah) A windowlike opening; an opening or open area, as in an anatomical structure.

fetus (fe'tus) The unborn offspring of any viviparous animal; the developing young in the human uterus after the end of the second month.

fibrosis (fi-bro'sis) The formation of fibrous tissue; fibroid degeneration.

fistula (fis'tu-lah) An abnormal passage or communication, usually between two internal organs, or leading from an internal organ to the surface of the body.

flex (fleks) To bend or put in a state of flexion.

follicle (fol'li-k'l) A very small excretory or secretory sac or gland.

foramen (fo-ra'men) A natural opening or passage; used as a general term in anatomical nomenclature to designate such a passage, especially one into or through a bone.

fornix (for'niks) A general term designating an archlike structure or the vaultlike space created by such a structure.

fossa (fos'sah) A trench or channel; used in anatomical nomenclature as a general term to denote a hollow or depressed area.

fracture (frak'tūr) 1. The breaking of a part, especially a bone. 2. A break or rupture in a bone.

frenum (fre'num) A restraining structure or part.

fundus (fun'dus) The bottom or base of anything; used in anatomical nomenclature as a general term to designate the bottom or base of an organ, or the part of a hollow organ farthest from its mouth.

G

gamete (gam'ēt) A reproductive element; one of two cells, male and female, whose union is necessary, in sexual reproduction, to initiate the development of a new individual.

ganglion (gang'gle-on) 1. A knot, or knotlike mass; used in anatomical nomenclature as a general term to designate a group of nerve cell bodies located outside of the central nervous system. 2. A form of cystic tumor occurring on an aponeurosis or tendon, as in the wrist.

gene (jēn) The biologic unit of heredity, self-producing and located in a definite position (locus) on a particular chromosome.

genetics (je-net'iks) The study of heredity.

glabella (glah-bel'ah) 1. The smooth area on the frontal bone between the superciliary arches. 2. The most prominent point in the midsagittal plane between the eyebrows; used as an anthropometric landmark.

gland (gland) An aggregation of cells, specialized to secrete or excrete materials not related to their ordinary metabolic needs.

glia (gli'ah) The neuroglia. Used as a word termination to denote a gluelike structure or tissue.

globulin (glob'u-lin) A class of proteins characterized by being insoluble in water, but soluble in saline solutions (euglobulins), or water soluble proteins (pseudoglobulins) whose other physical properties closely resemble true globulins.

glomerulus (glo-mer'u-lus) A tuft or cluster; used in anatomical nomenclature as a general term to designate such a structure, as one composed of blood vessels or nerve fibers.

glottis (glot'is) The vocal apparatus of the larynx, consisting of the true vocal cords (vocal folds) and the opening between them (rima glottidis).

gluteal (gloo'te-al) Pertaining to the buttocks.

gustatory (gus'tah-to″re) Pertaining to the sense of taste.

H

haustrum (haws'trum) A recess. Of the colon: sacculations in the wall of the large intestine produced by adaptation of its length to that of the tenia coli, or by the arrangement of the circular fibers.

haversian (ha-ver'shan) **system** A haversian canal and its concentrically arranged lamellae, constituting the basic unit of structure of compact bone.

hemangioma (hē-man″je-o'mah) A benign tumor made up of new-formed blood vessels.

hematoma (hem″ah-to'mah) A tumor containing effused blood.

hemoglobin (he″mo-glo'bin) The oxygen-carrying pigment of the erythrocytes, formed by the developing erythrocyte in bone marrow.

hernia (her'ne-ah) The protrusion of a loop or knuckle of an organ or tissue through an abnormal opening.

heterozygous (het″er-o-zi'gus) Possessing different alleles in regard to a given character.

hiatus (hi-a'tus) A gap, cleft, or opening.

hilus (hi'lus) or **hilum** (hi'lum) A depression or pit at that part of an organ where the vessels and nerves enter.

homeostasis (ho″me-o-sta'sis) A tendency to uniformity or stability in the normal body states (internal environment or fluid matrix) of the organism.

homozygous (ho″mo-zi'gus) Possessing an identical pair of alleles in regard to a given character or to all characters.

hydrolysis (hi-drol'ĭ-sis) The splitting of a compound into fragments by the addition of water, the hydroxyl group being incorporated in one fragment, and the hydrogen atom in the other.

hyoid (hi'oid) 1. Shaped like the Greek letter upsilon. 2. Pertaining to the hyoid bone.

hypermotility (hi″per-mo-til'ĭ-te) Excessive or abnormally increased motility.

hypersthenia (hi″per-sthe'ne-ah) Exalted strength or tonicity.

hypersthenic (hi″per-sthen'ik) Pertaining to or characterized by hypersthenia.

hypertension (hi″per-ten'shun) Abnormally high tension; especially high blood pressure.

hypomotility (hi″po-mo-til'ĭ-te) Deficient power of movement in any part.

hyposthenia (hi″pos-the'ne-ah) An enfeebled state; weakness.

hypothenar (hi-poth'e-nar) The ridge on the palm along the bases of the fingers and the ulnar margin.

I

ileum (il'e-um) The distal portion of the small intestine, extending from the jejunum to the cecum.

ileus (il'e-us) Obstruction of the intestines.

ilium (il'e-um) The expansive superior portion of the hip bone. It is a separate bone in early life.

incision (in-sizh′un) 1. A cut, or wound produced by cutting. 2. The act of cutting.

incisura (in-si-su′rah) A cut, notch, or incision; used in anatomical nomenclature as a general term to indicate an indention or depression, chiefly on the edge of a bone or other structure.

incisura angularis (ang′gu-lar-is) **ventriculi** The lowest part on the lesser curvature of the stomach, marking the junction of the cranial two-thirds and caudal one-third of the stomach.

incisura cardiaca (kar′de-ak-a) **ventriculi** The point on the stomach where it articulates with the esophagus.

incus (ing′kus) The middle of the three ossicles of the ear.

induration (in″du-ra′shun) An abnormally hard spot or place.

infarct (in′farkt) An area of coagulation necrosis in a tissue due to local anemia resulting from obstruction of circulation to the area.

inferior (in-fe′re-or) Situated below, or directed downward; in official anatomical nomenclature, used in reference to the lower surface of an organ or other structure.

inguinal (ing′gwĭ-nal) Pertaining to the inguen, or groin.

inhale (in-hāl′) To take into the lungs by breathing.

inion (in′e-on) The most prominent point of the external occipital protuberance.

inspiration (in″spĭ-ra′shun) The act of drawing air into the lungs.

insufflation (in″sŭ-fla′shun) The act of blowing a powder, vapor, gas, or air into a cavity, as into the lungs.

intercalate (in-ter′kah-lāt) To insert between.

internal rotation (in-ter′nal ro-ta′shun) The process of turning or rotating a part inwardly about its axis, as in rotating the humerus; opposed to external rotation.

intima (in′tĭ-mah) Innermost.

intussusception (in″tus-sus-sep′shun) A receiving within: specifically, the prolapse of one part of the intestine into the lumen of an immediately adjoining part.

inversion (in-ver′zhun) A turning inward, inside out, upside down, or other reversal of the normal relation of a part.

ischium (is′ke-um) The inferior dorsal part of the hip bone.

J

jaundice (jawn′dis) A syndrome characterized by hyperbilirubinemia and deposition of bile pigment in the skin and mucous membranes resulting in a yellowish appearance of the patient.

jejunum (je-joo′num) That portion of the small intestine that extends from the duodenum to the ileum.

K

karyokinesis (kar″e-o-ki-ne′sis) The phenomena involved in division of the nucleus, in the process of indirect cell division, or mitosis.

keloid (ke′loid) A new growth or tumor of the skin, consisting of whitish ridges, nodules, and plates of dense tissue.

keratitis (ker″ah-ti′tis) Inflammation of the cornea.

Kupffer's (koop′ferz) **cells** Large star-shaped or pyramidal cells that are attached to the walls of the sinusoids of the liver. They form a part of the reticuloendothelial system.

kyphosis (ki-fo′sis) A condition characterized by an abnormally increased convexity in the curvature of the thoracic spine as viewed from the side.

L

labium (la′be-um) a fleshy border or edge; used in anatomical nomenclature as a general term to designate such a structure; also called lip.

laceration (las″er-a′shun) 1. The act of tearing. 2. A wound made by tearing.

lacrimal (lak′rĭ-mal) Pertaining to the tears.

lacteal (lak′te-al) Any one of the intestinal lymphatics that take up chyle.

lacuna (lah-ku′nah) A small pit or hollow cavity; used in anatomical nomenclature as a general term to designate such a compartment within or between other body structures.

lambda (lam′dah) The point at the site of the posterior fontanel where the lambdoidal and sagittal sutures meet.

lamella (lah-mel′ah) A thin leaf or plate, as of bone.

lamina (lam′ĭ-nah) A thin flat plate, or layer; used in anatomical nomenclature as a general term to indicate such a structure, or a layer of composite structure.

larynx (lar′inks) The musculocartilaginous structure, lined with mucous membrane, situated at the top of the trachea and below the root of the tongue and the hyoid bone.

lateral (lat′er-al) 1. Denoting a position farther from the median plane or midline of the body or of a structure. 2. Pertaining to a side.

leukemia (lu-ke′me-ah) A fatal disease of the blood-forming organs, characterized by a marked increase in the number of leukocytes and their precursors in the blood, together with enlargement and proliferation of the lymphoid tissue of the spleen, lymphatic glands, and bone marrow.

leukocyte (lu′ko-sīt) Any colorless ameboid cell mass. Applied especially to one of the formed elements of the blood, consisting of a colorless granular mass of protoplasm, having ameboid movements, and varying size between 0.005 and 0.015 mm. in diameter.

leukocytosis (lu″ko-si-to′sis) An increase in the number of leukocytes in the blood.

ligament (lig′ah-ment) A band of tissue that connects bones or supports viscera.

lobe (lōb) A more or less well-defined portion of any organ, especially of the brain and glands. Lobes are demarcated by fissures, sulci, connective tissue, and their shape.

locus (lo′kus) Place. In genetics, the specific site of a gene in a chromosome.

lordosis (lor-do′sis) An abnormally increased concavity in the curvature of the lumbar spine as viewed from the side.

lumen (lu′men) The cavity or channel within a tube or tubular organ.

luxation (luks-a′shun) Dislocation.

lymph (limf) A transparent, slightly yellow liquid of alkaline reaction, found in the lymphatic vessels.

lymphosarcoma (lim″fo-sar-ko′mah) A malignant neoplasm arising in lymphatic tissue from proliferation of atypical lymphocytes.

lysis (li′sis) 1. Destruction, as of cells by a specific lysin. 2. Decomposition, as of a chemical compound by a specific agent. 3. Loosening of an organ from adhesions. 4. The gradual abatement of the symptoms of a disease.

M

macrophage (mak′ro-faj) Metchnikoff's name for a large mononuclear wandering phagocytic cell that originates in the tissues.

macula (mak′u-lah) A stain or spot; used in anatomical nomenclature as a general term to designate an area distinguishable by color or otherwise from its surroundings. Applied to a fairly well-pronounced opacity of the cornea.

malar (ma′lar) Pertaining to the cheek or cheek bone.

malignant (mah-lig′nant) Tending to become progressively worse and to result in death.

malingerer (mah-ling′ger-er) An individual who is guilty of malingering.

malingering (mah-ling′ger-ing) The willful, deliberate, and fraudulent feigning or exaggeration of the symptoms of illness or injury, done for the purpose of a consciously desired end.

malleolus (mal-le′o-lus) A rounded process, such as the protuberance on either side of the ankle joint.

mammary (mam′er-e) **gland** The breast.

mandible (man′di-b′l) The bone of the lower jaw.

manubrium (mah-nu′bre-um) A handlelike structure or part; often used alone to designate the manubrium sterni.

manus (ma′nus) The distal portion of the arm, or hand, including the carpus, metacarpus, and fingers.

mastectomy (mas-tek′to-me) Excision of the breast; mammectomy.

mastoid (mas′toid) 1. Nipple shaped. 2. The mastoid process of the temporal bone.

matrix (ma′triks) The groundwork on which anything is cast, or basic material from which a thing develops.

maxilla (mak-sil′ah) The irregularly shaped bone that with its fellow forms the upper jaw.

meatus (me-a′tus) An opening; in anatomical nomenclature, used as a general term to designate an opening to some passageway in the body.

medial (me′de-al) 1. Pertaining to the middle; closer to the median plane or the midline of a body or structure. 2. Pertaining to the tunica media.

mediastinum (me″de-as-ti′num) 1. A median septum or partition. 2. The mass of tissues and organs separating the two lungs, between the sternum in front and the vertebral column behind, and from the thoracic inlet above to the diaphragm below.

medulla (me-dul′lah) The middle, innermost part; used as a general term in anatomical nomenclature to designate the inmost portion of an organ or structure.

meiosis (mi-o′sis) A special method of cell division, occurring in maturation of sex cells, by means of which each daughter nucleus receives half the number of chromosomes characteristic of the somatic cells of the species.

membrane (mem′brān) A thin layer of tissue that covers a surface or divides a space or organ.

meninges (mĕ-nin′jēz) The three membranes that envelop the brain and spinal cord: the dura mater, arachnoid, and pia mater.

metabolism (mĕ-tab′o-lizm) The sum of all the physical and chemical processes by which living organized substance is produced and maintained, and also the transformation by which energy is made available for the uses of the organism.

metacarpus (met″ah-kar′pus) The part of the hand between the wrist and the fingers, its skeleton being, five cylindric bones (metacarpals) extending from the carpus to the phalanges.

metaphysis (me-taf′ĭ-sis) The wider part at the extremity of the shaft of a long bone, adjacent to the epiphyseal disc. During development it contains the growth zone and consists of spongy bone; in the adult it is continuous with the epiphysis.

metastasis (mĕ-tas′tah-sis) The transfer of disease from one organ or part to another not directly connected with it. It may be due either to the transfer of pathogenic microorganisms or to transfer of cells, as in malignant tumors.

metatarsus (met″ah-tar′sus) The part of the foot between the tarsus and the toes, its skeleton being the five long bones (metatarsals) extending from the tarsus to the phalanges.

microcyte (mi′kro-sīt) An abnormally small erythrocyte, i.e., one 5 microns or less in diameter.

micturate (mik′tu-rāt) Urinate.

midaxilla (mid″ak-sil′ah) The center of the axilla.

mitosis (mi-to′sis) A method of indirect cell division, consisting of a complex of various processes, by means of which the two daughter nuclei normally receive identical complements of the number of chromosomes characteristic of the somatic cells of the species.

mitral (mi′tral) 1. Shaped somewhat like a miter. 2. Pertaining to the mitral or bicuspid valve.

molecule (mol′ĕ-kūl) A very small mass of matter; an aggregation of atoms; specifically, a chemical combination of two or more atoms that form a specific chemical substance.

morbid (mor′bid) Pertaining to or affected with disease.

moribund (mor′ĭ-bund) In a dying state.

morula (mor′u-lah) The solid mass of blastomeres formed by cleavage of a fertilized ovum, filling all the space occupied by the ovum before cleavage.

mucin (mu′sin) A muculopolysaccharide or glycoprotein, the chief constituent of mucus.

multipara (mul-tip′ah-rah) A woman who has had two or more pregnancies that resulted in viable offspring, whether or not the offspring were alive at birth.

mutant (mu′tant) A sport or variation that breeds true.

mutation (mu-ta′shun) In biology, a permanent transmissible change in the characteristics of an offspring from those of its parents; also, an individual showing such change.

myelin (mi′ĕ-lin) The fatlike substance forming a sheath around certain nerve fibers.

myxedema (mik″se-de′mah) A condition characterized by a dry, waxy type of swelling, with abnormal deposits of mucin in the skin (mucinosis), and associated with hypothyroidism.

N

naris (na′ris) One of the openings of the nasal cavity.

nasal (na′zal) Pertaining to the nose.

nasion (na′ze-on) An anthropometric landmark, the point at which a horizontal line tangential to the highest points on the superior palpebral sulci is intersected by the midsagittal plane.

navel (na′vel) The umbilicus.

navicular (nah-vik′u-lar) Boat shaped.

necropsy (nek′rop-se) Examination of a body after death.

necrosis (ne-kro′sis) Death of a tissue, usually as individual cells, groups of cells, or in small localized areas.

neoplasm (ne′o-plazm) Any new and abnormal growth, such as a tumor.
nephron (nef′ron) The anatomical and functional unit of the kidney, consisting of the renal corpuscle, the proximal convoluted tubule, the descending and ascending limbs of Henle's loop, the distal convoluted tubule, and the collecting tubule.
nerve (nerv) A cordlike structure that conveys impulses between a part of the central nervous system and some other region of the body.
neural (nu′ral) Pertaining to a nerve or to the nerves.
node (nōd) A swelling or protuberance.
nucha (nu′kah) The back, nape, or scruff of the neck.
nucleus pulposus (nu′kle-us pul-po′sus) A semifluid mass of fine white and elastic fibers that forms the central portion of an intervertebral disc.

O

obese (o-bēs′) Excessively fat.
oblique (ob-lēk′, ob-līk) Slanting, inclined; between a horizontal and a perpendicular direction.
occiput (ok′sĭ-put) The back part of the head.
occlude (ŏ-klōōd′) To fit close together; to close tight, as to bring the mandibular teeth into contact with the teeth in the maxillae.
occult (ŏ-kult′) Obscure, concealed from observation; difficult to understand.
odontoid (o-don′toid) Toothlike, resembling a tooth.
olecranon (o-lek′rah-non) The proximal bony projection of the ulna at the elbow, its anterior surface forming part of the trochlear notch.
omentum (o-men′tum) A fold of peritoneum extending from the stomach to adjacent organs in the abdominal cavity.
oocyte (o′o-sīt) A growing or full-grown oogonial cell that has not yet completed its maturation process.
organ (or′gan) A somewhat independent part of the body that performs a special function.
orifice (or′ĭ-fis) 1. The entrance or outlet of any body cavity. 2. Any foramen, meatus, or opening.
os (os) 1. The anterior or proximal opening of the digestive apparatus. 2. Bone; used in anatomical nomenclature as a general term that is combined with the appropriate adjective to designate a specific type of bony structure or a specific segment of the skeleton.
osmosis (os-mo′sis) The passage of pure solvent from the lesser to the greater concentration when two solutions are separated by a membrane that selectively prevents the passage of solute molecules but is permeable to the solvent.
osseous (os′e-us) Of the nature or quality of bone; bony.
ossicle (os′sĭ-k'l) A small bone.
ossification (os″ĭ-fi-ka′shun) The formation of bone or of a bony substance; the conversion of fibrous tissue or of cartilage into bone.
osteitis (os″te-i′tis) Inflammation of a bone, involving the haversian spaces, canals, and their branches, and generally the medullary cavity, and marked by enlargement of the bone, tenderness, and a dull, aching pain.
osteoblast (os′te-o-blast) A cell that arises from a fibroblast and, as it matures, is associated with the production of bone.
osteoclast (os′te-o-klast) A large multinuclear cell associated with the absorption and removal of bone.
osteocyte (os″te-o-sīt′) An osteoblast that has become embedded within the bone matrix, occupying a flat oval cavity (lacuna) and sending, through apertures in its walls, thin cytoplasmic processes that directly connect with other osteocytes in developing bone.
osteogenesis (os″te-o-jen′e-sis) Formation of bone; the development of the bones.
osteology (os″te-ol′o-je) The scientific study of the bones; applied also to the body of knowledge relating to the bones.
osteomyelitis (os″te-o-mi″ĕ-li′tis) Inflammation of bone caused by a pyogenic organism.
osteoporosis (os″te-o-po-ro′sis) Abnormal rarefaction of bone due to failure of the osteoblasts to lay down bone matrix.
ovum (o′vum) A round cell about 0.1 mm. in diameter, produced in the ovary.

P

palatine (pal′ah-tīn) Pertaining to the palate.
palliate (pal′e-āt) To reduce the severity of; to relieve.
palm (palm) The hollow of the hand.

palmar (pal'mar) Pertaining to the palm.

paracentesis (par″ah-sen-te'sis) Surgical puncture of a cavity for the aspiration of fluid.

paralysis (pah-ral'ĭ-sis) Loss or impairment of motor function in a part due to lesion of the neural or muscular mechanism; also, by analogy impairment of sensory function.

paraplasm (par'ah-plazm) An abnormal growth.

paraplegia (par″ah-ple'je-ah) Paralysis of the legs and lower part of the body, both motion and sensation being affected.

parenchyma (par-eng'kĭ-mah) The essential elements of an organ; used in anatomical nomenclature as a general term to designate the functional elements of an organ, as distinguished from its framework, or stroma.

paroxysm (par'ok-sizm) A sudden recurrence or intensification of symptoms.

patella (pah-tel'lah) A triangular sesamoid bone, about 5 cm. in diameter, situated at the front of the knee in the tendon of insertion of the quadriceps femoris muscle. Called also the knee cap.

patent (pa'tent) Open, unobstructed, or not closed.

pathology (pah-thol'o-je) The branch of medicine that treats of the essential nature of disease, especially of the structural and functional changes in tissues and organs of the body that cause or are caused by disease.

pelvic strait (pel'vik strāt) The narrow passageway of the pelvis, consisting of the superior strait, the pelvic inlet, and the inferior strait, the pelvic outlet.

pericardium (per″ĭ-kar'de-um) The fibroserous sac that surrounds the heart, comprising an external layer of dense fibrous tissue (pericardium fibrosum) and an inner serous layer (pericardium serosum). The base of the pericardium is attached to the central tendon of the diaphragm.

perineum (per″ĭ-ne'um) The pelvic floor and the associated structures occupying the pelvic outlet. It is bounded anteriorly by the pubic symphysis, laterally by the ischial tuberosities, and posteriorly by the coccyx.

periosteum (per″e-os'te-um) A specialized connective tissue covering all bones of the body, and possessing bone-forming potentialities; in adults, it consists of two layers that are not sharply defined, the external layer being a network of dense connective tissue containing blood vessels, and the deep layer composed of more loosely arranged collagenous bundles with spindle-shaped connective tissue cells and a network of thin elastic fibers.

peristalsis (per″ĭ-stal'sis) The wormlike movement by which the alimentary canal or other tubular organs provided with both longitudinal and circular muscular fibers propel their contents. It consists of a wave of contraction passing along the tube.

peritoneum (per″i-to-ne'um) The serous membrane lining the abdominopelvic walls (parietal peritoneum) and investing the viscera (visceral peritoneum). It is a strong, colorless membrane with a smooth surface, and forms a closed sac except in the female, in whom it is continuous with the mucous membrane of the uterine tubes.

pH The symbol commonly used in expressing hydrogen ion concentration, the measure of alkalinity and acidity. It signifies the logarithm of the reciprocal of the hydrogen ion concentration in gram molecules per liter of solution.

phagocyte (fag'o-sīt) Any cell that ingests microorganisms or other cells and foreign particles. Phagocytes are either fixed (cells of the reticuloendothelial system) or free (polymorphonuclear leukocytes, macrophages).

phalanx (fa'lanks) Any bone of a finger or toe.

pharynx (far'inks) The musculomembranous sac between the mouth and nares and the esophagus.

phlebolith (fleb'o-lith) A calculus or concretion in a vein; a vein stone.

phren (fren) 1. The diaphragm. 2. The mind, as seat of the intellect, or the heart, as the seat of the passions.

pia mater (pi'ah ma'ter) The innermost of the three membranes (meninges) covering the brain and spinal cord, investing them closely and extending into the depths of the fissures and sulci.

placenta (plah-sen'tah) The cakelike organ within the uterus that establishes communication between the mother and child by means of the umbilical cord; listed in anatomical nomenclature it consists of a uterine and a fetal portion.

placenta praevia (pre′vĭ-ah) A placenta that develops in the lower uterine segment, in the zone of dilatation, so that it covers or adjoins the internal os.

plantar (plan′tar) Pertaining to the sole of the foot.

pleura (ploor′ah) The serous membrane investing the lungs and lining the thoracic cavity, completely enclosing a potential space known as the pleural cavity. There are two pleurae, right and left, entirely distinct from each other.

pneumoperitoneum (nu″mo-per″ĭ-to-ne′um) The presence of gas or air in the peritoneal cavity, sometimes deliberately introduced as an aid to examination and diagnosis.

pneumothorax (nu″mo-tho′raks) An accumulation of air or gas in the pleural cavity, which may occur spontaneously or as a result of trauma or a pathological process, or be introduced deliberately.

polymerization (pol″ĭ-mer″ĭ-za′shun) The chemical union of two or more molecules of a substance to form a new compound without the elimination of a secondary compound.

polyp (pol′ip) A morbid excrescence, or protruding growth, from mucous membrane.

posterior (pos-te′re-or) Situated in back of, or in the back part of, or affecting the back part of an organ; in official anatomical nomenclature, used in reference to the back or dorsal surface of the body.

primipara (pri-mip′ah-rah) A woman who has had one pregnancy that resulted in a viable child, regardless of whether the child was living at birth, and regardless of whether it was a single or multiple birth.

proliferate (pro-lif′er-āt) To grow by the reproduction of similar cells.

pronate (pro′nāt) To assume or place in a prone position

prone (prōn) Lying with the face downward.

prosthesis (pros′the-sis) 1. The replacement of an absent part by an artificial substitute. 2. An artificial substitute for a missing part, such as an eye, leg, or denture.

proteolysis (pro″te-ol′ĭ-sis) The hydrolysis of proteins into proteoses, peptones and other products by means of enzymes.

protoplasm (pro′to-plazm) The only known form of matter in which life is manifested. It is a viscid, translucent, polyphasic colloid with water as the continuous phase, and it makes up the essential material of all plant and animal cells.

proximal (prok′sĭ-mal) Nearest; closer to any point of reference; opposed to distal.

ptosis (to′sis) 1. Prolapse of an organ or part. 2. Drooping of the upper eyelid from paralysis of the third nerve.

pubis (pu′bis) The os pubis.

pyknic (pik′nik) Having a short, thick, stocky build.

pylorus (pi-lo′rus) The distal aperture of the stomach, through which the stomach contents pass into the duodenum; it is surrounded by a fold of mucous membrane enclosing a circular layer of muscle fibers.

pyogenic (pi″o-jen′ik) Producing pus.

Q

quadrate (kwod′rāt) Square or squared; four sided.

R

racemose (ras′e-mōs) Resembling a bunch of grapes on its stalk.

radius (ra′de-us) The bone on the outer or thumb side of the forearm.

ramus (ra′mus) A branch; used in anatomical nomenclature as a general term to designate a smaller structure given off by a larger one, or into which the larger structure, such as a blood vessel or nerve, divides.

raphe (ra′fe) A seam; used in anatomical nomenclature as a general term to designate the line of union of the halves of various symmetrical parts.

Reid's base line (rēdz) The base line of the skull; also called the infraorbitomeatal line and the anatomic base line; a line from the infraorbital ridge to the external auditory meatus and the middle line of the occiput.

renal (re′nal) Pertaining to the kidney.

renin (re′nin) A proteolytic enzyme liberated by ischemia of the kidney or by diminished pulse pressure that changes hypertensinogen into hypertensin.

rennin (ren′in) The milk-curdling enzyme obtained from the fourth stomach of calves.

retina (ret′ĭ-nah) The innermost of the three tunics of the eyeball, surrounding the vitreous body and continuous posteriorly with the optic nerve.

rickets (rik′ets) A condition caused by deficiency of vitamin D, especially in infancy and childhood, with disturbance of normal ossification.

rickettsia (rĭ-ket′se-ah) An individual organism of the genus *Rickettsia*.

rigor mortis (ri′gor mor′tis) The stiffening of a dead body, as a result of depletion of adenosine triphosphate in the muscle fibers.

ruga (roo′gah) A ridge, wrinkle, or fold, as of mucous membrane.

S

sacrum (sa′krum) The triangular-shaped bone formed usually by five fused vertebrae that are wedged dorsally between the two hip bones.

sagittal (saj′ĭ-tal) 1. Shaped like or resembling an arrow; straight. 2. Situated in the direction of the sagittal suture, said of an anteroposterior plane or section parallel with the long axis of the body.

sarcoma (sar-ko′mah) A tumor made up of substance like the embryonic connective tissue; tissue composed of closely packed cells embedded in a fibrillar or homogeneous substance. Sarcomas are often highly malignant.

scapula (skap′u-lah) The flat, triangular bone in the back of the shoulder.

Schmorl's nodule (shmorlz nod′ūl) A nodule seen in roentgenograms of the spine, due to prolapse of a nucleus pulposus into an adjoining vertebra.

sclerosis (skle-ro′sis) An induration, or hardening; especially hardening of a part from inflammation and in diseases of the interstitial substance.

scoliosis (sko″le-o′sis) An appreciable lateral deviation in the normally straight vertical line of the spine.

secrete (se-krēt′) To separate or elaborate cell products.

sella turcica (sel′ah tur′si-ka) A transverse depression crossing the midline on the superior surface of the body of the sphenoid bone, and containing the hypophysis.

semilunar (sem″e-lu′nar) Resembling a crescent, or half-moon.

sepsis (sep′sis) Poisoning that is caused by the products of a putrefactive process.

septum (sep′tum) A dividing wall or partition; used as a general term in anatomical nomenclature.

sequela (se-kwe′lah) Any lesion or affection following or caused by an attack of disease.

sequestrum (se-kwes′trum) A piece of dead bone that has become separated during the process of necrosis from the sound bone.

serrated (ser′āt-ed) Having a sawlike edge.

sesamoid (ses′ah-moid) Resembling a grain of sesame.

sesamoid bone A type of short bone occurring mainly in the hands and feet, and found embedded in tendons or joint capsules.

Sharpey's fibers (shar′pĕz fi′berz) Fibers that pass from the periosteum and embed in the periosteal lamellae.

sigmoid (sig′moid) 1. Shaped like the letter S, or like the Greek sigma. 2. The sigmoid flexure.

sinus (si′nus) 1. A cavity, or hollow space; used in anatomical nomenclature as a general term to designate such spaces as the dilated channels for venous blood, found chiefly in the cranium, or the air cavities in the cranial bones. 2. An abnormal channel or fistula permitting the escape of pus.

somatic (so-mat′ik) Pertaining to or characteristic of the body (soma).

spermatid (sper′mah-tid) A cell derived from a secondary spermatocyte by fission, and developing into a spermatozoon.

spermatocyte (sper′mah-to-sīt″) The mother cell of a spermatid.

spermatozoon (sper″mah-to-zo′on) A mature male germ cell, the specific output of the testes.

sphenoid (sfe′noid) Wedge shaped; designating especially a very irregular wedge-shaped bone at the base of the skull.

sphincter (sfingk′ter) A ringlike band of muscle fibers that constricts a passage or closes a natural orifice.

sphygmomanometer (sfig″mo-mah-nom′e-ter) An instrument for measuring blood pressure in the arteries.

spina bifida (spi′nah bif′ĭ-dah) A developmental anomaly characterized by a defect in the bony encasement of the spinal cord.

spina bifida occulta (ŏ-kul′tah) A spina bifida in which there is a defect of the bony spinal canal without protrusion of the cord or meninges.

spiralization (spi′răl-ĭ-zā′shŭn) The act of making a spiral, a coil, or a twist.

splenic (splen′ik) Pertaining to the spleen.

spondylitis (spon″di-li′tis) Inflammation of the vertebrae.

spondylolisthesis (spon″di-lo-lis′the-sis) Forward displacement of one vertebra over another, usually of the fifth lumbar over the body of the sacrum, or of the fourth lumbar over the fifth.

spondylolysis (spon″di-lol′i-sis) Dissolution of a vertebrae; a condition marked by platyspondylia, aplasia of the vertebral arch, and separation of the pedicle.

squamous (skwa′mus) Scaly, or platelike.

stasis (sta′sis) A stoppage of the flow of blood or other body fluid in any part.

stenosis (ste-no′sis) Narrowing or stricture of a duct or canal.

sterile (ster′il) 1. Not fertile; infertile; barren; not producing young. 2. Aseptic; not producing microorganisms; free from microorganisms.

steroid (ste′roid) A group name for compounds that resemble cholesterol chemically and contain a hydrogenated cyclopentenophenanthrene-ring system. Included in this group are bile acids, cardiac aglycones, sex hormones, sterols proper, saponins, toad poisons, and some cancerigenic hydrocarbons.

sternum (ster′num) A longitudinal unpaired plate of bone forming the middle of the anterior wall of the thorax and articulating above with the clavicles and along the sides with the cartilages of the first seven ribs. It consists of three portions, the manubrium, the body, and the xiphoid process.

sthenic (sthen′ik) Active; strong.

stomach (stum′ak) The musculomembranous expansion of the alimentary canal between the esophagus and the duodenum.

stroma (stro′mah) The structural elements of an organ; used in anatomical nomenclature as a general term to designate the tissue that forms the ground substance, framework, or matrix of an organ, as distinguished from that constituting its functional element, or parenchyma.

styloid (sti′loid) Resembling a pillar; long and pointed.

subarachnoid (sub″ah-rak′noid) Situated or occurring beneath the arachnoid.

subluxation (sub″luk-sa′shun) An incomplete or partial dislocation.

sulcus (sul′kus) A groove, trench, or furrow; used in anatomical nomenclature as a general term to designate such a depression, especially one of those on the surface of the brain, separating the gyri.

superior (su-pe′re-or) Situated above, or directed upward; in official anatomical nomenclature, used in reference to the upper surface of an organ or other structure, or to a structure occupying a higher position.

supination (su″pi-na′shun) The act of assuming the supine position, or the state of being supine. Applied to the hand, the act of turning the palm forward (anteriorly) or upward, performed by lateral rotation of the forearm.

supine (su′pin) Lying with the face upward.

suppuration (sup″u-ra′shun) The formation of pus, the act of being converted into and discharging pus.

sustentaculum tali (sus″ten-tak′u-lum ta′li) A process of the calcaneus that supports the talus.

suture (su′tur) A type of fibrous joint in which the opposed surfaces are closely united.

symphysis (sim′fi-sis) A site or line of union; used in official anatomical nomenclature to designate a type of cartilaginous joint in which the apposed bony surfaces are firmly united by a plate of fibrocartilage.

synapse (sin′aps) The anatomical relation of one nerve cell to another; the region of contact between processes of two adjacent neurons, forming the place where a nervous impulse is transmitted from one neuron to another.

synapsis (si-nap′sis) The pairing off and union of homologous chromosomes from the male and female pronuclei at the start of meiosis.

syndrome (sin′drom) A set of symptoms that occur together; the sum of signs of any morbid state; a symptom complex.

synovia (si-no′ve-ah) A transparent alkaline, viscid fluid, resembling the white of an egg, secreted by the synovial membrane, and contained in joint cavities, bursae, and tendon sheaths.

synovial (si-no′ve-al) Of, or pertaining to, or secreting, synovia.

systole (sis′to-le) The contraction, or period of contraction, of the heart, especially that of the ventricles. It coincides with the interval between the first and second heart sounds, during which the blood is forced into the aorta and the pulmonary trunk.

T

talus (ta'lus) The highest of the tarsal bones and the one that articulates with the tibia and fibula to form the ankle joint.

tarsus (tahr'sus) The region of the articulation between the foot and the leg.

tetanus (tet'ah-nus) 1. An acute infectious disease caused by a toxin produced in the body by the *Clostridium tetani*, with tonic spasm of the masseter muscles, causing trismus (lockjaw), followed by spasms of the back muscles, producing the characteristic opisthotonos. 2. Continuous tonic spasm of a muscle, steady contraction of a muscle without distinct twitching.

tetany (tet'ah-ne) 1. A syndrome manifested by sharp flexion of the wrists and ankles, muscle twitchings, cramps, and convulsions, sometimes with attacks of stridor. It is due to abnormal calcium metabolism and occurs in parathyroid hypofunction, vitamin D deficiency, alkalosis, and as a result of the ingestion of alkaline salts. 2. Tetanus, def. 2.

thalamus (thal'ah-mus) The middle and larger portion of the diencephalon, which forms part of the lateral wall of the third ventricle and lies between the hypothalamus and the epithalamus. It is the main relay center for sensory impulses to the cerebral cortex.

thenar (the'nar) The mound on the palm at the base of the thumb.

thoracentesis (tho"rah-sen-te'sis) Thoracocentesis. Surgical puncture of the chest wall with drainage of fluid.

thoracic (tho-ras'ik) Pertaining to the chest.

thoracoplasty (tho"rah-ko-plas'te) Plastic surgery of the thorax; operative repair of defects of the chest.

thrombus (throm'bus) A plug or clot in a blood vessel or in one of the cavities of the heart, formed by coagulation of the blood, and remaining at the point of its formation.

thymus (thi'mus) A ductless glandlike body situated in the anterior mediastinal cavity that reaches its maximum development during the early years of childhood.

thyroid (thi'roid) 1. Resembling a shield; scutiform. 2. The thyroid gland. 3. A pharmaceutical substance derived from thyroid glands obtained from domesticated animals used for food by man, the glands having been deprived of connective tissue and fat and then cleaned, dried, and powdered: used in replacement therapy.

thyroxine (thi-rok'sin) or **thyroxin** A crystalline iodine-containing compound possessing the physiologic properties of thyroid extract.

tissue (tish'u) An aggregation of similarly specialized cells united in the performance of a particular function.

tonus (to'nus) The slight, continuous contraction of muscle, which in skeletal muscles aids in the maintenance of posture and in the return of blood to the heart.

torsion (tor'shun) The act of twisting; the condition of being twisted.

trabecula (trah-bek'u-lah) A little beam; used in anatomical nomenclature as a general term to designate a supporting or anchoring strand of connective tissue, as such a strand extending from a capsule into the substance of the enclosed organ.

trachea (tra'ke-ah) The cartilaginous and membranous tube descending from the larynx to the bronchi.

tragus (tra'gus) The cartilaginous projection anterior to the external opening of the ear.

transverse (trans-vers') Placed crosswise; situated at right angles to the long axis of a part.

trauma (traw'mah) A wound or injury.

Trendelenburg's (tren-del'en-berg) **position** The patient on the back on a plane inclined 30 to 40 degrees, the legs and feet hanging over the end of the table.

trigone (tri'gōn) A triangular area.

trochlea (troch'le-ah) A pulley-shaped part or structure; used as a general term in anatomical nomenclature.

tubercle (tu'ber-k'l) 1. A nodule, especially a solid elevation of the skin, larger in size than a papule. 2. Any small, rounded nodule produced by the *Mycobacterium tuberculosis*. 3. A nodule, or small eminence, such as a rough, rounded eminence on a bone.

tuberosity (tu"ber-os'ĭ-te) An elevation or protuberance.

tuft (tuft) A small clump or cluster; a coil.

tumor (tu'mor) 1. A swelling, one of the

cardinal signs of inflammation; morbid enlargement. 2. Neoplasm. A mass of new tissue that persists and grows independently of its surrounding structures, and has no physiologic use.

U

ulna (ul'nah) The inner and larger bone of the forearm, on the side opposite that of the thumb.

umbilicus (um"bĭ-li'kus) The cicatrix marking the site of attachment of the umbilical cord in the fetus.

ureter (u-re'ter) The fibromuscular tube that conveys the urine from the kidney to the bladder.

urethra (u-re'thrah) The membranous canal conveying urine from the bladder to the exterior of the body.

uterus (u'ter-us) The hollow muscular organ in female animals that is the abode and the place of nourishment of the embryo and fetus. In the human, it is a pear-shaped structure, about 3 inches in length, consisting of a body, fundus, isthmus, and cervix.

uvula (u'vu-lah) A pendent, fleshy mass; used as a general term in anatomical terminology. Usually used alone to designate the *uvula palatina*.

V

vagina (vah-ji'nah) 1. A sheath, or sheathlike structure; used as a general term in anatomical nomenclature. 2. The canal in the female, extending from the vulva to the cervix uteri, which receives the penis in copulation.

varicose (var'ĭkōs) Of the nature of or pertaining to a varix; unnaturally swollen; said of a vein.

vascular (vas'ku-lar) Pertaining to or of vessels.

vein (vān) A vessel through which blood passes from various organs or parts back to the heart.

vena cava inferior (ve'nah ca'vah) The venous trunk for the lower extremities and for the pelvic and abdominal viscera, it begins at the level of the fifth lumbar vertebra by union of the common iliac veins, passes upward on the right side of the aorta, and empties into the right atrium of the heart.

vena cava superior The venous trunk draining blood from the head, neck, upper extremities, and chest; it begins by union of the two brachiocephalic veins, passes directly downward, and empties into the right atrium of the heart.

ventral (ven'tral) 1. Pertaining to the belly or any venter. 2. Denoting a position closer to the belly surface.

ventricle (ven'trĭ-k'l) A small cavity, such as one of the several cavities of the brain, or one of the lower chambers of the heart.

vertebra (ver'te-brah) Any one of the thirty-three bones of the spinal column.

vertebra prominens (prom'ĭ-nenz) The seventh cervical vertebra; so called because of the length of its spinous process.

vertex (ver'teks) A summit or top; used as a general term in anatomical nomenclature. Sometimes used alone to designate the top of the head.

vestigial (ves-tij'e-al) Of the nature of a vestige, trace, or relic; rudimentary.

villus (vil'lus) A small vascular process or protrusion, especially such a protrusion from the free surface of a membrane; used as a general term in anatomical nomenclature.

viscera (vis'er-ah) Plural of viscus.

viscus (vis'kus) Any large interior organ in any one of the three great cavities of the body, especially in the abdomen.

vitelline (vi-tel'in) Resembling or pertaining to the yolk of an egg or ovum.

vitreous (vit're-us) Glasslike or hyaline; often used alone to designate the vitreous body of the eye.

void (void) To cast out as waste matter.

volar (vo'lar) Pertaining to the palm or sole; indicating the flexor surface of the forearm, wrist, or hand.

vulva (vul'vah) The region of the external genital organs of the female, including the labia minora, labia majora, mons pubis, clitoris, perineum, and vestibule of the vagina. Called also the *pudendum femininum*.

W

wormian (wer'me-an) **bone** A small flat bone lying wholly within a suture. Ossa suturarum.

wound (wōōnd) An injury to the body caused by physical means, with disruption of the normal continuity of body structures.

X

X chromosome The differential sex chromosome carried by half the male gametes and all the female gametes in human beings.

xiphoid (zi'foid) 1. Shaped like a sword. 2. The xiphoid process.

Y

Y chromosome The differential sex chromosome carried by half the male gametes in human beings.

Z

zoology (zo-ol'o-je) The biology of animals; the sum of what is known regarding animals.

zygoma (zi-go'mah) The zygomatic process of the temporal bone. Also the malar bone.

zygote (zi'gōt) 1. The cell resulting from the fusion of two gametes; the fertilized ovum. 2. The individual developing from a cell formed by the union of two gametes.

Pronunciations and definitions in this glossary are taken by permission from *Dorland's Illustrated Medical Dictionary*, ed. 24, Philadelphia, 1965, W. B. Saunders Co., with slight alterations.

Index

A

Absorption
   of alcohol, 188
   of food, 188, 219-231
Acetabulum, 74, 123, 124
Acid
   amino, 15
   and base, 43
   carbonic, 45, 221, 320
   DNA, 14
   hippuric, 225
   hydrochloric, 221
   lactic, 362
   nucleic, 14
   phosphoric, 15
   pyruvic, 362
   RNA, 14
   uric, 225
Acromegaly, 397
Addison's disease, 402
Adipose tissue, 34
Ameba, 11
Amitosis, 17
Amnion, 51
Ampulla, hepatopancreatic, 208
Anaphase, 17
Anatomic landmarks
   of skull, 165
   of trunk, 124, 125
Anatomic location and direction, 2-5
Anatomic position, normal, 3
Anatomy, 1
   defined, 8
Angiocardiography, 273
Anions, 10, 18
Ankle, 78, 80
Aorta, 243, 244
Aortography, 274
Arachnoid membrane, 334

Arch
   longitudinal, 80
   transverse, 80
Areolar tissue, 34
Arterial circle, 245, 246
Arteries, 243-249
   coronary, 242
   of head and neck, 244-246
   hepatic, 215
   of lower extremity, 248, 249
   nutrient, 56, 78
   of trunk, 246, 247
   of upper extremity, 246
Arteriography, 274
Arterioles, 249
Articulations, 166-173
   of thoracic vertebrae with ribs, 111, 112
Astral rays, 16
Atlas, 90, 91
Atom, 36
   orbits of, 8, 9
   size of, 9
   stable, 8
Atomic number and weight, 10
Atrioventricular node of heart, 240
Autonomic nervous system, 330, 331
   craniosacral outflow, 331
   thoracolumbar outflow, 331
AV bundle, 240
Axis, 90, 91
Axons, 29, 329

B

Balance, homeostatic
   acids and bases, 43
   fluids and electrolytes, 42
Bile
   excretion, 226
   formation, 225, 226
Biliary system, 218, 219
Binary fission, simple, 17

Biologic effects
  of alpha particles, 49
  of beta particles, 49
  of gamma rays, 49
  of neutrons, 49
  of x-rays, 49
Biologic science, 8
Biology, 8
Blastocyst, 51
Blood, 32, 261, 262
  acid-base balance, 261, 262
  cells, 32, 33, 262-264
  clotting mechanism, 266
  fetal, 52
  flow of, 241, 242
  hydrostatic pressure, 250
  material, 52
  osmotic pressure, 250, 261
  percent of body weight, 32
  pH, 261
  plasma, 32, 264
  platelets, 33, 264
  pressure, 264-266
  serum, 264
  as tissue, 261
Blood-brain barrier, 329, 344
Blood-cerebrospinal fluid barrier, 344
Blue baby, 274
Body
  planes, 124, 125
  stalk, 53
  types, 201
Bone, 55, 175-186
  cancellous, 33, 57, 176
  classes, 34
  cortical, 33, 34, 56, 176
  depressions, 60, 61
  epipteric, 138
  functions, 34
  markings, 61
  marrow, 55, 56, 176
  osteogenetic functions, 55
  physiology, 173-187
  projections, 58, 59
  resorption, 400
  structure, 176
  tissue, 33, 34, 176
  trabeculae, 176
  types, 57
  typical, 55
Botany, 8
Bronchi, 312
Buffers, 44
Building blocks, 10

C

Calcium, 8, 11
  phosphorus ratio, 400
  storage, 55
  tetany, 400
Calyx, of kidney
  major, 291
  minor, 290
Canals
  anal, 212
  auditory, 142
  condyloid, 147
  haversian, 177
  hypoglossal, 147
  medullary, 57
  spinal, 86
  of Volkmann, 177
Capillaries, 249, 250
Carbohydrates, 16, 220, 227-229, 231
Carbon, 8
  atoms in protein molecules, 11
Carbon-dioxide transport, 320
Carbonic anhydrase, 320
Cardiac cycle, 241
Cardiac notch, 314
Carpals, 70-72
Carrying angle of elbow, 67
Cartilage, 34, 173-175
  arytenoid, 310
  corniculate, 310
  cricoid, 308
  cuneiform, 310
  elastic, 34, 174, 175
  epiglottis, 308
  fibrous, 34, 175
  hyaline, 34, 174
  thyroid, 307, 308
Cations, 10, 18
Cavities of body, 3
  abdominal, 197-199
  abdominopelvic, 3, 197-199
  buccal, 189, 190
  cranial, 3
  dorsal, 3
  mediastinal, 316, 317
  oral, 190-195
  pericardial, 237
  peritoneal, 197
  pleural, 316
  spinal, 3
  thoracic, 3
  ventral, 3
Cecum, 211
Cell, 2
  argentaffine, 207
  bone, 178-180

Cell—cont'd
  cancer, 17
  chief, 222
  connective tissue, 19, 31
  cytoplasm, 2, 14
  division, 17
  epithelial, 19
  germ, 12, 46
  Kupffer's, 225
  of Leydig, 385
  membrane, 2, 14
  microscopic examination of, 12
  mitosis, 14, 17-19
  muscle, 19
  nerve, 19, 29
  nucleus, 2, 14
  numbers, 2
  of Paneth, 207
  parietal, 221
  power plants, 17
  Schwann, 31
  serozymogenic, 207
  of Sertoli, 385
  size, 14
  somatic, 12, 19, 46
  specialization, 2, 12
  spermatogenic, 385
  structure, 14, 15
  totipotent, 46, 47
Cellular effects of ionizing rays, 49
Cellular respiration, 305
Central nervous system, 326-330
Centriole, 16
  and mitosis, 16
Centrosome, 16
Cephalic index, 127, 130
Cerebellum, 336, 338, 339
Cerebral angiography, 274
Cerebrospinal fluid, 343, 344
Cerebrum, 336, 337
Cervical ribs, 93
Chiasma
  in meiosis, 49
  optic, 134
Chloride shift, 320
Chlorine, 8
Chordae tendineae, 239
Chorion, 52
Choroid plexus, 339
Chromatid, 47
Chromosome
  and energy, 49
  number, 47
  number in human, 17
  number in species, 17
Chyle, 258
Chyme, 206, 221

Circulatory system, 235-286
  anatomy, 237-261
    arteries, 243-249
    arterioles, 249
    capillaries, 249, 250
    coronary vessels, 242, 243
    heart, 237-243
      cardiac cyle, 241
      chambers, 239
      special structures, 240
      structure, 238-240
      valves, 239, 240
    lymphatic system, 256-261
    veins, 250-256
    venules, 256
  blood, cells, fluids, and interrelated functions, 261-268
  divisions, 270-272
    coronary, 270
    portal, 271
    pulmonary, 270
    systemic, 271
  fetal circulation, 268-270
  function, 235-237
    heat-equilibration, 237
    transportation, 235, 237
    water-balance maintenance, 237
  lymphatic system, 256-261
  radiographic studies, 272-286
    direct, 273, 274
    heart margins, 277-286
    transportation mechanism, 274
Cisterna
  lumbar, 340
  manga, 338
Clavicle, 106, 107
Coccyx, 119
Coelem, 3
Colloid, 11, 36, 398
  osmotic pressure, 297
Colon, 203
  ascending, 211
  descending, 212
  iliac, 212
  sigmoid, 212
  transverse, 211, 212
Colostrum, 390
Compartments, fluid, 42
Concha, inferior, 152
Connective tissue, 31
Conus medullaris, 339
Corpus callosum, 336, 338
Costophrenic angle, 111, 313
Cranial floor, 142-144, 146, 147
  bones of, 142
  foramina of, 144-147
  fossae of, 142

Cretinism, 399
Cryptorchidism, 379
Crystalloids, 36
Crystals, 16
Cysterna chyli, 259
Cytology, 8
Cytoplasm, 16

D

Decomposition, 36
Deglutition, 220
Dendrites, 329
Dendrons, 29
Dermis, 23
Diabetes insipidus, 398
Diabetes mellitus, 227
Dialysis, 36
Diaphragm, 199, 315
  openings of, 196, 374
Diaphysis, 177
Diastole, 241, 357
Diencephalon, 337
Diffusion, 37, 38
Digestion, 219-231
  defined, 224
  steps of, 188
Digestive system, 188-234
  and absorption of food, 219-231
  anatomy, 189-219
    abdominal cavity, 197-199
    biliary system, 218, 219
    buccal cavity, 189
      parotid glands and ducts, 190
    esophagus, 195-197
    fauces, 194, 195
    function, 188
    gallbladder, 216, 217
    large intestine, 211, 212
    liver, 213-216
    oral cavity, 190-195
      sublingual glands, 192
      submandibular glands and ducts, 190-192
    pancreas, 218
    pharynx, 195
    salivary glands, 190-192
    small intestine, 205-210
    stomach, 199-205
    teeth, 193, 194
    tongue, 192, 193
      frenum of, 190
  bile
    excretion, 226
    formation, 225, 226
  deglutition, 220
  large intestine, 229
  liver, 224-226

Digestive system—cont'd
  metabolism, 229-231
  oral cavity, 219, 220
  pancreas, 226, 227
  portal circulation, 229
  radiography of, 213, 231-234
  small intestine, 224
  stomach, 220-223
Digits
  manual, 73, 74
  pedal, 84
Diploë, 57, 185
Disc, intervertebral, 86
Discography, 86
Diverticulum
  of duodenum, 205
  of esophagus, 197
    pulsion, 197
    traction, 197
  Meckel's, 53
DNA, 14
  control messages, 15
  molecule replication, 18
  supermolecule, 14
Duct
  common bile, 208, 219
  cystic, 218
  pancreatic, 208, 219
    accessory, 219
  salivary, 190
  vitelline, 53
Duodenal papilla
  major, 224
  minor, 218, 226
Duodenum, 205-208
Dura mater, 334
Dwarfism, 397
Dyad, 47

E

Ectoderm, 52
Elasticity, 357
Elbow, 67, 68
  carrying angle of, 67
Electrolysis, 36
Electrolytes, 42
Electrons, 8
  compared with protons, 10
Elements
  of body, 8
  of periodic table, 8
Embryo development, 51-53
Embryology, 19-46
Embryonal tissue, 35
Endocardium, 237

Endocrine system, 394-404
    compound glands, 402, 403
        liver, 403
        pancreas, 403
    endocrine glands, 394-402
        adrenal, 400-402
        parathyroid, 400
        pituitary, 395-398
        thyroid, 398-400
    functions, 394
    general interest, 404
    radiographic interest, 403, 404
Endometrium, 51
Endoplasmic reticulum, 16
Endosteum, 57
Endothelium, 23
Entoderm, 52
Enzyme, 2
    amylase, 219, 220, 227, 228
    carbonic anhydrase, 319, 320
    chymotrypsin, 227, 228
    enterokinase, 227, 228
    erepsin, 227, 228
    hyaluronidase, 385
    lactase, 227, 228
    lipase, 225, 227, 228
    lysozyme, 220
    maltase, 227, 228
    monamine oxidase, 401
    pepsin, 221
    rennin, 222, 228
    sucrase, 227, 228
    trypsin, 227, 228
Epicardium, 237
Epidermis, 23
Epidural space, 334
Epimysium, 359
Epiphyseal plate, 186
Epiphysis, 176
Epiphysitis, juvenile, 98
Epithelium, 22
    choroidal, 343
    ependymal, 343
Erythrocyte, 12
    manufacture of, 54, 55
    number in circulating blood, 262, 263
E.S.D. studies, 196
Esophagus, 195-197
Ethmoid bone, 134
Eye orbit, 163-165

F

Facets
    costotransverse, 96
    costovertebral, 96
    location of rib, 98
Falx cerebri, 334
Fasciculus, 359
Fatty substance, 11
Femur, 74-76
    angle of shaft, 74
    compared with humerus, 75
Fertility, number of spermatozoa for, 385
Fertilization, 47, 50-51
Fetal circulation, 268-270
Fetus, 51-53, 173
Fibrillation, atrial, 240
Fibrils, 16, 17
Fibroblasts, 55
Fibrocartilage, 86
Fibrous joint, 166, 167
Fibrous tissue
    white, 35
    yellow, 35
Fibula, 78, 80
Filtration, 38, 39
Fingers, 73
Fissure
    great transverse, 198
    superior orbital, 137, 146
Flatfoot, 81
Floating ribs, 110
Fluid compartments of the body, 42
Fluids of the body
    blood, 32, 42, 261-264
    cerebrospinal fluid, 343, 344
    interstitial fluid, 42, 267
    intracellular fluid, 42, 267, 268
    lymph, 266, 267
    plasma, 264
    serum, 264
Foot, 80, 84
Foramen
    caecum, 146
    caroticoclinoid, 147
    epiploic, 199
    interventricular, 399
    intervertebral, 104
    jugular, 147
    lacerum, 138, 146
    magnum, 139, 147
    obturator, 124
    optic, 137, 146
    ovale, 138, 146
    parietal, 140
    rotundum, 137, 138, 146
    sacral, 118
    spinosum, 146
    transverse, 91
    vena caval, 374
    vertebral, 86
Forearm, bones of, 67-70
Fovea capitis, 74
Frontal bone, 131-134

G

Gallbladder, 213
  location of, 213
Gamete, 40, 46, 47
  effect of ionizing radiations on, 49
Ganglion, of nerve, 330
Gas laws, 319
Gaseous exchange, 305, 319-322
Gigantism, 397
Gland, 2
  adrenal, 400-402
  Bartholin, 384
  bulbourethral, 381
  cardiac, 204
  ceruminous, 303
  Cowper, 378, 381
  defined, 2
  endocrine, 23, 394-402
  exocrine, 394
  fundic, 204
  gastric, 204
  parathyroid, 400
  parotid, 190
  pituitary, 395-398
  pyloric, 204
  sebaceous, 302
  sublingual, 192
  submandibular, 190, 192
  sweat, 303
  thyroid, 398-400
  of unknown function, 403
Glenard's disease, 203
Glisson's capsule, 214
Glycogen synthesis, 16
Glycolysis, 361
Goiter, 399
Golgi complex, 16
Gonad, 47, 378
  protection, 176
Graafian follicle, 49, 50
Granules, 16
Granulocyte manufacture, 56
Groove, carotid, 147
Gustatory hair of tongue, 193
Gustatory pore, 193

H

Hand, 72
Haustral markings, 211
Haversian systems, 178-180
Hearing, 349-353
Heart-block, 240
Helium, 9
Hemopoiesis, 54
Hepatopancreatic ampulla, 208

Hiatus
  aortic, 244
  hernia, 196
His, bundle of, 240
Histology, 8
Homeostasis, 2, 42, 54
  cardinal rule of, 42
Hormones
  ACTH, 397, 402
  ADH, 397
  adrenotrophic, 397
  cholecystokinin, 226, 403
  defined, 394
  enterocrinin, 403
  enterogastrone, 403
  epinephrine, 401
  estradiol, 388, 402
  estrone, 387
  FSH, 386
  gastrin, 403
  gonadotrophic, 397
  insulin, 227, 403
  lactogenic, 397
  norepinephrine, 401
  oxytocin, 397
  pancreozymin, 226, 403
  parathormone, 400
  progesterone, 388
  secretin, 226, 403
  STH, 397
  testosterone, 385, 402
  thyroxine, 398
  TSH, 397, 399
  vasopressin, 397
Humerus, 66, 67
Hydrogen, 8
Hydrogen-ion concentration, 43, 44
Hydrolysis, 36, 222
Hydrostatic pressure, 39
Hyoid bone, 157-159

I

Ileum, 53, 208-210
Ilium, 119, 120
Impulse conduction by striated muscle, 16
Inclusions, 16
Insulin, 218, 227
Interstitial fluid, 267
Intervertebral foramen, 334
Intestine
  large, 211-212
    anal canal, 212
    cecum, 211
    colon, 211, 212
    rectum, 212

Intestine—cont'd
  small, 205-210
    duodenum, 205-208
    ileum, 208-210
    jejunum, 208
Intracellular fluid, 267, 268
Iodine, 8
  in thyroid metabolism, 399
Iodine-starch reaction, 36
Ionizing rays, 10
  effect on cells, 17
  effect on gametes, 49
Ions, 10
Iron, 8
Ischium, 120-124
Islets of Langerhans, 218
Isotopes, 10

J

Jejunum, 208
Joint
  cartilaginous, 167
  fibrous, 166, 167
  ginglymus, 67
  synovial, 167-173
Jugular vein
  external, 255
  internal, 255, 256

K

Karyokinesis, 17
Karyoplasm, 15
Karyosome, 14, 15
Kidney, 288-293
  and fluid regulation, 45, 298, 299
Kinetic theory of gases, 37
Knee, 78
  knock, 78
Kyphosis, 84-86

L

Lacrimal bone, 149
Larynx, 307-311
Leukocytes, 12, 33, 263, 264
Ligamentum teres, 74
Lipids, 16
Lithium, 9
Liver, 213-216
  bare area, 214
  and gallbladder, 214
  Glisson's capsule, 214
  ligaments of, 215
  lobules, 216
  portal fissure, 198, 215
  sinusoids, 216
Lordosis, 84-86
Lower leg, 78-80

Lumbar cistern, 340
Lumbarization, 103
Lumbosacral articulation, 102
Lungs, 312-316
Lymph, 258, 259, 266, 267
Lymphangiography, 274
Lymphatic system, 256-261
  lacteals, 258, 259
  lymph nodes, 257, 258
  lymph vessels, 258, 259
  lymphatics, 258
  spleen, 260, 261
Lymphocytes, 33, 257
Lymphoid tissue, 35, 256

M

Macrophages, 199
Magnesium, 8, 11
Mandible, 154-157
Manubrium of sternum, 112
Marrow, red, 176
Matrix, 2, 174
Matter
  forms of, 8
  and ionizing rays, 17, 18
Maturation division, 50
Maxilla, 150-152
Meatus, external and internal auditory, 147
Mediastinum, 316, 317
Medulla oblongata, 338
Meiosis, 17, 19, 46-49
Membrane, 2, 35
  passage of substance across, 35-42
  plasma, 16
  true, 16
Meninges, 334, 335
Mesenchyme, 173, 174
Mesentery, 198
Mesocolon, transverse, 198
Mesoderm, 52
Metabolism, 229-231
  products of, 17
Metacarpals, 72, 73
Metaphase, 17
Metaphysis, 186
Metatarsals, 80, 81
Microglia, 329
Microscope, 12
  compound, 12-14
  power of magnification, 14
Midbrain, 337
Mitochondria, 16, 17
Mitosis, 14, 17
Molecule, 10, 36
  DNA, 15, 19
  RNA, 15, 16
Monocytes, 33

Morula, 51
Motor end plates, 360
Muscle
    attachment, 363
    biceps brachii, 66
    cardiac, 24, 29, 377
    deltoid, 67
    diaphragm, 374
        openings of, 374, 375
    functions, 24, 357-362
        activity, 361, 362
        movement and heat production, 358-361
        posture, 362
    gemelli, inferior and superior, 74
    gluteus, maximus and minimus, 74
    of mastication, 375, 376
    movement, 363-365
        antagonist, 365
        fixation, 365
        prime mover, 365
        synergist, 365
    names, 365-376
    nonstriated (visceral), 376, 377
    obturator internus, 74
    shape, 363
    striated (skeletal), 24, 28, 29, 363
        elasticity, 357
        endomysium, 359
        epimysium, 359
        extensibility, 357
        fasciculus, 359
        fiber, 358
        micella, 360
        myofibril, 360
        perimysium, 359
        tonus, 357
    of tongue, 193, 375
    tissue, 24
        cardiac, 24, 29, 377
        nonstriated, 24, 28, 29, 376, 377
        striated, 24, 28, 29, 363
    of tongue, 193, 375
    vastus, lateralis and medialis, 74
Muscular papillae of heart, 239
Muscular system, 357-377; *see also* Muscle
Myelin sheath, 30
Myelography, 86
Myocardium, 238, 357
Myofibrils, 360
Myxedema, 399

N

Nasal bone, 149
Nasal cavity, 162, 163
Nephron unit, 291, 292

Nerve, 29
    function, 29
    tissue, 291, 292
Nervous system, 326-356
    anatomy, 326-335
        brain, 327
        central nervous system, 326, 327
        nerve cells, 29, 30, 327
        protective coverings, 331, 335
            meninges, 334, 335
            osseous, 331-334
        spinal cord, 327
    autonomic nervous system, 330, 331
    cerebrospinal fluid, 343, 344
        circulating force, 344
        formation, 343, 344
    function, 326
        brain, 336-339
        cranial nerves, 337-338, 341
        spinal cord, 339-341
        spinal nerves, 341-343
        ventricles, 339
    peripheral nervous system, 330
    physiology, 343-346
    radiographic studies, 353-356
    special senses, 346-353
    specific anatomy, 335-343
    synaptic and neuromuscular impulse conduction, 344-346
        neurohumoral theory of transmission, 345, 346
        refractory period, 345
Neuroglia, 329
Neurohumoral theory of impulse transmission, 345, 346
Neurons, 29, 329
    afferent, 329
    bipolar, 329
    conducting cells, 29
    efferent, 329
    internuncial, 329
    multipolar, 329
    unipolar, 329
Neutrons, 8, 10
Neutrophils, 33
Nitrogen, 8
Nose, 306
Nuclear material, 10
Nucleolus, 14-16
Nucleons, 10
Nucleoproteins, 14
Nucleotide, 15
Nucleus
    of atoms, 8
    of cells and transmission of hereditary traits, 14
    segmentation, 50

Nucleus pulposus, 87
Nutrient artery, 56, 78

O

Occipital bone, 139, 140
Odontoid process, 90, 91
Omentum
 greater, 199, 203
 lesser, 198, 203
Ontogeny, 46
Oocyte, 49
Oogonia, 46
Ootid, 49
Organ, 2
 accessory of digestion in abdominal cavity, 212-219, 224
Organelle, 16
Osmosis, 39-41
Osseous tissue, 33
Ossification, 180-186
 endochondral, 180, 181
 intracartilaginous, 180
 intramembranous, 180
 periosteal, 184
Osteoblasts, 33, 55, 178
Osteoclasts, 178
Osteocytes, 33, 177
Ovum, 47
Oxygen, 8
 flow, 319, 320
 transport, 320, 321

P

Palatine bone, 152
Pancreas, 218, 219
Papilla
 major duodenal, 208
 minor duodenal, 226
 of tongue, 193
Parietal bone, 140
Parotid gland, 190
Patella, 75, 78
Pectoral girdle, 65, 105-110
 clavicle, 106, 107
 scapula, 107-110
Pelvic bone, 61, 114
Pelvic girdle, 65, 74, 114-124
 acetabulum, 123, 124
 coccyx, 119
 ilium, 65, 119, 120
 ischium, 65, 120-123
 obturator foramen, 124
 pubis, 65, 123
 sacrum, 118, 119
 symphysis pubis, 115
Pelvic straits, 117
Pelvimetry, 116, 117, 393

Pericardial cavity, 237
Pericardium, 237
Perineum, 379
Periosteum, 55
Peripheral nervous system, 330
Peristalsis
 of digestive system, 196
 of ureters, 288
Peritoneum, 197, 198
pH, 43-45
 control mechanisms, 44
 and fracture healing, 179, 180
Phagocytosis, 41
Phalanges
 manual, 73, 74
 pedal, 84
Pharynx, 195
Phosphorus, 8
Photomicrograph, 12
Phylogeny, 46
Physiology, 1
 defined, 8
Pia mater, 334
Pigments, 16
 melanin, 23
Pinocytosis, 41
Placenta, 51
Planes of the body, 4, 6, 7
 coronal, 4
 oblique, 7
 sagittal, 4
 transverse, 4
Platelets, 264
Pleura, parietal and visceral, 316
Pneumoarthrography, 169
Pneumoencephalography, 355
Polar bodies, 49
Pons, 337
Portal fissure, 198, 215
Portal vein, 210
 circulation, 206, 271, 272
Position
 anatomic, 5
 of patient, 7
Potassium, 8, 11
Power plants, 17
Pregnancy, 389, 390
 ectopic, 50
Pressure
 gradient of stomach emptying, 204, 223
 hydrostatic, 39
Primordium, 184, 386
Projection of central ray, 7
Pronucleus
 female, 50
 male, 50
Prophase, 17

Protein, 11, 16
  factories, 16
  molecules, 11
Proton, 8, 10
Protoplasm, 11
Pseudoperistalsis, 196
Pterion, 138
Pubis, 123
Pulmonary lobule, 313
Pulmonary ventilation, 322, 323
Purkinje's fibers, 240

R

Radiograph, 7
Radiography
  circulatory system, 272-286
  digestive system, 231-234
  endocrine system, 403, 404
  nervous system, 353-356
  reproductive system, 390-393
  respiratory system, 323-325
  skeletal system, 65-84, 124-126, 165, 166
  urinary system, 300-302
Radius, 67, 70
Reduction division, 47
Regions, abdominal, 199, 200
Renal calyces, 290, 291
Renal corpuscles, 292
Renal shutdown, 297
Renal tubules, 291, 292
Reproductive system, 378-393
  anatomy, 378-384
    female, 380-384
      ovaries, 381
      uterine tubes, 382
      uterus, 383
      vulva, 384
    male, 378-381
      ductus deferens, 380
      ejaculatory duct, 380
      epididymis, 380
      penis, 380, 381
      scrotum, 378
      testes, 379
      urethra, 380, 381
  radiography, 390-393
  structure and function of gonads, 384-390
    female, 386-393
      corpus luteum, 388
      graafian follicle, 386
      menstrual cycle, 387, 388
      ovulation, 388
      primordial follicle, 386
    female mammary gland, 388-390
    fertilization of the ovum, 388

Reproductive system—cont'd
  structure and function of gonads—cont'd
    male, 384-386
      cells of Leydig, 385
      cells of Sertoli, 385
Respiration types, 323
Respiratory center, 323
Respiratory system, 305-325
  anatomy, 305-307
    bronchi, 312
    larynx, 307-311
    lungs, 313, 314, 320, 321
    nasal cavity, 306, 307
    nose, 306
    pharynx, 195, 307
    trachea, 311, 312
    vocal cords, 310
  function, 305
  radiography, 323-325
  respiration, 317-323
    mechanics of, 317-319
    physiology of, 319-323
    respiratory center, 323
  thoracic cavity, 314
    mediastinum, 316, 317
    pleura, 316
Reticular tissue, 35
Reticulo-endothelial system, 41
Ribosomes, 16
Ribs, 111, 112
  articulations of, 111, 112
  cervical, 93
RNA, 14, 16
  messenger, 15
  ribosomal, 15
  transfer, 15
RNP, 16
Rugae, 201

S

Sacrum, 118, 119
Sarcolemma, 24, 359
Sarcoplasm, 24
Scapula, 107-110
Schmorl's nodule, 87
Scoliosis, 84-86
Segmentation, 51, 52
Sella turcica, 134, 135, 395
Sharpey's fibers, 55
Sialography, 192
Sight, 347, 349
  learning by, 2
Sinoatrial node of heart, 240
Sinus
  air of the skull, 159-162
  aortic, 242
  carotid, 245

Sinus—cont'd
  coronary, 243
  inferior sagittal, 334
  renal, 290
  sigmoid, 246, 255, 335
  straight, 246
  superior sagittal, 246, 256
  transverse, 246, 255
Skeletal system, 54-187
  functions of, 54, 55
  hyoid bone, 157-159
  lower extremity, 74-84
  pectoral girdle, 105-110
  pelvic girdle, 114-124
  skeleton, 61-65
  skull, 126-157
  thoracic cage, 110-114
  trunk, 124-126
  upper extremity, 65-74
  vertebral column, 84-105
    cervical, 89-95
    lumbar, 98-105
    thoracic, 95-98
Skin, 18-20, 302-304
  anatomy, 23, 303, 304
  functions, 23, 24, 302, 303
Skull, 126-166
  calvarium, 127
  cephalic index, 127, 130
  cranial bones, 131-134
  cranial floor, 142-144
  epipteric bones, 138
  ethmoid, 134
  eye orbit, 163-165
  facial bones, 144, 149-157
  frontal, 131-134
  inferior concha, 152
  lacrimal, 149
  landmarks, 131, 164-166
  mandible, 154-157
  maxilla, 150-152
  nasal, 149
  occipital, 139, 140
  palatine, 152
  parietal, 140
  shapes, 127, 130
  sinuses, 159-162
    ethmoidal, 134, 159-161
    frontal, 133, 134, 159
    mastoid, 142, 162
    maxillary, 151, 162
    nasal cavity, 162, 163
    sphenoidal, 134, 162
  sphenoid, 134-138
  supernumerary bones, 138
  sutures, 130, 131
  temporal, 140-142

Skull—cont'd
  vertex, 127
  vomer, 153
  wormian bones, 138
  zygoma, 150
Smell, 347
Sodium, 8, 11
Solution
  hypertonic, 40
  hypotonic, 40
  isotonic, 40
Spermatid, 47
Spermatogonia, 47
Spermatozoon, 47
  nucleus in fertilization, 50, 51
Sphenoid bone, 134-138
Sphincter
  cardiac, 204
  choledochal, 208
  pyloric, 201, 204
Spina bifida, 88
  occulta, 102
Spleen, 260, 261
Spondylolisthesis, 103
Starch, 11
Sterility in the male, 385
Sternum, 112-114
Steroid hormone synthesis, 16
Stomach, 199-205
  digestion in, 220-223
  emptying, 201, 203, 223
  types, 201-203
Stresses and alarm reaction, 402
Subarachnoid space, 335
Sublingual gland, 192
Submandibular gland, 190, 192
Substance, 10
Sugar, 11
Sulfur, 8
Supernumerary bones, 138
Surgical landmarks, 126
Sutures, 166
Symphysis
  mandibulus, 157
  pubis, 115
Synapse in nervous system, 345
Synapsis in meiosis, 47
Synovia, 169
System
  circulatory, 2, 235-286
  defined, 2
  digestive, 2, 188-234
  endocrine, 2, 394-404
  muscular, 2, 357-377
  nervous, 2, 326-356
  reproductive, 2, 378-393
  respiratory, 2, 305-325

System—cont'd
  reticuloendothelial, 41
  skeletal, 2, 54-187
  urinary, 2, 287-304
Systole, 241, 357

T

Table
  acid-base relation at 15 pH values, 44
  approximate appearance time of bones, etc, 182-185
  articulations, 165
  articulations of carpal bones, 70
  articulations of tarsal bones, 81
  body planes, 6, 7
  body positions, 5
  bone depressions, 60, 61
  bone projections, 58, 59
  classification of bones by region, 64, 65
  cranial nerves, 341
  digestive juices and actions, 228
  directional terms, 5
  major viscera of the nine regions of the abdomen, 200
  menstrual cycle, 387
  muscles that make significant radiographic shadows, 366-373
  openings of the cranial floor, 146, 147
Tarsals, 80-82
Taste, 193, 353
Telophase, 17
Temporal bone, 140-142
Tentorium cerebelli, 256
Terminology, 3
Terms
  anatomical, 3, 5
  directional, 5
  of opposite reference, 3
  position, 5
Tetrad, 47
Thebesian veins, 242
Thoracic cage, 110-114
  ribs, 111, 112
  sternum, 112-114
Thumb, 73
Tibia, 78-80
  angle of shaft of, 78
Tissue, 2
  connective, 19, 31-35
  epithelial, 19, 22-24
  muscle, 19, 24-29
  nerve, 19, 29-31
Toes, 84
Tongue, 192, 193
  muscles, 193
Tonus, 357
Touch, 353

Trabeculae, 34
Trachea, 311, 312
  angles with bronchii, 312
Trophoblast, 51
Tufts, 74
Tungsten, 9

U

Ulna, 70
Ulnar flexion, 72
Units, basic, 8
Urinary system, 287-302
  anatomy, 288-296
    blood supply to kidneys, 292, 293
    innervation, 293
    kidneys, 288-293
    nephron unit, 291-292
    ureters, 293, 294
    urethra, 295, 296
    urinary bladder, 294, 295
  functions, 287, 288
  radiographic interest, 300-302
  urine, 299, 300
  urine formation, 296, 297
    filtration, 296, 297
    reabsorption, 297-299

V

Vacuoles, 17
Valve
  of heart, 239, 240
  of Heister, 218
  ileocecal, 208, 210
  of Kerckring, 206
Vasa vasorum, 243
Veins, 250-256
  of head and neck, 255
  of lower extremity, 251-253
  of trunk, 253, 254
  of upper extremity, 254, 255
Vena cava
  inferior, 253
  superior, 253, 254
Ventriculography, 355
Vertebra
  atlas, 89, 90
  axis, 89, 90
  body, 87, 88
  cervical, 89-95
  column of, 84-105
  curvatures of column, 84, 85
  lamina, 88
  lumbar, 98-104
  pedicle, 88

Vertebra—cont'd
 pillar views, 94
 processes, 87, 88
 prominens, 89, 91, 92
 thoracic, 95-98
 typical, 86-89
Vertebral notch, inferior and superior, 334
Visceroptosis, 203
Vitelline duct, 53
Vocal cords, 310
Vomer bone, 153

W

Wolff's law, 55, 176, 181
Wrist, 70-73

X

Xiphoid process of sternum, 112, 114

Z

Zona pellucida, 50
Zoology, 8
Zygoma, 150
Zygote, 47, 50